ENERGY NETWORKS AND THE LAW

Energy Networks and the Law

Innovative Solutions in Changing Markets

Edited by
MARTHA M. ROGGENKAMP
LILA BARRERA-HERNÁNDEZ
DONALD N. ZILLMAN
IÑIGO DEL GUAYO

OXFORD
UNIVERSITY PRESS

OXFORD
UNIVERSITY PRESS

Great Clarendon Street, Oxford OX2 6DP
United Kingdom

Oxford University Press is a department of the University of Oxford.
It furthers the University's objective of excellence in research, scholarship,
and education by publishing worldwide. Oxford is a registered trade mark of
Oxford University Press in the UK and in certain other countries

First Edition published 2012

British Library Cataloguing in Publication Data
Data available

Library of Congress Cataloguing in Publication Data
Library of Congress Control Number: 2012931718

ISBN 978–0–19–964503–9

Typeset by Newgen Imaging Systems (P) Ltd, Chennai, India
Printed in Great Britain on acid-free paper by
CPI Group (UK) Ltd, Croydon, CR0 4YY

Preface

This book marks the sixth collaboration of the Academic Advisory Group (AAG) of the International Bar Association's Section on Energy, Environment, Resources, and Infrastructure Law (SEERIL) with Oxford University Press (OUP). Over the years, the academic lawyers of the AAG have shaped a systematic approach to our biennial 'big picture' studies of energy topics.

This collaboration has followed the prior format. At the biennial meeting of SEERIL, the product of the prior biennium's study is released in OUP hardcover book form. The AAG rests only briefly on its laurels. At the same SEERIL meeting, the AAG members vet possible topics for the next biennial study. At the April 2010 SEERIL meeting in Toronto, AAG members selected the topic of networks and their legal implications. Networks are the physical frame of the energy market and crucial for securing energy supply. The organization and administration of these networks is faced with important changes and challenges following market liberalization, climate change, and security of supply policies leading to the introduction of new or renewable energy sources. The topic of networks is also one linking the interests of members of SEERIL as it relates to, and has an impact on, environmental and infrastructure law.

Individual topics were selected by the participating members in the months that followed. A number of additional authors were selected to write chapters that would further expand the coverage of the book. Four editors volunteered to guide the process through to conclusion. The authors then began or continued their research. Most chapters were available and many authors were present at the University of Groningen (Rijksuniversiteit Groningen, or RUG) in the last week of May 2011 for an intensive three-day seminar. Officials of the gas transmission company N.V. Nederlandse Gasunie ('Gasunie') provided expert commentary on several papers and on the overall project. Members of the Groningen Centre of Energy Law of the RUG provided similar expert commentary. Gasunie also provided a highly informative tour of one of its gas compressor facilities and covered some of the cost of the meeting and the book.

We thank Gasunie, the Groningen Centre of Energy Law, and the RUG for their generosity in this collaboration. We also thank the International Bar Association and SEERIL for their encouragement and financial support for the work of our group. A more personal gratitude goes to the Chair of the Council of SEERIL—Arent van Wassenaer—for his support of this particular project and his attendance at the midterm meeting.

As always, the midterm gathering reaffirmed some assumptions about the book and changed others. Authors and editors returned to their word processors for the final writing and editing that allowed the completed manuscript to reach the editors at OUP in August 2011. We express our appreciation to Merel Alstein and Anthony Hinton of OUP for their assistance in bringing the project to completion.

The Editors
October 2011

Table of Contents

PART IV OVERVIEW AND CONCLUSION

List of Abbreviations

Chapter 1

AAG	Academic Advisory Group
IEA	International Energy Agency
NAFTA	North America Free Trade Association
NIMBY	Not in my back yard
OECD	Organization of Economic Cooperation and Development
OUP	Oxford University Press
UAE	United Arab Emirates
UK	United Kingdom

Chapter 2

CEAA	Canadian Environmental Assessment Act
COGO	Canada Oil and Gas Operations
NEB	National Energy Board
POGG	Peace, order, and good government

Chapter 3

CFE	Comisión Federal de Electricidad (Federal Electricity Commission of Mexico)
CRE	Comision Reguladora de Energia (Energy Regulatory Commission of Mexico)
EPR	Empresa Propietaria de la Red (Grid Proprietor of Guatemala)
ERCOT	Electricity Reliability Council of Texas
LGEEPA	General Act for Ecological Balance and Environmental Protection
NAAEC	North American Side-Agreement on Environmental Cooperation
NAFTA	North American Free Trade Agreement
NERC	North American Electric Reliability Corporation
PROFEPA	Attorney General Office for Environmental Protection
ROW	right of way
SIEPAC	Central American Electricity Interconnection System
WECC	Western System Coordinating Council

Chapter 4

AEA	Andean Energy Alliance
CAF	Andean Finance Corporation
CAN	Community of Andean Nations

CANREL	Comité Andino de Organismos Normativos y Organismos Reguladores de Servicios de Electricidad (Council of Ministers of Energy, Electricity, Hydrocarbons and Mines of the Andean Community)
DRS	Dispute resolution system
IDB	Inter-American Development Bank
IIRSA	Integration of the Regional Infrastructure in South America
MERCOSUR	Southern Common Market
PRT	Permanent Review Tribunal
SA	South America
UNASUR	Union de Naciones Suramericanas (Union of South American Nations)

Chapter 5

CCP	Chad–Cameroon Project
CCSRP	Collège de Contrôle et de Surveillance des Revenus Pétrolières (Petroleum Revenue Oversight and Control Committee)
COTCO	Cameroon Oil Transportation Company
CPSP	Committee for Supervision of Pipelines
CSO	Civil Society's Organizations
DRM	Dispute resolution mechanism
EA	Environmental Assessment
ECMG	EMP Compliance Monitoring Consultant
EIB	European Investment Bank
EMP	Environmental Management Plan
FEDEC	Foundation for Environment and Development
FGF	Fund for Future Generations
IBRD	International Bank for Reconstruction and Development
IDA	International Development Agency
IFC	International Finance Corporation (private sector arm of the World Bank Group)
IPP	Indigenous Peoples Plan
NGO	non-governmental organization
PRMP	Petroleum Revenue Management Program
PRML	Petroleum Revenue Management Law
ROW	right of way
SNH	Société National des Hydrocarbures (national hydrocarbon company of Cameroon)
TOTCO	Chad Oil Transportation Company
WB	World Bank
WCS	World Conservation Society
WWF	World Wide Fund for Nature

Chapter 6

ASCOBANS	Agreement on the Conservation of Small Cetaceans of the Baltic, North East Atlantic, Irish and North Seas
BAT	Best available techniques
BEP	Best environmental practices

BIT	bilateral investment treaty
BP	British Petroleum
BTC	Baku–Tblisi–Ceyhan
CS	continental shelf
COTCO	Cameroon Oil Transport Company
EBRD	European Bank of Reconstruction and Development
ECHR	European Convention on Human Rights
EEZ	exclusive economic zone
HGA	Host Government Agreement
IFI	international financial institutions
IGA	Inter-governmental Agreement
IO	International Organization
NGO	non-governmental organization
OPEC	Organization of the Petroleum Exporting Countries
TBP	transboundary pipeline
UNCLOS	1982 United Nations Convention on the Law of the Sea
UN/IFC	United Nations/International Finance Corporation

Chapter 7

ADCC	Algemene Directie Crisiscentrum (General National Crisis Centre of the Federal Belgium Ministry of Internal Affairs)
AIVD	Algemene Inlichtingen- en Veiligheidsdienst (Dutch General Intelligence and Security Service)
CEN	Comité Européen de Normalisation (European Committee for Standardization)
CENELEC	Comité Européen de Normalisation Electronique (European Committee for Electrotechnical Standardization)
DSO	distribution system operator
ECI	European critical infrastructure
ECIP	European critical infrastructure protection
EEC	European Economic Community
EU	European Union
IEM	Internal Energy Market
ISO	International Organization for Standardization
KLIC	Kabels en Leidingen Informatie Centrum (Dutch Network Information Centre)
KLIM	Kabels en Leidingen Informatie Meldpunt (National Belgian Information Centre on Networks)
KLIP	Kabels en Leidingen Informatie Portaal (Regional Flemish Information Centre on Networks)
MS	Member States
NAVI	National Advice Centre on Vital Infrastructures (the Netherlands)
NCTb	National Coordinator for Counterterrorism (the Netherlands)
OCAD	Coordinating Body on Threat Analysis in Belgium
OSP	operator security plan
SLO	Security Liaison Officer
SOVI	Strategic Consultation on Vital Infrastructure

TFEU	Treaty on the Functioning of the European Union
TSO	transmission system operator
WION	Act on Information Exchange Concerning Subsoil Networks

Chapter 8

ACER	Agency for the Cooperation of Energy Regulators
CHP	combined heat and power
COM	Communication from the Commission
DSO	distribution system operator
ENTSO	European Network of Transmission System Operators
EU	European Union
GHG	greenhouse gas
IEA	International Energy Agency
IPPC	Intergovernmental Panel on Climate Change
SET-Plan	European Strategic Energy Technology Plan
TFEU	Treaty on the Functioning of the European Union
TSO	transmission system operator

Chapter 9

CIP	Conservation Improvement Program
CSP	Curtailment Service Provider
DR	demand response
EIA	Energy Information Administration
EISA	Energy Independence and Security Act of 2007
FERC	Federal Energy Regulatory Commission
FPA	Federal Power Act 1935
FPC	Federal Power Commission
GW	gigawatts
IRP	integrated resource planning
ISO	Independent System Operators
NECPA	National Energy Conservation Policy Act
NERC	North American Electric Reliability Corporation
PJM	Pennsylvania–New Jersey–Maryland Interconnection
PUHCA	Public Utility Holding Company Act
PURPA	Public Utility Regulatory Policies Act
QF	qualifying facility
RTO	Regional Transmission Organization

Chapter 10

AC	alternating current
BOEMRE	Bureau of Ocean Energy Management, Regulation and Enforcement
CZMA	Coastal Zone Management Act
DC	direct current
DOI	Department of the Interior

EEZ	exclusive economic zone
EIS	Environmental Impact Statement
EPAct	Energy Policy Act of 2005
FERC	Federal Energy Regulatory Commission
FPA	Federal Power Act 1935
MSP	marine spatial planning
NEPA	National Environmental Policy Act
OCS	Outer Continental Shelf
ROW	right of way
RUE	rights of use and easements
TSO	transmission system operator
UNCLOS	United Nations Convention on the Law of the Sea 1982
UNESCO	United Nations Educational, Scientific and Cultural Organization

Chapter 11

EA	Energy Act
EEA	European Economic Area
MPE	Ministry of Petroleum and Energy
NDA	Nature Diversity Act
NWED	Norwegian Water Resources and Energy Directorate
PBA	Planning and Building Act

Chapter 12

NDRC	National Development and Reform Commission
PV	Photovoltaic

Chapter 13

AGTL	Alberta Gas Trunk Line Limited
AUC	Alberta Utilities Commission
CCS	carbon capture and storage
CO2	carbon dioxide
DECC	Department of Energy and Climate Change
EOR	enhanced oil recovery
ERCB	Energy Resources Conservation Board
EU	European Union
GHG	greenhouse gas
MS	Member states
Mt	million tonnes
NGTL	Nova Gas Transmission Ltd
OGCA	Oil and Gas Conservation Act
PUB	Public Utilities Board
SGER	Specified Gas Emitters Regulation
TBQ,	Transportation by Others
TPA	third party access

Chapter 14

ANP	Agência Nacional do Petróleo, Gás Natural e Biocombustíveis (Brazilian Oil, Natural Gas, and Biofuels Agency)
CADE	Administrative Council for Economic Defense
CPAC	Open Competition for Capacity Allocation
LPG	liquefied petroleum gas

Chapter 15

ECNZ	Electricity Corporation of New Zealand Ltd
HVDC	high-voltage direct current
IEA	International Energy Agency
TPM	transmission pricing methodology

Chapter 16

ABARES	Australian Bureau of Agricultural and Resource Economics and Sciences
AEMA	Australian Energy Market Agreement
AEMC	Australian Energy Market Commission
AEMO	Australian Energy Market Operator
AER	Australian Energy Regulator
AUD	Australian dollar
CSIRO	Commonwealth Scientific and Industrial Research Organisation
MCE	Ministerial Council on Energy
NEM	National Electricity Market
NER	National Electricity Rules
NGMC	National Grid Management Council
NTDP	National Transmission Network Development Plan
RIT-T	Regulatory Investment Test for Transmission
SENE	Scale Efficient Network Extension
SWER	single-wire earth return

Chapter 17

capex	capital expenditure
DNO	distribution network operator
EDNO	electricity distribution network operator
EU	European Union
GDNO	gas distribution network operator
IQI	Information Quality Incentive
LENS	Long-term Electricity Networks Scenarios
LCN	Low-carbon Networks
Ofgem	Office of Gas and Electricity Markets
opex	operating expenditure
RIIO	Revenue using Incentives to deliver Innovation and Output
RPI	retail price index

RPZ	registered power zone
TNO	transmission network operator
UK	United Kingdom

Chapter 18

EU	European Union
LNG	liquefied natural gas
NRA	National Regulatory Authorities
TPA	third party access
TSO	transmission system operator
UIOLI	use it or lose it

Chapter 19

ACER	Agency for the Cooperation of Energy Regulators
EASEE	European Association for the Streamlining of Energy Exchange
ED	Electricity Directive of 2009 (Directive 2009/72/EC)
ENTSO-E	European Network of Transmission System Operators for Electricity
ENTSO-G	European Network of Transmission System Operators for Gas
ER	Electricity Regulation (Regulation 713/2009)
GD	Gas Directive of 2009 (Directive 2009/73/EC)
GR	Gas Regulation (Regulation 715/2009)
GSR	Gas Security Regulation
ISO	independent system operator
ITO	independent transmission operator
LNG	liquefied natural gas
NRA	National Regulatory Authorities
NTYNDP	National ten-year network development plan
TSO	transmission system operator
TYNDP	Community-wide ten-year network development plan
UCPTE	Union for the Coordination of Production and Transmission of Electricity
UCTE	Union for the Coordination of Transmission of Electricity

Chapter 20

ACER	Agency for the Cooperation of Energy Regulators
ARegV	Anreizregulierungsverordnung (Incentive Regulation Ordinance)
BGBl.	Bundesgesetzblatt (Federal Gazette)
BNetzA	Bundesnetzagentur (Federal Network Agency)
BVerfG	Bundesverfassungsgericht (Federal Constitutional Court)
DENA	Deutsche Energie-Agentur (German Energy Agency)
EEG	Erneuerbare-Energien-Gesetz (Renewable Energies Act)
EIA	Environmental Impact Assessment
EnLAG	Energieleitungsausbaugesetz (Electricity Grid Expansion Act)
ENTSO-E	European Network of Transmission System Operators for Electricity

EnWG	Energiewirtschaftgesetz (Energy Act)
ISO	Independent System Operator
ITO	Independent Transmission Operator
NABEG	Netzausbaubeschleunigungsgesetz (Grid Expansion Acceleration Act)
ROG	Federal Regional Planning Act
TEN	trans-European energy networks
UPV	Umweltverträglichkeitsprüfung (Environmental Impact Assessment)
UVPG	Gesetz über die Umweltverträglichkeitsprüfung (Environmental Impact Assessment Act)
VwVfG	Verwaltungsverfahrensgesetz (Administrative Procedure Act)

Chapter 21

E&P	Exploration and Production
FAS	Federal Anti-monopoly Service
FTS	Federal Tariff Service
GTS	gas transmission system
LNG	liquefied natural gas
RAO	a Russian joint stock company
RAO UES	Russian Joint Stock Company of Electricity and Electrification
RSFSR	Russian Soviet Federative Socialist Republic
TN owners	transmission network owners
TPA	third party access
UGS	Unified Gas Supply system

Chapter 22

CCS	carbon capture and storage
NOC	National Oil Company
TPM	transmission pricing methodology
RIIO	Revenue using Incentives to deliver Innovation and Output

List of Contributors

Nigel Bankes is Professor of Law, The University of Calgary and holder of the Chair in Natural Resources Law, Calgary, Alberta, Canada; email: ndbankes@ucalgary.ca

Lila Barrera-Hernández is Adjunct Assistant Professor, Faculty of Law, University of Calgary, Canada, and Abogada, Buenos Aires, Argentina; email: Lila.kbh@gmail.com

Barry Barton is Professor of Law, University of Waikato, Hamilton, New Zealand; email: barton@waikato.ac.nz

Mohammed A. Bekhechi is Lead Counsel, Legal Vice-Presidency of the World Bank, Washington DC, USA, Member of the Constitutional Council of Algeria and Visiting Professor at the Universities of Paris V (René Descartes), France; email: Mbekhechi@worldbank.org

Anderson Souza da Silva Lanzillo is Assistant Professor, Department of Private Law, Federal University Rio Grande do Norte, Natal, Brazil; email: adv.andersonss@gmail.com

Yanko Marcius de Alencar Xavier is Full Professor, Department of Public Law, Federal University Rio Grande do Norte, Natal, Brazil; email: ymxavier@ufrnet.br

Iñigo del Guayo is Professor in Administrative Law, University of Almería, Spain; email: iguayo@ual.es

Lee Godden is Professor of Law, Director of the Centre for Resources, Energy and Environmental Law, Melbourne Law School, University of Melbourne, Australia; email: lcgodden@unimelb.edu.au

José Juan Gonzáles is Professor, Department of Law, Universidad Autónoma Metropolitana, Mexico, and Director of the Mexican Institute for Environmental Law Research, Mexico; email: jjgonzalez@gonzalezasociados.com

Ivan V. Gudkov is Deputy Head of Division in the Legal Department of OJSC Gazprom, and Associate Professor at MGIMO-University, Moscow, Russia; email: I.Gudkov@adm.gazprom.ru

Ulf Hammer is Professor, Scandinavian Institute of Maritime Law, University of Oslo, Norway; email: ulf.hammer@jus.uio.no

Anne Kallies is a PhD student, Centre for Resources, Energy and Environmental Law, Melbourne Law School, University of Melbourne, Australia; email: a.kallies@unimelb.edu.au

Gunther Kühne is Emeritus Professor, Technical University of Clausthal, Institut für Berg- und Energierecht, Clausthal Zellerfeld, Germany; email: gunther.kuehne@tu-clausthal.de

Alastair R. Lucas, QC, is Dean and Professor, Faculty of Law, University of Calgary, Alberta, Canada; email: alucas@ucalgary.ca

Aileen McHarg is Senior Lecturer in Public Law, University of Glasgow, United Kingdom; email: Aileen.McHarg@glasgow.ac.uk

Rick Nilson is Manager Commercial and Regional Development, Syncrude Canada Ltd, Calgary, Alberta, Canada; email: r.nilson@shaw.ca

LeRoy Paddock is Associate Dean for Environmental Law Studies, The George Washington University Law School, Washington DC, USA; email: lpaddock@law.gwu.edu

Johann-Christian Pielow is Professor in Economic Law, Director of the Institute for Mining and Energy Law, Ruhr University, Germany; email: Christian.pielow@rub.de

Catherine Redgwell is Professor of International Law, University College London, United Kingdom; email: c.redgwell@ucl.ac.uk

Martha M. Roggenkamp is Professor of Energy Law, University of Groningen, Director of the Groningen Centre of Energy Law, board member of the Groningen Energy and Sustainability Programme and Chair of the Dutch Energy Law Association, The Netherlands; email: m.m.roggenkamp@rug.nl

Anita Rønne is Associate Professor in Energy Law, Faculty of Law, University of Copenhagen, Denmark, Chairman of the Danish Energy Law Society and Chairman of the Valuation Committee under the Law of Renewables; email: Anita.Ronne@jur.ku.dk

Peter J. Schaumberg is Principal, Beveridge & Diamond, P.C., Washington DC, USA; email: pschaumberg@bdlaw.com

Sergey S. Seliverstov is Partner in Sokolov, Maslov and Partner and Associate Professor at MGIMO-University, Moscow, Russia; email: sergey.seliverstov@smplawyers.ru

Graham St. Michel is Associate, Beveridge & Diamond, P.C., Washington, D.C. USA; email: GStMichel@bdlaw.com

Tjarda van der Vijver is case handler at the Network Sectors & Media Unit, Netherlands Competition Authority, The Hague and external PhD researcher at Leiden University, The Netherlands; email: tjarda.vandervijver@nma.nl

Wang Mingyuan is Professor and Executive Director, Center for Environmental, Natural Resources and Energy Law, Tsinghua University, China; email: wangmy@tsinghua.edu.cn

Olivia Woolley is a post-doctoral researcher, Groningen Centre of Energy Law, Faculty of Law, University of Groningen, The Netherlands; email: o.a.woolley@rug.nl

Charlotte Youngblood is a Masters in Law (LLM) recipient at The George Washington University School of Law, Washington DC, USA; email: cyoungblood@law.gwu.edu

Donald N. Zillman is President, University of Maine at Presque Isle and Godfrey Professor of Law, University of Maine Law School; email: donald.zillman@umpi.edu

1

Energy Networks and the Law: Innovative Solutions in Changing Markets

Donald N. Zillman, Martha M. Roggenkamp, Lila Barrera-Hernández, and Iñigo del Guayo

I. Introduction

This book discusses the role of networks in the energy sector, how these networks have developed, are being regulated, and what the impact can be of future market developments on network regulation. The energy sector is a vital part of most national economies and the networks are crucial for transporting energy resources from producers to consumers. However, networks are just one particular means of transportation and their current important role is the result of several technical innovations. In contrast to ships, trucks, and trains, which need drivers, networks (ie cables and pipelines) represent a mode of transportation that can be operated without direct human intervention. The sections below illustrate how the world of energy and networks has changed in little over two centuries.

Historian Pauline Maier has vividly recreated an important landmark in legal history in her study of the popular conventions in the United States that debated and ratified the Constitution of the United States.[1] Maier brings to life some of the little-known citizens who served as delegates to the conventions. Maier also provides a reminder of daily life at the end of the eighteenth century—a world in which modern energy was not present.

Maier traces the effort of two Massachusetts delegates to reach the state ratification convention in Boston. She laconically notes: 'It was no easy thing to get from Bath, an old shipbuilding center on the Kennebec River in Maine to Boston in January of 1788.' The trip required a journey of about 225 kilometers. Today the journey could be made in less than a three-hour drive on which the major concern would be traffic congestion.

In winter 1787–88, the best option was on horseback and the trip took six days. The other transportation option was by sailing ship. Wind and winter weather

[1] P. Maier, *Ratification: The People Debate the Constitution 1787–88* (New York: Simon & Schuster, 2010).

made that a challenging alternative. Even on the land route, water was a problem. Several major rivers divided the close-to-the-coast roads. At the rivers, ferry boats allowed passage, powered by a combination of human, animal, wind, and wave energy. Carbon-based energy (coal, petroleum, or natural gas) and/or electricity were decades in the future.

The six days of winter travel also required stopovers for food and lodging. Our two delegates satisfied their needs by staying with friends or relatives or in roadside inns. Here their energy source was wood, cut from the plentiful forests of Massachusetts of the time. The wood provided fireplace heat and heat for cooking sufficient to make travel survivable, if not pleasant.

Communications were difficult. The two delegates (lacking modern cell phones and computers) lost contact with their homes upon departure. Advice about political matters from home or news from Boston could have reached them only by carefully planned fast delivery by horse. This would have been both difficult and expensive.

The Massachusetts traveler of 1788 would have taken for granted that all energy was local. The horses and people who assisted his travel had likely never traveled more than 100 mile from the place of their birth in their lives. The wood that benefited the humans had been cut locally. The food that fed travelers and horses came from neighboring farms. Other products that came from human endeavor—blankets, saddles, plates, clothing—were likely to be of local origin. Distribution channels from raw material gatherers to product makers to consumers were basic and short.

This was the world of 1788, in one of the more modern parts of the globe, a world of very limited energy. It was also a world that would have looked familiar to the citizens of advanced civilizations of AD 1500 or 1066. Animal power, human power, wind, sunshine, water, and wood were the resources. Coal, oil, and natural gas played a very limited role. Also lacking were the benefits of that essential secondary source of energy—electricity. It is also worth remembering that the energy options available to the advanced nations of 1788 are the same that remain to between two and three billion people on Planet Earth in 2012.

The nineteenth, twentieth, and twenty-first centuries have been the centuries of energy. A remarkable combination of scientific discovery, engineering innovation, financial creativity, and political and legal initiative created what would have been unimaginable to our delegates of 1788. Oh, brave new world that has shaped the energy resources and networks that connect the world today!

II. Energy Sources and Networks

We begin our study of energy networks and the law with some background to energy itself. Much of this will be familiar to energy specialists. Our work as lawyers and teachers, however, persuades us that large proportions of the population, including professionals working outside the energy field, have little sense of the history of energy and its present scope. Energy knowledge extends little beyond the switch that turns on the electricity or the gas pump that fuels the car. We hope these introductory pages will provide a useful perspective before we move to our discussion of the networks that allow energy to reach much of the population of the earth.

A. The history of modern energy

The world of 1788 depended mainly on renewable energy sources. Wood provided energy for heating and cooking then as it does today in many underdeveloped regions. The equipment available to produce these resources was often limited to spades and axes. Wind energy already required a more advanced technique as it involved the construction of a windmill. Renewable resources like wind, tidal, and solar energy have gradually been adapted to provide new uses by developments of science and engineering. The industrial revolution produced more advanced techniques, which enabled production in the (deep) subsoil. This led to a new energy era.

1. Coal

Modern energy progress started with our first great fossil fuel—coal. Coal is extracted by both underground and surface mining techniques. Once removed from the ground, coal needs further processing to remove impurities and to increase energy values.

While coal was known and used long before the Industrial Revolution, the needs of manufacturing, residential services, and transportation prompted a coal revolution over several generations in the late eighteenth to the mid-nineteenth centuries, with Great Britain leading the way. Coal replaced many uses of wood and wind. Transportation changed from sail to coal-fired vessels that could ignore wind directions or lack of wind altogether. Transatlantic travel times went from months to a week or two.

On land, the new invention, the railroad, was developed between the 1810s and 1840s and boomed within two decades. The horse was no longer the means of long-distance travel. Coal replaced wood as the locomotive's fuel. Eventually, coal replaced wood and falling water for the manufacture of products. Large factories powered by coal became symbols of industrial success. Wood further lost out in the residential and business heating and cooking realms.

By 1900 coal had become *the* modern fuel. A traveling urban politician of 1900—the heir of our 1788 delegates—would probably have left his coal-heated home to ride a coal-fired railroad to a large city prospering with coal-powered industries and lighted by coal gas-manufactured lighting. In the twentieth century other resources have to some extent replaced the primary position of coal. Coal's decline from the versatile fuel of 1900 has concentrated much of its use as the fuel to run industrial operations and electricity-generating plants. In advanced societies, the use of coal for transportation or residential and commercial heating has become rare.

Coal's undesirable attributes are considerable. Mining, whether underground or surface, is dangerous for workers and damaging to the surrounding environment. Burning of coal probably constitutes the most environmentally harmful use of a fossil fuel. Sulphur, particulate matter, and other releases from coal combustion are harmful to humans and other life forms at both near and far distances. The release of CO_2 and other greenhouse gases make coal combustion one of the worst contributors to global warming.

The growth of coal gave rise to the first energy networks. Coal was not located close to or easily extractable by the ordinary citizen. Human ingenuity and wealth and hard physical labour were needed to locate coal seams, to extract and process the coal, and to ship it to the point of use. Coal can be shipped by truck, rail, barge, ocean freighter, or slurry pipeline. Often half a dozen separate business enterprises might be involved between the discovery and the ultimate consumption of the coal itself. The lumps of coal may have travelled hundreds of miles in the process. The sheer weight of coal can often make its transportation as expensive as its extraction.

2. *Oil*

King Coal in 1900 had already found a powerful challenger in an even more energy-potent fossil fuel—oil. Liquid carbon-based products had long been known. But the last half of the nineteenth century and the first half of the twentieth century saw oil's commercial emergence. Various attractive properties of oil began to make it the fuel of choice for powering water and rail transport, for many industrial purposes, and for residential and commercial heating. Oil also quickly captured two enormous markets: automobile and aeronautic transport. By 1950, transport was overwhelmingly powered by oil.

Oil retains its versatility into the twenty-first century, though some traditional uses are shrinking. Oil retains dominance in transportation, especially for motor vehicles, planes, and ships. Some nations and regions continue significant use of oil for residential and commercial heating and cooking. Industrial processes call on oil both for energy generation and for product component. Oil remains useful for electric generation, although economics often favours other fuels.

Oil, like coal, has its environmental harms. The Gulf Coast oil leak of 2010 provided a vivid reminder of how harmful and widespread an oil spill can be. Extraction and processing of oil presents a wealth of human health and environmental harms. Combustion of oil has a long history of environmental harms that have included smog in major cities and climate change worldwide.

Oil demanded even more networks than coal. Discovery and extraction of crude oil begins the oil cycle. The crude oil that comes out of the ground or from under the ocean needs refining to be usable for human purposes. In many instances this will require networks to get the oil from well-head to refinery to places of use. Pipelines can provide this service on land and in some shallow water situations. Discovery of rich oil deposits tended to be in locations far from the places where large amounts of refined oil were needed—the populated industrial cities of powerful nations. How to get the raw oil from rural Texas or the Dutch East Indies or the Caucasus, or later the Middle East, to its place of use became very important. A variety of networks emerged, from pipelines to long-distance tanker routes. For long-distance ocean shipments (think of moving crude oil from the Middle East to places of use around the world), supertankers take over. For shipments of smaller quantities of refined oil on land, rail, truck, and barge can also contribute. Energy and energy networks became more international, with financial and legal problems often accompanying their creation.

3. Natural gas

A third hydrocarbon fuel—natural gas—was often found with liquid petroleum. For half a century or more, natural gas was regarded as more of a waste product, and a dangerous one at that, than as an attractive energy resource. The gas that was used for energy was manufactured from coal. However, as access to oil became more difficult, local gas fields became highly attractive. By the end of the Second World War natural gas had gradually carved out attractive markets for itself. These included residential and commercial heating and cooking, taking over roles previously served by wood, coal, and oil. Natural gas also plays a major role in industrial production of all sorts. As with oil, natural gas can be both energy fuel or product component. Electricity generation has increasingly adopted natural gas as the fuel of choice. Natural gas serves some transportation needs as well.

Natural gas is the emerging fossil fuel. While many discoveries continue to find natural gas associated with oil, many fields that produce only natural gas have been discovered and developed. Natural gas's reputation as the 'clean' fossil fuel has helped its attractiveness. Its production of pollutants (including CO_2) per unit of energy delivered is below either coal or oil. Nonetheless, combusted natural gas is a pollution source. Additionally, health and environmental problems arise from its extraction (particularly using new hydraulic fracturing technologies) and transportation (eg pipeline explosions).

The nature of a gas made its shipment and storage particularly challenging. Movement across land by pipeline created another networked fuel. In the last decades of the twentieth century, with the growth of environmental awareness, natural gas became a highly valued resource and pipelines made its use feasible. Natural gas is therefore the most networked energy raw material. Its form limits natural gas's ability to move in other than pipeline form. Most land-based gas movement from place of extraction to place of processing to place of end use is by pipeline. Where wide expanses of ocean need to be crossed, the natural gas can be converted to liquefied natural gas (LNG) and shipped to the place of use in special ocean tankers.

4. Electricity

The last decades of the nineteenth century saw the development of a wonderfully versatile energy source—electricity. Unlike the fossil fuels or other energy sources, electricity is not extracted from nature, ready to use for energy purposes with some processing. Electricity needs to be manufactured using one of the primary energy sources. Fossil fuel combustion, the uranium-based nuclear reaction, the power of falling water, sunshine, or wind that allows turbines and generators to operate producing the electricity that can provide heat or light, run an immense variety of industrial and business machines and residential appliances. The computer age that arose at the end of the twentieth century had its origins in the development of electricity at the end of the nineteenth century.

Fossil fuels provide much of the fuel for electric power. Although coal remains the primary source of electricity-generating capacity, natural gas has taken an increasing share of the electricity-generating market. Hydroelectric power—the capability of falling water to provide the motive power for electricity generation—is another most prominent source. The old waterwheel used to grind grain or pump water can now provide many megawatts of electric power. So too, the old windmill has found a modern use in onshore and offshore electricity generation.

The virtues of renewable and non-carbon releasing energy sources have increased the attraction of hydropower in recent years. The major use for falling water in the energy world is for electricity generation. Giant dams provide significant megawattage and produce a major portion of the electricity of some geographically favoured nations. However, new dam sites in areas needing large amounts of power are limited. And, despite its carbon-friendly nature, hydroelectric power imposes environmental burdens and risks that range from population disruptions to risks of dam collapse. Similarly, despite their 'green image', the construction of large-scale wind parks raise objections as they can be noisy and create horizontal pollution.

A major modern source for electricity generation is nuclear power. The discoveries of the first half of the twentieth century that led to the atomic bomb also made possible the generation of large amounts of electric power from the use of enriched uranium nuclear fuel. However, no energy source has raised more controversy over its human and environmental risks than nuclear power. Tough policy questions face investors, government officials, and the general public about the future of nuclear power. How should society balance nuclear energy's potential to generate large amounts of electricity largely free of CO_2 releases against the high consequences of accidents like Three Mile Island in the United States in 1979, Chernobyl in the Soviet Union in 1986, and Fukushima in Japan in 2011?

In general, electricity demanded a new kind of network—the long-distance transmission and local distribution lines that allow the controlled and safe transport of kilowatts of electricity from the place of generation to the place of use in home, office, or factory. Most of the renewable energy sources also need networks. Whereas biofuels that end up in liquid or gaseous form may need pipelines to get from place of creation to place of use, the primary modern uses of hydropower, wind, and tidal power are for electricity generation. Even though most renewable energy projects need to rely on electricity cables, it may well be that a different treatment of these cables is required, as renewable resources are often injected in the distribution grid instead of the transmission grid as usually is the case and thus potentially may lead to balancing problems.

B. Energy sources today

Let us now focus more comprehensively on the world of energy in the early twenty-first century. What are the raw materials of energy? How are they prepared for their end use as a source of energy? What are those end uses? What are the undesirable consequences of those processes of extraction, processing, and use? That big picture helps us to assess more specific network concerns.

The 2010 Key World Energy Statistics of the International Energy Agency (IEA) provides a useful starting point. Its Factsheet projects energy consumption to the year 2035—a date within the lifetimes of many readers of this book, or, at least, their children's. The new Policies Scenario of the IEA study projects an increase of 36 per cent in world primary energy demand from 2008 to 2035. The economically developed and energy-sufficient nations of the Organization of Economic Cooperation and Development (OECD) provide a small portion of that increase. Non-OECD countries account for 93 per cent of the projected increase. China alone provides 36 per cent of the total projected growth. Despite a variety of public policies directed at reducing the consumption of the three fossil fuels (coal, oil, natural gas), 50 per cent of the projected increase in energy consumption over the next quarter-century comes from those fuels.

1. *Hydrocarbons*

As of 2009 statistics, China is the dominant coal producer in the world by a wide margin. The United States ranks second but produces less than one-third of Chinese production. India, Australia, Indonesia, South Africa, and the Russian Federation are next in order. Both China and the United States consume most of the coal they produce. China, in fact, is the second leading importer of coal, trailing only Japan. Australia, Indonesia, and the Russian Federation are the leading exporters of coal.

Oil remains the dominant and the most versatile fossil fuel. Crude oil production (as of 2009 figures) takes place in most parts of the world. The Middle East leads, with a 30 per cent share of total production. The OECD nations follow at 22 per cent followed by the former Soviet Union (17 per cent), Africa (12 per cent), Latin America (9 per cent), and China (5 per cent). Within regions, however, there are wide disparities from country to country. The major producing nations in order are the Russian Federation, Saudi Arabia, the United States, Iran, China, Canada, Mexico, Venezuela, Kuwait, and the United Arab Emirates (UAE). Net exporters of crude oil are led by Saudi Arabia, followed by the Russian Federation, Iran, the UAE, Nigeria, and Angola. The United States (despite ranking third in production) is the top net importer of crude oil followed by Japan, China, Korea, Germany, Italy, France, Spain, and the Netherlands.

Perspectives on oil's future are provided by proved oil reserves—crude oil identified and available for future extraction and use. *Oil & Gas Journal* reports that as of 2010 the largest oil reserves are found in Saudi Arabia, Canada, Iran, Iraq, Kuwait, Venezuela, the UAE, and Russia. The five Middle Eastern countries contain over 50 per cent of total reserves, with Saudi Arabia alone holding nearly 20 per cent. By contrast, China holds 1.5 per cent and the United States slightly less than that.

Natural gas is found worldwide. The leading producers are the United States and the Russian Federation, which divide almost equally 38 per cent of total world production. They are followed by Canada, Iran, Norway, China, Qatar, Algeria, the Netherlands, and Indonesia. The Russian Federation is the leading exporter of natural gas, followed by Norway, Canada, Qatar, and Algeria. The leading importers of natural gas are Japan, Germany, the United States, Italy, and France. Natural gas shows significant growth. A considerable portion of the growth in both oil and natural gas comes in the form of such 'new' fossil fuel resources as oil shale, tar sands, shale gas, coalbed methane, and tight gas.

2. *Electricity*

The fossil fuels continue to provide much of the fuel for electric power. In 2012 coal remains the primary source of electricity-generating capacity. Natural gas has taken an increasing share of the electricity-generating market. Hydroelectric power continues to be a factor in the world's electricity generation mix. The leading producers of hydroelectricity are China, Canada, Brazil, the United States, and the Russian Federation. In Norway, Brazil, Venezuela, Canada, and Sweden hydropower provides the majority, or near majority, of the nation's electricity.

Nuclear power is an even larger contributor to electricity generation. About 30 nations have located nuclear electricity-generating plants in their territory. The leading producers of electricity from nuclear energy are the United States, France, Japan, Russia, Korea, and Germany. The United States licenses the most nuclear plants and produces the most nuclear electricity. France draws the greatest percentage of its electricity (77 per cent) from nuclear power, followed by Ukraine, Sweden, Korea, Japan, Germany, the United States, the Russian Federation, and Canada. At the front end of the nuclear cycle are the sources of the raw uranium that will be processed and refined for nuclear fuel. Among the leading sources of minable uranium are Australia, Canada, Kazakhstan, the Russian Federation, Uzbekistan, Brazil, the United States, Namibia, South Africa, Niger, the Czech Republic, and the United Kingdom.

In addition, the share of 'new' sources of renewable energy is increasing and in some geographic areas and technologies, increasing rapidly. The OECD estimates that all renewables provide about 12 per cent of the world energy supply, with hydropower being the major renewable. Wind, solar, tidal, and geothermal energy are growing contributors, mostly in the area of electricity generation. Certain countries rely significantly on renewable energy, especially hydropower (see figures above). However, other renewable energy sources are becoming more important. Brazil has moved aggressively towards biofuel use. Denmark draws 20 per cent of its electricity from wind power. Iceland is a major geothermal energy user. In the EU the goals set by the European Commission to reach a 20 per cent renewable energy target in 2020 will definitely stimulate the production and use of this type of energy.

3. *The role of networks*

To the extent that energy worries the average citizen, the concern is at the raw materials stage. Coalmines collapse. Petroleum reserves leak. Nuclear reactors malfunction. Petroleum prices escalate. Both short-term accidents and longer-term shortages become the lead story on the evening news. Yet, energy network issues can be just as crucial. Can a new fossil fuel deposit be extracted and shipped by pipeline to its place of processing or end use in a way that meets economic and legal requirements? Can a remotely located renewable energy source be linked to a new or existing electricity grid? Will uncertainties about future supply of a resource or about the regulatory scheme that will govern it a decade in the future discourage multibillion dollar investments?

III. Networks, Changing Markets and the Role of Law

The above historical and contemporary perspective on all aspects of energy has shown that the energy sector is constantly changing due to the availability of energy sources and technical developments. Similarly the role of the transportation network is changing. Two centuries of progress in energy have thus provided a world in which four billion people can call on oil for a wide range of transportation options. They look to natural gas, oil, and electricity to heat and cool their homes, places of work, commerce, and public business. They rely on electricity generated from fossil fuels, nuclear energy, hydropower, and other renewable sources to power the wealth of labour-saving and life-enhancing devices that they take for granted. When relying on these energy resources they also rely on their means of transportation. What type of transportation networks are we actually discussing?

A. The networks further identified

Earlier we have noted that networks can be broadly defined to include everything that occurs between the initial extraction or capture of an energy resource and its final use by a consumer. That definition would certainly include road, rail, and shipping lines as energy networks, even though they serve a wide number of other purposes. For example, the sea route that carries oil or LNG also carries other products and passengers. Sometimes, however, the choice between transportation modes is limited, as in the case of natural gas and electricity. These two energy sources are largely dependent on transportation via pipelines and cables, although other means of transportation are developed and used (LNG and batteries).

Long-distance transportation of electricity and gas is the result of technical developments. For the construction of pipelines use has been made of different types of materials, varying from wood (Ontario 1872) to glass (Maastricht around 1850). Steel has been used as of early 1900. The ability to weld steel pipelines together allows gas to be transported over longer distances. Similarly, the technique of converting high voltages to lower voltages is a crucial element in the development of electricity cables and long-distance electricity supply.

These networks linking producers and consumers are usually subdivided in several categories. The first category includes the pipelines and cables connecting production facilities with the main grid. Whereas these cables are generally considered as part of the installation, a separate category applies to these pipelines as they are referred to as upstream pipelines as a rule. The next category involves the main grid, ie high-pressure and high-voltage networks. These so-called transmission lines (pipelines) mostly cover the entire territory of a state and are considered as the 'national energy highways'. The cross-border connections are interconnections of transmission grids. These transmission lines are again connected to another—third—category of infrastructure, ie distribution lines which operate under lower pressure and/or voltage and have a direct connection to consumers. Each category of networks can be governed by a different type of legislation.

B. Changing markets

The organization of energy markets reflects the above developments. The first electricity and gas supply companies were established at the end of the nineteenth century by municipalities in order to supply local communities. As a result of new techniques the local networks of municipalities have been interconnected, resulting in regional supply companies. Since the Second World War, most developed countries have been able to create national grids and national energy companies. For a number of reasons these national grids were again interconnected. Such a reason could be security of supply or the need to organize a large-scale export of gas, as was the case in the Netherlands after the discovery of the Groningen gas field in 1959. The size of that field, combined with the expectation that nuclear energy would soon take over as the main energy source, led to a situation where gas export contracts were concluded with Belgium, Germany, France, and Italy, and subsequently a transnational European gas transmission grid was developed. Without the Groningen field there would not have been an EU gas market. Currently, gas is imported over even longer distances onshore as well as offshore.

Since the 1980s the energy sector has been faced with considerable challenges and changes all over the world. The energy liberalization/deregulation/privatization process has rendered energy networks more independent from the energy supply and production sector. Hence, the operation of the networks needs to take place without direct involvement/guidance of network users. In addition, the situation on the supply and demand side of the grid is changing as well. Fossil fuels and large hydro plants are increasingly located in distant areas and require the construction of major infrastructure often involving several countries and jurisdictions. By contrast, increasingly new (often renewable) resources are being developed and this again requires a different type of grid development and management. Last but not least, energy suppliers are not the only drivers influencing grid construction and management. Demand-side management also has an impact on grid operations. By controlling and limiting energy demand governments hope to be able to restrict the need for further expansion and development of networks. These changes need to be reflected in the laws governing the sector.

C. The role of law

Developing networks in the above described changing energy markets depends on a variety of legal requirements, which may depend on the type of network (electricity, gas, or oil), its position in the energy chain (upstream, midstream, or downstream) or its actual location (onshore, offshore, or cross-border).

Whichever network is developed, the construction always involves the application of planning and environmental laws. The impact of these laws is becoming increasingly onerous due to increased public participation (the 'not in my back yard', or NIMBY, effect), safety requirements, and more densely populated areas. Additionally, issues of planning and environmental law become more complicated when the network is more international or inter-jurisdictional and is crossing several borders.

Other changes result from an increased use of renewable energy sources. Sometimes this may be the result of clear and binding government targets, as is the case in the EU. The generation of renewable energy sources may be dealt with in a specific law on renewable energy but also in an Electricity Act, Gas Act, or Energy Act. Whichever law is chosen, it will have an impact on the networks and especially as increasingly renewable energy is injected in the distribution grids instead of the transmission grids, as used to be the case. Issues like balancing, congestion management, and grid expansion need to be dealt with and could possibly require innovative legal and regulatory solutions.

Similarly, liberalization processes have had an important impact on the existing legal frameworks. Electricity and gas laws are constantly changed and adapted in order to meet new market requirements. Liberalization does not result in deregulation but severe additional regulation. Governments are put at arms' length of the energy markets but are blamed if there is insufficient energy supply. The latter depends on the availability of reliable networks. What kind of regulatory regime or law is in place to guarantee that the proper networks are developed to transport electricity, gas or CO_2? These questions will be addressed in this book.

IV. The Book

A. Our approach

We have noted that networks can be broadly defined in order to include everything that occurs between the production of energy and its final consumption: transportation by road, rail, and ship, as well as pipelines and cables. In the chapters that follow we use network in a more precise sense. Our focus is on systems exclusively dedicated to the movement of energy resources from place of extraction or capture to place of final use by consumers. The familiar examples are oil and natural gas pipelines and electricity transmission and distribution lines.

What is needed to operate an existing energy network or to create a new one? Experienced energy project developers identify three elements to make an energy project successful.

First, the science and engineering must enable the necessary network features to work. Can transmission lines move generated electricity over hundreds or thousands of kilometres to the consumers? Will the natural gas pipeline safely contain the gas as it moves from field to consumer? Can wind-powered electricity generated offshore be moved to land and connected with existing power grids? Prior scientific discoveries and engineering creativity have answered many network questions in the affirmative. Research and development goes on to enable new network possibilities.

Second, how is the project to be financed? Economic learning is valuable in weighing the return on investment of any network project. Even if the science and technology of the project is assuredly 'doable', the project that has high installation and operational costs and a limited customer base to bear those costs is unlikely to be funded. Financing also raises the fascinating intersection of private and public sectors. The answer to infrastructure in a planned economy was often for the government to be financier, builder, and operator of the network project. Government would look at the economic return on the network project. But it frequently had other objectives in its network decisions. If network financing decisions are left to the workings of the free market, return on investment looms larger. Socially attractive projects (bringing electricity to poor and rural areas, helping to reduce carbon emissions) may not attract investors. A sociologist or political scientist's view of 'necessary' network investment may differ from that of an economist, banker, or finance expert. These tensions give rise to the private–public partnerships that are often essential for network development. The private sector may take the lead and provide a large portion of the investment. However, some government subsidy or assistance or assurance is essential to secure the private sector participants. A wide range of government involvement—direct subsidies, tax benefits, land grants, regulatory waivers, etc—have been used to bring network ventures to successful completion.

Third, what political, public policy, and legal hurdles must be overcome for the network project to succeed? This is often the most crucial inquiry. Many technically feasible and well-financed projects have failed because of government or citizen objection. As the chapters will reveal, law is involved in many ways in network decisions. Competition law, regulated industries law, land use law, environmental law, and cross-jurisdictional laws may impact a single network project. We will consider all these matters of law and more in the chapters that follow.

B. Organization of the book

The chapters lent themselves to a variety of organizational patterns. We selected a four-part division. Part I addresses cross-border energy infrastructure and supply security from a variety of perspectives. Each chapter centres on the fact that significant geographic distance separates the energy resource from the place of end use. What legal consequences arise when the resource—the coal, oil, or natural gas deposit, the hydroelectricity generation plant, the wind or solar energy location—is different from the place of consumer use of the fuel or the electric energy? At least, the laws of two nations may be at play. Principles of international law may also be relevant, particularly where no single national law may be controlling (the high seas, the polar regions, outer space). Even if all activity takes place in one country, that country's federal governmental structure may mean that the laws of a province or state or even municipalities must be considered along with the national law.

Part II examines new energy sources and innovative network management. The emphasis is on 'new'. Innovation is an important feature of contemporary energy. Two decades ago, development of tar sands, or shale gas by hydraulic fracturing, or attention to smart grid systems, or offshore electricity grids, or demand-side management, were in early exploratory stages. Yesterday's bold innovation becomes today's emerging industry. Each of these technological advances have an implication for networks—both existing and future. Can the appropriate network structure be devised and implemented? Is there a financial arrangement that will make it possible to implement the technology? Will legal and public policy issues advance or retard the project?

Part III studies market liberalization and challenges for network investments and planning. As the chapters make clear, the past three decades have seen the triumph of the free market in many energy realms. Socialist states, with the expectation that government would do everything in energy from resource discovery and exploitation to providing energy services to the populace (often at greatly discounted prices to the consumer), have vanished from many parts of the world. Free-market economies have also scaled back government's role in energy production and distribution. Privatization and liberalization have advanced. Yet, the total triumph of the free market is by no means certain. The economic theory and reality of natural monopolies shows that certain business ventures are at their best when a single or few business ventures provide all the product or services to the customers. Generations of economists have used the hypothetical horror of dozens of separate business ventures, stringing electric transmission wires or building dozens of natural gas pipelines that would serve only their product. The infrastructure costs would be enormous. The environmental disruption would be large (imagine a dozen sets of electric wires penetrating a crowded downtown or remote countryside). Harsh competition is likely to promptly reduce the field to a single survivor, who would then hold monopolistic power unless it is controlled by government in some form. Several chapters note that government policy, occasionally expressed in binding law, may seek more network development than the profit-driven free market is ready to provide. Consider the case of a geographic region lacking basic electricity or natural gas services. Government, for self-serving or altruistic motives, may want the services expanded to the region. The private sector does not see a profitable return on the investment. What are the government's options?

Part IV presents an overview of our findings and presents some of our ideas on network development, the innovative solutions required, and the role of law and legislators in shaping the law and the daily life of network operators.

C. The chapters

Twenty-nine authors and editors have contributed to this book. They are primarily based in fifteen nations, although many authors' educational and professional careers have intersected with multiple nations.

Alastair Lucas begins Part I by explaining cross-border issues in a single nation, Canada. He identifies the federal structure of the nation that gives individual provinces and First Nations considerable power over resources and networks. Canada is also a party of the North America Free Trade Agreement (NAFTA). This brings us to the next chapter in which José Juan Gonzalez provides a study of networks in Mexico and the impact of integration with North and Central America. The chapter examines the distinctive regulatory issues that are present as a result. Thereafter Lila Barrera-Hernández highlights regional arrangements in South America that advance energy integration in the region. Three significant multi-state agreements have the potential to provide the framework for moving energy from source to end user in South American nations. Next we move from the Americas to another continent in the Southern hemisphere: Africa. Mohammed Bekheshi's study of the Chad–Cameroon Pipeline provides a multinational case study. It also introduces some of the complexities of project financing, especially where agencies like the World Bank are involved. From Africa we continue with two chapters on Europe. Catherine Redgwell introduces us to arrangements supporting large European transboundary pipeline projects. She observes the mixture of treaty arrangements and private contractual agreements that often govern transnational network projects. Martha Roggenkamp concludes the first part with a European Union perspective. She highlights the crucial aspects of security of supply and the safe treatment of networks to achieve that objective.

Part II considers new energy sources and innovative ways of network management. We begin this part with chapters on innovative network management. Anita Rønne commences with an analysis of 'smart grids' or 'intelligent energy systems' from an EU perspective. Smart grids can play a large role in balancing energy supply and demand and thus in the envisaged large-scale introduction of renewable energy sources. The study of demand-side management continues with Lee Paddock's and Charlotte Youngblood's look at United States practices. The authors study both the technology and the supporting infrastructure development in this emerging area. The US also features in the next chapter by Olivia Woolley, Peter Schaumberg, and Graham St. Michel on offshore wind. Their study concentrates on wind power development in the North Sea (Europe) and US waters and the network structure necessary to make it feasible. These subsea cables are a typical example of new and innovative networks. Ulf Hammer brings the focus to another renewable energy resource requiring long-distance cables. He looks at Norway's heavily renewable electricity industry, which consists mainly of hydropower. He fits network development with the need for a more liberalized market. Renewable energy sources are also the focus of the next chapter by Wang Mingyuan. He highlights the large role of energy issues in China and, in particular, the role of networks in advancing renewable energy projects in China. Finally, Nigel Bankes and Rick Nilson examine a new technique that can be useful in reducing carbon emissions and limiting global warming, ie carbon capture and storage. Pipelines may prove to be essential in bringing the captured CO_2 to the subsoil reservoirs. When choosing a regulatory framework for developing and exploiting CO_2 pipelines, lessons can be learned from the natural gas sector.

Part III concentrates on market liberalization and challenges for network investment and planning and provides different national perspectives on the tensions between network financing and market liberalization. We note that liberalization policies have been introduced worldwide, Brazil being one of the more recent examples. Yanko Marcius de Alencar Xavier and Anderson Souza da Silva Lanzillo present a Brazilian perspective on liberalization issues. Their study of natural gas and natural gas networks in Brazil provides a good example of the ebb and flow of regulation and deregulation in a major emerging economy. This is followed by Barry Barton presenting us with an overview of energy network regulation in New Zealand. Both New Zealand's distinct geography and its variant enthusiasms for regulation remind us that similar legal issues can be differently treated in different nations. Australia has started slightly earlier on the path of energy market liberalization. Lee Godden and Anne Kallies examine Australia's new challenges in transmission network development. They note that control of carbon emissions and serious problems with bushfires caused by ageing networks have pushed the development of new networks. Next we move to Europe again, where energy market liberalization started some 30 years ago, beginning with the United Kingdom. The UK has engaged in some of the most creative thinking about paying for additional network development. Aileen McHarg reviews the sophisticated efforts to balance necessary regulation and adequate incentive for private investment. Network investment incentives also loom large in Tjarda van der Vijver's examination of the regime exempting investors from the general rules on third party access to energy networks. Such exemptions have been considered necessary to organize for sufficient investments in networks and network developments. Network investments also play a dominant role in the next chapter, where Iñigo del Guayo and Johann-Christian Pielow provide a European Union perspective on both electricity and gas infrastructure planning. The need for such planning is illustrated in the subsequent chapter, where Gunther Kühne highlights Germany's proposed massive shift to renewable generating sources for its electricity and the challenge of moving electricity generated by offshore wind parks to the major electricity-consuming regions in southern Germany. This shift requires massive investments in grid development. The last chapter brings us to Russia. Sergey Seliverstov and Ivan Gudkov consider both gas and electricity networks in the Russian Federation. Their study captures the uneasy balance of private sector and governmental action in the treatment of networks and land use.

Part IV presents a conclusion and overview. In the final chapter the editors put forward some observations on the rich lessons to be learned about the legal and policy treatment of networks in the early part of the twenty-first century.

D. The book and the Academic Advisory Group

This book marks the sixth collaboration of the Academic Advisory Group (AAG) of the International Bar Association's (IBA's) Section on Energy, Environment, Resources, and Infrastructure Law (SEERIL) with Oxford University Press

(OUP). One of the delights of the work of our 2010–12 cycle has been the discovery of connections between the networks project and our five previous OUP projects: *Property and the Law in Energy and Natural Resources* (2008–10), *Beyond the Carbon Economy: Energy Law in Transition* (2008), *Regulating Energy and Natural Resources* (2004–06), *Energy Security: Managing Risk in a Dynamic Legal and Regulatory Environment* (2004), and *Human Rights in Natural Resource Development: Public Participation in the Sustainable Development of Mining and Energy Resources* (2002). Earlier insights from those works have been confirmed or modified in our work here. We are confident that the networks project will enhance our 2012–14 project. Energy law is regularly changing and rarely dull.

PART I

CROSS-BORDER ENERGY INFRASTRUCTURE AND SUPPLY SECURITY

2

Canadian Energy Infrastructure and the Federalist Dilemma

Alastair R. Lucas

I. Introduction

Development of major energy network infrastructure in federal states faces a fundamental dilemma. On one hand, increasing size and scope of projects produces corresponding increase in national significance and national concern about issues raised by projects. This militates in favour of central government authority. On the other hand, because infrastructure is local, on-the-ground reality—from planning and construction to operation and maintenance—local authority is engaged amid cries for local autonomy and control. All of this may shine a spotlight on the federal system itself, reopening central–local fissures or creating flashpoints for broader local–central conflict. Thus infrastructure project developers face the equally unappealing prospects of dealing with intractable local issues or complex and far-reaching national issues. Whichever level is successful, there is a strong likelihood that the projects will become a source of local–central federalism disputes.

It gets worse. There may be additional levels of government in federal systems. Whether these are recognized formally, given tacit recognition, or regarded as potentially authoritative on certain matters, these levels of government may have to be dealt with by energy infrastructure developers. The most obvious is local or municipal governments. Another level, common and increasingly powerful in Canada, is First Nations governments. The latter may draw explicit authority from legislatively confirmed modern treaties, from historic treaties, and even from claimed but unconfirmed 'aboriginal rights' protected by the Canadian Constitution.[1]

[1] Constitution Act 1982, s 35, being Schedule B to the Canada Act 1982 (UK), 1982, c 11.

II. Federalism Constraints and Solutions

This chapter is about public regulation in an interjurisdictional setting. It asks whether there are fundamental constitutional federalism constraints that must be taken into account in energy infrastructure development. If so, what is the potential impact of these constraints?

For this purpose, case studies of two major Canadian energy infrastructure projects are undertaken. Both involve pipelines—one oil and the other natural gas. Electricity infrastructure, though it raises issues of interest, is not addressed. Both are private sector projects, a common form of major Canadian infrastructure development.

If there are federalism constraints that go beyond mere regulatory requirements and process, what arrangements or instruments can be used to mitigate constitutional constraints? Major issues highlighted by the case studies are facilities' approval and regulation, environmental, and socio-economic impacts, including effects on property and aboriginal rights of First Nations.

III. Theories of Federalism

Federalism is thought to be an almost organic element of human social and political organization. Strands can be glimpsed in the alliances and coalitions of the ancients and perhaps in the relationship of the mediaeval church and early nation states or their precursors.[2] It is a way to explain many types and variations of group political behaviour, and it has been a response to concentration of political power and a template for power sharing in its various forms.

Perhaps most commonly, federalism is associated with the American state—a model that incorporated the ideas of equal relationships and mechanisms to maintain the designed balance.[3] More recently, Europe produced a different federalism around the idea of a treaty-created union and the principle of subsidiarity.[4] The latter, at least in principle, situated government action as close as possible to the citizens likely to be most affected.

Canadian federalism developed in the period between the American Declaration of Independence and the European Union. Its roots in the mid-nineteenth century were pragmatic—security for the diverse British North American colonies, the beginnings of devolution of power from Britain, the need to reconcile the English

[2] D. Elazar, 'The Political Theory of Covenant: Biblical Origins and Modern Developments' (1980) 10 *Publius* 3; D. Elazar, *Federalism: An Overview*, Tuscaloosa: University of Alabama Press, 1987, 19.

[3] The Federalist Papers: Thomas (Library of Congress), available at: <http://thomas.loc.gov/> (accessed 25 September 2011).

[4] A. Moravcisik, 'Federalism in the European Union: Rhetoric or Reality', in K. Nicolaidis and R. Howse (eds), *The Federal Vision: Legitimacy and Levels of Governance in the US and the EU* (Oxford: Oxford University Press, 2001) 161.

and French language, culture, and political traditions, and the imperative to begin building a nation 'from sea to sea', and incidentally holding American Western North American expansion in check.[5] So we see the main currents, first the idea of a compact between the two founding peoples that included fundamental guarantees and the institutions and processes to deliver them, and second the imperatives of nation building in a difficult and demanding geographic and geopolitical context. A third thread can only be glimpsed at the time of Confederation in 1867: guaranteeing and protecting civil liberties.[6]

The compact theory of Canadian federalism has been articulated by scholars and to some degree by courts, at least as an aid to constitutional interpretation, in constitutional cases.[7] It sits closely with federalism theories like those of Daniel Elazar, that see federalism as a form of political organization that both distributes power among general and regional governments and recognizes and protects the authority and integrity of each government.[8] The federation is created by bargaining, and maintained and renewed by active intergovernmental relations.[9] Furthermore, it is anchored by a moral or ethical perspective that underpins commitments to justice in political relationships.[10]

IV. Canadian Confederation and the Formal Division of Powers over Energy Infrastructure

A. Land, resources, and energy

The Canadian judicial approach that has been taken to resolving federal–provincial jurisdictional disputes concerning energy resources may be explained by classical political organizational federalist theory. This describes constitutional arrangements involving sharing of powers between central and sub-units in a state according to an agreed division of powers.[11] It is the result of the pre-constitutional discussions among representatives of the federating colonies in the 1860s and produced sufficient agreement to permit English drafters to prepare the bill that the Imperial Parliament enacted as the British North America Act 1867.[12]

5 P. Hogg, *Constitutional Law of Canada* (5th edn), Toronto: Carswell, 2009, 52. 6 Ibid.

7 P. Russell, *Constitutional Odyssey* (3rd edn, Toronto: University of Toronto Press, 2004) 17–18; R. Vipond, 'Whatever Became of the Compact Theory?' (1989) in *Queen's Quarterly* 96 793. In *Re Resolution to Amend the Constitution*, [1981] 1 SCR 753, the Supreme Court of Canada said at p 816 that compact theories operate in the political realm and 'do not engage the law, save as they might have some peripheral relevance to actual provisions of the British North American Act and its interpretation and application'.

8 D. Elazar, 'Federalism vs Decentralization: The Drift from Authenticity' (1976) 6 *Publius* 9.

9 D. Elazar, *Exploring Federalism*, 16–17 (Tuscaloosa: University of Alabama Press, 1987).

10 Ibid at 146.

11 See A.V. Dicey, *Introduction to the Study of the Law of the Constitution* (London: MacMillan, 1985) 131.

12 See 'From Contact to Confederation', in Joel Bakan et al (eds), *Canadian Constitutional Law*, (3rd edn, Toronto: Emond Montgomery, 2003) 77–90 [The Constitutional Law Group].

The agreement that emerged from the colonial representatives' negotiations addressed division of legislative jurisdiction between the central (federal) government and the colonies which became provinces. But the representatives also had to consider ownership of the abundant public land and resources within the boundaries of each federating colony. The result—recognition that each province owns the public land and the resources within its boundaries[13]—correlates with the agreed provincial legislative jurisdiction in relation to property, local matters, and the 'management and sale of the public lands belonging to the provinces'.[14] Even more significant, this allocation of land and resource ownership to the provinces supported the division of legislative powers by ensuring that the provinces had not just a tax base but a public resource endowment that would generate at least a portion of the provincial revenues necessary to manage government operations. This loose federation, with provincial property ownership and broad exclusive provincial legislative powers, presents a strong analogy to international obligations of sovereign states.

Thus, at least a plausible theory begins with the idea of an agreement or compact among the federating provinces, and continues with the federal arrangements being supported at least in part by the public land and resources owned, managed, and potentially developed by each province. There was, of course, no equity in these resource endowments. Energy and other valuable resources are where one finds them. So nearly 150 years later, Canada has energy resource-rich and energy resource-poor provinces. The scale of energy resource wealth in some provinces led, not surprisingly, to federal attempts to obtain a share. In response, provinces took defensive action to thwart perceived federal intrusion. These disputes and the litigation that resulted[15] have contributed significantly to the judicial delineation of federal and provincial legislative jurisdiction in relation to energy resources. There are two main strands: energy and environmental protection, and jurisdiction over transportation infrastructure.

B. Energy and environmental protection

Consideration of constitutional jurisdiction over environmental protection, including that related to energy resources, brings in different perspectives—ideas of national vision, protection, and conservation. No one thought about environmental protection and management in 1867, though they did address some specific natural resource issues with direct environmental implications. The most obvious is fisheries, which made the federal list of exclusive subjects of legislative jurisdiction.[16] Though

[13] See Constitution Act 1867 (UK), 30 & 31 Vict, c 3, s 109, reprinted in RSC 1985, App II, No. 5; and Natural Resources Transfer Agreements, which placed the prairie provinces in the same position as the other provinces. See *Reference Re: Proposed Federal Tax on Exported Natural Gas, 1982*, 1 SCR 1004 ('Natural Gas Reference').

[14] Constitution Act 1867, s 91(5).

[15] Natural Gas Reference, above n 13; see A. Lucas, 'Energy Law: The Court and the Prosperity Bonus' in J. Swainger (ed), *The Alberta Supreme Court at 100*, Edmonton: University of Alberta Press and Osgoode Society, 2007, 227 at 241–4.

[16] Constitution Act 1867, s 91(12) (Sea Coast and Inland Fisheries).

this was seen in 1867 as concerning the fishing industry, fish nevertheless came to play a key role in federal environmental jurisdiction in relation to water quality.[17]

Environmental jurisdiction issues arose in the 1970s as both provinces and federal government moved to address emergent environmental protection issues. The approach was functional and pragmatic, involving prohibitions and contaminant discharge licensing systems. Initially, in the 1970s, the federal government confined itself largely to legislation that established national standards, along with consultation to encourage provinces to enact environmental legislation.[18] At the time, constitutional scholars argued that while there was overlap, federal environmental jurisdiction was largely limited to supporting the provinces.[19]

Provincial resource ownership and reliance on resource revenues caused apprehension about federal intrusion. Alberta, with its energy resource dependence is perhaps the best example. In Alberta, according to Grace Skogstad, 'environmental policy is energy policy'.[20] This pattern of federal–provincial dialogue and cooperation suggests that intergovernmental relations theories may be useful in explaining the development of federal and provincial environmental jurisdiction.[21] Scholars have agreed that if there ever was a pattern of federal environmental jurisdiction deference to the provinces in the 1970s this changed significantly thereafter. Deference and collaboration gave way to greater federal environmental assertiveness based on public (and therefore electoral) pressure and on international influences.[22] Citizen involvement, often through environmental group action, validated by judicial decisions such as in *Friends of the Oldman River Society v Canada (Minister of Transport)*, has arguably played a role in pressuring the federal government to assert its environmental jurisdiction.[23]

For energy infrastructure, the most significant federal environmental legislation concerned environmental impact assessment. After two decades of measured and deferential federal environmental assessment action, a national system was established in 1992 by the Canadian Environmental Assessment Act.[24] The Supreme Court of Canada held the Act to be constitutionally valid in *Oldman River Society*.[25]

[17] K. Webb, 'Pollution Control in Canada: The Regulatory Approach in the 1980s: A Study Paper Prepared for the Law Reform Commission of Canada' (1988) Ottawa: Law Reform Commission of Canada.

[18] G. Skogstad, 'Intergovernmental Relations and the Politics of Environmental Protection in Canada' in Holland et al (eds), *Federalism and the Environment* (Westport, Conn.: Greenwood Press, 1996) 103 at 108. For a broader review, see W. MacKay, 'Canadian Federalism and the Environment: The Literature' (2004) 26 *Georgetown International Environmental Law Review* 25.

[19] L. Lundquist, 'Do Political Structures Matter in Environmental Problems? The Case of Air Pollution Control in Canada, Sweden and the United States' (1974) 17 *Can. Pub. Admin.* 119 at 130.

[20] Skogstad, above n 18 at 109.

[21] C. Thompson, 'Intergovernmental Relations: A Legal Review' (2008) [unpublished, archived at University of Calgary Faculty of Law].

[22] K. Harrison, *Passing the Buck: Federalism and Canadian Environmental Policy*, Vancouver: University of British Columbia Press 5, 1996.

[23] N. Hawke, 'Canadian Federalism and Environmental Protection' (2002) 14 *Journal of Environmental Law* 185 at 188–9; *Friends of the Oldman River Society v Canada (Minister of Transport)* (1992) 1 SCR 3.

[24] Canadian Environmental Assessment Act, SC 1992, c 37.

[25] *Oldman River Society*, see n 23.

The result is overlapping federal and provincial environmental jurisdiction in relation to energy activities. For the most part, the federal and provincial governments have operated collaboratively.[26] These collaborative relations have been formalized by federal and provincial legislation empowering ministers to enter into intergovernmental agreements,[27] such as the Canada Wide Accord on Environmental Harmonization[28] and its sub-agreements[29] on environmental standards and environmental assessment. The latter has resulted in collaborative federal–provincial assessment of many major energy projects.[30] The next section sets the stage for a more doctrinal approach to federal and provincial energy and environmental law jurisdiction by identifying the main relevant heads of provincial and federal constitutional jurisdiction.

C. Powers in relation to energy infrastructure and environment

Focus on energy and environmental jurisdiction requires that proprietary rights and legislative powers be distinguished.[31] The Constitution Act 1867 distributes legislative powers between the federal government and the provinces. Under section 109, it also confirms provincial ownership of public lands and resources within their boundaries. As owners, provinces have the legal authority, based ultimately on the Royal Prerogative, to manage and sell this property.[32] Thus, in an otherwise archaic decision, the Privy Council held that while a province cannot legislate to prevent aliens from working in mines in the province,[33] it can, as owner, enforce provisions in Crown timber leases preventing employment of aliens.[34]

Provincial public property ownership must be respected by the federal government in exercising its legislative powers. The 1898 Fisheries Reference, in which the Judicial Committee of the Privy Council was asked to state which level of government had jurisdiction to issue exclusive fishing leases in provincial rivers and lakes, is illustrative. In its decision, the court concluded that fishery leases were exclusively within the power of the provinces.[35] Further, the court stated that this provincial power cannot be so constrained by federal legislation under its

[26] Though some scholars maintain that in most cases governments favour a competitive approach: see S. Kennett, 'Meeting the Intergovernmental Challenge of Environmental Assessment' in P. Fafard and K. Harrison (eds), *Managing the Environmental Union* (Regina: Institute of International Relations; Saskatchewan Institute of Public Policy, 2000) 107.

[27] See Canadian Environmental Assessment Act, SC 1999, c 33, ss 9–10 ('Agreements Respecting Administration' and 'Agreements Respecting Equivalent Provisions', respectively).

[28] Canadian Council of Ministers of the Environment, 'Canada Wide Accord on Environmental Harmonization', available at CCME <http://www.ccme.ca/assets/pdf/accord_harmonization_e.pdf> (accessed 25 September 2011).

[29] Ibid.

[30] Such as the joint federal–Alberta environmental assessment panels on the Shell Muskeg River Oil Sands Mine and Upgrader and the Imperial Resources Kearl Oil Sands facilities.

[31] *Canada (AG) v Ontario (AG)* (1898) AC 700 (Privy Council) ('Fisheries Reference'). See A. Lucas and C. Sharvit, 'Proprietary and Legislative Powers Distinguished' in *Canadian Environmental Law*, R. Cotton and A. Lucas (eds), 2nd edn, Toronto: LexisNexis, 1991) c 3 at 3.10–3.12.

[32] Constitution Act 1867, s 109; and Resources Transfer Agreements (Constitution Act 1930).

[33] *Union Colliery v Bryden* (1899) AC 850 (Privy Council).

[34] *Brooks-Bidlake & Whiltall v British Columbia (AG)* (1923) AC 450 (Privy Council).

[35] Fisheries Reference, see n 31.

fisheries management powers that the value of the provincial property is negated. Lord Herschell said:

> It must also be borne in mind that there is a broad distinction between proprietary rights and legislative jurisdiction. The fact that such jurisdiction in respect of a particular subject matter is conferred on the Dominion Legislature, for example, affords no evidence that any proprietary rights with respect to it were transferred to the Dominion. There is no presumption that because legislative jurisdiction was vested in the Dominion Parliament proprietary rights were transferred to it. The Dominion of Canada was called into existence by the British North America Act 1867. Whatever proprietary rights were at the time of the passing of that Act possessed by the provinces remain vested in them except such as are by any of its express enactments transferred to the Dominion of Canada.[36]

The consequence is that provinces can deal with their public land and resources like any other owners, including attaching conditions to leases or sales that if legislated would be beyond provincial jurisdiction.[37] In the absence of federal legislation that conflicts directly, such conditions are enforceable by provinces as matters of property and contract law. In principle, provinces may structure their public resource transactions by, for example, attaching environmental conditions and retaining ownership of energy resources to the point of export, so as to maximize provincial legislative powers.[38] From another perspective, federal environmental legislation, even legislation based on peace, order, and good government (POGG) power, should in its scale of impact on provincial powers and in its relative distinctness of subject matter leave appropriate room for provincial public resource management.[39]

D. Powers over transportation undertakings

The British North America Act expressly contemplated regulation of major transportation infrastructure, particularly railways, shipping lines, and telegraphs. So we see the idea of nation building, but very much from a nineteenth-century perspective. These powers were assigned to federal jurisdiction as exceptions to provincial jurisdiction over 'local works and undertakings'.[40] But the exceptions clause included a general category: 'other works or undertakings connecting the Province with any other or others, or extending beyond the limits of the Province'.[41] There was also a sweeping federal power in parliament to declare 'works' in provinces to be for the 'general advantage of Canada' and thus assure exclusive jurisdiction.[42]

The interconnecting undertakings power remains significant for interjurisdictional network infrastructure, including pipelines and electricity transmission lines. The declaratory power has not been used for 40 years, though the Supreme

36 Ibid at 709–10.

37 See D. Thring, 'Alberta, Oil and the Constitution' (1979) 27 *Alta L. Rev.* 69; W. Moull, 'Natural Resources: the Other Crisis in Canadian Federalism' (1980) 18 Osgood Hall LJ 1.

38 See Natural Gas Reference, above n 13.

39 See the POGG, 'national concern' test outlined in *R. v Crown Zellerbach Canada Ltd* (1988) 1 SCR 401.

40 Constitution Act 1867, s 92(10)(a).

41 Ibid.

42 Ibid at s 92(10)(c), read with s 91(29).

Court of Canada has explained that, when clearly made, these declarations are essentially immune from judicial review.[43]

E. Provincial energy infrastructure powers

Potential support for provincial energy infrastructure legislation is provided by the following provisions of the Constitution Act 1867 (formerly, British North America Act). Most important is s 92(13): 'Property and Civil Rights in the Province'. In *Bedard v Dawson and A-G Quebec*, Iddington, J. said:

> I have long entertained the opinion that the provincial legislatures have such absolute power over property and civil rights, as given them by section 92 of the BNA Act, item 13 thereof, that so long as they did not in fact encroach upon the powers assigned by the said Act to the Dominion Parliament it would be almost impossible to question any such exercise of power...[44]

Other provincial powers include:

- s 93(16) 'Generally all Matters of a Local or Private Nature in the Province';
- s 92(10) 'Local Works and Undertakings...';
- s 92(5) 'The Management and Sale of the Public Land and Resources belonging to the Province...';
- s 92A 'Conservation and Management of Non-Renewable Natural Resources, Forestry Resources and Sites and Facilities for the Generation and Production of Electric Energy'.

This latter provision, added by the Constitution Act 1982, was intended to confirm settlement of the federal–provincial 'energy wars' of the 1970s. Courts have given it little effect.[45]

- s 109 'Natural Resources Transfer Agreements'

Jurisdictional implications of provincial ownership of public natural resources within their boundaries are discussed in *Land, Resources, and Energy*, above.

F. Federal energy infrastructure powers

On the federal side, the following Constitution Act, 1867 provisions are relevant:

- s 91(1A) 'The Public Debt and Property'

This power permits federal spending, with conditions relevant to energy and environment infrastructure matters.[46]

- s 91(2) 'The Regulation of Trade and Commerce'

[43] Hogg, see n 5. [44] *Bedard v Dawson and Quebec (AG)* (1923) SCR 681 at 683.
[45] See Lucas, above n 15 and Moull, above n 37.
[46] *Winterhaven Stables Ltd v Canada (AG)* (1988), 62 Alta LR (2nd) 266 (CA).

This includes economic and potential environmental effects of energy import and export facilities and activities.[47]

- s 91(3) 'The Raising of Money by any Mode or System of Taxation'

This power authorizes establishment of taxes that may be designed to impact energy infrastructure matters.

- s 91(10) 'Navigation and Shipping'

Energy products moved by ships may be regulated under this power.[48]

- s 91(12) 'Sea Coast and Inland Fisheries'

This power authorizes legislation to protect water quality, and aquatic habitat for fish.[49] Energy exploration, development, production, transportation, and use have potential to harm aquatic environments.

- s 91(24) 'Indians and Lands Reserved for the Indians'

This power supports legislation consistent with a federal Crown fiduciary obligation that arguably includes ensuring sustainable energy development and infrastructure on aboriginal lands.[50]

- s 91(27) 'The Criminal Law'

This conceptual power is wide and consequently controversial, as discussed below. It authorizes laws that establish prohibitions and penalties directed at a 'legitimate public purpose', that could, in principle, include actions related to energy infrastructure that cause societal harm.[51]

- 'Peace, Order, and Good Government' (POGG)

Generally conceived as a residual power, POGG is based on the opening clause of s 91. There are (temporary) national emergency and (non-temporary) national concern elements. It can apply only to subject matters characterized as single distinct and indivisible and not be of a scale that would have a significant impact on matters of provincial concern. Inability of a province to deal effectively with a matter is considered to be a prime indicator.[52]

- s 91(29) 'Subjects Expressly Excepted From Those Assigned to the Provinces'

For greater certainty, this clause confirms allocation to the federal government of matters, such as interconnecting undertakings (eg pipelines and railways) under

[47] *Westcoast Energy Inc. v Canada (National Energy Board)* (1998) 1 SCR 322.

[48] *R. v Canada Steamship Lines Ltd* (1961) 127 CCC 205 (ON Co Ct).

[49] *R. v Northwest Falling Contractors* (1980) 2 SCR 292.

[50] See Hogg, above n 5 at c 28. But provincial laws of general application, such as heritage conservation, legislation that may for example affect pipeline and powerline infrastructure, apply to Indians and Indian Lands: *Kitkatla Band v British Columbia (Minister of Small Business, Tourism and Culture)* (2002) 2 SCR 146.

[51] *R. v Hydro Quebec* (1997) 3 SCR 213.

[52] *R. v Crown Zellerbach Canada Ltd* [1988] 1 SCR 401.

s 92(10)(a), that are expressly excepted from broader provincial local works and undertakings powers in s 92.

V. Overlap and Conflict—Limits on Federal and Provincial Powers

According to well-established jurisprudence, the constitutionality of both federal and provincial energy and environmental legislation can be challenged on three main grounds. The first is the *validity* of the statute, namely that its subject matter is outside the enacting legislature's authority and thus within the authority of the other level of government.[53] A second type of challenge is to the *applicability* of legislation. A law can be fundamentally valid, but may have to be limited or 'read down' so as not to apply to certain activities in order to avoid entering the exclusive jurisdiction of the other legislature. Doctrinally, this is referred to as interjurisdictional immunity.[54] The third ground, *operability*, comes into play when both federal and provincial statutes apply to the same subject, but nevertheless, as a matter of validity, each can be upheld as in pith and substance in relation to federal and provincial powers. This situation has been labelled 'double aspect'.[55] Each statute is valid and can operate, unless they come into conflict. This can happen either where there is conflict in operation so that compliance with one statute results in breach of the other, or where federal legislative purposes would be 'frustrated'.[56] In either event, the doctrine of federal paramountcy results in the provincial law being declared inoperative to the extent of the conflict.[57]

VI. Other Governments

A. Municipalities

Municipalities are governmental institutions subject to exclusive provincial jurisdiction under s 92(8) 'Municipal Institutions in the Province'. They operate under provincial municipal government statutes that authorize municipalities to carry out their essential local government legislative, regulatory, and operational functions. Residual 'good government' provisions are given a purposive interpretation—empowering municipalities, for example, to regulate pesticide use and application,[58] and planning powers authorizing by-laws to regulate certain energy activities such as fuel sales by service stations, and propane and compressed natural gas distributors.

[53] See *Hydro Quebec*, above at n 51.
[54] See *Kitkatla*, above n 50 at paras 65–8.
[55] *Canadian Western Bank v Alberta* (2007) SCC 22 [CWB].
[56] *Multiple Access v McCutcheon* (1982) 2 SCR 161; Natural Gas Reference, above n 13; ibid.
[57] *Bank of Montreal v Hall* (1990) 1 SCR 121.
[58] *114957 Canada Ltee (Spraytech, Societe d'arrosage) v Hudson (Town)* (2001) SCR 241.

B. First Nations

Under section 91(24), the federal government has legislative power over 'Indians and lands reserved for the Indians'. This includes reserve lands, lands recognized by the Proclamation of 1763 as 'reserved' for the Indians (all land specified in the Proclamation held at that time by the Indians and not ceded to the Crown), as well as lands held under 'aboriginal title'.[59] The s 91(24) legislative power does not give the federal government proprietary rights in these Indian lands.[60]

Generally, provincial energy and environmental laws apply to Indians and Lands reserved for the Indians. But aboriginal cultural values and practices, including certain hunting, fishing, and trapping activities are, since 1982, constitutionally protected and cannot be impaired without compelling justification.[61] Nor can provincial laws remove or infringe rights to take game and fish for food granted by the Natural Resources Transfer agreements which were confirmed by the British North American Act 1930.

The result is that energy operations on reserve lands within provinces have been managed by the federal government for the benefit of the First Nations. Though never challenged, elements of Alberta's energy regulatory regime (though not energy resource rights ownership provisions), particularly infrastructure approval and public safety and environmental operating requirements managed by the Energy Resources Conservation Board, have been applied to these Indian lands.[62]

Provincial environmental laws of general application also apply to Indians and Indian lands, subject to the qualifications above.[63] However, provincial energy and environmental laws of general application cannot infringe 'aboriginal and treaty rights' protected under s 35 of the Constitution Act 1982.[64] But the reconciliation theory developed by the courts contemplates the possibility of aboriginal rights qualification by provincial or federal legislation, subject to 'justification' consistent with upholding the 'honour of the Crown' in its dealings with First Nations.[65] A significant justification element is consultation of aboriginal people affected by action taken under otherwise valid provincial energy and environmental legislation.[66] The parameters of this consultation, which may be by the federal or provincial Crown, and which has become a critical element in the assessment, review, and approval of energy infrastructure projects that affect First Nations or their people, have been outlined in a series of Supreme Court of Canada decisions.[67] The result of these constitutionalized rights, including

[59] *Delgamuukw v British Columbia* (1997) 3 SCR 1010.

[60] *St Catherines Milling v The Queen* (1888) 14 App. Cas. 46 (Privy Council).

[61] See *Dick v The Queen* (1985) 2 SCR 309.

[62] In cooperation with Indian Oil and Gas Canada, a federal agency: Energy Resources and Conservation Board, 'ERCB EnerFAQs 10: Public Health and Safety: Roles and Responsibility of Agencies that Regulate Upstream Oil and Gas', available at <http://www.ercb.ca/> (accessed 2 October 2011).

[63] Hogg, above n 5 at para 28.2; *Dick v The Queen*, above n 61.

[64] *R v Badger* (1996) 1 SCR 771.

[65] *R. v Sparrow* (1990) 1 SCR 1075.

[66] *Haida Nation v British Columbia* (2004) 3 SCR 511.

[67] See especially *Haida Nation*, ibid; *Mikisew Cree First Nation v Canada* (2005) 3 SCR 388.

Treaty rights, consultation requirements, and limited self-government rights, is that First Nations do for many purposes have 'sub-national' status within the Canadian federation.

VII. Jurisdiction over Environmental, Public Health, and Socio-economic Effects of Particular Energy Activities

A. Conventional oil and gas

There is little doubt that the most direct societal impacts of conventional oil and gas activities are essentially within provincial jurisdiction. This includes the access roads, gathering pipelines, and upgrading facilities that are fundamental elements of hydrocarbon exploration. It is based primarily, as indicated above, on ss 92(13) (16) (5) (10) and s 92A. In addition, as suggested above, some measure of practical authority flows from provincial ownership of the resources themselves. Alberta (as opposed to private persons), for example, owns over 80 per cent of the oil and gas reserves within its boundaries.

B. Oil and gas network infrastructure

Focusing now on network infrastructure, jurisdiction over pipeline gathering and transportation facilities, including their direct environmental impacts, is complex. International and interprovincial pipelines are federally regulated (by the National Energy Board (NEB)) under s 92(10)(a)—interjurisdictional works and undertakings. Pipeline-gathering facilities, including gas-processing plants within a province, are within provincial authority unless these facilities are owned and operated by an interprovincial transmission company and dedicated to exports from a province.[68] Connection of a new gas processing plant with an interjurisdictional pipeline may bring the plant and its associated gathering lines under federal jurisdiction.[69]

However, in principle, provincial environmental legislation of general application should still apply to these federally regulated pipelines. Barring direct conflict in operation (compliance with one law results in breach of the other), both provincial and federal environmental requirements should operate concurrently.[70] Operational conflict, which would result in paramountcy of the federal law, seems unlikely, since meeting the most stringent requirements should permit compliance with both laws.[71] Unrealistic or unduly burdensome provincial requirements aside, it is not likely that federal legislative objectives in regulating interjurisdictional

[68] *Westcoast Energy*, above n 47. However, similar upstream gathering facilities owned by companies separate from the interprovincial transmission company were held to be within provincial jurisdiction: *Canadian Hunter Exploration Ltd v Canada (National Energy Board)* (1999) FCJ No. 460 (CA).

[69] Ibid.

[70] See *Bank of Montreal*, above n 57.

[71] *Ross v Registrar of Motor Vehicles* (1975) 1 SCR 5.

pipelines would be frustrated.[72] Similarly, should the doctrine of interjurisdictional immunity be engaged, the environmental standards for air, water, and land pollution, and compliance processes in provincial legislation, would not seem to 'impair' the vital or essential functions of the interprovincial pipelines under federal regulatory jurisdiction. This would be consistent with *Ontario v Canadian Pacific Ltd*,[73] in which the Supreme Court of Canada held that Ontario environmental legislation concerning air contamination applied to railway right of way burning by Canadian Pacific.

C. Oil sands

The analysis of jurisdiction over environmental effects of conventional oil and gas activities applies almost completely to oil sands operations. One difference is the significant contribution of oil sands operations to greenhouse gas (GHG) emissions relative to conventional oil and gas activities.[74]

One issue concerning infrastructure jurisdiction is whether the federal government can prohibit bitumen export on environmental grounds. During the autumn 2008 General Election Campaign, Prime Minister Harper indicated that federal legislation would be proposed to prohibit export of bitumen to countries that have less stringent greenhouse gas emission standards than Canada.

Such legislation could plausibly be based on the same federal powers—POGG, Criminal Law, and Regulation of Trade and Commerce—that might support federal greenhouse gas emissions legislation. Bringing such a scheme under POGG would have the effect of largely allocating the subject to the federal government and transferring potentially overlapping provincial powers to the federal side. If a bitumen export restriction were specifically made part of such a valid federal scheme, its functional relationship to the core goal of greenhouse gas emission reduction would very likely place the legislation within federal jurisdiction. However, a Criminal Law rationale for an overall greenhouse gas regime that included a bitumen export prohibition arguably strengthens federal jurisdictional claims because it preserves scope for provincial legislation to address the issue.[75]

Federal Trade and Commerce justification might be advanced as a possibility for bitumen export prohibition, whether part of a more general greenhouse gas scheme, or as a standalone law. If the latter, the issue would be one of characterization. Would the leading feature of the legislation be a trade matter—export of bitumen—or would it be an environmental objective, namely greenhouse gas emission reduction? While there is not yet legislation to analyse, the external context of

[72] See *Campbell-Bennett v Comstock Midwestern* (1954) SCR 207. Provincial mechanics lien legislation under which a pipeline could have been sold piecemeal to different purchasers held inapplicable.

[73] *Ontario v Canadian Pacific Ltd.* (1995) 2 SCR 1031.

[74] In 2008, oil sands contributed 15 per cent of total Albertan GHG emissions; coal, oil, and gas contributed 24 per cent. Government of Alberta, Environment, Regulating Greenhouse Gas Emissions, available at <http://environment.alberta.ca> (accessed 25 September 2011).

[75] Perhaps by attaching provincial processing conditions based on provincial property jurisdiction to bitumen leases.

the prime minister's remarks, namely environmental sustainability and specifically climate change impacts of the oil sands subsector, suggests that the leading feature of the law would be greenhouse gas emission reduction by upgraders and refiners in importing countries. If bitumen export prohibition is part of a broader climate change regime, the issue would be whether the broader scheme would be in relation to trade and commerce. While emission trading can be seen as trade in 'goods', namely emission reduction credits, the core objective of the trading scheme, and the broader climate change regime, including the mandatory emission limits that establish the base line for reduction credits, would even more clearly be characterized as in pith and substance in relation to GHG emission reduction and not trade.

Another related issue concerns jurisdiction over nuclear plants to generate electricity and to generate steam for oil sands production operations. Facilities of this kind have been proposed to address energy needs for steam-assisted gravity drainage (SAG-D) bitumen production. These would replace and release natural gas for higher order domestic and commercial use.

VIII. Federalism and Pipeline Case Studies

Major federalist theories centre on the idea of contract.[76] But agreements are never without ambiguity, particularly as the circumstances of the parties change over time. So the objectives of federalist covenants and the interests of the parties to federalist arrangements become the focus of attention. This means minority or local rights, as well as nation building and pursuit of global community goals. Management of the tensions between these interests is at the heart of federal constitutional instruments, cooperative arrangements, and practices.

Powers concerning pipeline, particularly market pipeline, infrastructure present federalist challenges because they concern physical structures, not just arrangements that by their nature bridge local and central interests. They are intended to transport locally produced energy resources, but the transportation, beyond local movement to treatment and processing facilities, extends beyond the boundaries of provinces (subordinate federal units) and often beyond national boundaries. The powerful local interests involved in accessing national or international markets, completes the loop linking local exploration development and conservation with extra-jurisdictional markets.

Yet the fundamental federalist issues presented by this kind of pipeline infrastructure are by no means new. In the 1860s, when Canada's constitution was under development and negotiation, railways and other interjurisdictional transportation modes such as merchant shipping and ferry lines were well known and received considerable attention in the constitution building process.[77]

Thus, as a counterpart to section 109, public land allocation to provinces, there is the s 92(10)(a) exclusive federal jurisdiction in relation to extra-provincial works

[76] The compact theory of federalism is discussed in *III. Theories of Federalism*, of this chapter.
[77] See Constitution Act 1867, 3rd Schedule.

or undertakings. Though not envisaged in 1867, trunk pipelines nevertheless present a good fit for the extra-provincial undertakings power and this has received judicial confirmation.[78] The difficulty was in the detail—technological, business, and political.

Beginning in the 1950s, it was recognized that provincial laws that could empower federally regulated pipelines in their core operations did not apply.[79] But, it took much longer to resolve the question of how far this federal pipeline jurisdiction extended back toward the well head. The Supreme Court confirmed in *Westcoast Energy Inc. v Canada (National Energy Board)*[80] that federal power extended to plants processing natural gas for delivery to interprovincial or international pipelines. It took even longer for federal National Energy Board regulation over Alberta's Nova gas collection system, initially developed through a de facto provincial Crown corporation, to be recognized.[81]

It is apparent that judicial constitutional interpretation played an important role in resolving issues arising between provinces and the central government. It is also apparent that the result has reinforced and extended federal jurisdiction. Nation building, in the sense of developing national economic activity,[82] and more recently in facilitating access to the global economy, has played very strongly in the doctrinal constitutional analysis with its core pith and substance determination and classification to the most appropriate heads of constitutional power.

The last part of this chapter involves two case studies. These are the proposed Mackenzie Gas Project, which would transport gas produced in Canada's Western Arctic to southern Canadian and United States markets, and the proposed Northern Gateway Pipeline project, intended to transport bitumen from the Athabasca Oil Sands to the British Columbia coast for shipment to Asian markets. In each case, constraints arising from Canada's federal system will be identified and the significance for infrastructure regulation assessed. Instruments and processes that have either addressed or potentially can address these constraints will be discussed. Three functional areas at the heart of necessary regulatory processes will be examined. These are (1) facility approval and regulation; (2) environmental and socioeconomic impact assessment and effects on property; and (3) First Nations aboriginal rights.

[78] *Westcoast Energy*, above n 47; *Campbell-Bennett*, above n 72.

[79] *Campbell-Bennett*, above n 72.

[80] *Westcoast Energy*, above n 47.

[81] In 2009, on the application of TransCanada Pipelines Limited, the Federal National Energy Board (NEB) issued a declaratory order and certificate GC-113, placing the Alberta transmission system under NEB jurisdiction. TransCanada Alberta System Jurisdiction, available at <http://www.transcanada.com> (accessed 25 September 2011).

[82] See eg *Caloil Inc. v Canada (AG)* (1971) SCR 543. The Supreme Court of Canada upheld a regulation authorizing the National Energy Board to issue licences for the import of oil if it 'is of the opinion that importation of oil...will be consistent with the development and utilization of Canadian indigenous oil resources...'.

A. The Mackenzie Gas Project

This massive 16 billion (Canadian) dollar project involves a 1,196km pipeline from the Beaufort Sea in the Western Arctic (the Northwest Territories) up the Mackenzie River Valley to Northeastern Alberta, three onshore natural gas fields and related facilities.[83] The intention is that this pipeline will stimulate Arctic natural gas development (following earlier discoveries) and exploration. An application for approval by the NEB was filed by the project consortium in 2004. Board hearings, including sessions in fifteen Arctic communities, began in 2006. Approval, subject to 264 specific conditions concerning engineering, environment, and other matters, was granted in December 2010.[84]

A unique feature of the regulatory approval of this project was an initial cooperation agreement concerning review and assessment. This Cooperation Plan, made in 2002 among ten federal, territorial, and First Nations agencies with some type of regulatory or consultative processes, established a blueprint for environmental review and regulatory processes relevant to the Mackenzie Gas Project.[85] This led to the April 2004 Agreement for Coordination of the Regulatory Review of the Mackenzie Gas Project.[86]

The National Energy Board's regulatory hearings convened first in January 2006, closely followed by Joint Federal–Territorial–First Nations Environmental Review Panel hearings beginning in February 2006. However, though the NEB heard evidence until December 2006, it adjourned to await the Joint Review Panel Assessment Report. This report informed the argument part of the NEB proceedings, which commenced in 2010. A tangible link between the two processes was provided by a member of the National Energy Board, who sat as a member of the Environmental Review Board, then as a member of the NEB hearing panel.[87]

In jurisdictional terms, the project involved federal pipeline approval and regulatory authority, as well as federal public land ownership in the Northwest Territories. Additional federal legislative authority included environmental

[83] See Mackenzie Gas Project available at <http://www.mackenziegasproject.com> (accessed 25 September 2011).

[84] Canada, National Energy Board, Mackenzie Gas Project—Reasons for Decision, vol 1–2 (2010, Calgary: National Energy Board) available at National Energy Board <http://www.neb-one.gc.ca> (accessed 25 September 2011).

[85] Canada, Northern Pipeline Environmental Assessment and Regulatory Chairs' Committee, 'Cooperation Plan for the Environmental Assessment and Regulatory Review of a Northern Gas Pipeline Project Through the Northwest Territories' (National Energy Board, 2002) ('Cooperation Plan') available at < http://www.neb.gc.ca> (accessed 25 September 2011).

[86] 'Agreement for an Environmental Impact Review of the Mackenzie Gas Project' (2004, Northern Gas Project Secretariat) available at <http://www.ngps.nt.ca>, accessed 25 September 2011). This agreement was concluded between the Mackenzie Valley Environmental Impact Review Board (established under the Mackenzie Valley Resource Management Act, SC 1998, c 25, s 99(1) to implement the Gwich'in, and Sahtu Dene and Metis Comprehensive Land Claim Agreements), the Inuvialuit first nation, represented by the Inuvialuit Game Council, and the Federal Minister of the Environment, 3 August 2004.

[87] The NEB Member, Rowland Harrison, was appointed under s 15 of the National Energy Board Act, RSC 1985, c N-7.

impact assessment,[88] aboriginal people,[89] certain wildlife matters, particularly endangered species and migratory birds,[90] and navigation and shipping.[91] The territorial government administers certain lands and has delegated authority over matters including wildlife and water resources.[92] The province of Alberta has authority over certain non-site-specific project impacts,[93] and owns public land through which the right of way passes.[94] First Nations have aboriginal and treaty rights to hunt, fish, and trap, surface access controls, and limited surface and mineral land ownership.[95] Municipalities affected are subordinate bodies, established and empowered by their territorial or provincial governments. Thus the levels of government with authority or concerns in relation to the Mackenzie Gas Project were federal, provincial, territorial, municipal, and (multiple) First Nations.

1. Facilities approval

Facilities approval and regulation was essentially federal, exercised by the National Energy Board. This covered all facilities, including the Beaufort Sea gas fields, the pipeline, and terminal facilities in the Territories and in Northwestern Alberta. Under the National Energy Board Act,[96] economic regulation of the pipeline, including tariffs and tolls, is also within federal authority. But, while direct environmental effects must be taken into account by the NEB, there is a federal environmental impact assessment process triggered by factors including disposition of federal lands. This process, now under the Canadian Environmental Assessment Act,[97] has, as it is framed, been held by the Supreme Court of Canada to be within federal legislative jurisdiction.[98] But because there is an environmental impact assessment (EIA) interface with provinces and with First Nations under land settlement agreements, both constitutional law and political imperative made cooperative arrangements a necessity for the Mackenzie Gas Project. This led to the ten-party 'Cooperation Plan' to establish a joint process under the Canadian

[88] Under Canadian Environmental Assessment Act, above n 24. Predecessor federal environmental assessment legislation was held to be constitutionally valid in *Oldman River Society*, above n 23, to support environmental assessment in relation to federal heads of power under s 91 of the Constitution Act 1867, or alternatively as a process supporting federal decision-making under the residual federal 'peace, order and good government' power.

[89] Constitution Act 1967, s 91(24).

[90] The Migratory Birds Convention Act, RSC 2000, c 22. implemented originally under Constitution Act 1867, s 132 concerning 'Empire Treaties'. The 1909 Migratory Birds Convention between Great Britain (for Canada), the US and Mexico, and the Species at Risk Act, RSC 2000, c 29 is very likely valid in relation to federal lands, including provisions affecting provincial lands, under the POGG or Criminal Law (s 91(27)) powers.

[91] Northwest Territories Act, RSC 2000, c N-27.

[92] Under Constitution Act 1867, s 92(10) (Local Works and Undertakings) and s 92(13) (Property and Civil Rights in the Province).

[93] Constitution Act 1867, s 109.

[94] Constitution Act 1982, s 35(1), 'The existing aboriginal and treaty rights of the aboriginal peoples of Canada are hereby recognized and affirmed'.

[95] *Delgamuukw*, above n 59.

[96] National Energy Board Act, above n 87 s 52.

[97] Canadian Environmental Assessment Act, above n 24.

[98] *Oldman River Society*, above n 23.

Environmental Assessment Act, with environmental assessment fully undertaken, but linked to the National Energy Board's regulatory approval process.[99]

Federalism tensions developed along several fault lines. One concerned federal NEB pipeline regulatory jurisdiction, versus separate federal jurisdiction (Indian and Northern Affairs Department) over oil and gas exploration and development under the Canada Oil and Gas Operations (COGO) Act.[100] This did not engage separate federal units, but the COGO authority concerns local operations that occur in the Northwest Territories, a federal governmental sub-unit,[101] that nevertheless has certain local powers.

A motion by a group of Mackenzie Delta gas explorers, urging the NEB to regulate the Mackenzie gas-gathering system as part of the overall pipeline system was denied by the Board.[102] The matter in issue was tolling advantage on the system as between the explorer group, which would produce the gas, and the project sponsors. Leave to appeal was granted by the Federal Court of Appeal, but the appeal did not proceed.

A second federalism fault line concerned First Nations versus federal (and provincial) governments, This division erupted with a 2005 judicial review application to the Federal Court by the Dene Tha' First Nation, whose lands would be traversed by the pipeline. This is discussed below under 'First Nations'.

In a separate proceeding before the NEB, the Dene Tha' requested directions concerning NEB jurisdiction over the interconnection of proposed Nova Gas Transmission facilities in Northern Alberta, with the Mackenzie Valley Pipeline. Thus an issue of federal–provincial jurisdiction was raised. The Board, following a procedural hearing and consideration, decided to treat this matter as separate from the Mackenzie Valley application.[103] This is discussed below.

2. *Environmental Impact Assessment*

Because the project involved federal lands, environmental impact assessment requirements under the Canadian Environmental Assessment Act (CEAA) were triggered. As noted above, CEAA provides for interagency and interjurisdictional agreements to establish joint assessment panels.[104] Here, the plan for this was established by the Cooperation Agreement and confirmed by an agreement among the Mackenzie Valley Environmental Impact Review Board, the Inuvialuit First Nation, and the Federal Minister of the Environment.[105] These parties represented respectively the environmental assessment board under the federal legislation[106] implementing the Mackenzie Valley comprehensive land settlement with the Dene Tha' First Nation, the aboriginal people of the

[99] Cooperation Plan, above n 85.

[100] Canada Oil and Gas Operations Act, RSC 1985, c O-7.

[101] Northwest Territories Act, above n 91.

[102] National Energy Board, Mackenzie Gas Project, Hearing Order GH-1-2004, Ruling #16.

[103] National Energy Board, Letter Concerning Request for Directions, 2 February 2007.

[104] Canadian Environmental Assessment Act, above n 24.

[105] See 'Agreement for an Environmental Impact Review of the Mackenzie Gas Project', above n 86.

[106] Mackenzie Valley Resource Management Act, above n 86.

Mackenzie Delta region, and the federal Canadian Environmental Assessment Agency that administers CEAA.

The seven-member review panel, nominated by these parties, was appointed under sections 40 and 41 of the Canadian Environmental Assessment Act. The National Energy Board nominee, who linked the environmental assessment and the energy regulatory processes, was appointed under s 15 of the National Energy Board Act. The Joint Review Panel process commenced in the summer of 2006, with the Panel's report released in December 2009.[107]

It is significant that the route extends into northwestern Alberta, where the pipeline will connect with existing transportation facilities. During the environmental review process, no federal–provincial issues were raised.

Federalism strains manifested in the environmental review process were introduced above. These pitted First Nations against the federal government. A main event was the Dene Tha' judicial review[108] concerning the failure by the federal government to consult on the Cooperation Plan that not only mapped out the dual environmental and regulatory processes, but set out the framework for a joint Canadian Environmental Assessment Act review panel and the terms of reference for that environmental review. The result of this litigation confirmed procedural rights of the Dene Tha' in the review process. However, it also contributed to the overall length of what was already a very lengthy regulatory process.

Release of the Joint Review Panel's report in December 2009 set the stage for arguments in the NEB energy regulatory proceeding. While the NEB focused on technical and economic feasibility, tolls and tariffs, and land issues, the Joint Review Panel's recommendations were also considered.

3. *First Nations*

Affected First Nations have been heavily involved in the Mackenzie Gas Project. This is based on recognized First Nations' rights. In particular, existing aboriginal and Treaty rights are guaranteed by s 35 of the Constitution Act 1982. These include hunting, fishing, and trapping rights on traditional lands. The Mackenzie Valley region is subject to Treaties involving comprehensive land settlements, with the Gwich'in, Sahtu, and Inuvialuit peoples. Section 35, as judicially interpreted, provides that *prima facie* infringement of these aboriginal and treaty rights must be justified by the government in its fiduciary role. This involves full and effective consultation; and where, as here, settlement lands are taken up and used, includes fair compensation.[109] Justification, though it may involve parties such as the Mackenzie Gas Project proponents, is ultimately the responsibility of government.

The Mackenzie region land settlements have been implemented by federal legislation. This establishes land, water, and wildlife management agencies for each

[107] Joint Review Panel for the Mackenzie Gas Project, 'Foundation for a Sustainable Northern Future: Report of the Joint Review Panel for the Mackenzie Gas Project' (Canada: Minister of Environment, 2009) available at Canadian Environmental Assessment Agency—Review Panels—Completed <http://www.ceaagc.ca> (accessed 25 September 2011).

[108] *Dene Tha' First Nation v Canada (Minister of Environment)* (2006) FC 1354.

[109] See *Sparrow*, above n 65; *Delgamuukw*, above n 59.

settlement area.[110] The effect is to create for these purposes regional governments for each First Nation. The initial negotiations and the resulting Cooperation Plan involved Federal and Territorial government agencies, as well as authorized agencies in these First Nations settlement areas. This led in turn to the Joint Environmental Review Agreement, involving the Inuvialuit First Nation and the Mackenzie Valley Environmental Impact Review Board representing the other First Nations. First Nations were well represented among the seven Joint Review Panel members.

Another matter of First Nation's engagement was potential equity participation in the pipeline. An Aboriginal Pipeline Group representing First Nations economic development agencies was offered partnership in the project—up to one-third ownership. In addition, the project proponents made commitments to inform, train, and provide business opportunities for the First Nations people.

First Nations are clearly affected by the Mackenzie Gas Project. Exploration has occurred on traditional lands, the subject of an earlier land settlement, and the proposed pipeline will be built through those lands, affecting land, water, and wildlife, as well as communities.

A central problem in establishing the joint panel environmental review and the coordinated environmental and energy regulatory processes for the project was to ensure that affected First Nations were consulted and in agreement. While there was considerable cooperation, not all First Nations were in agreement. This became apparent when the Dene Tha' filed their judicial review application with the Federal Court in 2005.

They claimed that the federal government had infringed s 35 of the Constitution Act 1982 under which 'existing aboriginal and treaty rights are recognized and affirmed'. In particular, they invoked the Supreme Court of Canada's jurisprudence that established a 'duty to consult' on the part of the government as part of its fiduciary obligations in relation to First Nations in the event of a *prima facie* infringement of the s 35 right. The Dene Tha' pointed to the failure to involve them in the beginnings and the foundations of the regulatory review process for the project—the original Cooperation Agreement that led to the integrated environmental assessment and the NEB processes. They argued that they had not been consulted in the formulation of these cooperative processes. The Dene Tha' motion was granted in 2006 by the Federal Court.[111] A duty to consult was confirmed, and a conclusion reached that the duty was not met. The result was that the Joint Review Panel environmental hearings were stayed as they concerned Dene Tha' lands. Remedies granted would impact the scope and timing of the NEB hearings.

This decision was appealed by the federal government to the Federal Court of Appeal, which ultimately dismissed the appeal.[112] Meanwhile, the parties had settled the dispute and the dual environmental and regulatory processes proceeded.

110 See n 106.

111 *Dene Tha' First Nation v Canada (Minister of Environment)* (2006) FC 1354.

112 *Dene Tha' First Nation v Canada (Minister of Environment)* (2008) FCA 20.

4. Lessons

Dominant features of the Mackenzie Gas Project regulatory review have been complexity arising from the number of jurisdictions and decision authorities and the length of the regulatory proceedings. This occurred notwithstanding innovative cooperative regulatory arrangements to address federalism constraints. The process took so long that during its course the Canadian and global natural gas industries changed fundamentally. Rapid development of technologies to support shale gas and other unconventional gas development have cast doubt on the economic viability of the Mackenzie Gas Project.

B. The Northern Gateway Project

Enbridge Inc.'s Northern Gateway pipeline and marine terminal project involves transporting oil, particularly diluted bitumen and upgraded 'synthetic crude', from Alberta's Athabasca oil sands region to a terminal to be built on Canada's West Coast near Kitimat, British Columbia. Included is an import pipeline to move condensate, needed for diluent in bitumen transportation.[113] This system will enable Canada to diversify from the single US market to include refiners in Asia.

A Preliminary Application Package for this project was filed by Northern Gateway with the National Energy Board and the Canadian Environmental Assessment Agency in November 2005.[114] Under an agreement between the NEB and the federal Minister of the Environment (acting under the Canadian Environmental Assessment Act),[115] a Joint Review Panel was established following negotiations and public comments, including comments from Aboriginal groups.[116] The agreement with terms of reference and scope of assessment of factors identified and process to be followed, was finalized in December 2009. Other federal regulatory approvals required include those under the Fisheries Act, the Indian Act (for crossing reserve lands), Transport Canada (for the marine terminal and river crossings), and Environment Canada concerning necessary sea disposal and dredging.

Certain non-site-specific provincial approvals are required for project components such as access roads. Easements for use of private and public lands in the provinces of Alberta and British Columbia are also necessary. Municipal lands may also be required, but municipal approval authority arises ultimately under provincial statutes.

Agreements are necessary for right-of-way and ancillary facilities located on First Nations reserves. Aboriginal rights to traditional uses claimed over additional lands must also be addressed.[117]

Thus federal, provincial, First Nations, and municipal authorities are involved in the assessment, approval, and monitoring of the project.

[113] 'Enbridge Northern Gateway Project, Section 52 Application' (May 2010), available at <http://www.northerngateway.ca> (accessed 25 September 2011), section 2, 'Project Description'.

[114] Ibid. [115] Ibid at 6.1. [116] Ibid at 6.1.1.

[117] As a result of s 35(1) of the Constitution Act 1982.

1. *Facilities approval and regulation*

Federal regulatory jurisdiction over this extra-provincial pipeline and coastal marine terminal project is supreme. The principal agency will be the National Energy Board, acting under the National Energy Board Act to determine whether the project is consistent with present and future public convenience and necessity.[118] In addition to the s 92(10)(a) extra-provincial undertakings power, federal marine waters environmental jurisdiction is supported by the 'peace, order, and good government' power. Regulation of the marine terminal and tankers falls under the federal Navigation and Shipping power.[119]

2. *Environmental impact assessment*

Use of federal lands triggers the environmental impact assessment requirement under the Canadian Environmental Assessment Act. Provincial environmental assessment jurisdiction is not engaged by a federally regulated project of this kind. In this case, as noted, a joint NEB–CEAA assessment process and review panel was established by interagency agreement.

3. *First Nations*

The project proposal recognizes the necessity for appropriate First Nations involvement in several ways. First, Enbridge has offered aboriginal communities the opportunity to invest in ten per cent of the project.[120] In addition, aboriginal employment, training, and business development opportunities are to be made available.[121]

A second point of aboriginal engagement involves the application information and the aboriginal economic opportunity commitments.[122] This is part of the consultation that the federal government, through the National Energy Board process, and the proponents' actions, is required to carry out. This requirement is based on the aboriginal rights guaranteed by s 35 of the Constitution Act 1982. If rights are infringed, justification that meets the conceptual tests established by the Supreme Court of Canada must be met. As discussed above, this includes full and effective consultation.[123]

4. *Lessons*

The Northern Gateway Project, while still under regulatory review, demonstrates textbook cooperative arrangements to deal with potential federalism constraints. It underlines the significance of the well-established provisions in Canada's Constitution that allocate legal authority over physical and operational matters concerning interjurisdictional energy infrastructure. This creates conditions for

[118] National Energy Board Act, above n 87, s 52. [119] Constitution Act 1867 s 91(10).
[120] Enbridge, see n 113 at 1–6. [121] Ibid at 1–7. [122] Ibid.
[123] *Haida Nation*, see n 66.

sub-unit issues, in this case First Nations, to be dealt with collaboratively under a legally supported consultative process.

5. *Conclusions*

There is little doubt that decentralized federations like Canada's place constraints on interjurisdictional energy infrastructure development. Constitutional division of powers create classic local versus central conflicts, or at least uncertainties concerning legal powers of central and sub-unit governments to authorize and regulate energy infrastructure that crosses jurisdictional boundaries. These are not merely policy constraints. As legal limitations, they create the hard constraints that must be carefully assessed and respected by infrastructure developers. They also produce uncertainty, since intergovernmental or even private disputes can lead to judicial constitutional decisions that demolish the regulatory assumptions of governments and citizens.

In Canada, a complicating and perhaps surprising federalism feature is the constitutional recognition and significance of First Nations as sub-national units. Tensions are not limited to federal government versus provinces.

However, the constitutional analysis and the pipeline case studies show that there is not a stalemate. Relatively clear federal jurisdiction over interjurisdictional works and undertakings supports regulation of major hydrocarbon transportation infrastructure. But this is not a complete answer. Interjurisdictional cooperation and agreement is essential. First, federal–provincial cooperation is required concerning major infrastructure project strategy. There must be at least tacit agreement on Northern natural gas development or offshore bitumen export. Otherwise, skirmishes at jurisdictional pressure points, including the interface between provincial land and resource ownership and development, and production regulation and federal interjurisdictional infrastructure regulation, could fatally delay projects. The same is true of First Nations' cooperation, as the Mackenzie Gas case study illustrates.

Interjurisdictional agreements providing for joint regulatory processes are a potent instrument for mitigating federalism constraints. These are not merely formal intergovernmental agreements. Issue-specific arrangements like the coordinated environmental and facilities review processes for the Mackenzie Gas Project may be necessary. Yet Mackenzie Gas also shows that there may be levels of complexity caused by multiple parties, and processes, and process length that defy effective collaboration. Party-to-party bargaining and accommodation on matters, such as the social and employment benefits and equity participation offered to First Nations by pipeline sponsors, may also be significant.

At a minimum, the prescription for federalism constraints on interjurisdictional hydrocarbon infrastructure involves cooperative federalism, agency facilitation, rights accommodation, and a workable legal foundation.

3

Law and Regulation Governing Electricity Networks in Mexico in the Context of Regional Integration with North and Central America

José Juan González

I. Introduction

This paper discusses the legal and regulatory regime that governs ownership, building, and operation of public and private energy networks in Mexico and explains how Mexico's economic integration with North and Central American markets has influenced the recent transformation of that regime as well as factors that discourage private investment in this area. The research includes an analysis of regulations governing electricity networks as well as regulations regarding environmental impacts associated with such infrastructure.

II. Mexican Electricity Networks' Law

In Mexico, building and operation of electricity networks are regulated by the Federal Constitution, the North American Free Trade Agreement (NAFTA),[1] the Electricity for Public Service Law of 1975[2] as amended in 1993,[3] and its 1993 regulation[4] as amended in 2003,[5] the Ecological Equilibrium and Environmental

[1] The North American Free Trade Agreement (NAFTA) was signed in three separate ceremonies in the three capitals on 17 December 1992, by USA President Bush, Mexican President Salinas, and Canadian Prime Minister Mulroney. This framework agreement proposed to eliminate restrictions on the flow of goods, services, and investment in North America. It took effect on 1 January 1994. The full text of NAFTA is available at <http://www.international.gc.ca> (accessed 25 September 2011).

[2] Diario Oficial de la Federación, 22 December 1985.

[3] Diario Oficial de la Federación, 22 December 1993.

[4] Diario Oficial de la Federación, 31 May 1993.

[5] Diario Oficial de la Federación, 24 May 2003.

Protection Act of 1988[6] as amended in 2011,[7] and its Regulation on Environmental Impact Assessment,[8] as well as other federal and local environmental laws.[9]

As will be discussed in this paper, according to the legal reforms passed by Federal Congress in the early 1990s, currently in Mexico, energy networks can be built, owned, and operated by the Federal Electricity Commission (CFE),[10] as well as by private national or foreign companies. These networks may have the purpose of transmitting energy into the national territory to satisfy public service needs, as well as of transporting electricity for both import and export purposes.

According to the legal regime currently in force, private parties interested in building, owning, and operating energy networks are basically required to obtain an authorization from the Energy Regulatory Commission (CRE)[11] and an environmental impact authorization from the Ministry of Environment.[12] Notwithstanding that, it cannot be asserted that in Mexico there exists an open market for electricity. Private energy networks have a very limited objective; they may be used only for import and export purposes.[13] When electricity transmission is aimed at providing public electricity services it can only be transmitted through the National Electricity Network under the property and control of CFE.

The legal reforms that liberalized the energy sector in 1992 were influenced by the process of energy integration with North and Central America that started at the end of the last century. Nevertheless, energy integration of Mexico with the USA and Guatemala resulted from two quite different processes. In the case of the USA, energy integration was not new but received extra stimuli from NAFTA's signature.[14] In contrast, although Mexico has signed international trade agreements with Costa Rica (1995)[15] and Nicaragua (1998),[16] as well as with El Salvador, Guatemala, and Honduras (2000),[17] energy integration with Central America is very recent and it is not the result of those agreements but of the so-called 'Plan Puebla Panama'[18]

[6] Diario Oficial de la Federación, 28 January 1988.

[7] Diario Oficial de la Federación, 28 January 2001.

[8] Diario Oficial de la Federación, 30 May 2000.

[9] Eg the General Act for Sustainable Development of the Forest of 2003, the National Waters Act of 1992 as amended in 2008, and local laws and regulations on land use.

[10] CFE is Mexico's state-owned electricity company. Until 1992, when the Electricity for Public Service Law of 1975 was amended, it had the monopoly of almost all activities related to electricity generation, transmission, and distribution.

[11] The Energy Regulatory Commission was established in 1993 by Executive Order as a technical semi-autonomous office in charge of enforcing the law under article 27 of the Federal Constitution regarding the Energy Sector. See Diario Oficial de la Federación, 4 October 1993.

[12] A number of additional permits and authorizations may be required from federal and local governments, such as land use permits or federal zones concessions.

[13] See Miguel G. Breceda-Lapeyr, 'Inversión privada en el sector eléctrico de México' (2002), available at <http://www.cec.org> (accessed 25 September 2011).

[14] According to one study, NAFTA influenced the transformation of the electricity industry in Mexico. See Sarahí Angeles Cornejo, 'Los efectos del TLC en los cambios de la organización de la industria eléctrica en México' (2003), available at <http://www.depfe.unam.mx> (accessed 25 September 2011).

[15] Available at <http://www.sice.oas.org> (accessed 25 September 2011).

[16] Available at <http://www.sice.oas.org> (accessed 25 September 2011).

[17] Available at <http://www.sice.oas.org> (accessed 25 September 2011).

[18] The Plan Puebla Panama is the result of the initiatives proposed by eight countries on 15 June 2001 in El Salvador. The Plan aims at the economic integration of Central America from the

and of the Treaty for the Central American Electricity Market signed in 1998 by Costa Rica, El Salvador, Guatemala, Honduras, Nicaragua, and Panama.[19] The signature of this Treaty made it possible to negotiate the interconnection of the Central American network with the Mexican Electricity Network,[20] in the context of the so-called 'Tuxtla Mechanism'.[21]

A. Constitutional provisions

Article 27 of the Mexican Constitution points out that it pertains exclusively to the state to transport and distribute electric power for public service and it asserts as well that in such matters no concessions shall be granted to private persons. This constitutional provision has been interpreted by federal legislation in the sense that private individuals or corporations are not allowed to build, own, or operate electricity networks when energy transmission or distribution is aimed at providing a public service. In that case, electricity networks can only be built, owned, or operated by the state. In any event, it does not mean that private persons or corporations are forbidden to build, own, and operate electricity networks when electricity transmission or distribution is directed to a different purpose than providing a public service. In this regard, article 3 of the Electricity for Public Service Law describes the activities related to the energy sector that are not considered a public service. Such activities include import and export of electricity. Moreover, as will be analysed in the following sections of this paper, modifications of this provision passed after NAFTA became enforceable made possible private investment in building and operating energy infrastructure.

Thus, according to the Mexican Constitution and its corresponding regulations, building, owning, and operating energy networks for export and import purposes are allowed to both public and private-sector parties.

B. The North American Free Trade Agreement provisions on energy networks

In accordance with the most recent Supreme Court's decisions on the hierarchy of international agreements, NAFTA's provisions are held to be hierarchically superior to federal laws.[22]

Mexican state of Puebla to Panama. It consists of eight initiatives that include highway integration, trade interchange, energy interconnection, telecommunication services, sustainable development, tourism promotion, human development, and prevention and mitigations of disasters.

[19] The full text of the Treaty for Central American Electricity Market is available at <http://untreaty.un.org> (accessed 25 September 2011).

[20] See OLADE–ACDI/CIDA–University of Calgary, 'Competition in Energy Markets: An Assessment of Restructuring of Energy Markets in the LAC Region' (2004), available at <http://temp2.olade.org/documentos/LAC%20Energy%20Market%20Report.pdf> (accessed 20 October 2011).

[21] The Tuxtla Mechanism of Dialogue and Negotiation was adopted at the Summit of Presidents of Central America and Mexico held in January 1991 in Tuxtla, Mexico. It represents a forum for consensus building, political dialogue, and cooperation. It includes Belize, Costa Rica, El Salvador, Guatemala, Honduras, Nicaragua, Panama, Mexico, and the Dominican Republic.

[22] Amparo en revisión 1475/98. Sindicato Nacional de Controladores de Tránsito Aéreo. 11 May 1999. Register No. 903483. Appendix 2000, Vol. I Supreme Court of Justice.

Chapter 6 on energy and basic petrochemical goods of NAFTA regulates cross-border energy trade between Mexico and its commercial partners under the agreement, USA, and Canada.[23] The three basic principles that govern regional energy trade set forth in article 601 of NAFTA are:

(1) The Parties confirm their full respect for their Constitutions;
(2) The Parties recognize that it is desirable to strengthen the important role that trade in energy and basic petrochemical goods plays in the free trade area and to enhance this role through sustained and gradual liberalization;
(3) The Parties recognize the importance of having viable and internationally competitive energy and petrochemical sectors to further their individual national interests.

As a consequence of the above-mentioned first principle, although NAFTA generally deregulated trade in partner countries,[24] major exceptions were negotiated for Mexico, in accordance with the constitutional requirement of state control of the energy sector. In this regard, Annex 602.3 of NAFTA defines the scope of the energy materials and activities over which the Agreement holds sway in the case of Mexico.[25] The restrictions established in the Annex are compatible with the Mexican Constitution. Provision 1 of the Annex reserves for the Mexican state a number of strategic activities, including investment in such activities and the provision of services in such activities.[26]

However, regarding international trade of electricity the restriction refers only to 'the supply of electricity as a public service in Mexico, including (...) the generation, transmission, transformation, distribution, and sale of electricity'.[27] There are exemptions to such restriction. Opportunities can be found for private investment in Mexican electricity generation in the case of production for own use (self-generation), co-generation,[28] independent power production,[29] and small generators. According to Annex 602.3, self-producers, co-generators, and independent producers are allowed to acquire, establish, and/or operate electrical infrastructure, including networks.

[23] Available at <http://www.sice.oas.org> (accessed 25 September 2011).

[24] When NAFTA became effective on 1 January 1994, it eliminated many but not all of the energy trade barriers that had existed between Canada, Mexico, and the USA. In fact, the kind of arrangements prevailing for energy under the Free Trade Agreement between USA and Canada were not accepted by Mexico under NAFTA.

[25] Annex 602.3: Reservations and Special Provisions, available at <http://www.iadb.org> (accessed 25 September 2011).

[26] The list of restrictions includes: (a) exploration and exploitation of crude oil and natural gas; refining or processing of crude oil and natural gas, and production of artificial gas, basic petrochemicals and their feedstock and pipelines; (b) foreign trade; transportation, storage and distribution, up to and including the first-hand sales of the following goods: (i) crude oil, (ii) natural and artificial gas, (iii) goods covered by this chapter obtained from the refining or processing of crude oil and natural gas, and (iv) basic petrochemicals; and (d) exploration, exploitation, and processing of radioactive minerals, nuclear fuel cycle, generation of nuclear energy, transportation and storage of nuclear waste, use and reprocessing of nuclear fuel, and regulation of their applications for other purposes and the production of heavy water.

[27] NAFTA, Annex 602.3, proviso 1.

[28] Self-generators and co-generators are obliged to sell all surpluses to CFE.

[29] Independent producers have to sell all their production to CFE.

Given that according to the Mexican Constitution and legislation, generation, transmission, distribution, and sale of electricity is viewed as a public service and a 'strategic area reserved to the state',[30] private investment is not permitted in 'reserved' Mexican energy activities and any cross-border trade in them is confined to contracts approved by the Mexican authorities (Annex 602.3). However, contract approval is not required for non-reserved activities.

In this regard, the rules established by Annex 602.3, proviso 5, are:

a) Production for Own Use: An enterprise of another Party may acquire, establish, and/or operate an electrical generating facility in Mexico to meet the enterprise's own needs.
b) Co-generation: An enterprise of another Party may acquire, establish, and/or operate a co-generation facility in Mexico that generates electricity using heat, steam or other energy sources associated with an industrial process. Owners of the industrial facility need not be the owners of the co-generating facility.
c) Independent Power Production: An enterprise of another Party may acquire, establish, and/or operate an electricity-generating facility for independent power production (IPP) in Mexico.

As discussed in the next section, NAFTA negotiations during the period 1992–1994 influenced the reform of the Electricity for Public Service Law in a way that the mentioned NAFTA principles were incorporated within the 1992's reform of such law.

C. National law on electricity networks

Until very recently, the traditional model of operation of the Mexican electricity sector had been a vertically integrated state monopoly. From 1960 to 1992 the CFE was in charge of operating the electricity system and it had exclusivity in providing the public service throughout the Mexican territory. Besides, the Electricity for Public Service Law as passed by the Federal Congress in 1975 did not regulate any aspect of international trade in electricity.[31] The original text of the Act was basically addressed to regulating generation, transmission, transformation, distribution, and supply of electricity for the public service, but not for international trade. According to articles 3 and 4, with the only exemption of self-generation of energy, all energy sector activities were considered to have a public service purpose. Consequently, building and operating energy infrastructure, including energy networks, was an area reserved to the constitutional monopoly of the state aimed at providing the public service of electricity. However, CFE was not forbidden to export its surpluses of electricity to satisfy the USA's demand or even import it to satisfy the national demand. As a result, when NAFTA went into force and the Electricity for Public Service Law reform was passed in 1992, a significant electricity trade market between Mexico and the USA was already in place and networks

[30] Annex 602.3, proviso 1(c).

[31] Diario Oficial de la Federación, 22 December 1975.

that connect the Mexican Electricity System with the USA Electricity System had already been built by CFE.

NAFTA's provisions on energy trade were adopted by the Mexican energy law amendment of 1992.[32] Article 3 of the Electricity for Public Service Law was amended to include a reference to import and export of electricity and to introduce the modality of independent producers of electricity that were allowed to generate electricity only to be exported or to be sold to CFE.[33] The amendment incorporated also the modalities of co-generators and small producers that were allowed to both export their surpluses and to sell them to CFE. The reform introduced two important changes in relation to networks.

First, according to the reform, generation of electricity by private parties[34] to be exported is not considered a public service.[35] In the same way, the import of electricity by private persons or corporations is not considered public service if such electricity is to be exclusively used for their own consumption. In addition, the amendment does not forbid CFE to import energy for public service purposes or to export the electricity generated by its own electricity plants.

Second, as a consequence of the 1992 amendment, the Mexican electricity system passed from a full state monopoly to the single buyer model, where private parties—basically independent producers—build electricity plants that increase the capacity of generation of the national system and CFE acquires the generated electricity and distributes it through the National Electricity Network.[36] According to paragraph II of the mentioned article 3 as amended in 1992, generating electricity by independent producers to be sold to CFE is not considered as a public service. In the same vein, the reform allowed small generators and co-generators to sell their surpluses to CFE. Thus, under the 1992 reform, CFE lost the monopoly of generation but retained the monopoly of transmission and distribution of electricity when it is aimed at providing a public service. Import and export of electricity are not part of such monopoly.

It is possible to assert that the incorporation of rules governing energy networks for purposes of international electricity trade is a direct consequence of free trade

[32] One author argues that: 'NAFTA provisions on electricity were not consistent with the Electricity for Public Service Law in force up to December 1992, nor with the contents of Article 27 of the constitution.' See Jacinto Viqueira Landa, 'Electric Power Regulation in Mexico' (1994) Energy Studies Review, Vol. 6: Iss. 3, article 2 at 250.

[33] Diario Oficial de la Federación, 23 December 1992.

[34] It is important to mention that, according to the amended article 3 of the Electricity for Public Service Law, private parties include not only independent producers but also co-generators and small generators.

[35] The text passed by Congress in December 1992 established article 3. The following are not considered public services: (1) electric power generation for self-supply, cogeneration, or small-scale production; (2) energy generation effected by independent producers to sell to the Federal Electricity Commission; (3) electric power generation for export, derived from cogeneration, independent production and small-scale production; (4) electric power imports by private persons or societies for their exclusive use; and (5) electric power generation to be used for emergencies due to public service electricity supply interruptions.

[36] According to the reform, private generators are allowed, but must sell their production through long-term power purchase agreements to CFE, unless energy produced is used for export or self-supply.

commitments aimed at allowing national and foreign private investment in some areas of the energy sector. Since NAFTA came into force and the Electricity for Public Service Law was amended, North American investors—established in Mexico as co-generators, independent producers, or small generators—can generate electricity on the Mexican side to be exported to the USA and as a consequence they may build, own, and operate energy networks for such purpose or they can export the generated electricity through the National Electricity Network owned and operated by CFE. Additional legal requirements for those activities are not too complex and both import and export activities are subjected to prior authorization from the Ministry of Energy through the CRE (articles 116 and 120).[37]

Official data[38] shows that, by 2011, the CRE had issued 955 permits for private generation[39]—720 for self-supply; 83 for co-generators; 28 for independent production to be sold to CFE; nine for export (this number includes co-generators, independent producers, and small generators generating electricity for export); 42 for import; six for small generation and 68 for continuous own uses.[40] Permits for import and export of electricity include the authorization for building and operating the necessary infrastructure.[41] Similarly, independent producers may build and operate their own transmission lines when they generate electricity for import or export purposes. According to article 73 of the mentioned regulation, those authorized to carry on such activities are also allowed to transmit, transform, and supply electricity. Notwithstanding that, article 123 points out that when importers are not interconnected to the National Electricity System, they are obliged to build and operate their own facilities with their own employees and resources.[42]

[37] In the case of electricity exports, to obtain the authorization the applicant has to attach the energy purchase contract or a letter of intention to that effect (article 117 of Regulations to the Electricity for Public Service Law). [38] See:< www.cre.gob.mx> (accessed 20 October 2011).

[39] This means that private generation capacity authorized at the national level is about 168,370.2 GW/h which is 48.58 per cent of total national generation. See <http://www.fte-energia.org/sdp/2010/6365.pdf> (accessed 5 December 2011).

[40] The so-called 'permits for continuous own uses' were granted before the 1992 reform under article 36 of the Electricity for Public Service Law as passed in 1975. This article held that: 'The Ministry of Industry and Trade in consultation with the Ministry of National Patrimony and CFE shall grant permits for self-supply of electricity aimed at satisfying the corporations' or individuals' own needs in those cases where CFE does not have the capacity to provide them with public electricity service.' This article was modified in 1992 and permits for continuous private use were substituted by permits for self-generation, but the reform did not cancel the permits previously granted. See CRE/GTZ, 'Guía para trámites con la Comisión Reguladora para permisos de generación e importación de energía eléctrica con energías renovables, cogeneración y fuente firmé (2010), available at <http://www.gtz.de> (accessed 25 September 2011).

[41] A similar situation may be found in the USA, where energy generation corporations interested in building an international transmission line are required to get a permit from the Department of Energy. Regarding lines crossing Mexico's south border, Guatemala's legal regime is quite different. According to the General Act of Electricity of that country, neither electricity generation nor electricity transmission require any authorization from the state. Nevertheless, authorization is required when such activities entail using public resources. See 'Invest in Guatemala: Energy Sector' (2008), available at <http://www.investinguatemala.org/index> (accessed 20 October 2011).

[42] In contrast, in the case of exporters the regulation does not impose a duty to build their own infrastructure.

III. Regional Energy Integration

Due to its geographical position, the Mexican economy is naturally integrated with both North America's[43] and Central America's.[44] In addition, as a result of NAFTA, entered into in the 1990s with Canada and the United States of America, as well as of the integration plans with Central[45] and South America,[46] Mexico has increased its exports of energy across both its borders. However, as will be analysed in the following sections, regional integration of electricity markets is not the main factor that has influenced the expansion of the National Electricity Network in Mexico.

A. Mexican electricity system

The Mexican network of transmission and distribution of electricity is basically under the control of CFE. By 2010 the network of electricity transmission operated by CFE had a longitude of 713,259km—49,250km of transmission network; 638,000km of distribution network, and 26,000km of optic fiber network.[47] This network covered 97 per cent of the country's territory and provided electricity to 93.84 per cent of the Mexican population.[48]

In this context, private companies—co-generators, independent producers, or small generators—that generate electricity for international trade purposes are allowed to either build and to operate their own energy networks or to connect to the National Electricity System. However, when independent producers produce energy just to be sold to CFE or when small generators and co-generators sell their surpluses to this entity, they are obliged to connect to the CFE's network and in consequence they are forbidden to own energy transmission lines. In addition to the mentioned prohibition and despite the fact that Mexican legislation allows private

[43] Mexico and the United States of America share a common border of about 1,310km.

[44] The Mexico–Guatemala border became the concrete reference point for the increasing discontinuities and asymmetries between both countries and more broadly, with the neighbouring region. Nevertheless, in local and regional terms, the border maintained a relatively dynamic trade relationship, albeit one that was persistently favorable to Mexico's interest. See Manuel Angel Castillo, 'The Mexico–Guatemala Border: New Controls on Trans-border migrations in View of Recent Integration Schemes?' (2003) available at <http://aplicaciones.colef.mx> (accessed 25 September 2011).

[45] The Mesoamerican Integration and Development Project (Plan Puebla Panamá) includes energy integration.

[46] In 1999, 12 South American countries conceived the Initiative for Integration of the South American Regional Infrastructure (IIRSA) that has, among others, the objective of increasing the security of energy supply, reducing the cost of investments, and developing countries in a peaceful and equitable way. See World Energy Council, 'Regional Energy Integration in Latin America and the Caribbean. Executive Summary' (2008), available at <http://www.worldenergy.org> (accessed 25 September 2011).

[47] See Eric Zenon and Juan Rosellon, 'The Expansion of Electricity Networks in North America: Theory and Application' (2010), available at <http://mpra.ub.uni-muenchen.de> (accessed 25 September 2011).

[48] See CFE, 'The Mexican Electric System: Investment Opportunities' (2010), available at <http://www.banobras.gob.mx> (accessed 25 September 2011).

investment in electricity transmission for international trade purposes, private electricity lines can be unnecessary given that CFE's network already covers most of the Mexican territory, and as a result a private enterprise would not have an incentive to undertake this investment.[49] Moreover, as will be analysed in the next sections, the National Electricity System under the control of CFE is already connected to both North American and Central American Networks.

B. The North American energy market

Electricity trade between the USA and Mexico is a two-way arrangement—from Mexico to the USA and from the USA to Mexico—but Mexican exports are currently higher than its imports. According to the US Energy Information Administration, by 2009 the USA imported 1,082,093 megawatt hours from Mexico, whereas it just exported 647,720 megawatt hours to the same country.[50]

According to Avila and Sarmiento, the level of exchange of electricity between the USA and Mexico has been influenced by a variety of geographic, economic, and political factors, including:[51]

- the increase of energy demand on the USA side because of demographic and economic conditions;
- the establishment in 1967 of the Western System Coordinating Council (WECC);[52]
- the signing of a 220 MW firm power purchase agreement between the CFE[53] and San Diego Gas & Electric and South California Edison in 1984;
- the incorporation of CFE as member of WECC in 1985.

[49] Glenn P. Jenkins, Henry B.F. Lim, and Gangadhar P. Shukl, *Evaluation of an Expansion of the Electricity Transmission System in Mexico* (Harvard University: Harvard Institute for International Development, 1999) 68.

[50] See US Energy Information Administration, 'Electric Power Industry—U.S. Electricity Imports from and Electricity Exports to Canada and Mexico', January 2011, available at <http://www.eia.gov> (accessed 25 September 2011).

[51] Miguel Angel Avila Rosales and Hector G. Sarmiento, 'Integrating the Electricity Markets in United States–Mexico–Central America' (2011), available at <http://www.scribd.com> (accessed 25 September 2011).

[52] According to its by-laws, WECC is a Utah-based non-profit corporation with a mission to: (1) maintain a reliable electric power system in the Western Interconnection that supports efficient competitive power markets ('Reliability Mission'); and (2) assure open and non-discriminatory transmission access among Members and provide a forum for resolving transmission access disputes between Members consistent with FERC policies where alternative forums are unavailable or where the Members agree to resolve a dispute using the mechanism provided in Section 11 (Transmission Access Mission). WECC is geographically the largest and most diverse of the eight Regional Entities that have Delegation Agreements with the North American Electric Reliability Corporation (NERC). WECC's service territory extends from Canada to Mexico. It includes the provinces of Alberta and British Columbia, the northern portion of Baja California, Mexico, and all or portions of the 14 Western states in between. See Western Electricity Coordinating System, 'Empowering Western Connection' (2011), available at <http://www.wecc.biz> (accessed 25 September 2011).

[53] The Federal Commission of Electricity is fully under the authority of the Ministry of Energy and is in charge of the state's electricity monopoly.

Apart from the mentioned factors, the signature of NAFTA in 1994 gave extra stimuli to regional North American energy integration in two ways.

First, as analysed, NAFTA negotiations influenced the reform of the legal framework for electricity making it possible for private national and foreign corporations to generate electricity in Mexico and deliver it to the North American market through the Mexican electricity network. Second, the idea that Mexican environmental and labour regulations were less burdensome that those of the USA stimulated foreign companies—not only from the USA but also from France and Spain—to build power plants in northern Mexico to generate electricity to satisfy rapid demand growth from industrial and residential users in the South of the USA.[54]

Overall, it seems that the USA is not interested in exporting electricity to Mexico but, rather, in the possibility of investing in this sector on the Mexican side of the border and thus allowing the USA to take advantage of the benefits of NAFTA while avoiding the heightened environmental cost of electricity generation and transmission in the USA. Among the environmental costs associated with electricity generation it is possible to consider the increase of fossil fuel consumption in the side of the exporter country that implies higher greenhouse gas emissions,[55] whereas transmission of electricity implies deforestation along the route on the transmission lines.[56] Nevertheless, only a few years were necessary to demonstrate that the benefits of NAFTA for USA's Electricity Corporations were apparent. First, Mexico's deficit of electricity generation due to economic and demographic considerations obliged CFE to focus more on the national market than on exporting electricity.[57] Second, whereas NAFTA rules apply just to Canada, Mexico, and the USA, the 1992 reform of the Mexican law opened the market of electricity generation to all foreign investors. Consequently, whereas investment of USA corporations in electricity generation in Mexico has not been considerable, official data shows that NAFTA has made it possible for European corporations to access USA's electricity market through the Mexican network.[58] Third, as discussed below, signature of NAFTA's environmental side-agreement and the process of modernization of Mexico's environmental legislation that followed NAFTA negotiations demonstrated that Mexico was not a polluters' heaven.

Accordingly, NAFTA did not substantially modify the bilateral electricity trade that was already in place before this agreement was negotiated. In fact, CFE has several

[54] See Global Energy Networks Institute, 'National Energy Grid: Mexico' (2010), available at <www.geni.org> (accessed 25 September 2011).

[55] Gloria Soto Montes de Oca, 'Quantifying NAFTA Environmental Impacts: Energy and Agriculture' (2002), available at <www.cserge.ac.uk> (accessed 25 September 2011).

[56] Economic Consulting Associates, 'The Potential of Regional Power Sector Integration: Central American Electricity Interconnection System (SIEPAC) Transmission and Trading Case Study' (2010), available at <http://www.esmap.org> (accessed 25 September 2011).

[57] Martin J. Pasqualetti (ed), 'The Geography of Energy at the US-Mexican Border, in Trade, Energy, and the Environment: Challenges and opportunities for the Border Region, Now and in 2020' (2003), *Southwest Center for Environmental Research and Policy*, Monograph Series No. 7,133 at167, available at <http://scerp.org/pubs/ENERGY%20BOOK.pdf> (accessed 25 October 2011).

[58] Out of 26 permits for independent producer of electricity granted by CRE, 10 were allocated to Spanish corporations (six to Union Fenosa and four to Ibedrola) and three to the French Corporation Electricité de France.

interconnections with electricity systems of various utilities along the USA–Mexico border.[59] Some of these ties have been used for permanent interchanges of energy while others are used only for emergency purposes.[60] Energy trade between Mexico and the USA takes place through Mexico's CFE and two USA operators of regional transmission systems: WECC and ERCOT. The main flows of electricity import/export between both countries are channelled through the interconnections CFE–WECC given that ERCOT interconnections are designed to be used in emergencies.

WECC connects with CFE in Baja California through two substations located in California (Miguel and Imperial Valley) in a synchronized and permanent way. It has a capacity of 800 MW. It also connects with CFE in Ciudad Juarez through the substations Insurgentes and Rivereña, connecting with two substations on the USA side at El Paso, Texas. This last connection has a capacity of 200 MW. It is synchronic and only operates during emergencies.

As mentioned, the CFE–ERCOT connection is mainly for emergencies, with the exception of an unsynchronized interconnection between substations in Piedras Negras, Coahuila, and Eagle Pass in Texas, which can operate permanently.

In addition to the above, there are two private transmission lines—Intergen and Sempra—connected to two gas-based independent producer plants placed on the Mexican side of the border.[61]

C. The Central American market

Cross-border electricity trade between Mexico and Central America is an important part of the 'Project for Mesoamerican Integration and Development'[62] that substituted the Plan Puebla Panama.[63] The Mesoamerican Project represents a high-level political space where the cooperation, development, and integration efforts of 10 countries (Belize, Colombia, Costa Rica, El Salvador, Guatemala, Honduras, Mexico, Nicaragua, Panama, and Dominican Republic) are discussed. It also facilitates the management and implementation of projects aimed at improving the quality of life in the region. It was officially launched during the Tenth Summit Meeting of the 'Tuxtla Mechanism' held in Villahermosa,[64] Tabasco, Mexico, on 28 June 2008. Specifically, the Project of Development and Integration of Mesoamerica has the following objectives:[65]

[59] There are currently three high-tension interconnections between the US and Mexico and several other transmission projects have been studied and proposed, particularly in the Arizona–Sonora corridor. See Edward A. Hoyt et al, 'Environmental Implications of Increased US–Mexico Electricity Trade' (1999), available at <http://www.sciencedirect.com> (accessed 25 September 2011).

[60] See CFE/ERCOT, 'Interconnections Study' (2003), available at <http://www.puc.state.tx.us> (accessed 25 September 2011).

[61] INTERGEN is an English corporation, whereas SEMPRA is a USA company. Both have permits as independent producers granted by the Mexican Authority.

[62] Available at <http://en.reingex.com> (accessed 25 September 2011).

[63] Available at <http://www.proyectomesoamerica.org/> (accessed 25 September 2011).

[64] See above, n 21.

[65] See Economic Consulting Associates, 'The Potential of Regional Power Sector Integration. Central American Electricity Interconnection System (SIEPAC): Transmission & Trading Case Study' (2010), available at <http://www.esmap.org> (accessed 25 September 2011).

a) to improve competitiveness in the Mesoamerican region through strengthening its energy sector, with a focus on electricity generation, electricity interconnection, renewable energies, energy savings and energy efficiency;
b) to satisfy regional energy demand;
c) to build interconnection infrastructure that allows the interchange of energy among the countries of the region;
d) to consolidate a Regional Electricity Market.

In order to meet these goals, the Project of Integration includes three main regional energy sub-projects, as described below.

1. The Central American Electricity Interconnection System (SIEPAC)

This is the most important sub-project, given that it provides the foundation for the other two energy sub-projects in the region.[66] It contemplates a series of electricity networks that integrates 1,800km of transmission lines and 15 sub-stations, 103km of the interconnections of the Mexico–Guatemala system and 614km of the Panama–Colombia connection.[67] It will facilitate trading up 300 MW of power and the establishment of a regional electricity market.[68]

The builder and owner of the system is a consortium comprised of private and public companies known as 'Empresa Propietaria de la Red—Grid Proprietor' (EPR). The consortium was established in 1988 in accordance with the Treaty for the Central American Electricity Market.[69] The company has the participation of Guatemala's National Institute of Electrification; the Hydroelectric Executive Commission of the Rio Lempa (El Salvador); Nicaragua's National Corporation of Electricity Transmission; Costa Rica's Electricity Institute of Costa Rica; Panama's Corporation for Electricity Transmission; Colombia's Electricity Interconnection; the Spanish Energy Corporation ENDESA; and the Federal Commission of Electricity (Mexico).

In the case of the six signatory parties of the Treaty for Central American Electricity Market, each government is required to grant a 30-year concession across its territory to the transmission line company.[70] In Mexico no concession was granted given that the transmission line connects with the national electricity

[66] It received support from the Inter-American Development Bank. The total investment for this project was US$500 million.

[67] Jeremy Martin, 'Central America Electric Integration and the SIEPAC Project: From a Fragmented Market Toward a New Reality' (2010), available at <http://www6.miami.edu> (accessed 25 September 2011).

[68] See Economic Consultant Associates, 'The Potential of Regional Power Sector Integration: Central American Electric Connection (SIEPAC) Transmission and Trading Case Study' (2010), available at <http://www.esmap.org> (accessed 25 September 2011).

[69] Article 15 of the Treaty establishes that 'Each Government shall designate a public agency in its country to participate in a public or private capital enterprise in order to develop, design, finance, build and maintain an initial regional transmission system that will interconnect the electrical systems of the six countries. None of the members shall have direct or indirect control of the said enterprise. The enterprise shall be called the Grid Proprietor (EPR); it shall be governed by private law and legally domiciled in a Central American country.'

[70] See article 17 of the Treaty.

system owned by CFE that is not part of the regional network. In the same vein, separate environmental impact authorizations granted by each government were required.

2. *Mexico–Guatemala interconnection*

The second sub-project connects the Mexican Electricity System with the SIEPAC. It consists of a transmission line of 103km in length (32km in Mexico and 71km in Guatemala) of 400 kV, as well as the expansion of two sub-stations, one in Tapachula, Mexico and the other in Los Brillantes, Retalhuleu, Guatemala.[71] This sub-project has already been concluded, enabling Mexico to export power to the entire Central American region. In March 2010, in accordance with a contract signed between both countries, Mexico started to export electricity to Guatemala.[72]

3. *Panama–Colombia interconnection*

The third sub-project consists of building a transmission line of 300 MW and approximately 614km between the substations of Cerromatoso in Colombia and Panama II in Panama.[73] The interconnection includes a maritime section of 55km that supposes environmental and social benefits by minimizing such impacts in Comarca Kuna Yala and the Serranía del Darien, which are the fundamental blocks of the so-called Mesoamerican Biological Corridor, where ten per cent of the world's biodiversity is located.[74]

Unlike what happened in the case of Mexico's integration with the North American electricity market, in the process of integration of the Mexican electricity system with the Central American Network, CFE has played the double role of exporter and investor in infrastructure. Given that according to the General Act of Electricity of Guatemala, there are no restrictions to foreign investment in building and operating electricity transmission networks,[75] the Mexican pub-

[71] The project's cost was US$56 million and was funded by an Inter-American Development Bank loan to Guatemala and by the Federal Commission of Electricity.

[72] The electricity interconnection between Mexico and Guatemala was officially opened in October 2009, when the Mexican President, Felipe Calderon, visited Guatemala.

[73] The cost of the project is estimated in US$300 million. The project will be built on the basis of an Inter-American Development Bank technical cooperation non-reimbursable loan of US$2.6 million to complete the basic engineering and pre-design of the project, and the environmental impact statement and analysis of regulatory harmonization.

[74] The Mesoamerican Biological Corridor was established in 1997 by the governments of the countries in the Mesoamerican region (Belize, Costa Rica, El Salvador, Guatemala, Honduras, Nicaragua, Panama, and Mexico) as a system of land planning consisting in four types of natural areas: (a) core areas which are exclusively for conservation of ecosystems and species and in which human activities are prohibited; (b) buffer zones, which are of restricted use by themselves; (c) corridors, which are areas that facilitate movement, dispersal, and migration species, and in which human activities are of low impact; and (d) multiple-use areas, which may include areas devoted to various activities such as agriculture, livestock, forest management, etc. The objectives of the Mesoamerican Biological Corridor are to maintain biological diversity, reduce fragmentation, and improve the connectivity of the landscape and ecosystems, and promote sustainable production processes that improve the quality of life of the local human populations who utilize, manage, and conserve biodiversity.

[75] See n 42.

lic corporation expanded its presence beyond Mexico's borders with the Central American region.

IV. Environmental Implications of Energy Networks

Electricity transmission lines are lineal facilities that could affect natural and sociocultural resources. The impacts of short lines are local but the longest ones—as, for instance, SIEPAC—can produce regional impacts. Most of such impacts occur on or near the right of way, that is to say, the area of land where the structures supporting the line are placed, including clearance allowances. The higher the voltage, the higher the impacts, and[76] bigger support structures are required, with the resulting increase in the corresponding environmental impacts.[77]

The negative environmental effects of transmission lines are caused by their construction, operation, and maintenance. The main causes of such impacts associated with transmission lines include land clearing of the right of way, building of access ways, and placement of towers and substations. In addition, maintenance work includes chemical or mechanical control of vegetation over the right of way. Finally, the impacts of transmission lines on the landscape are also considerable.

At the national level, environmental legislation addresses the control or mitigation of all transmission lines' negative impacts. However, in cases of regional markets such as North or Central American markets, when environmental legislation is not homogeneous in all the countries the apparent economic benefits of electricity exports may hide tremendous environmental impacts for the country where the electricity is produced or where the transmission networks are placed.[78]

A. Environmental protection in the North American region

Barely had NAFTA negotiations been initiated than USA environmental groups voiced three concerns about the free trade area:[79]

a) first, that NAFTA would encourage US businesses to move to Mexico to take advantage of lower environment standards or lax enforcement;[80]

[76] For instance, the negative effects of the magnetic field are stronger for lines of 1000 kV than for those of 69 kV.

[77] See n 59.

[78] In 2002 the environmental groups Earth Justice and Wild Earth Advocates, representing the Border Power Plant Working Group, filed a lawsuit against the US Government challenging permits granted to two companies planning to build electrical transmission lines from Baja, Mexico into California. The lines were built to take power from two electrical generation plants being built three miles inside of Mexico to supply power to the United States. The plaintiff held that 'Approving transmission lines for power plants under construction in Mexicali, without ensuring that these plants are built to minimize air and water quality impacts, will cause unnecessary harm to local US and Mexican communities.' See Earth Justice, 'Environmentalists Sue over Transmission Lines from Mexico to US', 19 March 2002, available at <http://earthjustice.org> (accessed 25 September 2011).

[79] F. Mayer, 'Negotiating NAFTA: Political Lessons for the FTAA' (2001), available at <http://www.pubpol.duke.edu> (accessed 25 September 2011).

[80] Gallagher holds that 'Although the majority of firms that move to Mexico do not move there because of low environmental standards, this does not imply that when firms move to Mexico they

b) second, that a free trade agreement with Mexico would exacerbate the already terrible environmental problems of the border region;
c) third, the free trade agreement could be used to challenge US environmental regulations.

That criticism provoked the following effects:

1) NAFTA was 'greened' with a number of rules related to environmental protection.[81]
2) Signature of NAFTA by Canada and the USA was conditioned to the adoption of a North American Side-Agreement on Environmental Cooperation (NAAEC).[82]
3) A plan to deal with environmental problems at the border was adopted by Mexico and USA.[83]
4) The Mexican government created the Attorney General Office for Environmental Protection (in Spanish PROFEPA) as an administrative agency with the aim of improving Mexico's national strategy for environmental law enforcement.[84]
5) The General Act for Ecological Balance and Environmental Protection (in Spanish LGEEPA) was amended for the first time, among other reasons with the objective of complying with commitments assumed by Mexico in both NAFTA and its Environmental Side-Agreement.[85]

are model environmental corporations. In fact, the World Bank conducted a survey of over 200 firms in Mexico and found that, contrary to prevailing assumptions, foreign firms were no more likely than domestic firms to comply with Mexican environmental law.' K. Gallagher, 'Free Trade and Environment: Mexico, NAFTA and Beyond' (Stanford University Press, 2004).

[81] A. Mol, *Globalization and Environmental Reform: The Ecological Modernizations of the Global Economy* (Cambridge, Mass: MIT Press, 2001).

[82] In the words of Bailey, NAAEC 'was designed to assuage fears that Mexico would became a heaven for pollution once NAFTA took effect drawing all industry south the border where environmental enforcement was weak'. K. Bailey, 'Citizen Participation in Environmental Enforcement in Mexico and the United States: A Comparative Study' (Winter 2004) *Georgetown International Environmental Law Review*, available at <http://findarticles.com> (accessed 25 September 2011). Trade officials from Canada, Mexico, and the United States began negotiations in March 1993. On 14 September 1993, the three countries concluded negotiations on environmental and labour side accords that include provisions to address problems with enforcement of environmental and labour laws.

[83] 'In separate negotiations, the United States and Mexico agreed to establish a Border Environmental Cooperation Commission (BECC) and a North American Development Bank (NADBank) to provide financing for environmental investments. NAFTA implementing legislation passed Congress in November, 1993, and the agreement entered into force on January 1, 1994.' Susan Fletcher and Mary Tiemann, *Trade and Environment: GATT and NAFTA* (Congressional Research Service, Library of Congress,1994).

[84] See 'Reglamento Interior de la Secretaría de Desarrollo Social, Diario Oficial de la Federación' 4 July 1992 and 'Acuerdo que regula la organización y funcionamiento interno del Instituto Nacional de Ecología y de la Procuraduría Federal de Protección al Ambiente, Diario Oficial de la Federación', 17 July 1992, available at <http://www2.ine.gob.mx/publicaciones/gacetas/208/ine-profepa.html> (accessed 20 October 2011).

[85] Published in *Diario Oficial de la Federación*, 13 December 1996 (Government of Mexico).

6) The Mexican government issued a number of standards in order to comply with NAFTA's requirements on regulatory harmonization.

In this context, the signature of the Environmental Side-Agreement to NAFTA[86] (hereinafter NAAEC) was aimed at avoiding the negative environmental impacts of international trade. In that sense some of the most important commitments assumed by signatory parties are:

a) Each country shall ensure that its laws and regulations provide for high levels of environmental protection and shall strive to continue to improve those laws and regulations (article 3).
b) Each Party shall effectively enforce its environmental laws and regulations through appropriate governmental action (article 5).

When NAFTA negotiations started, Canada and the USA had already established agencies in charge of enforcing environmental regulations.[87] In 1992 the Attorney General Office for Environmental Protection (PROFEPA in Spanish) was established by Mexico's Government. PROFEPA was given the power to enforce environmental laws in an attempt to deter foreign investors from seeing Mexico as a polluter's heaven. PROFEPA's creation represented an important step in the process of institutional harmonization required by the economic integration of the three countries. Additionally, NAAEC contemplates a special complaint procedure aimed at guaranteeing the effective enforcement and compliance of environmental laws in the territory of the three signatory parties. Thus, in theory, the applicable requirements for environmental impact assessment and compliance of energy plants and infrastructure for transmission networks are similar.

B. Environmental protection and electricity networks in Mexico

In the Mexican case, environmental legislation obliges those interested in building and operating electricity infrastructure to meet a series of requirements. First, in accordance with article 28, paragraph II of the Ecological Equilibrium and Environmental Protection Act (LGEEPA), all activities related to the electricity sector are required to obtain an environmental impact authorization from the Ministry of the Environment following the corresponding environmental impact assessment.[88] The regulations on environmental impact assessment specifically

[86] Signed in 1993 and enforceable since 1994. The Agreement creates a framework to better conserve, protect, and enhance the North American environment through cooperation and effective enforcement of environmental laws. See <http://www.naaec.gc.ca> (accessed 25 September 2011).

[87] See José Juan González (ed), *La naturaleza jurídica de la evaluación del impacto ambiental. Análisis de Derecho Comparado* (Mexico: Instituto Mexicano de Investigaciones en Derecho Ambiental, 2011).

[88] In the USA, the issuance of a connection permit for export purposes requires consideration of (a) the impact of the new transmission line on the reliability of US Electricity System; and (b) the environmental impact of the project. The procedure to obtain the EIA in the USA is quite different from the Mexican one. See above n 83. By contrast, in Guatemala there is no legal provision that specifically requires the environmental authorization of energy networks. See Grethel Aguilar et al, *Evaluación del Impacto Ambiental Transfronteriza en Centro America*, Lineamientos Generales (Costa Rica: UICN, 2006).

include building and operating infrastructure for electricity transmission.[89] Second, when transmission lines cross forest lands, article 58 of the General Act for Sustainable Development of the Forest[90] requires obtaining prior authorization for land use change. According to the mentioned provision, the Ministry of Environment could grant the land use change only by exception. Finally, if the transmission line is built on agricultural lands, a land use authorization has to be issued by local governments.[91]

Neither the environmental impact authorization nor the land use change authorization is easy to obtain. Normally both of them involve very costly environmental studies and a very exhaustive revision by the competent authorities. Additionally, in a number of recent cases important projects have confronted the opposition of environmental groups.[92] As a result, network developers may face costly and lengthy litigation processes.[93]

In addition to the mentioned environmental authorizations, investors are required to buy or to rent the land where the structures are placed, ie the right of way (ROW).[94] The ROW can be established on private, public, and *ejidos* or community lands.[95] This ROW may be obtained through three mechanisms: a) an easement from the property owner; b) a purchase and sale agreement with the property owner; and c) a rent agreement with the property owner. In all these cases, according to articles 1097, 2248, and 2398 of the Civil Federal Code, landowners on whose land the line is placed have the right to receive an economic compensation.

Sometimes acquiring the right of way is not easy given that in almost all Mexican agricultural lands as well as in those owned by *ejidos* there are a number of conflicts

[89] See article 5, sub-paragraph K-III of Regulation under LGEEPA on Environmental Impact Assessment.

[90] Diario Oficial de la Federación, 25 February 2003.

[91] According to article 35 of the General Act of Human Settlements, it corresponds to municipal governments to regulate land use in their territory.

[92] In November of 2003 the CFE had to modify the course of the transmission line Potrerillos–Aguascalientes–Potencia, in Guanajuato, to avoid environmental impacts on the natural protected area of the Silva's Dam given the strong opposition of the environmental groups, which asserted that the transmission line could affect the wildlife living in that area.

[93] For instance, the 'Accountability Counsel' filed complaints on 30 November 2010 on behalf of the villages of Paso Canoa and Santa Ursula, and on 17 January 2011 on behalf of the neighbouring indigenous community of Cerro de Oro. The complaints detail harm caused by the Overseas Private Investment Corporation (OPIC)-supported Cerro de Oro Hydroelectric Project. The Project entails converting the Cerro de Oro reservoir into a hydropower project through the construction of a water intake and conduction tunnel, powerhouse, voltage elevation substation, tailrace channel, and transmission lines. The Council held that the complainants had not received information about the project, had not been consulted, impacts on indigenous groups had not been considered, and that there were insufficient plans to address and mitigate social and environmental impacts, including destruction of important waterways that communities depend on for household use, consumption, and fishing. Complainants also note problems with land acquisition and the lack of a required, local grievance mechanism. See <http://www.accountabilitycounsel.org> (accessed 25 September 2011).

[94] An electricity line right of way (ROW) is a strip of land that an electricity utility uses to construct, maintain, repair, or replace an overhead or underground power line. The ROW allows the utility to provide clearance from trees, buildings, and other structures that could interfere with the line installation, maintenance, and operation.

[95] The non-mandatory standard NRF-014-*CFE* issued in 2001 recommends that the right of way is 52.48 feet wide.

over property rights, so the first problem consists of identifying who has the legal capacity to grant the easement or to sell or rent the land.[96] On the other hand, since the Civil Code does not establish any criteria to determine the value of the economic compensation for the easement and in consequence it has to be established by negotiation between the constructor and the landowner, landowners may demand a very high price for rent or sale.

Notwithstanding the above, what is missing in Mexican environmental law is a regulation on the transboundary environmental impact of both a) regional networks of electricity transmission; and b) energy plants that generate electricity to be transported and distributed beyond the frontiers of the country where such facilities are placed. In fact, the original Plan Puebla Panamá contemplated a number of environmental strategies to be implemented alongside the regional development plans such as the harmonization of environmental impact assessment systems across Central American Countries. However such strategies have not been implemented yet.[97] This gap in legislation prevents adequate and full evaluation of the environmental impacts resulting from Mexico's integration with both North and Central American electricity markets.

V. Conclusions

Despite constitutional provisions, Mexico's energy legislation allows participation of private investors in building and operating electricity infrastructure. However, private electricity networks have a limited objective. Private networks are only allowed to be built for import and export purposes. In addition, there are many disincentives to private investment in energy networks:

1. When NAFTA went into force and Mexican energy law was consequently amended to allow private investment in energy networks, the national energy transmission system under the control of CFE had already covered almost the entire Mexican territory.
2. Whereas the Electricity for Public Service Law allows importers and exporters to transmit electricity through the national energy system owned by CFE, building their own electricity networks could represent a number of problems to private investors, such as conflicts with environmental groups or difficulties in obtaining the required authorizations.

[96] Fifty-five per cent of Mexico's land area is agricultural land. Nearly 94 per cent of these lands belong to *ejidos* and *comunidades*. On *ejidos* and *comunidades*, disputes related to inheritance and parcel boundaries are common. As a result, normally it is not easy to identify who holds property rights of these lands. See USAID, 'Country Profile: Property Rights and Resource Governance, Mexico' (undated), available at <http://usaidlandtenure.net> (accessed 25 September 2011).

[97] Comisión Centroamericana de Ambiente y Desarrollo, 'Transversalidad del tema ambiental en el Plan Puebla Panamá- Analisis y recomendaciones' (2004), available at <http://www.incae.edu/es/clocds/publicaciones/pdf/cen718.pdf> (accessed 5 December 2011).

In sum, private investors do not have real incentives to build or operate their own electricity networks. Although a few permits for exporting energy have been granted to independent producers in the north of the country, producers prefer to be connected to national energy network. As a result, CFE operates as the single buyer and distributor of electricity while private investment is focused more in electricity generation than in electricity transmission.

Thus even when private networks are permitted by Mexican legislation, investing in this area could seem unattractive for private companies. Consequently, the Mexican electricity model has passed from a full state monopoly model to a model of state monopoly on transmission and distribution of electricity.

Mexico's energy integration with the North and with the South has different characteristics. In the case of the Mexico–USA energy market, when international trade took off, each country already had a developed national network. In the Mexico–Guatemala case, it was necessary to build electricity infrastructure to send electricity to that country but because of the limitations on the Mexican side no private investment participated in building this infrastructure. However, neither the process of integration with the North American Market nor Mexico's participation in the Central American electricity networks have substantially influenced the development of the national electricity system.

4

South American Energy Network Integration: Mission Possible?

Lila Barrera-Hernández

I. Introduction

South America (SA) is rich in a diverse mix of energy resources. To speak only of traditional resources, the region accounts for roughly 11 per cent of global reserves of oil and nine per cent of gas reserves.[1] In addition, despite its widespread use as a source of primary energy,[2] SA's hydropower potential is still significant, exceeding 600 GW.[3] Though abundant overall,[4] resources are unevenly distributed.[5] Natural gas's reserves-to-production ratio exceeds 130 years in Bolivia, Ecuador, Peru, and Venezuela, but is only ten years for the case of Argentina and zero for the case of Uruguay.[6] Hydropower potential is concentrated in the Amazon but is also abundant in the Plata basin, already the site of three major bi-national hydropower dams.[7] Both resource basins have the potential to complement each other, as well as other sources of power, as they are subject to different hydrological patterns and climate conditions.

Despite the abundance of energy resources, even where they are plentiful, sustainable access to sufficient and reliable energy supplies that are economically and socio-environmentally viable may be threatened by underinvestment, unfavourable climatic conditions, or political interference. Throughout the region, existing energy infrastructure is increasingly unable to handle existing and growing

[1] Reserves of coal are at around 5 per cent. C.A. Florez, OLADE, 'Latin America and the Caribbean Energy Integration', presentation, Montreal (13 September 2010).

[2] Hydropower accounts for 84 per cent of power production in Brazil.

[3] Brazil is also the second largest ethanol producer in the world. World Energy Council, 'Regional Energy Integration in Latin America and the Caribbean' (2008), available at <http://www.worldenergy.org> (accessed 26 September 2011).

[4] Latin America and the Caribbean only use 26 per cent of overall energy potential. F. Burgos, 'Regional Electricity Cooperation and Integration in the Americas' [undated], available at <http://www.oas.org> (accessed 26 September 2011).

[5] D. Otero Prada et al, 'Análisis de la integración energética en América Latina y de su importancia para Suramérica', August 2009, available at <http://www.ucentral.edu.co> (accessed 26 September 2011).

[6] World Energy Council, see n 3.

[7] The Plata basin includes Argentina, Bolivia, Brazil, Paraguay, and Uruguay.

demand, as evidenced by the frequent outages experienced across SA.[8] As a result, energy security—defined as the 'condition in which a nation and all, or most, of its citizens and businesses have access to sufficient energy resources at reasonable prices for the foreseeable future, free from serious risk of major disruption of service'[9]—remains elusive.

Because of the region's enormous potential for complementarity, regional energy integration as a concerted multilateral effort involving strategic, political, and economic convergence, and resulting in an integrated network system that goes beyond the transfer of energy from one country to another (interconnection), has long been championed as the best means to achieve energy security across South America. Efficiency, diversification, environmental impact mitigation, and economies of scale are also brought to bear in favour of network integration.[10] In turn, integration is believed to have the potential to be a significant motor of economic development and is therefore considered of vital importance for the region.[11]

Central to energy integration are network facilities that are developed and operated on a multinational scale. Given the enormous costs of developing and operating transport infrastructure,[12] and the pendular shifts from one end of the ideological spectrum to the other that are common in South American countries,[13] for energy integration to become a reality, the existence of clear and predictable rules to support the development and operation of a multinational network system is paramount. Moreover, those rules must be compatible and need to be supported by a governance system that is committed to stability in upholding the rule of law and that can resist, as far as possible, political manipulation.[14]

Experience shows that domestic laws in SA may often fall short of providing the clarity, predictability, and stability that is required for long-term development enterprises, including multinational energy networks, to succeed.[15] As explained below, even bilateral agreements have proven unable to guarantee continuous

[8] 'Don't Mention the B-Word: Hungry for Power', *The Economist*, 10 February 2011, available at <http://www.economist.com/node/18114669> (accessed 4 October 2011).

[9] 'Introduction' in B. Barton et al (eds), *Energy Security: Managing Risk in a Dynamic Legal and Regulatory Environment* (Oxford: Oxford University Press, 2004) 5.

[10] M. Mendes da Fonceca and L.E. Duque Dutra, 'Energy Integration in South America: Driving Force for Regional Integration Process?', October 2007, available at <http://www.crei.cat> (accessed 26 September 2011).

[11] ECLAC, 'Latin America and the Caribbean in the World Economy 2008–2009' (2009), available at <http://www.cepal.org> (accessed 26 September 2011).

[12] The International Energy Agency estimates that Latin America will need US$1.3 billion in new investment in the energy sector before the year 2030 in order to meet new demand. F. Rojas Aravena, 'Energy Integration in Latin America: Limits and Possibilities' in Arnson et al (eds), *Energy and Development in South America: Conflict and Cooperation* (2008), available at <http://www.flacso.org> (accessed 26 September 2011).

[13] See eg L. Stanley, 'Natural Resources & Foreign Investment: A Tale of Three Andean Countries' (April 2008), Working Group on Development and Environment in the Americas, Discussion Paper Number 16, available at <http//ase.tufts.edu> (accessed 26 September 2011).

[14] See eg R. Araneda K., GasAtacama, 'Desafíos para inversión en sector energético de Latinoamérica', 6 October 2004, available at <http://arpel.org> (accessed 16 September 2011).

[15] Lila Barrera-Hernández, 'The Andes: So Much Energy, Such Little Security' in B. Barton et al (eds), *Energy Security: Managing Risk in a Dynamic Legal and Regulatory Environment* (Oxford: Oxford University Press, 2004) 217.

flows of energy from one South American country to another. Regional cooperation agreements such as the Union of South American Nations (UNASUR), the Southern Common Market (MERCOSUR), and the Community of Andean Nations (CAN) are ideally positioned to fill that gap. Can those regional agreements supply an adequate legal platform for multinational energy networks to develop and operate successfully in SA? What are their strengths and weaknesses? These are the questions that the following sections attempt to answer.

II. Background

Historical relations amongst South American countries have not always been conducive to collaboration. Persisting animosity between Bolivia and Chile over the former's loss of ocean access to Chile in a war fought more than one hundred years ago has often resulted in Bolivia's refusal to sell gas to Chile. Chile and Uruguay are wary over increasing their dependence on Argentina after their neighbour cut off gas supplies and defaulted from its energy export commitments in 2004.[16] In 2006, Brazil also fell victim to Bolivia's oil and gas nationalization and had to renegotiate much-needed gas imports. Venezuela and Colombia, also linked via pipeline, have had numerous clashes over the past couple of years,[17] while Peru's relations with neighbouring Ecuador and energy-thirsty Chile have not been without problems.[18]

Despite the countries' dysfunctional relations, SA is laced with a number of interconnection facilities for gas and electricity.[19] However, existing cross-border infrastructure is the result of an ad hoc approach to energy trade between neighbouring countries and not a result of policies and rules designed to facilitate long term cooperation and network development. For the most part, the development of existing interconnected facilities has been a state-sponsored undertaking backed by individual agreements over the sale of power or gas.[20] The first cross-border gas pipeline in SA was built in the seventies between Bolivia and

[16] O. Landerretche, 'Chile's Choices: Maintaining Growth and Securing Supply' in C.J. Arnson et al, see n 12.

[17] See eg K.M. Tullos, 'More than Just Words?: The Relations between Venezuela and Colombia and UNASUR Intervention in Light of the Defense Cooperation Agreement between the United States and Colombia' (Summer 2010) 16 *L. & Bus. Rev. Am.* 559.

[18] For an account of South American relations in connection with energy integration see: J. Rios Sierra, 'Los Multiples Encuentros y Desencuentros de la Integracion Energetica Suramericana' (2010), available at <http://www.urjc.es> (accessed 26 September 2011); and A. Ruiz-Caro, 'Puntos de conflicto de la cooperación e integración energética en America Latina y el Caribe' (2010), CEPAL, Serie Recursos Naturales e Infraestructura #148, available at <http://www.eclac.org/cgi-bin/getProd.asp?xml=/publicaciones/xml/8/39408/P39408.xml&xsl=/drni/tpl/p9f.xsl&base=/drni/tpl/top-bottom.xsl> (accessed 4 October 2011).

[19] One oil pipeline connects Ecuador and Colombia; it is devoted to Caribbean exports. See L. Mewett, 'South America Snapshot' September 2009, available at <http://pipelinesinternational.com> (accessed 26 September 2011).

[20] A.F. Guerrero Bermeo, 'La integración energética como instrumento para el fortalecimiento del proceso de integración Suramericano UNASUR' (2010), available at <http://www.iadb.org> (accessed 26 September 2011). See also N.J. de Castro et al, 'La Integración del Sector Eléctrico

Argentina. The 1990s saw the development of 10 additional pipelines linking Argentina with Chile (seven), Brazil (one), and Uruguay (two). Though the latter were conceived as export connections, dwindling supplies in Argentina have led to speculation about their potential use as import lines though linkages to other producers.[21] Other gas interconnections include pipelines linking Bolivia with southern Brazil, and Colombia with Venezuela. In addition, there are nine electricity interconnection lines throughout SA.[22] As is the case with the cross-border pipelines, all electrical interconnections operate on a bilateral basis and have yet to be integrated into a multilateral network.[23]

Furthermore, physical interconnection received a boost in 2000 with the launch of the region-wide Initiative for the Integration of the Regional Infrastructure in South America (IIRSA). IIRSA's goal is to promote the development of transport, energy, and communications infrastructure throughout the region to stimulate the integration and development of isolated sub-regions. It is backed by financial assistance from the Inter-American Development Bank (IDB) and the Andean Finance Corporation (CAF).[24] So far, however, its focus has been on developing infrastructure on an individual project basis rather than as part of a concerted network development plan.

Notwithstanding the infrastructure already on the ground, unilateral country action due to changing local circumstances, such as unanticipated scarcity in the supplier country or changes in domestic policy usually following a significant shift in ideology, as was the case of Bolivia's nationalization of oil and gas assets, can disrupt the normal operation of interconnection facilities and is a threat to their viability and continued development. Rather than ad hoc interconnection arrangements, the stability and security required for achieving energy security and promoting sustained development throughout the region may only be achieved through network integration, including the adoption of overarching energy policy and regulation that levels the playing field, simplifies procedures, and minimizes the risk of unilateral action. What follows is an analysis of existing regional and sub-regional integration initiatives in relation to energy network development.

III. Regional Integration Initiatives

A. Union of South American Nations

Over the past couple of decades, the governments of South America have repeatedly pledged to increase cooperation to leverage development opportunities.

en Sudamérica: Características y Beneficios' (2009), available at <http://plataformaenergetica.org> (accessed 26 September 2011).

[21] D. Mansilla, 'Integración Energética y Recursos Naturales en América Latina' *La revista del CCC*, January/April 2011, available at <http://www.centrocultural.coop> (accessed 26 September 2011).

[22] C.A. Florez, see n 1.

[23] D. Mansilla, see n 21.

[24] For information on IIRSA see <http://www.iadb.org> (accessed 26 September 2011). M. Mendes da Fonceca and L.E. Duque Dutra, see n 10.

Increasingly, energy integration has been at the centre of cooperation-related initiatives. Thus, in the 2007 *Declaration of Margarita*, South American countries committed to energy integration as a means to promote growth and to combat poverty in the region.[25] The simultaneous creation of the Union of South American Nations (UNASUR—Union de Naciones Suramericanas), including Argentina, Bolivia, Brazil, Colombia, Chile, Ecuador, Guyana, Paraguay, Peru, Suriname, Uruguay, and Venezuela, also had energy integration at its centre.[26] Despite the fact that this cooperation agreement has only recently entered into force following Uruguay's ratification,[27] signatory countries have been remarkably active in moving towards energy integration.

Through the work of UNASUR's Energy Council,[28] some important advances in the road to integration have been reported.[29] In May 2010, at a meeting in Los Cardales, Argentina, UNASUR heads of state approved the *Outline of the South American Energy Strategy*[30] and the *Outline for the Action Plan for Regional Energy Integration*.[31] Though declaratory, both documents draw attention to the importance of infrastructure development for achieving a level of energy security and quality of supply that is compatible with the region's sustainable development. Countries also vow to take full advantage of complementarities and to diminish energy asymmetries while prioritizing internal supply and intra-regional trade. To this end, they commit to legal and regulatory harmonization and call on their leaders' awareness of the need for sustained political commitment and engagement. The countries' pledges are supported by a long list of action items that go all the way from vowing to undertake research into needs, gaps, and, potentials, to exploring potential infrastructure funding opportunities. Most importantly, at the same meeting, the *Outline of the South-American Energy Treaty* (the Outline) was also approved.[32] Although the Outline approved is little more than a table of contents,

[25] Declaración de Margarita (17 April 2007), available at <http://uniondelsur.menpet.gob.ve> (accessed 26 September 2011).

[26] Tratado Constitutivo de la Unión de Naciones Suramericanas, arts 2 and 3 (y), available at <http://www.pptunasur.com> (accessed 26 September 2011); 'Decisiones del Diálogo Político entre los Jefes de Estado y de Gobierno' I Cumbre Energética Suramericana, 16 and 17 April 2007, available at <http://www.comunidadandina.org> (accessed 26 September 2011).

[27] UNASUR entered into force on 11 March 2011. Current ratifications include Argentina, Bolivia, Chile, Ecuador, Guyana, Peru, Surinam, Venezuela, and Uruguay.

[28] The Energy Council (Consejo Energético de Suramérica) was established under the Declaration of Margarita and adopted under the UNASUR Charter as an organ of UNASUR. See Tratado Constitutivo de la Unión de Naciones Suramericanas, art 5, available at <http://www.pptunasur.com> (accessed 26 September 2011).

[29] Venezuela has been an important promoter of energy integration as a form of South–South collaboration which could contribute to a better distribution of power between the developed and developing nations of America. Under that vision, national energy companies were given an important role in facilitating integration by joining forces and working together. However, none of the big regional energy conglomerates proposed—Petroamérica, Petroandina, or Petrosur—have effectively taken shape.

[30] UNASUR, Lineamientos de la Estrategia Energética Suramericana (4 May 2010) [on file with the author].

[31] UNASUR, Lineamientos del Plan de Acción para la Integración Energética Regional (4 May 2010) [on file with the author].

[32] Declaración de Los Cardales, (4 May 2010).

it clearly centres on facilitating cross-border energy exchanges on a long-term, sustainable basis. After a section dedicated to principles, objectives, and provisions of general application, Section II, the core of the Outline, focuses on the following:

- necessary infrastructure for integration;
- regulatory framework, including treatment of transnational and supranational corporations;
- energy security, including rules on emergency management;
- energy trade, including free transit and cross-border issues;
- cooperation tools such as information systems, regional capacity-building, etc;
- environment; and
- investment.

The Outline's footnotes provide a glimpse into the issues that may require the most attention during the treaty's negotiation. Palpable in them are the countries' differences in political alignment and the consequent struggle to find an optimum balance between differing development models, as well as between sovereignty and integration. Thus, among other things, one footnote cautions that regulatory harmonization, an essential element for integrated network development, will be limited 'by the particularities of oil and gas producing and exporting countries and by recognition of the differences in regulatory frameworks' and will 'avoid, at all times, subordination of some [countries] to others',[33] Indeed, perhaps most important among the 'particularities' that may stand in the way of seamless harmonization are some countries' strict constitutional limits to private participation in the energy business. Such is the case, for example, of the 2009 Bolivian Constitution. Under the new Bolivian Constitution, any cross-border hydrocarbon project involving that country will have to include the national oil and gas company (YPFB) as a majority partner participant on the Bolivian side. Recourse to arbitration in disputes involving foreign investors is also banned under Bolivian constitutional law.[34] The Ecuadorian and Venezuelan foundational charters include similar constraints.

Also in the footnotes, Bolivia and Venezuela, two countries that are explicitly averse to foreign investment in energy resource development, state their aspiration that contracts under the treaty be made subject to renegotiation or cancellation whenever substantial changes in the underlying conditions occur. Venezuela goes even further to include a footnote to explicitly mention that application of the rule of law must not go against the principle of sovereignty over natural resources as recognized by the United Nations.[35]

[33] UNASUR, Anteproyecto de Estructura del Tratado Energético Suramericano, Versión Consensuada, (25 March 2010), footnote 4.

[34] D. Parravicini et al, 'International Investment and Development' (Spring 2010), 44 *Int'l Law* 283.

[35] Caveats such as these and those mentioned above, may be more or less destabilizing for integration, depending on the final shape of the Energy Treaty's and UNASUR's dispute resolution systems and whether UNASUR decision-making bodies gain supranational stature.

The differences mentioned above are within the realm of what can be expected given the existing ideological divide between the model proposed by Venezuelan leader Hugo Chávez and his allies, and the liberal-leaning countries in the region. Unfortunately, the fact that issues of ideology and political alignment continue to divert the countries' attention from making further tangible progress cannot be ignored, despite the countries' apparent support of the integration process, and the Energy Treaty in particular.[36] The reality that some countries, including Brazil,[37] have yet to ratify UNASUR's constitutive charter, further undermines the viability of its outputs. Moreover, existing asymmetries in terms of economic development and in the countries' capacity to host and manage new and old networks, need to be factored in as posing special challenges to integration.[38] The Energy Treaty's support framework, ie UNASUR's institutional and legal framework, is far from providing a robust platform for treaty implementation as its organs are weak or non-existent and lack supranational power.

To overcome the obstacles mentioned above, as a first step,[39] some South American countries projected to come together in a network of dedicated gas pipelines and interconnections that would function under a single set of rules, ie the Red de Gasoductos del Sur. The project would have the advantage of bypassing most of the barriers to full integration, while avoiding the piecemeal approach of single interconnection projects, and serving as a stepping stone towards true integration.[40] Though the proposal made good progress during its initial stages, no advance has been reported since 2007. Thus, in order to decide if SA's energy integration has any future, it may be useful to turn to the main sub-regional systems such as the South American Common Market: ie MERCOSUR and the Andean Community of Nations or CAN. Both MERCOSUR and CAN may be able to act as stepping stones towards greater regional integration.

B. Southern Common Market

The Southern Common Market (MERCOSUR) was created in 1991 with the signature of the *Treaty of Asunción* (the Treaty)[41] by Argentina, Brazil, Paraguay, and Uruguay. Venezuela recently joined as a full member,[42] while Bolivia, Chile, Colombia, Ecuador, and Peru enjoy associate status.

[36] Declaration of the Council of Heads of State and Government of the Union of South American Nations, 4th Ordinary Meeting of the Council of Heads of State and Government of the Union of South American Nations, Georgetown, Cooperative Republic of Guyana, 26 November 2010.

[37] Other hold-outs are Colombia and Paraguay.

[38] ECLAC, see n 11.

[39] Consistent with the Strategy and the Action Plan, the Structure's footnotes reaffirm parties' right to forge bilateral or multilateral energy agreements. Indeed, both the Strategy and the Action Plan encourage sub-regional agreements as potential stepping stones to full regional integration.

[40] MERCOSUR/SGT No 9, Acta No 2/2007, XLV Reunión del Subgrupo de Trabajo No 9 'Energía', Anexo VI, 30 October 2007, available at <http://www.mercosur.int> (accessed 26 September 2011).

[41] Amended by the Treaty of Ouro Preto, 1995.

[42] Signed membership agreement in 2006; received final ratification (Paraguay) in December 2010.

The objectives of the Treaty are: (1) free transit of goods, services, and factors of production through elimination of customs duties, non-tariff restrictions, and any other measures with similar effects; (2) establishment of a common external tariff; (3) adoption of common trade policies with regards to non-member states and coordination of positions in international commercial and economic fora; (4) coordination of macroeconomic and sector policies; and (5) legal harmonization to strengthen integration. MERCOSUR is based on the understanding that the Treaty's objective must be attained by means of an effective use of resources, environmental preservation, and enhancement of interconnections, as well as policy coordination and economic complementation. The parties take account of the asymmetries amongst them and vow to be guided by the principles of gradualism, flexibility, and balance in the implementation of the Treaty's objectives.[43] Though energy was initially excluded from the Treaty's application, several steps have been taken towards creating an integrated energy market within MERCOSUR.

Indeed, back in 1998, a Memorandum of Understanding on electricity integration[44] promised to deepen existing cooperative arrangements[45] by committing MERCOSUR countries to abide by 'minimum symmetry principles' directed at guaranteeing free market conditions including, for example, non-discrimination amongst producers and consumers from member countries; allowing direct purchases by large consumers as well as open access to excess transport capacity; and refraining from imposing any restrictions to the free movement of electricity. A similar understanding was arrived at in 1999 with regards to gas.[46] These commitments were reaffirmed in the Joint Declaration of Ministers and Secretaries of Energy from MERCOSUR and Chile, 29 June 2000,[47] and in subsequent declarations. The rhetoric in favour of energy integration and the need to deploy the necessary network infrastructure was particularly loud when energy blackouts throughout the region, coupled with a significant raise in the price of oil, opened the door to Venezuela's posturing as the energy provider of first resort.[48] Venezuela's plan included developing a pipeline to traverse the region from North to South (Gran Gasoducto del Sur) to provide Venezuelan energy to its MERCOSUR neighbours to the South.[49] It also included the creation of a supranational energy company led by PdVSA, Venezuela's own oil and gas company.[50] The supranational company would serve as a vehicle to sort out differences between neighbouring participating countries in cross-border undertakings, effectively bypassing the

[43] MERCOSUR, Tratado de Asunción, 26 March 1991, available at <http://www.mre.gov.py> (accessed 26 September 2011).

[44] MERCOSUR, CMC/DEC 10/98.

[45] L. Barrera-Hernandez, 'Institutions in Energy Trade: What is Going on in Latin America' (Fall 1999), *Natural Resources & Environment*, 87.

[46] MERCOSUR, CMC/DEC 10/99.

[47] MERCOSUR/GMC/RES 33/01.

[48] In 2006 Venezuela decided to resign its membership in the Andean Community and sought incorporation into MERCOSUR, first as an associate and later as full member.

[49] D. Otero Prada, see n 5.

[50] Venezuela's project was to create a supra-national government controlled oil and gas-holding company: Petroamerica, which would oversee three sub-regional government-controlled companies: Petrocaribe, Petroandina, and Petrosur. Venezuela would participate in all three sub-regional companies through its national oil and gas company, PdVSA. See D. Otero Prada, see n 5.

need to play by market rules and harmonizing regulatory requirements in building and operating the network. However, no real progress in support of integration may be reported to date either under the market model embraced by the 1998 and 1999 memoranda or under the one advanced by Venezuela. Instead, energy cooperation has veered towards efficiency and renewables, particularly biofuels,[51] highlighting Brazil's increasing prominence and Venezuela's diminishing influence in the region, as Brazil moves aggressively towards increasing energy independence and reasserts its importance in the alternative fuel field.[52]

Beyond energy-specific advancements to support integrated network development, MERCOSUR's efforts regarding rule-harmonization in other important areas for network deployment are also wanting. Such is the case of harmonization of environmental impact rules and requirements. So far, parties can show very little progress beyond a 2001 Framework Accord on Environment.[53] The Accord is limited to general statements of principle and provides minimal guidance for project implementation and development.

Notwithstanding MERCOSUR's failure to take substantial material steps towards facilitating energy integration, some authors suggest that network developers may find incentive in the block's gradual strengthening of its institutional and legal framework in support of a common free market generally.[54] As explained below, the evidence backing such a claim is weak, at best.

Those authors generally point at the fact that MERCOSUR's institutional framework includes several organs with the ability to issue decisions that are binding on all countries as a source of legal stability for prospective investors. True, according to the Ouro Preto Protocol (the Protocol)[55] all decision-making bodies—the Common Market Council, which is the top political decision-making agency of the trade block, the Common Market Group, which is MERCOSUR's executive body, and the Trade Commission, which oversees the application of common policies and trade instruments—have binding decision-making authority in matters within their jurisdictions. However, there are two major drawbacks to the way MERCOSUR's institutional framework operates. On the one hand, under article 37 of the Protocol, all decisions must be taken by consensus. As the lack of progress in relation to energy integration demonstrates, the requirement amounts to very slow and scarce progress in those areas that require enhanced collaboration and harmonization, particularly if broader issues of national security or ideology are at stake. On the other hand, in order to become effective, most decisions

[51] Directrices Generales de Eficiencia Energética 01/09; Directrices Generales sobre Energías Renovables 02/09; Decisión 049/2007, Plan de Acción del MERCOSUR para la Cooperación en Materia de Biocombustibles; available at <http://www.mercosur.org.uy> (accessed 26 September 2011).

[52] Over the past decade Brazil has doubled its petroleum reserves. Brazil's leading role in bioenergy development is well known. See M. Mendes da Fonceca and L.E. Duque Dutra, see n 10.

[53] MERCOSUR, Acuerdo Marco sobre Medio Ambiente, available at <http://ambiente.gov.ar> (accessed 5 October 2011).

[54] C.A. Rodriguez Yong, 'Providing Legal Certainty in South America: Can MERCOSUR Help?' (October 2010), 2 No. 3 *Pace Int'l L. Rev. Online Companion* 1.

[55] MERCOSUR, Protocolo de Ouro Preto, 17 December 1994, available at <http://www.mre.gov.py> (accessed 26 September 2011) .

emanated from the bodies mentioned above, though binding in principle, have to be internalized and implemented at the country level (art 38). Internalization and implementation might require passage of new norms, including ratification by the countries' legislatures, or modification of existing laws and regulations, procedures that in some instances could take years and often require internal wrangling and negotiations. Only when MERCOSUR decisions have been internalized and implemented in all member countries, as notified to the block's Secretariat, do those become simultaneously binding across the sub-region (art 40). In some cases, the process can be delayed for decades, utterly negating the so-called 'supremacy' of MERCOSUR institutional framework and its body of laws. Good examples of the system's ineffectiveness are Decisions 11/93 and 11/94 on foreign investment referred to below.

Indeed, in what looked like very promising developments, Decision 11/93 (Protocol of Colonia)[56] and Decision 11/94[57] set out to guarantee investors no less favourable treatment than national or third party investors. They also banned arbitrary or discriminatory expropriation, which was only allowed to proceed for reasons of public interest and upon payment of just compensation. In addition, no restrictions on the transfer of capital or profits abroad could be imposed. The Decisions define steps for resolving disputes between states and, most importantly, provide a more accessible system to resolve disputes brought by private parties against states. Unfortunately, these Decisions have not been ratified by all members and are not yet enforceable.[58]

Notwithstanding the above, if it is true that embracing a robust dispute resolution system, particularly one that is open to direct investor claims, 'transform[s] into "credible commitments" to treat foreign investors fairly',[59] hope for increased network development investments may still derive from the progressive enhancement of MERCOSUR's dispute resolution system (DRS) from one that relied strongly on diplomacy, to a system where rules and institutions prevail.[60] Any reassurance provided to investors interested in cross-border infrastructure development by way of the DRS, may be particularly important in light of the region's mounting antipathy to the system set up under the *Washington Convention on the Settlement of Investment Disputes between States and Nationals of Other States*.[61] However, it

[56] Available at <http://www.sice.oas.org> (accessed 26 September 2011). Decision 11/93 also includes country-specific exemption clauses, some of which relate to energy investments.

[57] Available at <http://www.sice.oas.org> (accessed 26 September 2011).

[58] T.A. O'Keefe, 'Dispute Resolution in MERCOSUR' (June 2002), *Journal of World Investment*, Volume 3, No. 3, pp 507–20, also available at <http://mercosurconsulting.net> (accessed 26 September 2011).

[59] I.A. Vicentelli, 'The Uncertain Future of ICSID in Latin America' (Summer 2010), 16 *L. & Bus. Rev. Am.*, 409, at 410. Contra: E. Gillman, 'The End of Investor-State Arbitration in Ecuador? An Analysis of Article 422 of the Constitution of 2008' (2008), 19 *Am. Rev. Int'l Arb.* 269; L. Stanley, see n 13; Stanley notes that foreign direct investment (FDI) in Bolivia, Ecuador, and Venezuela increased despite the countries' FDI-averse policies.

[60] L. Biukovic, 'Dispute Resolution Mechanisms and Regional Trade Agreements: South American and Caribbean Modalities', Spring 2008, 14 U.C. Davis *J. Int'l L. & Pol'y*, 255.

[61] Ibid; Gillman, see n 59.

is doubtful that network investors will find the kind of refuge in MERCOSUR's DRS found under the Washington Convention and ICSID.[62]

MERCOSUR's DRS is governed by the Olivos Protocol, which replaced the 1993 Brasilia Protocol. As in its predecessor, under the Olivos Protocol private parties can only access the system by submitting a claim to the National Chapter of the MERCOSUR state where they have their usual residence or place of business for a decision on admissibility. Only affirmative actions of states, and not their omissions, may give rise to a claim.[63] Claims cannot be brought against MERCOSUR bodies.[64] If the claim is admissible, the National Chapter will represent the private party in all proceedings (art 40). Thus, not only is there no direct access to an adjudicating body, but also, non-MERCOSUR investors may be at a disadvantage as their home state, ie the actual capital-exporting state as opposed to the MERCOSUR state where the claimant has legal domicile, is not allowed to take part in the proceedings, while the state of residence may not want to bring an action against one of its MERCOSUR partners.[65]

A further cause of concern is that the process takes place entirely within the diplomatic sphere.[66] Once a National Chapter accepts a claim, it must try to resolve the matter through direct negotiations with the National Chapter representatives of the infringing party. If a claim cannot be resolved directly, it goes to the Common Market Group which relies on a 'group of experts' to issue its decision. The group of experts' pronouncement must be unanimous (arts 41 and 42). If the experts decide that the claim is inadmissible or if they cannot reach unanimity, the Common Market Group must close the proceedings.[67]

[62] Under the Washington Convention the investor has standing to bring a suit directly against the host state, and does not have to persuade the capital-exporting state to represent him or her. By 2006 40 per cent of all cases against Latin American countries were energy-related. See A. Brunet and J.A. Lentini, 'Arbitration of International Oil, Gas, and Energy Disputes in Latin America', Spring 2007, 27 *Nw. J. Int'l L. & Bus.* 591.

[63] T.A. O'Keefe, see n 58.

[64] Ch. Leathley, The Royal Institute of International Affairs, Chatham House, MERCOSUR Study Group, The Mercosur Dispute Resolution System, 27 September 2002, available at <http://www.iadb.org> (accessed 26 September 2011). The author notes that claimants are free to bring action for a state's omissions under MERCOSUR to the local courts, but that the results have been inconsistent and unpredictable.

[65] T.A. O'Keefe, 'Latin American and Caribbean Trade Agreements: Key to a Prosperous Community of the Americas' (2009). As an alternative, private parties can file their claims with the Trade Commission under the Ouro Preto Protocol. Though technical in nature, as mentioned above, the Trade Commission is a diplomatic rather than an adjudicatory body. According to the Ouro Preto Protocol, if the dispute cannot be resolved at that level, it is referred to the Common Market Group, which will apply the procedure set under the Olivos Protocol, with identical results and drawbacks.

[66] Ch. Leathley, see n 64.

[67] The state backing the claim can re-instate it through the procedures set out under the Protocol of Olivos for disputes amongst states (art 44). The rules governing the state-to-state DRS have the advantage of allowing the proceedings to take place in a more apolitical sphere through the intervention of ad hoc arbitral panels. However, the likelihood of a MERCOSUR member investing its time and resources to defend a non-MERCOSUR investor against another member may not be very high and may, once again, depend on diplomacy and the amount of pressure that the capital-exporting country is willing to put on the MERCOSUR host.

Compliance and enforcement represent the Achilles heel of most international DRSs.[68] MERCOSUR is not the exception, as decisions and awards must be recognized and enforced by local courts.[69] Past experience in this area has been poor[70] and the region's increased rejection of investor-friendly fora such as ICSID[71] does not bode well for decisions rendered outside local courts.

Finally, the system's predictability and reliability is hindered by the lack of express precedent setting the value of its adjudicatory bodies' decisions.[72] At least in part, this is meant to be addressed by the creation of the Permanent Review Tribunal (PRT) under the Olivos Protocol. The PRT has authority to confirm, modify, or revoke decisions of the ad hoc arbitral panels on matters of law and legal interpretation (art 17), bringing them in line with its own interpretation, thereby introducing consistency and predictability to the DRS and MERCOSUR's management in general.[73] Upon request, the PRT can issue advisory opinions (art 3). Unfortunately, the PRT's advice is not binding, a fact that, as highlighted by the PRT itself, undermines its authority and ability to fulfil its role.[74]

In sum, energy integration seems to be a moving target within MERCOSUR, depending on which country is driving the group's agenda. Even if one looks beyond energy-specific initiatives and rules to find support for secure and sustainable cross-border network development, it is obvious that, at this juncture, the Treaty does not provide an adequate legal platform for network development. If MERCOSUR integration ever moves ahead, reinforcing the weak institutional and legal framework will be required to ensure security and stability for multinational network development and operation.

C. Andean Community of Nations

Bolivia, Colombia, Ecuador, and Peru are united under the Cartagena Agreement which creates the Andean Community of Nations (CAN).[75] The agreement's objective to 'promote the balanced and harmonious development of member Countries

[68] See eg A. Boralessa, 'Enforcement in the United States and United Kingdom of ICSID Awards Against the Republic of Argentina: Obstacles that Transnational Corporations May Face', Summer 2004, 17 *N.Y. Int'l L. Rev.* 53.

[69] The enhanced enforcement provisions of the Olivos Protocol are concerned with interstate disputes.

[70] See eg A. Brunet and J.A. Lentini, 'Arbitration of International Oil, Gas, and Energy Disputes in Latin America', Spring 2007, 27 *Nw. J. Int'l L. & Bus.* 591.

[71] I.A. Vicentelli, see n 59.

[72] Precedent is not mentioned amongst the sources of law listed under art 34 of the Olivos Protocol. Conf: Ch. Leathley, see n 64.

[73] R. Blackwood, 'Southern Common Market', August 2009, available at <http://www.asil.org> (accessed 26 September 2011).

[74] Opinion Consultiva 1/2007, re: Norte S.A. Imp. Exp. c/Laboratorios Northia Sociedad Anónima, Comercial, Industrial, Financiera, Inmobiliaria y Agropecuaria s/Indemnización de Daños y Perjuicios y Lucro Cesante (April 2007), available at <http://www.mercosur.org.uy> (accessed 26 September 2011).

[75] Acuerdo de Integracion Sub-regional Andino, 'Acuerdo de Cartagena', available at <http://www.comunidadandina.org> (accessed 26 September 2011). Venezuela renounced its membership in 2006, subsequently joining MERCOSUR; Chile, Argentina, Brazil, Paraguay, and Uruguay are associate members.

under equitable conditions' (art 1) is to be accomplished through gradual harmonization of law and policy, as well as industrial and infrastructure integration (ch IV), including energy interconnections (art 104).[76] The agreement's emphasis on integration of infrastructure and on leveling the playing field through harmonized rules has proven to be a powerful motor for overcoming existing barriers to cross-border energy network development.[77] In fact, the Andean Community has already taken numerous steps to implement energy network integration.

A first step was the signature in September 2001, of an agreement on electricity interconnection between Colombia, Ecuador, and Peru, whereby those countries committed to promote legal and regulatory harmonization to deepen electricity integration.[78] The agreement led to Decision 536, *General framework for subregional interconnection of electric power systems and intra-Community exchange of electricity*, a landmark network integration agreement which set down the general rules for electricity integration.[79] After adopting, among others, the principles of non-discrimination, free access, and economic dispatch, Decision 536 set down rules on permitting cross-border interconnection planning, pricing, short-term contracting, and legal harmonization. A regulatory body was created and charged with proposing and promoting those rules that may be required to achieve the Decision's objectives, as well as with following up on the countries' commitments regarding legal harmonization. That body is the Comité Andino de Organismos Normativos y Organismos Reguladores de Servicios de Electricidad, generally known as CANREL. Ironically, it was CANREL's active role in ensuring the scheme's sustained viability that led to the suspension of Decision 536 on electricity integration.[80]

Indeed, although some countries, namely Ecuador and Colombia, are fairly advanced in the implementation of Decision 536, CANREL proposed and obtained a two-year suspension of that Decision in November 2009. The suspension is intended to give parties time to deal with emerging issues around rent-sharing of congestion pricing in international transactions,[81] where Decision 536 provides no further guidance than saying that the owner of the cross-border transmission line cannot be assigned the rent due to congestion (art 1.10). While working on congestion pricing rules, parties may also be tackling other potentially

[76] CAN, Decisión 563, Codificación Del Acuerdo De Integración Subregional Andino (Acuerdo De Cartagena), 25 June 2003, available at <intranet.comunidadandina.org> (accessed 26 September 2011).

[77] L. Barrera-Hernández, see n 45.

[78] Comunidad Andina, Acuerdo para la Interconexión Regional de los Sistemas Eléctricos y el Intercambio Internacional de Energía Eléctrica, <http://www.comunidadandina.org> (accessed 26 September 2011).

[79] 19 December 2002, available at <http://www.comunidadandina.org> (accessed 26 September 2011). The Decision became binding on Bolivia in 2006 with Decision 639.

[80] CAN, Decisión 720, Sobre la vigencia de la Decisión 536 'Marco General para la interconexión subregional de sistemas eléctricos e intercambio intracomunitario de electricidad', 4 November 2009, available at <http://www.comunidadandina.org> (accessed 26 September 2011). A 'Transitory Agreement' was adopted to apply between Colombia and Ecuador during the suspension.

[81] A. Ruiz-Caro, 'Puntos de conflicto de la cooperación e integración energética en America Latina y el Caribe', March 2010, CEPAL, Serie Recursos Naturales e Infraestructura #148, available at <http://www.eclac.cl> (accessed 26 September 2011).

problematic issues not addressed by Decision 536, such as emergency procedures in cases of shortage of supply.

Notwithstanding the temporary suspension of Decision 536, the Andean Community is firmly on the path to energy network integration. This was made clear in 2003, with the creation of the Andean Council of Energy, Electricity, Hydrocarbons, and Mines Ministers (the Energy Council) to provide an adequate institutional vehicle to energy integration efforts within the Community.[82] The Energy Council's Action Plan reaffirms the Community's direction with regards to electricity integration and adds the intention to 'incorporate the interconnection of natural gas as a new element of Andean energy integration policy'.[83]

In addition to the above, in 2005, the bases for an Andean Energy Alliance (AEA) were defined. First on the list was '[c]onstruction of energy (electricity and gas) markets integrated through harmonized physical systems and regulatory frameworks'.[84] Though, per se, the AEA seems to have lost steam, reportedly due to the withdrawal of Venezuela from CAN,[85] the Community continues to be the most advanced example of energy integration in the region. This is particularly true if one takes into account that in addition to any specific steps taken by the Community towards effective energy integration, CAN's operation and the achievement of its goals are supported by a well-oiled and truly supranational institutional framework.

CAN's institutional framework is shaped mainly by the Cartagena Agreement and the Protocol of Trujillo, which introduced important changes to the original set up of CAN.[86] Since the adoption of the Protocol of Trujillo in 1997, the CAN is headed by the Andean Presidential Council. The Presidential Council presides over the Andean Integration System (as the Community calls its institutional framework), provides political guidance, and ensures implementation of its directives and objectives through the system's organs. The main legislative and decision-making body is the Andean Community Commission. Together with the Foreign Affairs Council, it is in charge of designing, implementing, and assessing the Community's integration policy, and of adopting all necessary measures to achieve the Community's goals. The Andean Council of Foreign Affairs has normative power over matters of foreign affairs. Commission and Council 'Decisions' are considered legally binding on all parties.[87] The Andean Secretariat is the main executive body and can issue binding 'Resolutions'. The Secretariat is also the first instance of recourse in cases of breach of community law and regulations, and can

[82] CAN, Decisión 557, Creación del Consejo de Ministros de Energía, Electricidad, Hidrocarburos y Minas de la Comunidad Andina, 25 June 2003, available at <http://www.comunidadandina.org> (accessed 26 September 2011).

[83] Action Plan of the Council of Andean Community Ministers of Energy, Electricity, Hydrocarbons and Mines, 19 June 2003, available at <http://www.comunidadandina.org> (accessed 26 September 2011).

[84] Ibid.

[85] A. Ruiz-Caro, see n 18.

[86] Available at <http://www.comunidadandina.org> (accessed 26 September 2011).

[87] The Foreign Affairs Council also issues non-binding 'Declarations'. The Andean Parliament can participate in the legislative process, but has no decision-making power and cannot issue rules of general application.

act ex officio to correct a breach when it finds that a member has failed to fulfil its obligations under Andean law.[88]

Andean law is defined in the treaty setting up the Court of Justice of the Andean Community[89] as including: (a) the Cartagena Agreement, its protocols, and ancillary instruments; (b) the treaty setting up the Andean Court and amending protocols; (c) Decisions; (d) Resolutions; and (e) industrial complementarity agreements and other agreements convened between member countries as part of the sub-regional integration process. CAN Decisions and Resolutions are binding, apply immediately and directly in the member countries, and would only require incorporation through local laws and regulations when explicitly indicated in the CAN instrument (arts 2 and 3).[90] Any local laws and regulations that contradict or are irreconcilable with Andean law will automatically cease to apply.[91] As affirmed by the Andean Court in a case that struck down Venezuelan regulations found to be in contravention of Community rules, the supremacy of Andean law is the cornerstone of the integration agreement and requires absolute obedience on the part of member countries and Community organs.[92] Member countries must take all necessary measures to ensure local application of Community law and must abstain from taking any measures or actions that may be against Community law or that may hinder its application (art 4). In applying Andean law, local courts may ask the Andean Court to interpret its provisions in relation to specific cases (arts 32–35). The procedure is devised to ensure harmonious application of Andean law across the region and the resulting interpretation is binding.[93]

Beyond the supranational character of Andean Law, another feature that sets the CAN system apart from MERCOSUR is the existence of a supranational DRS headed by the Court of Justice of the Andean Community (Andean Court).[94] According to its foundational charter, the Andean Court can hear cases relating to: 1) the annulment of a CAN Decision, Resolution, or of an industrial complementarity agreement and other agreements convened between member countries as part of the sub-regional integration process; 2) a breach of Community law; and,

[88] CAN, Decisión 472, Codificación del Tratado de Creación del Tribunal de Justicia de la Comunidad Andina, art 23, available at <http://www.comunidadandina.org> (accessed 26 September 2011).

[89] Ibid, art 1.

[90] Andean law supremacy has been confirmed by the Andean Court on several occasions. See eg Tribunal de Justicia de la Comunidad Andina, Proceso 0001-IP-1987, available at <http://www.tribunalandino.org.ec> (accessed 26 September 2011); Proceso 0002-IP-1990, available at <http://www.tribunalandino.org.ec> (accessed 26 September 2011); and, Proceso 0004-AI-1998, available at <http://www.tribunalandino.org.ec> (accessed 26 September 2011).

[91] Tribunal de Justicia de la Comunidad Andina, Proceso 0002-IP-1988, available at <http://www.tribunalandino.org.ec> (accessed 26 September 2011).

[92] Tribunal de Justicia de la Comunidad Andina, Proceso 0016-AI-1999, available at <http://www.tribunalandino.org.ec> (accessed 26 September 2011).

[93] In 1989 the Court ruled that its interpretation of Andean law may vary according to the circumstances of a case, or that the Court may change its opinion, and that existing interpretation does not exempt local courts from requiring a new interpretation. Tribunal de Justicia de la Comunidad Andina, Proceso 0007-IP-1989, available at <http://www.tribunalandino.org.ec> (accessed 26 September 2011).

[94] Id. <http://www.comunidadandina.org> (accessed 26 September 2011).

3) administrative inaction. Exceptionally, the Court will review its own decisions, but only in breach cases (art 29).

Amendments introduced to the Andean Court's statute by the Protocol of Cochabamba[95] of 1996, are significant in terms of empowering the private sector to sue at the Community level. Private parties can request annulment of Community rules whenever they are directly or indirectly affected. Also, while in the past private parties could only sue in breach before domestic courts,[96] they can now take their suit directly to the Andean Court.[97] Where the Andean Court finds that a country is in breach of its obligations under Andean law, the country has three months to fulfil its obligations (art 25).[98] Private plaintiffs can submit the ruling to local courts for immediate assessment of damages (art 30).

Private parties that are directly or indirectly affected by the inaction of CAN's decision-making organs can also bring a case to the Andean Court to force those bodies into action (art 37). Both the Andean Court and the Secretariat can act as adjudicators in arbitral proceedings involving private parties that request their assistance (arts 38 and 39).

Despite having a solid supranational system and although the Andean Community has gone a long way in terms of harmonizing laws and regulations in some areas, there are others that still pose a challenge to energy integration. Specifically, the Community has very little to show in terms of harmonized environmental and social impact rules and procedures that may guide network development. Coherent and compatible rules to guide management of environmental and social impacts may be key to the success of energy integration in the Andean Community, given the unique features of the region. There, most, if not all, network development projects will have to traverse the bountiful Amazon ecosystem, a natural focus of attention for environmentalists across the globe and home to indigenous populations that may have more in common amongst themselves than with the rest of the inhabitants of their respective countries. A robust system of harmonized laws and regulations to guide network developers and managers in relation to social and environmental impact management would go a long way in terms of securing successful integration and network development.

[95] Available at <http://www.comunidadandina.org> (accessed 26 September 2011). The Protocol entered into force in 1999.

[96] Tribunal de Justicia de la Comunidad Andina, Proceso 0001-AI-1996, available at <http://www.tribunalandino.org.ec> (accessed 26 September 2011).

[97] Though a private party's standing to bring an action in breach is less clearly defined in the Court's statute, which refers generally to 'affected' (private) parties, the rules of the court expand on the concept and include those private parties that are directly and indirectly affected by the breach. CAN, Desición 500, art. 49, available at <http://www.comunidadandina.org> (accessed 26 September 2011). The option to bring an action in breach before the local courts of the member country that has failed to take all necessary measures to ensure local application of Community law, or that has taken any measures or actions contravening Community law or that hinder its application is still available (art 31).

[98] Conf. Tribunal Andino de Justicia, Proceso 0001-AI-1996, available at <http://www.tribunalandino.org.ec> (accessed 26 September 2011).

IV. Conclusion

Notwithstanding past differences, the region could be moving towards greater convergence.[99] Sharp ideological differences between leftist (pro-Chávez) and right-leaning governments seem to be destined to fade as cash-strapped countries like Argentina and Bolivia struggle to attract investment to their energy sectors,[100] the Venezuelan leader gradually loses his grip on power, and strict adherence to neo-liberal economic recipes is abandoned for more balanced alternatives. The demise of Washington Consensus policies and the enhanced presence of the state in the energy sector through newly empowered national energy companies may bring around a uniquely South American model of energy integration and network development. It is yet to be seen whether that integration will move in the direction of a free regional energy market where national oil companies are at a par with private companies or whether it will favour a strong presence of the state(s) in developing and operating the network(s). In fact, that may be the first issue of joint energy policy that needs to be tackled for integration to become a reality.

In addition to defining the legal and regulatory basis for network integration per se, another item that looms large throughout the region's agenda to facilitate network development is achieving common ground in environmental management for cross-border projects, including regulatory harmonization. Managing the environment and any associated social impacts of infrastructure development may prove crucial to network sustainability in South America.

[99] P. Isbell, 'El Gas: Una Cuestión Conflictiva en América Latina' (April 2006), available at <http://www.realinstitutoelcano.org> (accessed 26 September 2011).

[100] Investment shortages have induced Argentine leaders to loosen their grip on energy price caps. Bolivia attempted a similar move in December of 2010 but had to backtrack in the midst of civilian upheaval.

5

The Chad–Cameroon Pipeline Project: Some Thoughts about the Legal Challenges and Lessons Learned from a World Bank-financed Large Infrastructure Project

*Mohammed A. Bekhechi**

I. Introduction

The Chad–Cameroon Project was designed to develop oil fields at Doba in southern Chad and to construct a 1070km pipeline from the Doba oil fields[1] to oil-loading facilities off Cameroon's Atlantic coast in Kribi, including all ancillary facilities, pumping stations, and other related infrastructures. It was, at the time of its preparation, the single largest oil and infrastructure project developed in that part of the world. It was sponsored by a consortium of petroleum companies[2] which committed to finance its development and operation. In addition to direct financing from the various shareholders of the consortium, the project involved financing from members of the World Bank Group, the European Investment Bank (EIB), export credit agencies,[3] and commercial banks. The estimated total project development cost was at the time of negotiation about US$3.5 billion, of which $2 billion (including financing costs, ie interest during construction and lenders' fees) corresponded to the downstream facilities (pipeline and associated facilities including the offloading facilities). The International Bank for Reconstruction and Development (IBRD, generally referred to as the World Bank, hereinafter WB) contributed (i) US$53.4m to the Government of Cameroon to finance its five per cent of Cameroon Oil Transportation Company's shares; and (ii) US$39.5m to the Government of Chad to finance its eight per cent of Chad Oil Transportation Company's shares and three per cent of Cameroon Oil Transportation Company's

* The opinions expressed in this chapter are solely those of the author, and cannot be attributed to any of the institutions for which he has worked.

[1] This paper will not discuss the oil permit granted by Chad to the Consortium for the development of the Doba oil field.

[2] See n 13 below for the membership and constitution of the Consortium.

[3] Mainly the US EXIMBANK and the French COFACE.

shares. In addition, the International Finance Corporation (the private sector arm of the WB Group—IFC) has contributed two separate loans for a total of US$200m. The International Development Agency (IDA) financed two credits to Chad and Cameroon respectively, to assist in their capacity development for management and monitoring of the Chad–Cameroon Pipeline (CCP).

II. Legal Framework for the Inception and Operation of the CCP

For the project to be feasible, Chad, as a landlocked country, was obliged to negotiate and secure access to the Atlantic shores of Cameroon to export its oil through a pipeline and an offloading facility built in the territory and territorial waters of Cameroon.[4] For that purpose, Chad and Cameroon entered into a bilateral international agreement on 8 February 1996 related to the 'Construction and Operation of a Transportation System of Hydrocarbons by Pipeline' (Chad–Cameroon Agreement).[5] Under this Agreement, it was envisioned that Chad, Cameroon, and a Consortium of petroleum companies would develop and operate under separate agreements the pipeline in both the Chadian and Cameroonian territories and the oil export terminal facilities to be built and installed off-shore and connected to the Cameroon Atlantic coastline through an 11km submarine pipeline.[6]

The Chad–Cameroon Agreement was and is still being implemented smoothly without any occurrence of dispute between the two countries. It is seen as 'an interesting example of constructive cooperation in Africa'[7] to allow Chad, a landlocked country, to access the sea off the shores of Cameroon, a transit country. The Agreement makes reference in its preamble to: (i) the General Agreement on Tariffs and Trade;[8] (ii) the provisions of the New York Convention of 8 July 1965 relating to the transit trade of landlocked countries;[9] and (iii) the United Nations Convention on the Law of the Sea of 10 December 1982, in particular Part X governing, on the one hand, the right of access to and from the sea in favour of landlocked states and free transit, and, on the other hand, the right of the transit states in their exercise of their full sovereignty, to safeguard their legitimate interests.[10]

The Chad–Cameroon Agreement provided the needed legal ground for establishing companies in Chad, the Chad Oil Transportation Company (TOTCO), and in Cameroon, the Cameroon Oil Transportation Company (COTCO), to build, own, and operate the respective portion of the Chad–Cameroon Pipeline (CCP).[11]

[4] For a full description of the CCP project see < http://www.worldbank.org> (accessed 27 September 2011).

[5] *Accord Tchad–Cameroun du 8 Février 1996 relatif à la construction et l'exploitation d'un système de transport des hydrocarbures par pipeline*. In furtherance of this agreement, Cameroon adopted Law 96–14 of 5 August 1996 on the legal regime of transportation by pipeline of hydrocarbons from other countries.

[6] Kishor Uprety, 'The Transit Regime for Land-Locked States' in *International Law and Development Perspectives* (Washington DC: World Bank Publications, 2006).

[7] Ibid.

[8] Preamble, Section 3.

[9] Preamble, Section 4.

[10] Preamble, Section 5.

[11] In French: 'Convention d'Établissement'. Initially, this Convention was entered into by, on one side, ESSO Exploration and Production Chad, Société SHELL Tchadienne de Recherches

These companies' shareholders include, respectively, the Governments of Chad and Cameroon, and the members of the Consortium of Oil Companies comprising Esso Exploration and Production Chad Inc (ESSO),[12] Chevron Petroleum Chad Company Ltd, and Petronas Carigali Chad EP (the Consortium).[13] The two governments have minority shareholdings in each company. Once formed, TOTCO and COTCO entered into Concession Agreements[14] with Chad and Cameroon[15] respectively, in order to define the legal regime applicable to their activities and protection of their investments.

III. Environmental and Social Aspects of the CCP

The CCP was designed to provide transportation for the oil developed in the Doba oil fields exclusively. The pipeline was designed for a maximum throughput of 250,000 barrels, which was the maximum production identified from the Doba fields but which was unlikely to be reached, creating opportunity for adding oil produced in fields other than the Doba fields. Therefore, the issue of whether oil developed in fields other than Doba could be transported through the CCP was one of the most sensitive issues raised by stakeholders during the preparation and negotiations of the financing of the project. It was clearly stated by the WB and other financiers that any oil to be transported by the CCP must be developed in compliance with the same principles, processes, and standards applied to the development of the CCP and the Doba oil fields. In other words, any oil proposed to be transported by the CCP must be developed in compliance with stringent environmental and social standards.[16]

Supporting and financing a project which could be very disruptive of the biophysical and social environment of Chad and Cameroon was not an easy decision to make for the WB Group. In making that decision, the WB Group had

et d'Exploitation et ELF Hydrocarbures Tchad and the Government of Cameroon on the other side; see Official Gazette of the Government of Cameroon dated 1 October 1997, pages 1214–83. SHELL and ELF were later, in 1998, replaced as shareholders by PETRONAS and CHEVRON. The Convention d'Établissement with Chad and Cameroon respectively, were amended accordingly in 1998. Upstream facilities are wholly owned by affiliates of the sponsors.

12 ESSO is a fully owned subsidiary of EXXON Inc USA.

13 The two governments do not own any shares in the upstream facilities (built for the purpose of developing and operating the Doba oil field), which are wholly owned by the Consortium members, through their respective affiliates.

14 These Concession Agreements are referred to as 'Convention d'Etablissmenent' in the original French version. The author will refer to them in this paper as 'Concession Agreements' or COTCO or TOTCO Convention, as the case may be.

15 See n 16.

16 Section 4.10 of the Loan Agreement between Chad and the WB reads as follows: 'The Borrower shall ensure that any oil developed outside the Doba Basin Oil Fields which is proposed to be transported through any part of the Transportation System in Chad is developed in accordance with the principles set forth in the EMP with respect to environmental analysis and protection, consultation, information disclosure, resettlement and compensation and with the equivalent legal and administrative approval processes and information disclosure as applied with respect to the oil developed in the Doba Basin Oil Fields.' The same language was included in Section 4.05 of the Loan Agreement between Cameroon and the WB.

to consider the opportunity to help further the fight against poverty and lift the two countries to a higher level of development without underestimating the weak governance structures and quasi absence of rule of law in both countries, and high level of corruption and political instability in Chad during the 1990s. The critical path to the final decision was characterized by the imposition of numerous conditions on the Consortium, mainly in terms of compliance with WB environmental, social, and safety standards, and on the respective governments on financial, governance, environmental, and social standards.

These aspects were subject to deep, wide, and sometimes heated discussion. While Civil Society's Organizations (CSOs)[17] argued that use of donors' resources was not appropriate for such a project,[18] the WB and other stakeholders suggested that their involvement would offer opportunities to help the two countries generate additional revenues necessary to fight poverty and promote reform and development, including sound environmental and social management of the project.

This paper focuses on some of the most important legal issues raised by the construction and operation of the Chad–Cameroon Pipeline, including those related to the conclusion and implementation of concession agreements under terms acceptable to the respective governments and the companies established for the purpose of building and operating the CCP and the financing of the CCP project. These substantive legal issues include: (i) lack of predictability and liability as a result of non-applicability of national laws; (ii) non-applicability of international treaties, conventions and agreements, especially those related to environmental protection and human rights; (iii) wide powers granted to the Consortium to act as an 'official' state organ without due regard for applicable laws and regulations; and (iv) lack of predictability and liability as a result of ambiguous provisions.

A. Reference to applicable law in the Concession Agreement

As mentioned above, the construction and operation of the CCP was subject to two concession agreements entered into by the Government of Chad and TOTCO and the Government of Cameroon and COTCO, respectively. Cameroon entered into a concession agreement with COTCO dated 23 March 1998[19] and Chad and TOTCO entered into a similar agreement dated 10 July 1998.[20] Both agreements define the rights and obligations of the parties with respect to the status of the

[17] In addition to local CSOs and religious groups in Chad and Cameroon, the project was subject to high levels of scrutiny from international NGOs, including IUCN, CIEL (Center for International Environmental Law), EDF (Environmental Defense Fund), to name a few. The Harvard Human Rights Program prepared a comprehensive legal review of the project and shared it with the WB and other stakeholders.

[18] See, for example, the analysis of the project by Amnesty International, 'Contracting Out Human Rights. The Chad–Cameroon Pipeline Project' 2005, available at <http://www.amnesty.org/en/library/info/POL34/012/2005> (accessed 27 September 2011).

[19] Amending a 1996 Concession Agreement approved by Law 097/016 of 7 August 1997, confirmed by Presidential Decree of 26 August 1999.

[20] Amending a 1996 Concession Agreement approved by Law 015//PR/98 of 17 August 1998, approved by Presidential Decree of 16 October 1999.

pipeline in Cameroon and Chad respectively, including those rights and obligations concerning the construction, operation, and maintenance of the CCP.

The concession agreements were negotiated by the parties in a manner consistent with their respective interests. As financier and stakeholder, the WB conducted a comprehensive due diligence on the two concession agreements in light of its proposed loans to the countries to finance their respective contribution to the equity of COTCO and TOTCO. The WB found the concession agreements to be satisfactory on the whole, since they had all the features commonly used in concession agreements for transboundary pipelines and similar projects in the petroleum and natural gas industry.[21] In general, the respective COTCO and TOTCO concession agreements contain provisions specifying the rights and obligations of COTCO and TOTCO, including but not limited to those related to the tax regime, use of foreign and local labour forces, customs, applicable law and stability of conditions, force majeure, and dispute resolution mechanisms. The concession agreements also indicate the commitments of the respective countries on several issues including land use rights, one of the most sensitive and complex issues connected to the project.

Regarding land use rights, the respective legislation of both countries provides that building a system of transportation of hydrocarbons by pipeline will be deemed a public interest project and therefore any land can be expropriated under their laws on expropriation of land in the public's interest (ie using eminent domain doctrine). Accordingly, Cameroon's Law 96/14 of 5 August1996, on the Transportation by Pipeline of Hydrocarbons from Other Countries, provides in its article 28 that 'the works related to the construction, operation and maintenance of a System of Transportation by Pipeline shall be declared to be for public purpose by the state under conditions and implementation terms of the expropriation procedures provided for in the land and real property legislation in effect'. The same is reflected in Chad's legislation governing pipeline construction and operation including: (i) Ordinance 07/PC/TP/MH of 3 February 1962, and its implementing Decree of 10 May 1967; and (ii) Decree 416/PR/99 of 5 October 1999 declaring the construction, operation, and maintenance of the TOTCO Transportation System for a public purpose. The issue with these laws was that they do not fulfill the standards of compensation set by the WB's Policy on Involuntary Resettlement.[22] This gap was filled through conditions and covenants included in the respective loan agreements between Cameroon and Chad on one side and the WB on the other side, and the project agreement between respectively TOTCO and COTCO on one side and the WB on the other side, mainly by measures required in the Environmental Management Plan (EMP). Section 4.01(iv) of the loan agreement between Cameroon and the WB states that Cameroon shall 'carry out the Land Acquisition Plan [. . .] and implement the Compensation Plan, set forth in Volume

[21] The same model of concession agreement was used for the purpose of the agreements related to the Eko Fisk Pipeline, the Frigg Gas Pipeline, the Siberian Pipeline, and the Trans-Mediterranean Pipeline (Algeria–Tunisia–Italy).

[22] Operational Directive 4.30 on Involuntary Resettlement, available at <http//:www.worldbank.org> (accessed 27 September 2011).

3 of the Cameroon Portion of the EMP, in coordination with COTCO for each segment of the Transportation System [ie the pipeline]...requiring compensation payment so that the compensation payments to affected individuals are made simultaneously by it and COTCO and acknowledged simultaneously by the recipients as described in such Volume 3 of the Cameroon Portion of the EMP'. Also, the Loan Agreement between Chad and the WB provided a similar provision but took into account the specifics of the resettlement action plan in the Chad portion of the pipeline. Section 4.01(c) states that the borrower (Chad) shall 'use all means available to it by law or contract not to permit construction to start on any segment of Part A of the Project (represented by an Alignment Sheet[23] as defined in Volume 6 of the Chad Portion of the EMP) until such time as all compensation payments specified for individuals in the Compensation & Resettlement Plan set forth in Volume 3 of the Chad Portion of the EMP shall have been paid as regards to that segment and all claims [...] in relation to such segment have been settled in accordance with the provisions of the Compensation & Resettlement Plan'. These provisions close the gaps in the respective legislation of the two countries which do not provide an adequate level of compensation (Cameroon) and payment of the compensation before the start of the works (Cameroon and Chad).

B. The Environmental Management Plan

Generally, the EMP, dated 1999, is part of the Environmental Assessment (EA) consisting of: (i) the Environmental Assessment Executive Summary and Update; (ii) the Environmental Management Plan—Chad Portion; (iii) the Environmental Management Plan—Cameroon Portion; (iv) the Project Description and Decommissioning Portions of the Supporting Documents (Volume 1); Supporting Documents (Volumes 2, 3, 4, 5, and 6);[24] and (v) the General Oil Spill Response Plan, which together describe the measures and actions that will be implemented by Chad and the operators of the oil field and pipeline during the design, construction, and operation of the project to eliminate, mitigate, reduce, or address identified biophysical, socio-economic, socio-cultural, health, and other similar issues and their impact, to acceptable levels defined in the EA and to monitor the same, as may be amended or supplemented from time to time. The EMP included a comprehensive description of the compensation process for land acquisition and involuntary resettlement.[25]

[23] An Alignment Sheet represents a section of the CCP as described in the EMP (Volume 6). Each section of the CCP is represented in a specific alignment sheet and the whole set of the alignment sheets representing the CCP from the Doba oil fields (Chad) to the offloading facility in Kribi (Cameroon) is part of the design and EMP documentation of the project.

[24] The Supporting Documents are the background studies and baseline that served for the purpose of designing the EMP for the project. These documents relate to: Project description, decommissioning (Volume 1), alternative analysis (Volume 2), Consultation and public review (Volume 3), Oil spill response plan (Volume 4), Chad biological diversity studies, Cameroon biological diversity studies (Volume 5), and Chad public health study and Cameroon public health study (Volume 6).

[25] See EMP, Volume 3 'Compensation and Resettlement Plan'. The loan agreements included a clear definition of the land (expropriation) decrees, and the level of protection which was adopted

However, the language used in the EA/EMP documents may not be entirely clear. For example, it is not the land itself which will be 'bought', but rather land use rights in accordance with common practice in many African countries. Thus, contrary to an opinion largely held by the public at large at the time of preparation of the CCP, the Consortium was granted land use rights in the right of way (ROW) of the pipeline but did not acquire ownership rights to the land. Payments for the use of land were based on the length of time used, similar to a lease whose present value is paid up front. There are basically two categories of land takings in Chad and Cameroon: one is temporary, for only a few months during construction activities, for which the compensation corresponds to the value of a year's production. The other is 'permanent', ie for the life of the project, and the compensation is calculated as the present value of future production over the life of the project. The modalities for returning the land after the project has been completed were also not fully worked out, but it was clear and well understood by affected persons that land taken for temporary construction activities would be rehabilitated and returned to the original user, with records to establish that the Consortium's obligations had been fulfilled. The Consortium's use rights on the land that is taken 'permanently' for the project expire after decommissioning, and that land will revert to the government and most likely to its original users.

C. Oil revenue management in Chad: legal aspects and challenges

Beyond the issue of compensation for land (use) acquisition and involuntary resettlement, all WB safeguard policies were made applicable to the project and duly identified in the respective loan agreements for the Chad and Cameroon portions of the CCP.[26] The loan agreements are similar in substance to those customarily utilized by the WB in similar natural resources and infrastructure operations. As mentioned, they include conditions and covenants on environmental and social protection with reference to project-specific mitigation measures. In addition, they include some specific conditions and covenants the impacts of which reach beyond the limited scope of the project, such as the obligation to adopt and implement a Petroleum Revenue Management Program (PRMP) for oil-generated revenues in Chad,[27] the establishment of a foundation to finance a biological diversity offset programme and an Indigenous Peoples Plan (IPP) in Cameroon, and the implementation of an independent environmental compliance-monitoring mechanism. This innovative approach followed by the WB in financing the CCP Project was

through the EMP and found to meet Bank standards on socio-economic impacts which include involuntary resettlement.

[26] See definition of safeguard defined as 'Relevant Guidelines and Policies' in Section 1.01 para (ww) of the Loan Agreement for Chad; and, Section 1.01 para (vv) of the Loan Agreement for Cameroon, available at <http://www-wds.worldbank.org/external/default/main?pagePK=64187835&piPK=64187936&theSitePK=523679&siteName=WDS&menuPK=64187283&callBack=&projectId=P051059,P044305&siteName=WDS&menuPK=64187283&callBack=> (accessed 13 October 2011).

[27] Chad Petroleum Revenue Management Law No. 0001/PR/99 dated 11 January 1999 governing the management of petroleum revenues.

influenced by the nature of the state of Chad in the early 1990s as a fragile state emerging from a long and deadly civil war paying little or no attention to the rule of law. These issues were at the heart of the CSOs' mobilization during the preparation and first phase operation of the CCP[28] and were the subject of numerous publications.[29]

The main objectives of the PRMP as described in the Petroleum Revenue Management Law (PRML) were to: (i) ensure that petroleum revenues generated by the production and export of the oil produced in Doba would be utilized solely for the purpose of alleviating poverty in Chad by focusing on investments in priority areas including education, health, and rural development; (ii) establish and finance a Fund for Future Generations (FGF); and (iii) manage petroleum revenues in a sound and transparent manner through a governance mechanism including representatives from both government and a civil society organization, CCSRP (Collège de Contrôle et de Surveillance des Revenus Pétroliers , also known as the Petroleum Revenue Oversight and Control Committee).[30] As per the agreement with the World Bank, the PRML could not be amended, abrogated, modified, or waived.

No other financing agreement for an infrastructure project had ever stipulated that revenue generated by use of natural resources connected to such infrastructure would be managed through specific rules or used for determined purposes. However, at the time, the risk of non-compliance was foreseen and did subsequently materialize. As petroleum revenues increased and the government faced political challenges, it decided to revise substantially the PRML and modify its main features which were among the conditions the WB requested the government to adopt and maintain in order for the WB Group to finance part of the cost of the CCP. In 2005, the government of Chad decided to amend the PRML to:

(i) extend the scope of the PRML to all oil fields, including those to be put into exploitation outside Doba;

(ii) abolish the FGF;

(iii) broaden the priority areas to which oil revenues are principally allocated, such as the energy and petroleum sector, justice, territorial administration and security, with the option to further modify such list by government regulation;

[28] This Project was subject to two requests for investigation by the Inspection Panel of the WB. The two cases are publicly disclosed at <http://web.worldbank.org/WBSITE/EXTERNAL/EXTINSPECTIONPANEL/0,,contentMDK:21680496~menuPK:64129250~pagePK:64129751~piPK:64128378~theSitePK:380794,00.html> (accessed 13 October 2011).

[29] African forum and network on Debt and Development, 'The Contribution of Oil to Debt and Under-Development in Africa: The Case of the Chad–Cameroon Oil Pipeline Project' (2007) available at <http://afrodad.org> (accessed 27 September 2011); Korinna Horta, Samuel Nguiffo and Delphine Djiraibe, 'The Chad–Cameroon Oil and Pipeline Project. A Project Non-Completion Report', April 2007, available at <http://apps.edf.org/documents/6282_ChadCameroon-Non-Completion.pdf> (accessed 13 October 2011) and on the same subject see <http://apps.edf.org/documents/728_ChadCameroon_pipeline.pdf> (accessed 13 October 2011).

[30] An English translation would be 'Group for the Monitoring and Surveillance of the Petroleum Revenues'.

(iv) reallocate direct revenues from dividends and royalties in the following manner:

(a) reduction of royalties and dividends allocated to priority sectors (royalties from 80 to 65 per cent and dividends from 85 to 70 per cent);

(b) increase of royalties and dividends going to the general budget from 15 to 30 per cent, with the time limit for such general budget allocation to be extended from 2007 to the date at which revenues from corporate taxes for the corresponding exploitation would equal the modified percentage; and

(v) modify the membership of the CCSRP and extend the appointment of members of the CCSRP from three to nine years renewable by one-third.

As a result of the amendments the WB threatened suspension of its participation in the project and with termination when the government decided not to maintain the PRML as agreed upon during project preparation. When the government of Chad decided to abrogate the 1999 Law on Revenue Management, the WB raised strong opposition but was not able to enforce the relevant loan condition. In the end, the government of Chad decided to terminate the loan agreement and pay back the full loan amount.[31]

IV. Legal Issues Raised and Faced by the CCP

Notwithstanding its challenges, the implementation of the CCP project was an opportunity to learn lessons on project finance and ways to improve WB involvement and financing in projects involving high risks but also presenting opportunities for high rewards in terms of fighting poverty and promoting sustainable development strategies and programmes.[32]

A. To what extent does domestic law and regulation apply to the CCP?

It is undisputed in international business that the parties to an international agreement enjoy the freedom to choose the applicable law governing their business-related relationships. Because the CCP project was a unique and unlikely replicable activity in the two countries, it is normal that the legal framework be provided within one unique instrument which provided for the applicable rules and conflict resolution mechanisms. It is worth mentioning that the concession agreements in Chad and Cameroon were approved by the respective parliaments which found

[31] For additional information see: <http://www.bicusa.org/en/Article.2490.aspx> (accessed 27 September 2011) and <http://www.worldbank.org/WBSITE/EXTERNAL/NEWS/0,,contentMDK:20696758~menuPK:34464~pagePK:34370~piPK:34424~theSitePK:4607,00.html> (accessed 27 September 2011).

[32] Another large infrastructure project was financed by the WB, which benefited from lessons learned from the CCP Project.

them to be in compliance with their respective constitutions.[33] As such they were enacted and became law in each country, and superseded prior existing national laws with respect to matters specifically addressed by them.

The International Law Institute mentioned expressly, in its 1979 Resolution, that 'contracts between a state and a foreign private person shall be subjected to the rules of law chosen by the parties', and that 'the rules of law chosen in accordance with the preceding provisions shall govern the incidence of contractual liability between the parties, in particular those raised by the state's exercise of its sovereign powers in violation of any of its commitments towards the contracting partner'.[34] This is the case of the two concessions agreements related to TOTCO and COTCO. Article 30 of the COTCO concession agreement expresses this principle of freedom of choice of the governing law and defines the applicable law. The instrument provides also for the so-called 'stabilization clause' or 'intangibility clause', the purpose of which is 'to reach a compromise between the prerogatives of the state involved and the legitimate quest of the private party for stability of status consistent with sound business judgment'.[35] Article 30 states that: '(1) . . . the Parties acknowledge that the provisions of this Convention[36] which are contrary to or inconsistent with Law 96–14 are provisions which reiterate the commitments made by the Republic of Cameroon before the enactment of Law 96–14 and that therefore, such provisions shall prevail over the provisions of the said law and its implementing instruments which are contrary thereto or inconsistent therewith, . . . and (2) . . . all ordinary law provisions of the Republic of Cameroon which are not contrary to nor inconsistent with the provisions of this Convention apply to activities undertaken under this Convention'. Although the wording may differ, similar examples may be found in other conventions related to oil and mining concessions and are common in international industry practice.[37] This provision was agreed upon by Cameroon and COTCO to ensure that any future implementing regulation of Law 96–14 or interpretation of such law is not opposed to a provision of the Concession Agreement.

The COTCO–Cameroon agreement imposes on the Government the obligation not to modify any law or tax after the date of the agreement without COTCO's prior consent, if such modifications might adversely affect the rights and obligations of COTCO, the Consortium Members, lenders, and others. The COTCO–Cameroon Agreement in its Article 24.2 states:

> With regard to the activities undertaken under this Convention, the Republic of Cameroon shall not modify such legal, tax, customs, and exchange control regime in such a way as to

[33] See n 13.

[34] See International Law Institute, Resolution on the Proper Law of States Contract, Athens' Session, 1979. Yearbook, Vol. 58, T.2, at 193.

[35] Abdullah Faruque, 'Validity and Efficacy of Stabilisation Clauses. Legal Protection vs Functional Value' (2006), *Journal of International Arbitration* 23(4): 317–36.

[36] Convention here refers to the 'Convention d'Etablissment' or Concession Agreement. See n 14.

[37] George Delaume, 'Transnational Contracts: Applicable Law and Settlement of Disputes. Law and Practice' (July 1983) Oceana Publication, Booklet 8, at 39.

adversely affect the rights and obligations of COTCO, shareholders, Affiliates, Contractors, Sub-Contractors, Shippers or Lenders arising from this Agreement and no legislative, regulatory or administrative measure contrary to the provisions of this Convention shall apply to the persons mentioned above without COTCO's prior written consent.

As for the case of Chad, Article 34.3 of the TOTCO concession agreement provides:

During the period of validity of this Agreement, the Government of Chad shall ensure that it shall not apply to the Consortium, without prior agreement of the Parties, any future governmental acts with the duly established effect of aggravating, directly, as a consequence, or due to their application to the Consortium Members, the obligations and charges imposed by the provisions of this Convention, or with the effect of undermining the rights and economic advantages of the Consortium or its shareholders, specified by this Convention, including the effect duly established and reflected over the Consortium of the aggravation of the charges of the Affiliated Companies or of the Contractors due to these acts. Only the Consortium shall be able to cite this stability clause, which is offered to it to the exclusion of any third party to this Agreement. This shall notably apply to the following elements: (a) exemption from rights, taxes and levies; (b) obligations in the matter of royalties and taxes on profits; (c) the right to keep and repatriate abroad funds and foreign currencies; (d) non-discrimination for charges by way of services rendered by the Government in connection to those applied by the Government for similar services provided in the public domain.

In addition, the Convention of 1988 on Exploration, Operation, and Transportation of Hydrocarbons between the Republic of Chad and the Consortium provides in Article 34.3 similar provisions.[38]

These stabilization clauses were the focus of criticism from CSOs because they were seen as preventing governments in Chad and Cameroon from enforcing environmental and social laws against the consortium members to address any environmental and social damage or harm during the CCP operation phase.[39] However, the WB's due diligence review on the subject found that the stabilization clauses included in the two Agreements were consistent with widely accepted international practice.[40] Moreover, as mentioned above, the respective TOTCO and COTCO Concession Agreements define the rights and obligations of COTCO

[38] Approved by Ordinance 041/PR/88 of 19 December 1988 as amended in 1993 and 1997.

[39] See Amnesty Report, above, note 18.

[40] Examples include: (i) article 22 of the Agreement between The Sultan of Oman and the Sun Group, which states: 'The Sultan shall not annul this Agreement by general or special legislation or by administrative measures or by any other act...'; 2 (ii) article 21 of the Concession Agreement between the Government of Liberia and Liberia Iron and Steel Corporation which stipulates: 'This Concession Agreement shall be governed, construed and interpreted in accordance with the laws of the Republic of Liberia excluding however, any enactment passed or brought into force in the Republic of Liberia before or after the date of this Concession Agreement which is inconsistent with or contrary to the terms thereof' 2; and (iii) article 4 of the Convention d'Établissement entre la République Islamique de Mauritanie et AGIP Spa, Getty International, Inc, Hispanica de Petroleos and Phillips Petroleum International Corp. which states 'Le Gouvernement garantit... pour la durée de la Convention, la stabilité des conditions générales, juridiques, économiques, financières et fiscales dans lesquelles la deuxieme partie exercera son activité, telles que ces conditions résultent de la législation et de la règlementation en vigueur à la date de signature de la Convention ainsi que les dispositions de la dite Convention.'

and TOTCO as approved by the respective Parliament of Cameroon and Chad and promulgated by their respective heads of state. As far as environmental, social, and disclosure issues are concerned, the Concession Agreements clearly specify the obligations of TOTCO and COTCO via an express link to the EMP required by the WB. Thus Article 13 of the COTCO instrument and 14 of TOTCO's specifically include provisions which require COTCO and TOTCO to implement the EMP as approved by the respective country and submitted to and endorsed by the WB Group. In this respect, it should be noted that, in addition to being in compliance with the relevant provisions of each of the two countries involved, the EMP was developed, approved, and designed to be implemented in compliance with the WB policies and procedures related to environmental and social safeguards.[41] These include the applicable WB and IFC social and environmental safeguard policies which, at the time of the CCP project's processing, were used as benchmarks for other international financial institutions. In fact, IFC standards were adopted by numerous commercial banks under the so-called 'Equator Principles', which are voluntary standards to ensure the banks finance only those projects that are socially responsible and reflect sound environmental management practices. However, those standards were also criticized as not ensuring that the legal framework governing large infrastructure projects such as the CCP project respect human rights.[42] The WB did not dispute that statement because it is not equipped with a formal human rights policy that would have been triggered for the CCP project. Instead, it took all the steps necessary to ensure that the Consortium agreed to establish in both Chad and Cameroon grievance procedures mechanisms[43] to allow affected persons to file claims and get redress. In addition, communication and disclosure measures were implemented to afford opportunities to affected persons and the public at large to raise issues of concern with the Consortium members. Consortium members in both Chad and Cameroon were required to log and report all grievances to the government and the WB. It should be noted, however, that grievance procedures and mechanisms are focused on the project's impacts and do not address the wider issue of governance and human rights protection.[44]

Another layer of protection provided to affected persons under the WB's financing framework is the recourse to the Bank's Inspection Panel whose main mission is to monitor whether WB-financed projects achieve compliance with applicable WB environmental and social safeguards and other policies.[45] Affected groups requested an Inspection Panel investigation. As a result, the Investment Panel found some situations of non-compliance, prompting Bank management to

[41] Above n 24 and <http//:www.worldbank.org/safeguards> (accessed 27 September 2011).

[42] See Amnesty International Report, above n 18, at 27.

[43] EMP, Volume 3 Chapters 8–9 through 8–11.

[44] See below re. assistance to strengthen the legal and regulatory framework for environmental management and social development.

[45] The mandate of the Inspection Panel does not cover monitoring compliance with WB's procurement policies and guidelines. Cf <http://www.worldbank.org/inspectionpanel> (accessed 27 September 2011).

recommend an action plan with additional mitigation and/or corrective measures which were subsequently implemented.[46]

B. Are international treaties, conventions, and agreements applicable to the CCP project?

Many critics of the CCP have mentioned that article 41 of the COTCO concession agreement leaves out international treaties, conventions, and agreements from the enumeration of the sources that must be taken into account in the interpretation of the concession agreement and that international legal instruments cannot be subsumed under the category of 'general principles of international law', referred to in article 41 of the COTCO Convention. To respond to such criticism, lawyers who conducted the due diligence on the legal framework for implementing the CCP project argued that under their respective constitutions, Cameroon and Chad[47] recognize that international treaties, conventions, or agreements ratified by or adhered to by them become part of the law of the land by virtue of the acts of ratification or accession, and are therefore to be implemented and enforced in their respective territory and cannot be opposed by domestic legislation.[48] As a result, any reference to domestic legislation includes those international treaties, conventions or agreements that have been ratified, or adhered to, by the two countries. The reference to the general principles of international law should be understood as in addition[49] to the applicable domestic law.

In addition, neither concession agreement prevents Chad or Cameroon from taking action under relevant applicable international law principles and norms consistent with their obligations under the respective concession agreement. For example, they do not prevent each of the two countries from exercising any rights they have in respect of natural and other resources in the interest of their respective people. They also provide for the use of international arbitration procedures in case of disputes between the parties about the implementation and interpretation of the concession

[46] Inspection Panel, Investigation Report following the Request related to Cameroon Portion of the CCP #25734, 2 May 2003; and Management Response, Report and Recommendations #26446, 28 May 2003.

[47] Constitution of Cameroon, 1972, as amended 1996, available at: <http://www.prc.cm/instit/consti.htm> (accessed 27 September 2011); Constitution of Chad, 1995, available at <http://www.presidencetchad.org/Constitution_Tchad.pdf> (accessed 13 October 2011).

[48] Article 222 of the Constitution of Chad (31 March 1996) provides: 'Les traités ou accords régulièrement ratifiés ont, dès leur publication, une autorité supérieure à celle des lois, sous réserve pour chaque accord ou traité de son application par l'autre partie.' Article 45 of the Constitution of Cameroon of 2 June 1972 as amended on 18 January 1996, states: 'Duly approved or ratified treaties and international agreements shall, following their publication, override national laws, provided the other party implements the said treaty or agreement.' These provisions confirm the principle of international law that once a duly ratified treaty is in force in the territory of a state, that state may not, justify non-observance by reference to any impediment of its domestic law.

[49] This reference to 'general principles of international law' in international agreements is intended to fill gaps in treaty law and international customary law. In case of a claim, and if the agreement does not provide explicit rules to be applied, the judge may use international legal principles and principles that are common to most of the existing legal systems and are widely accepted by the jurisprudence and business practice.

agreements. Finally, the contractual relationships between the Consortium members and the countries on the one hand, and the WB on the other, is governed by the CCP project and loan-related legal agreements which, by nature, are international agreements governed by international law.

Moreover, the WB Group ensured that no international obligations under any conventions, treaties, and agreements would be violated because of the project's approval and implementation.[50] This is the result of routine due diligence by the WB Group's staff who, since the adoption of the Operational Manual Statement (OMS) 2.36 in 1984 and Operational Policy (OP) 4.01 in 1989, have been taking steps to help countries implement their obligations under environmental treaties and agreements. Operational Manual Statement 2.36 (OMS 2.36 of May 1984) states that 'the Bank... will not finance projects (e) that contravene any international environmental agreement to which the member country concerned is a party... [and] (g) which would significantly modify natural areas designated by international conventions as World Heritage sites or Biosphere Reserves, by national legislation as national parks, wildlife refuges or other protected area'.[51] The provision in OMS 2.36 is now embodied in Operational Policy 4.01 (OP 4.01 of January 1999) on Environmental Assessment, which expressly states that the environmental assessment of any Bank-funded and/or assisted project should take 'into account... the country's overall policy framework, national legislation, and institutional capabilities related to the environment... and obligations of the country, pertaining to project activities, under relevant international environmental treaties and agreements. The Bank does not finance project activities that would contravene such country obligations, as identified during the EA.'[52]

More specifically, it should be noted that all the treaties, conventions, and agreements related to biological diversity, including (i) the Convention on the Conservation of Migratory Species of Wild Animals (1979); (ii) the Convention on the Protection of the World Cultural and Natural Heritage (1972);

[50] The WB's Operational Policy OP 4.01 mandates Bank staff supervising a project to ensure that the Environmental Assessment includes a review of the legal framework of the country in which the project is to be implemented. Obligations of the country pertaining to project activities under relevant international environmental treaties and agreements must be assessed. OP 4.01 clearly states that 'The Bank does not finance project activities that would contravene such country obligations, as identified during the EA.'

[51] No natural habitats protected under the RAMSAR or the World Heritage Sites Conventions were identified and/or impacted. One of the potential major impacts identified was the risk of pollution of the Mbere River which flows into other rivers before reaching Lake Chad, which is a habitat of regional if not global importance. The project would cross the Mbere River and several of its tributaries. This river flows along the border between Cameroon and the Central African Republic (CAR). Lake Chad is bordered by Niger and Nigeria, as well as Cameroon and Chad. Similarly, the project facilities and pipeline in the oilfield area in southern Chad cross tributaries of the Logone River, which again eventually empties into Lake Chad. In 1998, CAR, Niger, and Nigeria were notified of the project in accordance with the WB's OP 7.50. CAR and Niger endorsed the project; the third riparian did not oppose the project. All Lake Chad riparians are members of the Commission for the Lake Chad Basin, headquartered in N'Djamena, which provides a vehicle to address potential impacts on the Lake Chad. Chad also informed the Commission of the proposed project. Moreover, the project is designed (including physical design of the pipeline, strategic placement of block valves, monitoring and response systems) so as not to cause appreciable harm to Lake Chad riparians.

[52] Available at <http://www.worldbank.org> (accessed 27 September 2011).

(iii) the Convention on Biological Diversity (1992); (iv) the Convention on the Conservation of Nature and Natural Resources (1968); and (v) the Convention on Wetlands of International Importance Especially as Waterfowl Habitat (1971), were fully reviewed through a thorough internal due diligence review which included field visits and consultation with all stakeholders, including potentially affected persons and CSOs[53] and taken into account in the development of the above-mentioned WB Group's OP 4.01 and 4.36. As a result, one of the positive aspects of the World Bank involvement in the project has been to ensure the effective protection of all legally protected areas, as well as of ecologically sensitive sites not under any legal protection (eg Mbere Rift Valley). The pipeline route through the coastal forest was optimized on the ground to pass through areas already degraded by agricultural and other human activities. The few remaining pristine and untouched areas of forest were avoided and a major biodiversity offset was put into place in Cameroon to more than compensate for any damage that could be done by pipeline construction.[54] This involved the effective protection of a much larger and more pristine area of Atlantic coastal forest. With Bank assistance, the Government of Cameroon has already enacted two decrees on 12 January 2000, to establish two new national parks in M'Bam and Djerem and Campo Ma'an for which management plans were prepared respectively by the World Conservation Society (WCS) and the World Wild Fund for Nature (WWF), with financing provided by the Foundation for Environment and Development (FEDEC).[55]

As for the conventions dealing with the marine environment including (i) the United Nations Convention on the Law of the Sea (1982); (ii) the Convention for Co-operation in the Protection and Development of the Marine and Coastal Environment of the West and Central African Region (1981); (iii) the Protocol concerning Cooperation in Combating Pollution in Case of Emergency (1981); and (iv) the International Convention on Civil Liability for Oil Pollution Damage (1969) with its 1976 Protocol, those instruments were considered by WB Group's staff during its review of the protection of the marine environment in the Kribi region and the oil spill response plan. The Oil Spill Response Plan, as part of the EMP, was found satisfactory by all applicable international standards. It was released for public review on 15 October 1999, and made available in the two countries. The marine terminal was required to have a tailor-made specific oil

[53] Available at <http://www.worldbank.org> (accessed 27 September 2011).

[54] FEDEC is a foundation established as an endowment fund for the purpose of providing financing for an IPP and a biological diversity offset (see EMP Cameroon Portion, Volume 4). During the preparation of the project, the WB requested the Consortium to consider all potential alternative corridors. Eight options were assessed; alignment of the pipeline within the preferred corridor was optimized from cost, technical, safety, environment, and social perspectives, in addition to aligning the pipeline to follow existing infrastructure and/or traverse degraded land. While reviewing the options, the WB, the Government of Cameroon, and the Consortium agreed to avoid a ROW that would impact the Deng-Deng forest area, which is one of the few remaining habitats for gorillas in Cameroon. Minor impacts would occur in small areas within the ROW. Mitigation measures were designed and included two environmental offsets—one for the semi-deciduous forest (M'bam and Djerem National Park) and one for the Atlantic littoral forest (Campo Ma'an National Park). EMP. Volume 4 Part 1.

[55] Ibid.

spill response plan satisfactory to the WB Group in place before the oil started to flow in the pipeline. Under the technical assistance projects mentioned below, the two governments were also committed to develop their own oil spill response plans that would comprise the activities that the governments have to undertake to comply with their own commitments in addition to those of the Consortium. The Cameroon National Oil Spill Response Plan was developed and adopted officially in 2008. It was widely disseminated and shared through an international conference with riparian countries of the Gulf of Guinea in 2010.[56] By doing so and defining what needs to be done to prevent and clean up potential oil spills, the project is helping Cameroon and Chad to meet their obligations under the relevant international conventions. For example, a minor oil spill occurred on 15 January 2007 off the coast of Kribi in Cameroon and was addressed by COTCO in accordance with the area oil spill response plan for the offloading facility site. Following that oil spill episode, it was further agreed between COTCO and the government of Cameroon to further revise the oil spill response plan to include communication with and public awareness education of neighbouring communities and local authorities which were lacking from the originally agreed oil spill response plan.[57]

C. Are wide powers recognized to the CCP operators a threat to host States and a challenge to their legal and regulatory functions?

Numerous observers mentioned the fact the Consortium was empowered with a wide mandate to implement the project in the territory of the two countries and in some cases without even consulting or obtaining a permit and/or authorization from the host state.[58] Observers highlighted article 27.12 of the COTCO Convention, and article 23.15 of the TOTCO Convention, which allow the two companies to access any private or public land 'in case of emergency' for the 'purpose of investigating the causes or remedying the emergency or situation of danger, without prior authorization and with the possible assistance of the public or private emergency services'. This provision envisages fire, destruction of installations, and other events and circumstances beyond the reasonable control of COTCO and TOTCO,[59] which may have a substantial harmful impact on the environment

[56] In addition to WB assistance, Cameroon obtained assistance from the International Maritime Organization to develop and adopt its National Oil Spill Response Plan and share it with neighbouring countries. See <http://www.giwacaf.org/activities.asp> (accessed 13 October 2011).

[57] An independent review of the oil spill incident and consequences was undertaken by a local scientific group called ENVIREP (based in the University of Yaoundé) under a WB-financed operation. The review determined that impacts were minor and recommended measures to further protect the area fisheries and water quality monitoring. (The ENVIREP report was not published but was made available to the author of this article.)

[58] IUCN, 'Liability for Environmental Damage and the WB's Chad–Cameroon Oil and Pipeline Project', 25 February 2000, Selected papers of the NC–IUCN Symposium, S.A. Bronkhorst (ed) The Hague, 2000.

[59] Although a very thorough analysis of the prevailing situation in the two countries was conducted and risk of sabotage minimized, such risk was considered during the CCP Project preparation and the likelihood of sabotage was understood as closely connected to the degree of satisfaction of the local people with the treatment they receive from the Consortium and the two governments.

and the safety of the local population. On its plain construction, this means that the companies are under an obligation to act promptly to take all preventive and remedial measures. The companies, however, remain responsible for indemnifying property owners and dwellers who are adversely affected by their actions in accordance with relevant legislation in the two countries.

In its legal due diligence, the WB concluded that this was a common practice in international and national operations in the petroleum sector. It illustrates the recognition of such right of intervention by private operators in emergency situations to preserve the environment and protect the safety and interests of the population.[60] The rights of the Consortium members were not found to create any restriction on the rights of the dweller population or a violation of the sovereign rights of the state. If these rights are used in violation of applicable laws and regulations, the Consortium members may be sued for compensation by the impacted population and its government.

It should be mentioned further, that article 17 of the COTCO Convention states that 'pursuant to the provisions of Section 23.4 of Law 96–14, COTCO shall be bound to make reparation, in accordance with ordinary law, for the damages incurred by any person, due to the construction, operation and maintenance of the Cameroon Transportation System, other than the compensation resulting from the making available of the land provided for in Article 27'. The same rule is provided for in Article 20.2.aa of the TOTCO Convention. Liabilities are well defined in case of damages incurred by people. Moreover, as mentioned above, the Consortium members are bound by the applicable WB policies and standards related to environmental and social safeguards, including compensation for any harm or substantial negative impact resulting from the implementation of the CCP project and operation of the pipeline.

Furthermore, the financing agreements entered into by the Consortium members, the countries, and the WB clearly and strongly specified that no waiver, change, or modification could be made to the EMP, except those that were consistent with WB policies and standards. Also, all project proponents were obliged under the legal agreement to report periodically on the implementation of the EMP, including facilitating the work of an independent EMP Compliance Monitoring Consultant (ECMG) who was jointly contracted by the WB Group and the other lenders to monitor the implementation of all mitigation measures and report to them. The reports of the ECMG were made public on a regular basis and discussed in various fora by stakeholders, including CSOs and affected people's representatives.[61] In

For example the Chad Revenue Management Program was designed by the Government with assistance from the WB to produce tangible benefits for the populations of the Doba Region and also countrywide. The Consortium has also developed a communication strategy to address concerns expressed by the affected populations. Finally, a prevention policy was developed to prevent any easy access to the pipeline and risk of sabotage.

[60] Letter from the Task Team Leader (Mr Ph. Benoit) in charge of the CCP Project to the Policy Director of IUCN–Netherlands (Mr Wouter Veening), dated 12 May 2000.

[61] See the WB website, which provides a full description of the monitoring system put in place for the CCP and the reports and other documents related to such system, at <http://www.worldbank.

this regard, it should be noted that the CCP was supervised and monitored through various layers of processes and bodies including: (i) supervision by WB Group staff in charge of the project; (ii) supervision and monitoring by an independent body, ie the ECMG; and (iii) monitoring by an Independent Advisory Group led by a former Prime Minister from Senegal and comprising international experts from the Netherlands and Canada.[62]

Because the CCP was developed and meant to be implemented under concession agreements that refer to domestic law as the law governing the activities of the Consortium members, TOTCO and COTCO, numerous questions were raised by CSOs about the capacity of the two countries to enforce their own legislation in case the Consortium members, TOTCO and/or COTCO did not comply with the provisions of each of their respective agreements. The WB considered this issue seriously and agreed with the two countries to help develop and fund two technical assistance projects (one for Cameroon and one for Chad) to strengthen the legal and regulatory frameworks of the two countries, including enhancing the judiciary's capacity to address environmental and social issues under the Project and training of civil servants, judges, lawyers, and non-governmental organizations (NGOs) in order to equip them with the necessary knowledge and tools to monitor project-related activities and to inform the population, and especially the affected people, of their rights under the project.

The two technical assistance projects were implemented and completed with mixed results after the construction of the pipeline started, while under normal circumstances they should have been carried out and completed before any construction. Although the two capacity development projects have delivered some capacity and good results on the ground, they were subject to criticism from many CSOs and other observers for not delivering all of their promised outcomes in terms of building capacity and equipping the countries with the needed legal and regulatory frameworks to control and monitor large infrastructure projects and enforce related laws.

D. Are the agreements between the Consortium and the two countries hampered by ambiguous provisions that can be a source of lack of predictability and liability?

This issue was mainly and primarily raised in connection with the environmental and social mitigation measures that were identified and approved to mitigate, reduce, and offset negative impacts of the project on the natural and human environment.[63] Articles 13.4 of the COTCO Convention and 14.1.d of the TOTCO Convention, mention clearly the obligation of COTCO and TOTCO to comply

org> (accessed 27 September 2011). See also ECMG Reports on the IFC website, such as the 2001 Report, <http://www.ifc.org> (accessed 27 September 2011).

[62] As mandated by its policies, the WB conducted an evaluation of the project by the Independent Evaluation Group (IEG). Its outcome is available at: <http://www.worldbank.org> (accessed 27 September 2011).

[63] IUCN Report above n 58; responded to in the Letter mentioned in n 60.

with, and implement, the EMP in a manner satisfactory to the WB Group.[64] Under the legal agreements entered into between the WB Group and respectively the Government of Chad, the Government of Cameroon, COTCO, TOTCO, Esso Exploration and Production Chad Inc, Chevron Petroleum Chad Company Ltd, and Petronas Carigali (Chad EP) Inc, an obligation to implement and comply with all the terms of the EMP was defined and accepted by all the project's proponents. As mentioned, no waiver, change, or modification could be made to the EMP, except those that were consistent with the WB Group safeguard policies and guidelines. The WB Group and other lenders had the right to suspend disbursements or accelerate the reimbursement of the loans if these commitments were not met.

The legal agreement with Chad included provisions pursuant to which Chad had to carry out, in a timely manner, any action required to be performed by it under the EMP, taking all necessary action to enable TOTCO to carry out the resettlement and compensation plan described in the EMP. Chad was also obliged to comply with the obligations relating to the implementation of the compensation plan and disclosure of information related to the EMP. Similar provisions were included in the legal agreements with Cameroon. In addition, Cameroon was required to (i) establish the national parks of Campo Ma'an and M'Bam and Djerem as a biodiversity offset for the loss of biological diversity due to the selection of the route of the pipeline; and (ii) carry out an IPP and related compensation plan, all of which are described in the EMP.

TOTCO and COTCO were obligated to carry out, in a timely manner, all actions required to be performed by them under the EMP. Specific obligations related to the carrying out of the compensation plan and the oil spill response plans were stated in the legal documents. The requirements for Chad, Cameroon, and the Consortium to develop oil spill response plans were spelled out in the respective agreement, and included area oil spill response plans to be developed by the Consortium at different places of the pipeline's right of way, including the marine area off the coast of Cameroon. The two countries were also required to develop and approve their respective national oil spill response plans to clarify their roles and responsibilities versus those of the Consortium.

Notwithstanding the measures taken to ensure adequate environmental and social impact management as described above, in its investigation report the Inspection Panel of the WB reported that many aspects of the EA of the CCP were characterized by weaknesses that impacted the identification of the necessary mitigation measures and the overall implementation of the EMP. Among these weaknesses, the Inspection Panel identified the following: (i) incomplete baseline data; (ii) absence of analysis of environmental costs and benefits; (iii) no review of the adequacy of the EA and EMP by an independent panel of experts as mandated by the relevant WB's Operational Policy 4.01; (iv) no analysis of cumulative impact

[64] This condition was also included in financing agreements of other lenders, including EIB and IFC.

assessment; and (v) lack of baseline survey outside the pipeline's right of way with regard to forest conservation and use.

E. Dispute resolution mechanisms with regard to impacts of the CCP on affected persons and communities

The main issue here was to define the competent forum to settle disputes that could arise between the CCP builder and/or operator and any affected person or community as a consequence of the construction and operation of the CCP. Under the concession agreements and the laws of the two countries, all disputes between respectively the Consortium members, TOTCO, and COTCO on one hand, and affected peoples and communities, on the other, are to be tried before local courts.[65] The same applies to disputes between the government and its citizens as a consequence of the pipeline's construction and/or operation.[66] This is the regular way of settling disputes of this kind. An additional mechanism may undermine the local legal system and, in particular, the role of the countries' judiciary. However, in addition to the right to claim before the courts, the respective land acquisition and involuntary resettlement plan described in the EMP for the Cameroon and Chad portions of the CCP and endorsed by the WB Group has established a grievance mechanism that provides affected persons and communities with an additional opportunity to make their concerns and cases heard and discussed with the government and the Consortium members.[67] The grievance mechanism provides a remedy to affected persons and communities for any land and/or other assets' loss or negative impact on livelihood during the preparation and construction phase of the CCP. Liaison officers were contracted to maintain direct links with affected persons and communities to facilitate the relationships with the CCP owner and operator, including for the case of filing and addressing complaints. Local CSOs and government bodies in charge of supervision and

[65] Grievance procedures and mechanisms established under the EMP do provide for an upstream process to deal with people's grievances before any recourse to courts.

[66] It should be mentioned that at least in Cameroon the *actio popularis*, as known in many French-speaking African countries, is recognized by law. Individuals and associations have the right to sue when vested interests are at stake. For example, article 8.2 of the Environmental Code of Cameroon (Law 96–12 of 5 August 1996) provides that 'authorized grassroots communities and associations contributing to...environmental protection may exercise the rights of standing with regard to facts constituting a violation of the provisions of this law and causing direct or indirect harm to the common good they are intended to defend'. In addition, many of such CSOs will benefit from the above-mentioned technical assistance projects to improve and enhance their capacity to enforce their rights under this provision. However, no *actio popularis* was undertaken by any grassroots organizations which instead preferred, for example, to make requests for review of the project before the WB Inspection Panel.

[67] In Cameroon, local NGOs such as FOCARFE and CED were very active in assisting affected persons and communities to raise their concerns and claim additional compensation from COTCO. COTCO, the Government of Cameroon through the Pipeline Steering and Monitoring Committee and local NGOs including CED, Global Village, and FOCARFE, organized site visits and consultations with affected persons and communities to ensure that the resettlement action plan and compensation plan were fully and adequately implemented.

monitoring played an important role in ensuring that complaints were treated efficiently, diligently, and with fairness.[68]

The training, information, education, and communication components of the two technical assistance projects have helped develop, although not up to the initial expectations, people's awareness and judicial and administrative capacity to address any such disputes. The technical assistance projects have achieved part of that objective, including providing affected peoples and communities with resources to obtain all necessary compensation for impacts suffered, including impacts that were unforeseen during project preparation. One such impact was the impact on the fishermen community in Cameroon in the Kribi area. This community suffered loss of revenues during the construction of the pipeline off the coastline of Kribi. However, it has benefited from the technical assistance project in Cameroon which made it possible for the community to claim and obtain compensation through a participatory and consultative process. The same technical assistance project provided indigenous peoples in Cameroon with assistance to fully benefit from the IPP provided for under the EMP.[69] But in hindsight, the benefits would have been greater if the technical assistance had been completed before the start of the construction.

V. Lessons Learned from the CCP Project

The WB's loan instruments entered into by the Governments of Chad and Cameroon, COTCO and TOTCO, as well as the Consortium members,[70] refer to the implementation of the EMP which embodies a whole set of measures designed to address all the environmental and social issues described in the Environmental Assessment reports. The implementation of these measures by the governments of Chad, TOTCO, COTCO, and the Consortium members was to be monitored through different procedures, including governmental procedures. However, it was known that both governments lacked capacity to effectively monitor the implementation of the legal obligations of the builders and operator of the CCP. Clearly, international ad hoc mechanisms are not a panacea and cannot replace government agencies responsible for protecting the environment and the people in any country. It is desirable and important that countries develop and establish their own effective and efficient monitoring mechanisms before launch-

[68] The Amnesty International Report mentioned above (see n 18, 39, and 42) has criticized this grievance mechanism. The Report concluded that it 'does not provide an effective remedy in human rights terms'.

[69] Even the most critical CSOs, such as FOCARFE in Cameroon, recognized that through the grievance mechanism consultation happened between the COTCO and affected communities and many complaints were handled quite effectively. See FOCARFE, 'The Chad–Cameroon Pipeline: Where Do We Stand Today?', April 2006, available at <http://pwypdev.gn.apc.org/sites/pwypdev.gn.apc.org/files/chadcamtoday_0.pdf> (accessed 27 September 2011).

[70] These are two Loan Agreements between the IBRD and, respectively, Chad and Cameroon; two Credit Agreements between IDA and, respectively, Cameroon and Chad, and project Agreements with, respectively, TOTCO, COTCO, Esso Chad, Petronas, and Chevron.

ing large infrastructure projects with a significant environmental and social footprint. Relying on international financial institutions and ad hoc mechanisms is not a solution, nor can it be a replacement for national ownership of processes and outcomes of such projects. For the CCP, it is recognized that while TOTCO and COTCO were able to deliver the construction and operate their respective portion of the pipeline in due time, both governments lacked capacity to monitor the environmental and social performances of the two companies and to ensure that the promised outcomes were reached.

Connected to the above, is the full description of the environmental and social obligations of the project companies in the concession documentation and their precise itemization in the relevant environmental impact assessment and EMP. The EMP prepared and adopted for the CCP was characterized by too many general statements and by a lack of specific definition of implementing measures for mitigation and compensation measures. It will be useful and desirable for any future projects of CCP's nature to include in their governing instruments provisions and annexes describing with all the needed detail any mitigation and compensation measures, the timeline for their implementation, the budget allocated to that end, and even categories of mitigation measures and compensation that may be limited by cost or result. That was not the case of the CCP, under which the only cost-limited measure was the financing of measures to maintain a biological diversity offset and plans for assisting indigenous peoples in Cameroon through an independent foundation: FEDEC. The foundation was established to provide sustainable financing for the two offsets in Cameroon (Campo Ma'an and M'bam & Djerem) and the implementation of an IPP for the Bakola–Bagyieli communities impacted by the CCP. The US$3million allocated to FEDEC were found to be insufficient to implement the two measures, and a solution for a sustainable financing plan is yet to be adopted.

International financial institutions such as the WB will be well advised to ensure that Environmental Management Plans and compensation measures for potentially affected peoples are fully budgeted in a transparent manner before any work starts on large infrastructure projects such as the CCP. For the CCP, although the EMP was well designed, no budget was ever disclosed. Also, actual payment to compensate impacts and for mitigation measures implementation were not always transparent.[71] A good practice was developed subsequently in connection with a large infrastructure project in which the WB and other international financial institutions are involved, the Nam Theun 2 Dam and Hydropower Project in Laos People's Democratic Republic (Laos PDR). For that project the concession agreement between the developer and operator of the project and the Government which entered into effect in 2005 included a specific schedule defining with great detail all mitigation measures for environmental and socioeconomic impacts, along with a dedicated budget and a timeline for implementation and a monitoring plan. In addition, the concession agreement provided for additional financial resources in

[71] With the exception of the US$3 million endowment for the FEDEC and budget allocated to mitigate impact on archeological resources in Cameroon during the construction of the pipeline.

case the dedicated budget did not cover unforeseen impacts and/or impacts that were not properly assessed during project design. That schedule 4 to the Concession Agreement between the Nam Theun 2 Power Company (NTPC) and the Laos PDR Government is widely disclosed on the web, along with all implementation monitoring reports.[72]

Concerning human rights, many CSOs, including Amnesty International, raised issues connected to compliance with the 1966 Covenants on: (i) Political and Civil Rights; and (ii) Economic, Social and Cultural Rights, as well as the African Charter on Human and Peoples' Rights (1981) and other legal instruments related to human rights protection and promotion. These CSOs did not dispute the principle that 'the Bank does not interfere in the political affairs of its members, including their positions on the political rights of their citizens'. On the other hand, the Bank's involvement in human rights issues goes far beyond the political arena where it is prohibited from interfering. It encompasses such fundamental issues as the alleviation of poverty, the fulfillment of basic human needs for nutrition, safe water, education, health and housing, the concern for resettlement of people affected by large development projects and of tribal peoples, the role of women in development, and the avoidance of the negative impact of development on the environment.[73] Although the WB Group has no mandate to implement political human rights instruments, the project in many aspects has been instrumental in improving the rights of the citizens and specific groups of citizens within the two countries. Accordingly, an important consultation[74] process took place during the preparation of the project to (i) share project information with the affected communities; (ii) determine project land needs, establish entitlements, and agree on a valuation system with affected peoples; (iii) agree on a framework for mitigation of adverse effects; and (iv) elaborate a regional development plan for the project-related areas. This process was implemented through specific projects not directly linked to the CCP project, such as PROADEL,[75] which is in its second phase and is promoting decentralization and self-government of local communities.

Finally, on a very specific issue related to Cameroon's institutional set up for participation in and monitoring of the CCP, the Government of Cameroon decided that its fully owned oil company, the National Hydrocarbon Company, or SNH,[76]

[72] See <http://www.namtheun2.com/index.php?option=com_content&view=article&id=48&Itemid=53> (accessed 27 September 2011).

[73] I.F.I. Shihata, 'The World Bank and Human Rights: A Presentation Before the 1993 UN World Conference on Human Rights' (Vienna: 17 June 1993) in *The World Bank Legal Papers* (Leiden: Martinius Nijhoff Publishers, 2000) pp 815–17; and for further details on the subject, see I.F.I. Shihata, *The World Bank in a Changing World*, Volume 1, Chapter 3 (1991); Volume 2, Chapters 18 and 19 (1995); and Volume 3, Chapter 4 (Leiden: Martinius Nijhoff Publishers, 2000).

[74] Numerous meetings took place to discuss the EA findings and EMP recommendations which were given to the affected people for review and consultation with the project sponsors and governments. Information was disseminated in local languages and, where necessary, WB Group's staff got involved in improving the quality of the consultation process, including building capacity for constructive engagement of local CSOs.

[75] PROADEL is in phase 2 and was assessed as having an effective positive impact on local communities. See <http://www.worldbank.org> (accessed 27 September 2011).

[76] In French, Société National des Hydrocarbures, or SNH.

would be its representative and share holder within COTCO, while at the same time establishing a Committee for Supervision of Pipelines (CPSP). The CPSP was meant to be an inter-ministerial committee representing all ministries and agencies having a mandate to monitor CCP's construction and operation. In reality it was staffed with SNH's staff and headed by SNH's CEO. This led to concerns of conflict of interests and lack of effective willingness to monitor environmental and social performance of COTCO. Although empirical evidence showed that the CPSP showed some independence, it was nonetheless perceived as acting in many instances as a shareholder and not a compliance-monitoring body. This kind of institutional set up must be avoided at any cost to ensure that a government's mandate to monitor construction and operation of large infrastructure projects will be effectively undertaken in an independent and efficient manner.

6

Contractual and Treaty Arrangements Supporting Large European Transboundary Pipeline Projects: Can Adequate Human Rights and Environmental Protection Be Secured?

Catherine Redgwell

I. Introduction

The purpose of this chapter is to analyse the complex interplay of private contractual and treaty arrangements supporting large energy infrastructure projects and the extent to which, and how effectively, these address the environmental and human rights impacts[1] of such projects. A case study approach will be adopted, with two examples of transboundary pipeline (TBP) 'mega-projects' in particular serving to highlight how human rights and environmental protection is secured through supranational mechanisms. The case studies selected are the Baku–Tblisi–Ceyhan (BTC) and the Nord Stream pipeline projects, discussed in detail in Parts II and III below.[2] Both are designed to ensure that non-OPEC Russian

[1] While the focus of this contribution is upon the impairment of human rights to which contractual and treaty arrangements (eg stabilization clauses) constraining host state action may give rise, it is acknowledged that there are instances where foreign investors have simultaneously pursued investment treaty arbitration and a claim for violation of their human rights (usually relating to due process and/or the right to property) where there is a forum to do so, eg before the European Court of Human Rights. It is rare for human rights claims to arise directly in investment treaty arbitration itself: for recent discussion see B. Simma, 'Foreign Investment Arbitration: A Place for Human Rights?' (2011) 60:3 *ICLQ* 573–96, critiquing *inter alia* the view that the current system of investment arbitration 'seems to be leaning toward separation of human rights and investor's rights like oil and water' citing M. Toral and T. Schultz, 'The State, a Perpetual Respondent in Investment Arbitration? Some Unorthodox Considerations' in M. Waibel, A. Kaushal, K.-H. Chung and C. Blachin (eds), *The Backlash Against Investment Arbitration: Perceptions and Reality*, (Amsterdam: Wolters Kluwer, 2010) 577–602.

[2] There are other illustrations which could be used here: eg the human rights implications of the project agreement between COTCO (the Cameroon Oil Transport Company) and Cameroon in the context of the Chad–Cameroon Pipeline. See Amnesty International, *Contracting Out of Human Rights: the Chad–Cameroon Pipeline Project* (2005), available at <http://www.amnesty.org/en/library/info/POL34/012/2005> (accessed 19 October 2011). Notwithstanding criticism, the project agreement between COTCO and Cameroon was not amended. In addition, the World Bank provided a

petroleum is able to access European markets either overland—the BTC pipeline from the Caspian region through Georgia to Turkey—or at sea—the Nord Stream pipeline from Russia to Germany, via the continental shelves of Finland, Sweden and Denmark.

Both are transboundary pipelines, though with key distinctions between them, most notably that the Nord Stream project is a transboundary undersea gas pipeline while the BTC pipeline is a land-based oil pipeline. One consequence of this geographic distinction is that human rights are a much more prevalent concern in respect of the land-based BTC pipeline, a factor exacerbated by imperfect observance of the rule of law and good governance, especially in the Caucasus region.[3] On the other hand, the fragile environment of the Baltic Sea area has raised acute concerns regarding the impact of the pipeline, in particular of its construction.[4] For Nord Stream these impacts transcend the project participants alone with assessment of the impacts of the project on all nine Baltic littoral states. A notable feature of this project has been the role of procedural human rights under the regional 1991 Espoo Convention on Environmental Impact Assessment in a Transboundary Context ('Espoo Convention') and its harmonizing influence in the face of diverse national planning and environmental regulations.

Potentially serious environmental issues arise with both projects owing to the length of the pipelines, their routes and their impact on land[5]/seascape, species and habitat, and the potential for spillage and/or 'explosions'.[6] Both are illustrative

small part of the funding for the project, and it was the subject of an inspection panel report with deficiencies found on environmental assessment and mitigation measures in particular: see IBRD/IDA, Management Report and Recommendation In Response to the Inspection Panel Investigation Report, Cameron Petroleum Pipeline Project (Loan No. 7020-CM) and Petroleum Environmental Capacity Enhancement (CAPECE) Project (Credit No. 3372-CM), Report No. INSP/R2003-0003, 28 May 2003, available at <http://siteresources.worldbank.org/EXTINSPECTIONPANEL/Resources/CMMgmtReporttoINVReportMay28.pdf>, (accessed 19 October 2011). The Bank's involvement effectively ended in 2008 and the loans were repaid. For background see S. Pegg, 'Chronicle of a Death Foretold: The Collapse of the Chad–Cameroon Pipeline Project' (2009) 108/431 *African Affairs* 311; see also P. Cameron, *International Energy Investment Law: The Pursuit of Stability* (2010) 398–401, and chapter V of this volume.

3 See, for example, C.P.M. Waters, 'Who Should Regulate the Baku–Tbilisi–Ceyhan Pipeline?' (2003–04) 16 *Georgia International Environmental Law Review* 403 and Amnesty International, 'Human Rights on the Line: The Baku-Tblisi-Ceyhan Pipeline Project' (2003), available at <http://www.amnesty.org.uk> (accessed 2 October 2011).

4 E. Karm, 'Environment and Energy: The Baltic Sea Gas Pipeline' (2008) 39 *Journal of Baltic Studies* 101; and B. Whist, *Nord Stream: Not Just a Pipeline—An Analysis of Political Debates in the Baltic Sea Region Regarding the Planned Gas Pipeline from Russia to Germany* (Fridtjof Nansen Institute, 2008). See further discussion in Part III of this chapter.

5 See, for example, Water's criticism of the routing of the BTC pipeline through the Borjomi Gorge in Georgia: above n 3 at 415. Hill relates that BTC Co. hired a team of archaeologists to excavate and record data on ancient sites discovered during construction of the BTC pipeline in Georgia: F. Hill, 'Pipelines in the Caspian: Catalyst or Cure-all?' (2004) 5 *Georgia Journal of International Affairs* 17 at 23.

6 Indeed, one of the reasons suggested (albeit with journalistic hyperbole) for the Estonian refusal to permit the Nord Stream pipeline to cross its maritime territory was concern regarding detonation of the highly pressurized gas with explosive force 'equal that of about 50 Hiroshima bombs': 'Estonia says "No" to Nord Stream', *The Baltic Times*, 20 September 2007, available at <http://www.baltictimes.com> (accessed 2 October 2011). The Estonian authorities refused permission to survey the

of the self-evident fact that, 'as distances from producing fields to consuming markets increase, so will the frequency with which large-scale pipeline developments straddle international boundaries'.[7] And so too will the frequency increase with which 'bespoke' legal arrangements[8] are concluded, resulting in a complex blend of domestic, regional, and international law, in addition to the project-specific contractual arrangements. The need for such bespoke arrangements is underscored by the absence of pipeline-specific generic regulation at the regional or international level: there is no general treaty on the construction and use of pipelines at international law.[9] Nonetheless, there is a number of international treaties that may apply to condition where, and how, such projects may be carried out particularly with respect to environmental impact and human rights. The focus of this chapter is on how, in response to pressures from a variety of stakeholders, the range of public and private law instruments applicable to the project are brought to bear to secure basic (indeed, inalienable) human rights guarantees and environmental protection.[10]

II. The BTC Pipeline

A. Description of the project

The Baku–Tbilisi–Ceyhan (BTC) pipeline project is designed to transport oil from the Caspian Sea to markets in Europe via a 1,090-mile pipeline crossing three states—Azerbaijan, Georgia, and Turkey—and without going through the Persian Gulf or Russia.[11] The project is supported by a number of related agreements of a

continental shelf, for the purposes of assessing this potential route for the pipeline. See further Part III of this chapter.

[7] W. Browning and T. Dimitroff, 'Transboundary Pipeline Development and Risk Mitigation' in G. Picton-Turbervill (ed), *Oil and Gas Law: A Practical Handbook* (London: Globe Business Publishing, 2009) at 93.

[8] As early as 1995, Jeremy Carver and Greg Englefield wrote of the need to establish 'an international pipeline authority to build and manage a major oil and gas pipeline network' to deliver oil and gas from land-locked frontier regions in Russia and Central Asia to European markets because of the absence of a legal regime to protect international pipelines from regional instability: J. Carver and G. Englefield, 'A New Regime for International Pipelines from Central Asia' in G.H. Blake et al (eds), *The Peaceful Management of Transboundary Resources* (London: Graham & Trotman, 1995) 97–109.

[9] This is a descriptive rather than a normative statement; it is far from clear what the utility of such a general instrument would be, and in any event a variety of bespoke legal arrangements will inevitably be necessary for a large pipeline project, as the case studies below serve to illustrate.

[10] This is a difference of emphasis from Cameron, whom in his chapter 8 on 'Stability in the Context of Environmental and Human Rights Issues' uses three case studies—the Chad–Cameroon pipeline, the Sakhalin II gas development, and the BTC pipeline—'to examine how investors have developed and adapted legal regimes for large energy infrastructure projects in ways that attempt to mitigate human rights risks to the long-term stability of the investment'. Above n 2 at 367.

[11] Geopolitics as well as commercial considerations thus had a large role to play in determining the existence, and route, of the pipeline. See A.S. Reyes, 'Protecting the 'Freedom of Transit of Petroleum: Transnational Lawyers Making (Up) International Law in the Caspian' (2006) 24 *Berkeley Journal of International Law* 842 at 843; S. Stern, 'Turning Towards Turkey: Its Importance as an Energy Distributor and Ally in Post-9/11 Stabilization' (2004) 28:1 *Fletcher Forum of World Affairs* 201; and T.W. Waters, '"The Momentous Gravity of the State of Things Now Obtaining":

treaty or contractual character,[12] including in particular, an Inter-Governmental Agreement (IGA) among the three states;[13] Host Government Agreements (HGAs) between each of the participating states and the project investors (a BP-led oil consortium, the BTC Company),[14] and the project commercial agreements. Each HGA is appended to the IGA,[15] which pursuant to Article II(2) IGA constitute an integral part of the intergovernmental agreement.[16] This is an 'incorporation by reference' of private contractual commercial arrangements into a treaty framework.[17] Additionally, a Code of Practice was subsequently appended to the HGAs setting forth certain principles, standards, and agreements that are applicable to project activities, including on environment and health and safety, thus incorporating by reference binding and non-binding standards within the treaty framework.[18]

B. The project agreements: securing adequate human rights and environmental guarantees?

The IGA is an international agreement governed by the law of treaties pursuant to which the three participating states undertake certain responsibilities, including giving full force and effect to the agreement in domestic law. Of particular relevance for present purposes is IGA Article IV on technical, safety, and environmental

Annoying Westphalian Objections to the Idea of Global Governance' (2009) 16:1 *Indiana Journal of Global Legal Studies* 25 at 43.

[12] This is a considerable simplification of a highly complex transboundary project linked to the upstream development of the Azerbaijan's Caspian Sea area (eg there are 208 finance documents): see further <http://www.caspiandevelopmentandexport.com> (accessed 2 October 2011) ('BP Caspian' site, under 'legal agreements').

[13] *Agreement Among The Azerbaijan Republic, Georgia and The Republic of Turkey Relating to the Transportation of Petroleum Via the Territories of The Azerbaijan Republic, Georgia and the Republic of Turkey Through the Baku-Tbilisi-Ceyhan Main Export Pipeline*, concluded 18 November 1999 at Istanbul, Turkey. On the legal definition of such agreements see generally M. Dulaney and R. Merrick, 'Legal Issues in Cross-Border Oil and Gas Pipelines' (2005) 23 *JERL* 247 at 251.

[14] BTC Co. comprises 11 co-venturers: BP (UK) (30.1 per cent); SOCAR (Azerbaijan) (25 per cent); Unocal (US) (8.9 per cent); Statoil (Norway) (8.71 per cent); TPAO (Turkey) (6.53 per cent); ENI (Italy) (5 per cent); Total (France) (5 per cent); Itochu (Japan) (3.4 per cent); INPEX (Japan) (2.5 per cent); ConocoPhillips (US) (2.5 per cent); Amerada Hess (US) (2.5 per cent).

[15] For the HGA between Turkey and the MEP [main export pipeline] participants, for example, see Appendix 2 of the IGA of 18 November 1999.

[16] Here internationalization of the agreements is by virtue of their incorporation by reference in the IGA, rather than by contract internationalization through reference to the governing law of the contract as international law, or by a combination of international law and stipulated domestic law (as per *Texaco v Libya* (1978)17 ILM 1). The governing law of the BTC project commercial agreements is English law. See D. Nougayrede, 'Binding States: A Commentary on State Contracts and Investment Treaties' (2005) 6 *Business Law International* 373.

[17] In turn, the preamble of each of the commercial agreements states that the three participating states 'have entered into the Intergovernmental Agreement to give the Project's legal and commercial conditions the support and framework of international law'.

[18] For the Code of Practice attached to Turkey's HGA, see Appendix 5 of the IGA of 18 November 1999. As Pitts notes, referencing of voluntary standards in host agreements ensures that these become part of the 'prevailing legal regime', explicitly citing the BTC pipeline HGAs as examples: J. W. Pitts, 'Business, Human Rights, & the Environment: The Role of the Lawyer in CSR & Ethical Globalization' (2008) 26 *Berkeley J. Int'l L.* 479 at 487.

standards. This imposes on the state parties the obligation to cooperate and coordinate with each other, and the relevant project investors, in formulating uniform technical, safety, and environmental standards applicable to the construction, operation, repair, replacement, capacity extension or extension and maintenance of the facilities. Such operations are to be carried out in accordance with 'international standards and practices within the Petroleum pipeline industry' but with these benchmarked to EU standards[19] and to the requirements set forth in the HGAs. These explicitly prevail over any standards and practices set forth in the domestic law of each state. The IGA also underscores the commercial nature of the BTC project in the express warranty that the project 'is not intended nor required to operate in the service of the public benefit or interest in [the relevant host state's'] territory'.[20] Last but not least for present purposes, priority is given to the IGA arrangements in that each IGA state party expressly warrants that it is not a party to any domestic or international agreement or commitment, or obliged to observe or enforce any domestic or international law or regulation, that conflicts with the IGA and the pipeline project.[21]

The HGA is also termed an 'international agreement' which, while not transforming it into a separate international treaty[22]—it creates only a commercial contract between the host state and the investor—ensures that as a matter of domestic constitutional law, in the event of conflict between the HGA and an ordinary (ie not constitutional) domestic law, present or future, it is the provisions of the agreement which prevail.[23] Each host state grants to BTC Co. the right to build and operate the pipeline and to provide various kinds of support to it. The term of the HGA is forty years, renewable twice for a ten-year period each time. The host state is prohibited from unilateral change of the concession structure for the duration of the HGA.[24] As Waters views it, this 'essentially constitutes an opt-out for large oil interests from some aspects of the standard legislative regime for oil and gas exploitation and environmental protection'.[25] A more measured approach is found in Cameron, who suggests it is a question of the interpretation of the stabilization clause in each case.[26] Regardless of the doctrinal view taken, the fact remains that such provisions may serve to insulate the project from the application of ordinary

[19] IGA article IV states that such standards and practices 'shall in no event be less stringent than those generally applied within member states of the European Union'.

[20] Article II(8). This clause was evidently inserted to address problems in Turkish law and to avoid a legal classification of the HGA essentially nullifying its effects: Cameron above n 2 at 404 n 116, citing H. Boyd-Carpenter and W. Labadi, 'Striking a Balance: Intergovernmental and Host Government Agreements in the Context of the Baku–Ceyhan–Tblisi Pipeline Project' (EBRD, 2004), available at <http://www.ebrd.com/pubs/legal/lit042e.pdf> (accessed January 2009).

[21] IGA article II(6).

[22] An international treaty is a written agreement between states (or between states and International Organizations (IOs) or between IOs) governed by international law. The mere fact that one party to an agreement is a state does not transform the agreement into an international treaty where the other party is a private actor (as here, the Project Investors), even where a state supports the company in its contract negotiations: see Anglo-Iranian Oil Company case (1952) *ICJ Rep.* 93.

[23] See eg section V, ch X, art 151 of the Constitution of the Republic of Azerbaijan; ch 1, art 6(2) of the Constitution of Georgia.

[24] See eg article 7.2(x) of the Turkish HGA.

[25] Waters above n 3 at 405.

[26] Cameron above n 2.

domestic law and to internationalize the Project by benchmarking conduct to international standards.[27] This is not, of course, unique in the energy context—one of the functions of BITs is to provide international standards and thus stability for investor protection[28]—but what is striking here is the degree of autonomy from otherwise applicable domestic law in 'all keys areas of the project (environment, taxation and customs, land rights, etc)'.[29] In consequence, as Bridgeman and Hunter observe, bespoke legal arrangements such as for the BTC pipeline 'can have the same deregulatory effect as investment agreements'.[30]

The stabilization clause varies slightly between the HGAs,[31] and is in the form of an economic equilibrium clause[32] which (a) requires economic equilibrium to be restored where, for example, disrupted by changes in tax, health, safety, or environmental laws; and (b) creates a right to compensation of investors if newly introduced legal measures adversely affect the value (ie the economic equilibrium) of the project. The host state has an obligation to take all appropriate measures promptly by whatever means may be necessary to restore the economic equilibrium established under the concession structure. Additionally, there is a non-interference provision common to the HGAs, article 5 ('Government Guarantees'), which states, *inter alia*:

(iii) that the state authorities shall not act or fail to act in any manner that could hinder or delay any Project Activity or otherwise negatively affect the Project or impair any rights granted under the Project Agreement (including any such action or inaction predicated on *security, health, environmental or safety considerations* that, directly or indirectly, could interrupt, impede or limit the flow of Petroleum in or through the Facilities, *except* under

27 Ibid. The HGAs build on the example of production-sharing agreements already extant in the region, which amongst other things benchmark compliance with domestic environmental measures to international petroleum industry standards: see, for example, the *Agreement on the Joint Development and Production Sharing for the Azeri and Chirag Fields and the Deep Water Portion of the Gunashli Field in the Azerbaijani Sector of the Caspian Sea*, concluded 20 September 1994.

28 See generally Cameron above n 2 and M. Erkan, *International Energy Investment Law: Stability Through Contractual Clauses* (New York: Kluwer Law International, 2011).

29 Nougayrede above n 16 at 394.

30 N.L. Bridgeman and D.B. Hunter, 'Narrowing the Accountability Gap: Toward a New Foreign Investor Accountability Mechanism' (2008) 20 *Georgetown International Environmental Law Review* 187 at 198; see also K. Tienhaara, 'Unilateral Commitments to Investment Protection: Does the Promise of Stability Restrict Environmental Policy Development?' (2006) 17 *Yearbook of International Environmental Law* 139.

31 For example, the Azeri HGA adopts the 'modern' balancing approach and requires state authorities to 'take all action available to them to restore the Economic Equilibrium established under the Project Agreements if and to the extent that Economic Equilibrium is disrupted or negatively affected, directly or indirectly, as the result of any change... ' (art 7(2)(x)). 'Economic Equilibrium' is defined as 'the economic value to the Project Participants of the relative balance established under the Project Agreements at the applicable date between the rights, interests, exemptions, privileges, protections and other similar benefits provided or granted to such Person and the concomitant burdens, cost obligations, restrictions': see Annex I to the HGAs.

32 This is not intended to suggest that there is only one form of stabilization clause, which may take the form of an intangibility, freezing, fiscal, or tariff stabilization, or economic stabilization, clause: see L. Cotula, 'Foreign Investment Contracts' (IIED Briefing Paper No. 4, 2007); see also Erkan, above n 28 at 101–43 and Cameron, above n 2 at paras 2.19–2.56. For sample stabilization clauses employed in the petroleum industry see—ibid., and also J. Bishop, W. Crawford, and M. Reisman (eds), *Foreign Investment Disputes* (New York: Kluwer Law International, 2005) 286–307.

circumstances in which continued operation of the Facilities without immediate corrective action *creates an imminent, material threat to public security, health, safety or the environment* that renders it reasonable to take or fail to take, as the case may be, such action and, then, only to the extent and for the period necessary to remove that threat).[33]

In essence, this non-interference provision prevents the host state from intervening on environmental, health, or safety grounds except as a temporary and exceptional measure where reasonable to do so. Since the trigger for such action is an imminent and material threat, this reduces the scope for precautionary action and arguably sets a higher threshold than required under the European Convention on Human Rightst (ECHR), and most environmental treaties, for example.[34] It has also been suggested that such non-interference provisions may inhibit the ability of the host state to comply with its obligations to respect, protect, and fulfil fundamental human rights guarantees and to regulate in the public interest since such action may give rise to the obligation to pay compensation.[35] Whether indeed this has led to 'regulatory chill' is empirically a complex task to demonstrate,[36] and much will turn on the interpretation of the agreement in question. As the tribunal in *Methanex* observed:

> [A]s a matter of general international law, a non-discriminatory regulation for a public purpose, which is enacted in accordance with due process and which affects, *inter alia*, a foreign investor or investment is not deemed expropriatory and compensable unless specific commitments had been given by the regulating government to the then putative foreign investor contemplating investment that the government would refrain from such regulation.[37]

This final caveat may be made all the more forcefully here, where the non-interference commitment is enshrined in a bespoke legal agreement between the parties.

33 Emphasis added.

34 See generally P. Birnie, A. Boyle, and C. Redgwell, *International Law and the Environment* (3rd edn, Oxford: Oxford University Press, 2009).

35 J.L. Cernic, 'Corporate Human Rights Obligations under Stabilization Clauses' (2010) 11:2 *German Law Journal* 210 at 221; see also the research commissioned by the United Nations (UN)/International Finance Corporation (IFC), Stabilization Clauses and Human Rights (2009); UNCTAD, Selected Recent Developments in IIA Arbitration and Human Rights, IIA Monitor No. 2 (2009); and, generally, L.E. Peterson, 'Human Rights and Bilateral Investment Treaties: Mapping the Role of Human Rights Law within Investor-State Arbitration (Rights and Democracy, International Centre for Human Rights and Democratic Development)', available at <http://www.dd-rd.ca/site/_PDF/publications/globalization/HIRA-volume3-ENG.pdf> (accessed 19 October 2011).

36 See generally L. Cotula, 'Reconciling Regulatory Stability and Evolution of Environmental Standards in Investment Contracts: Towards a Rethink of Stabilization Clauses' (2005) 1 *Journal of World Energy Law* 7 Business 158; ibid., 'Stabilization Clauses and the Evolution of Environmental Standards in Foreign Investment Contracts' (2006) *Yearbook of International Environmental Law* 236; A.S. Reyes, 'Protecting the "Freedom of Transit of Petroleum": Transnational Lawyers Making (Up) International Law in the Caspian' (2006) 24 *Berkeley J. Int'l L.* 842; and T. Ishikawa, 'Third Party Participation in Investment Treaty Arbitration' (2009) 59:2 *ICLQ* 349.

37 *Methanex Corp. v United States*, Jurisdiction and Merits (NAFTA chapter 11 arbitral tribunal, 3 August 2005), Part IV, chapter D, at 4, para 7.

C. Controversy and project agreement adjustments

The human rights and environmental issues in the project gave rise to 'a perfect storm',[38] with concerns expressed particularly by non-governmental organizations (international and local) regarding the human rights and environmental impacts of the BTC project[39] The stabilization clause in particular proved a lightning rod for criticism, not only because of the apparent ousting of domestic regulatory sovereignty;[40] but also because such freezing effect fails to take account of the inherent dynamism and evolutionary nature of human rights and environmental law norms. Amnesty International highlighted the potential trumping effect of the 'no-conflict' warranty of IGA Article II(8) over human rights obligations should, for example, Turkey's international human rights obligations conflict with the pipeline's construction or operation (eg with respect to workers, indigenous communities, and property rights).[41]

The response to these criticisms by the host states and BTC Co. was the issuing of a Joint Statement in 2003 which contained two essential elements. The first element addressed the standards which would apply to the project, and indicated a 'basement' for the assessment of changes in the law and a 'ceiling' for the limitation of the effects of future legislation. This was done by an interpretation of the IGA[42] which 'commits each state to the application of environmental standards that are no less stringent than those generally applied within member states of the European Union from time to time. The HGAs and other BTC Project Agreements give effect to this commitment, and provide a dynamic benchmark that will evolve as EU standards evolve.' The ceiling came in the modification to the stabilization clauses, excluding previously agreed clauses from application to legal changes in the fields of health, safety, and the environment made over the life of the project so long as such changes are 'in a manner (1) reasonably required by international labour and human rights treaties to which the relevant Host Government is a party from time to time; and (2) otherwise as required in the public interest in accordance with domestic law in the relevant project state from time to time, provided that such domestic law is not more stringent than the highest of [these international standards]'.

The second element was the addition of a new Code of Practice to the HGAs setting forth internationalized standards to be applied to the project regarding environmental and social issues to which BTC Co. must conform. Here the 'basement' (ie 'no less stringent than') is the relevant standards and practices applicable to comparable projects in the Netherlands or, in the case of a lacuna and for mountainous or earthquake-prone areas, in Austria. The choice of these two EU states with similar terrain and likely to keep abreast of changes in the technical and

[38] So characterized by Cameron above n 2 at 402.

[39] See, for example, Amnesty above n 3.

[40] Reyes graphically refers to the agreements as creating 'a thousand-mile swath of militarized corporate sovereignty running from Azerbaijan's Caspian shore to Turkey's Mediterranean': above n 11 at 842.

[41] Ibid at 16.

[42] The Joint Statement is expressly stated to be part of the Project Agreements.

environmental standards pertinent for pipeline projects introduced a degree of certainty to what otherwise could be a vague, flexible, and costly standard.[43]

The response by BTC Co. was not to insist on the extensive guarantees of stability contained in the Agreements but rather to limit the scope of these guarantees by an express 'Human Rights Undertaking' which was made publicly available.[44] However, the legal effect of this Undertaking does not appear to have been tested. Additionally, BP and Amnesty International concluded a nonbinding 'Memorandum of Understanding in relation to Human Rights' which disapplies the suspension of new legislative provisions in relation to human rights protection, anti-discrimination, and health and safety requirements. However, as Watchmen argues, such factors are not relevant to the financial basis of the project—thus the 'cost' for the consortium is negligible—and implies that these are not matters on which BP should have sought special protection or exclusion from in the first place.[45]

III. The Nord Stream Pipeline

A. Description of the project

Part of the Trans-European National Gas Network,[46] Nord Stream is a submarine natural gas pipeline running from Portovaya Bay, near Vyborg in Russia, across the Baltic Sea to Lubmin, near Greifswald in Germany, where it connects via existing European pipeline infrastructure with key European gas markets.[47] It transits five Baltic states from the point of departure (Russia) to landing (Germany)[48]

[43] There are exceptions for liability (where the Agreement applies) and the administrative/regulatory regime for implementation of environmental laws (where Georgia is the comparator country).

[44] A UN/IFC Report endorses transparency through publication of such undertakings: UN/IFC, *Stabilization Clauses and Human Rights* (2009). In this connection it should be noted that the HGAs apply the Extractive Industries Transparency Initiative (available at <http://eiti.org>, accessed 2 October 2011) and that BTC Co. created a publicly accessible website with documentation on the project (see <http://www.caspiandevelopmentandexport.com>, accessed 2 October 2011). The IFC, which provided some project finance, also published information on its website at <http://www.ifc.org> (accessed 2 October 2011). In consequence, Richardson concludes that the BTC project broadly met the procedural, decision-making requirements of the Equator Principles, while failing on human rights and environmental guarantees: see further below n 65 at 288.

[45] P. Watchman, 'Banks, Business and Human Rights' (2006) *Journal of International Banking and Financial Law* 46.

[46] In 2000, the European Commission approved a northern natural gas pipeline through the Baltic Sea as part of this network, confirmed by the European Parliament and Council in 2006. However, the specifically EU law dimensions of the project are not considered further here.

[47] See E. Zolfagharifard, 'Nord Stream: The World's Largest Gas Pipeline', *The Engineer* 4 May 2010, available at <http://www.theengineer.co.uk> (accessed 2 October 2011); and N. Cho and F. Geelhoed, 'The Nord Stream Pipeline Project—A Brief Overview of its Legal and European Relevance for Supply Security' in *European Energy Law Report VI*, M.M. Roggenkamp and Ulf Hammer (eds) [(Cambridge: Intersentia, 2009) 227–47.

[48] The pipeline is consequently not wholly submarine, with land portions in Russia and Germany.

via the continental shelves of Finland, Sweden, and Denmark.[49] The pipeline is the joint project of Nord Stream AG owned by four companies: Gazprom (51 per cent), Wintershall (20 per cent), E.ON (20 per cent) and Gasunie (9 per cent). Construction commenced in April 2010 and when completed in 2012 it will run 1,224 kilometres, in two parallel lines, the first with a transmission capacity of 27.5 billion cubic metres per year and the other increasing annual capacity to 55 billion cubic metres by 2012. It is designed to last for 50 years, with maintenance and monitoring for the duration of its lifespan carried out using advance pipeline technology which eliminates the need for an offshore service platform.[50]

B. Overview of the environmental impacts of the project

Owing to its location in the Baltic Sea area, a marine region recognized as of particular environmental sensitivity,[51] consideration of the environmental impact of the construction, operation, and decommissioning of the pipeline has been a major element of the project. In addition to concerns regarding the impact on small cetaceans in the Baltic Sea area,[52] there is also the impact on other legitimate uses of the marine area—especially for fishing and navigation—and on sites of cultural[53] and archaeological value.[54] The applicable law is national, regional, and international; as for the BTC pipeline, there is no bespoke pipeline treaty among the participating states and hence no 'one-stop shop' for regulation of the pipeline. In the case of the BTC pipeline, the interlocking project agreements and their incorporation

[49] In terms of maritime zones, the pipeline traverses the territorial seas of Russia, Denmark, and Germany, and the exclusive economic zones of all five states.

[50] For further details of the pipeline see <http://www.nord-stream.com> (accessed 2 October 2011).

[51] It is so recognized in the preamble to the Convention on the Protection of the Marine Environment of the Baltic Sea Area 1992 (as amended), text available at <http://www.helcom.fi> (accessed 2 October 2011).

[52] See the 1991 AGREEMENT on the Conservation of Small Cetaceans of the Baltic, North East Atlantic, Irish, and North Seas (ASCOBANS) concluded under the auspices of the 1979 Convention on Migratory Species (Bonn Convention). The Baltic littoral states are party, but not Russia (a key range state). The Convention obliges parties to engage in habitat conservation and management, surveys and research, pollution mitigation and public information. The Conservation and Management Plan annexed to ASCOBANS further requires prevention of 'other significant disturbance, especially of an acoustic nature' to which eg pipeline construction and decommissioning might give rise. On ASCOBANS see further M. Bowman, P. Davies, and C. Redgwell, *Lyster's International Wildlife Law* (Cambridge: Cambridge University Press, 2010).

[53] Of potential relevance is the 1972 World Heritage Convention, which protects both cultural and natural heritage, and to which all five Baltic states are party: see, generally, F. Francioni and F. Lenzerini (eds), *The World Heritage Convention* (Oxford: Oxford University Press, 2007). Petroleum projects have been recognized adversely to impact world heritage, with the first delisting of a site in 2007—a sanctuary for the Arabian oryx in Oman—in consequence of petroleum licensing activities. While the World Heritage List of protected sites includes marine and coastal heritage (eg High Coast/Kvarken archipelago, a joint Swedish and Finnish site), the route of the pipeline does not cross any designated world heritage sites in the Baltic region, nor is there any indication that any such sites were indirectly affected. There is also a 2001 treaty specifically for the protection of underwater cultural heritage, but no Baltic states are yet party to it.

[54] There is a number of munitions dumps in the Baltic Sea, with the route of the pipeline designed to avoid these, as well as to minimize impacts on marine protected areas and busy shipping lanes.

by reference to binding and voluntary standards perform a harmonizing and internationalizing function. A notable feature of the Nord Stream project is the extensive general legal framework for the 'internationalized' space of the marine environment applicable to the project, and its harmonizing influence given the diversity of national laws potentially applicable to the project. In particular, the Espoo Convention, considered further below, performed an important harmonizing function with respect to the five national legal systems applicable to parts of the project.[55]

C. Submarine pipelines: the applicable international law

The general jurisdictional framework for the regulation of offshore pipelines is provided by the 1982 United Nations Convention on the Law of the Sea (UNCLOS), to which all nine Baltic littoral states are party. Coastal states enjoy sovereignty over their territorial sea, and sovereign rights over the resources of the continental shelf (CS) and exclusive economic zone (EEZ). While coastal state consent is clearly required for the laying of pipelines in the territorial sea[56]—there is no right to do so—with respect to the CS/EEZ beyond there is the generally recognized right of all states to lay submarine cables and pipelines. Article 79 makes clear that this is a regulated right of immersion enjoyed by 'all states' and that it is subject to the right of the coastal state to take reasonable measures, *inter alia*, for 'the prevention, reduction and control of pollution from pipelines'. In so doing 'the coastal state may not impede the laying or maintenance of such cables and pipelines'.

Unlike in the territorial sea, for pipelines on the CS/EEZ coastal state consent is thus required only for the delineation of the course of the pipeline. There is no legal basis under UNCLOS for a state to prohibit entirely the laying of a submarine pipeline on its CS/EEZ, with Article 79 making clear that the right of immersion enjoyed by all states is subject to the right of the coastal state only to take *reasonable* measures for the exploration of its CS and the exploitation of its resources, for pollution control, and to consent to the delineation of course of the pipeline. In the case of the Nord Stream project, refusal to grant permission to conduct surveys of the continental shelf as part of assessment of the feasibility of the Estonian route for the pipeline had the effect of ensuring that alternate routes, not so hampered, were pursued.[57] However, UNCLOS does not provide for direct recourse by Nord Stream against Estonia—this would have been a matter to be pursued at

[55] The extent to which transboundary challenges of domestic permitting conditions can be mounted depends also on domestic law. For example, the Estonian Naturalist Society filed a complaint in Denmark challenging the permit to construct the pipeline awarded pursuant to the Danish Continental Shelf Act, resulting in the first decision of the Danish Energy Board of Appeal of 31 May 2010 (upholding the permit to construct the pipeline). Here the domestic legislation was interpreted in the light of article 3.9 of the 1988 Aarhus Convention on Access to Information, Public Participation in Decision-Making and Access to Justice in Environmental Matters (the case turned wholly on environmental and standing issues).

[56] And such consent was obtained by Nord Stream from the three states where the pipeline transits the territorial sea (Russia, the state of export; Germany, the landing state, and Denmark, a transit state).

[57] See further Koivurova and Polonen below n 61.

Estonian law, for example—nor have any of the consequently affected Baltic states, hosts of the revised route, invoked inter-state dispute settlement proceedings under UNCLOS.

In addition to providing the general jurisdictional framework, the 1982 UNCLOS contains extensive treatment of protection of the marine environment in Part XII with provisions of potential application to the construction, operation and decommissioning of the Nord Stream pipeline, with obligations of states including to: protect and preserve the marine environment; prevent, reduce, and control pollution of the marine environment from any source, including measures necessary to protect and preserve rare or fragile ecosystems as well as the habitat of depleted, threatened, or endangered species and other forms of marine life;[58] prevent damage by pollution to other states arising from incidents under their jurisdiction and control; and notification, monitoring, and assessment obligations (though linked to 'substantial pollution of, or significant and harmful changes in, the environment').

These general rules are supplemented by the *lex specialis* for the Baltic region, the Helsinki Convention, which applies to the whole of the Baltic Sea Area, including the internal waters, territorial sea, and EEZ of the nine littoral states. Substantively it applies, *inter alia*, to the prevention of dumping (defined as 'deliberate disposal of... man-made structures at sea') and to prevent pollution from exploration and exploitation of its part of the seabed and subsoil or 'from any associated activities thereon'.[59] The Convention requires parties to prevent and eliminate pollution of the marine environment of the Baltic Sea Area caused by harmful substances from all sources, in order to promote the ecological restoration of the Baltic Sea Area and the preservation of its ecological balance (the Baltic Action Plan 2007 sets a restoration target date of 2021). 'Harmful substances' are defined in Annex I by reference to their intrinsic properties and characteristics liable to cause pollution, with a 'priority group' list including 'oils and hydrocarbons of petroleum origin'. 'Oil' is defined to mean 'petroleum in any form' and thus clearly includes gaseous petroleum. The Helsinki Convention requires the application of modern environmental principles such as the precautionary principle, best environmental practices (BEP), and best available techniques (BAT), and imposes the widely recognized obligation to prevent transboundary pollution. There are also assessment, notification and consultation, and reporting provisions. Additionally, the regulatory framework for environmental protection of the Baltic requires consideration of impacts not only upon the participating states but on all Baltic Sea littoral states party to the Helsinki Convention with Article 7 requiring notification and consultation where transboundary impacts in the Baltic Sea Area may occur. The Helsinki Convention thus clearly requires coordination with the Espoo Convention; in

[58] This obligation is reinforced by the marine and coastal biodiversity conservation provisions of the 1992 Convention on Biological Diversity, to which all five Baltic states, and the EC, are party.

[59] The Helsinki Convention thus provides regional reinforcement both of the general marine environmental provisions of UNCLOS and of the global prevention of dumping regulation of the 1972 London Convention, to which the five Baltic states are party. The 1972 London Convention is replaced by a 1996 Protocol, but only with respect to the parties to it (Finland and Russia are not).

turn, Article 8 of the latter explicitly envisages bilateral and multilateral arrangements in this regard.[60]

D. Environmental impact assessment

The Nord Stream project was the first occasion for consideration of a 'complex megaproject' under the 1991 Convention on Environmental Impact Assessment in a Transboundary Context ('Espoo Convention').[61] The Convention requires parties,[62] individually or jointly, to take all appropriate and effective measures to prevent, reduce, and control significant adverse transboundary environmental impact from activities subject to decision by a competent authority of the state party in accordance with national procedures (Article 2). Appendix I lists activities likely to cause such adverse effect, including large-diameter oil and gas pipelines. Appendix II sets out the assessment procedures to be applied, including public participation, preparation of the assessment documentation, and consultation requirements. Environmental assessment was thus required not only as a matter of the domestic law of all five Baltic states, but also as a requirement under Espoo, where the impacts on other littoral states, and of the project as a whole, were taken into account.

The Espoo process required cooperation over 3 ½ years, and generated an assessment document of over 2,500 pages.[63] One of the difficulties posed by the project was the implicitly limited transboundary notion at the heart of Espoo, ill-adapted to an offshore multi-jurisdictional project. An example is the Article 5 requirement of consultation regarding project alternatives, including no action, which if applied in isolation at the national level would be highly impracticable where the pipeline route crosses multiple jurisdictions.[64] The Parties chose to take a holistic, 'community' approach to the impacts across the region, and Nord Stream AG prepared an omnibus assessment covering all five jurisdictions with certain common elements—eg water quality, seabed sediment, hydrography and seabed topography, benthic flora and fauna, and cultural heritage impacts—as well as factors pertinent to only one or two states—eg landscape impacts in Germany and Russia. Here the Espoo Convention appears to have acted as 'an efficient instrument for intergovernmental cooperation', not to mention the harmonizing effect of domestic implementation of its provisions directly, and through the prism of

[60] Espoo Appendix VI stipulates the matters to be addressed by such arrangements, including impact assessment and monitoring, and the establishment of critical loads of transboundary pollution. This harmonization is particularly important for a sensitive regional ecosystem such as the Baltic regulated by special agreement, to ensure that the benchmarks employed under Espoo are not lower than those required of Parties under the Helsinki Convention.

[61] See generally <http://www.unece.org/env/eia> (accessed 19 October 2011). For detailed discussion see T. Koivurova and I. Polonen, 'Transboundary Environmental Assessment in the Case of the Baltic Sea Gas Pipeline' (2010) 25 *International Journal of Marine and Coastal Law* 151.

[62] Russia is a signatory, but not a party, to the Espoo Convention, but agreed to apply its terms in the impact assessment process so long as compatible with its domestic law.

[63] See <http://www.nord-stream.com> (accessed 2 October 2011).

[64] Exhaustive critique of the process is found in Koivurova and Polonen above n 61; see also Cho and Geelhoed, above n 47.

EU implementation for member states. A joint report by the Swedish and Danish Ministries of Environment concluded that 'without Espoo, large transboundary pipeline projects would suffer from chaotic, uncoordinated public participation and unclear permitting processes'.[65] Indeed, the Espoo Working Group on Environmental Assessment facilitates the exchange of good practices on large-scale TBPs.[66]

IV. Conclusions: A Tale of Two Pipelines

The legal rules applicable to TBPs are a complex mix of 'local' and international law.[67] Non-state actors—largely human rights and environmental non-governmental organizations (NGOs)—can play a significant role in highlighting concerns regarding the human rights and/or environmental impacts of the projects. Although not a dominant factor in either of the case studies considered here, external pressure may also come from international financial institutions (IFIs)[68] increasingly alive to issues such as the potentially negative impact of stabilization clauses on human rights[69] and the adverse environmental consequences of large energy infrastructure projects.

In the BTC case study, it is clear that one of the concerns of NGOs, and ultimately of the project investors, was state compliance with human rights obligations (whether vis-à-vis the investor—particularly economic and social rights—or the general population). Simma's proposed general solution to the perceived tensions between human rights and investment is a 'human rights audit'. The host state and investor, as part of due diligence, would conduct 'a proper assessment of the host state's pre-establishment regulatory information', which would build on voluntary corporate responsibility initiatives.[70] The audit would include a survey of the host state's human rights treaty commitments and methods for implementation

[65] Danish and Swedish Ministries of Environment, Experiences of Nordstream.

[66] See <http://www.unece.org/ea> (accessed 19 October 2011).

[67] See Dulaney and Merrick above n 13.

[68] International obligations may be imported indirectly through loan conditions eg of the World Bank Group. The EBRD and IFC provided finance to the BTC pipeline project, and there are suggestions that IFIs influenced the response by the host states and BTC Co. to criticisms of the original IGA/HGAs.

[69] See, for example, the research commissioned by the UN/IFC, Stabilization Clauses and Human Rights (2009); see also UNCTAD, *Selected Recent Developments in IIA Arbitration and Human Rights, IIA Monitor No. 2* (2009); and, generally, L.E. Peterson, *Human Rights and Bilateral Investment Treaties: Mapping the Role of Human Rights Law within Investor–State Arbitration* (Rights and Democracy, International Centre for Human Rights and Democratic Development).

[70] Above n 1 at 594. He distinguishes this audit from the more ambitious reach of 'human rights impact assessments': see, for example, S. Walker, 'Human Rights Impact Assessments of Trade-Related Policies' in *Sustainable Development in World Trade Law*, M.W. Gehring and M.C. Cordonier-Segger (eds), (Montreal: CISDL, 2005) 217–56 and T.F. Maassarani, M.T. Drakos, and J. Pajkowska, 'Extracting Corporate Responsibility: Towards a Human Rights Impact Assessment' (2007) 40 *Cornell International Law Journal* 135. On the distinct legal question of the human rights obligations of non-state actors, see A. Clapham, *Human Rights Obligations of Non-State Actors* (Oxford: Oxford University Press, 2006), in particular ch 6 (Corporations and Human Rights); see also P. Muchlinski, *Multinational Enterprises & The Law* (2nd edn, Oxford: Oxford University Press,

of such commitments (eg whether through transformation or incorporation[71] or processes of interpretation[72]) with access by the investor to the host state's periodic reports to human rights monitoring bodies and to their observations on state performance.[73] In an innovative blend of public international treaty, and foreign investment contractual provision, the objective of this audit is to ensure that the human rights obligations of the state become part of the applicable law. This in turn feeds into the scope of the legitimate expectations of the investor in terms of regulatory measures the state might take to implement human rights obligation, and ensures that host states 'would not be unduly constrained from defining their public policy agenda as a result of investment protection guarantees within foreign investment contracts and their corresponding treaties'.[74] It would also sit alongside the soft law norms of the voluntary Equator Principles[75] which project lenders—IFIs and large banking groups—may adopt[76] whereby they undertake to refrain from lending to projects where the investor/borrower will not, or is unable to, comply with stipulated international standards with respect to the environment and health and safety.[77]

Last but not least, there should be some form of ongoing internal monitoring and control mechanism 'such as regulatory compliance boards and voluntary contractual undertakings not to invoke broad stabilization clauses when these would prevent the host state from adopting regulations that implement obligations under

2007), which addresses both 'the social dimension' (labour relations, human rights, and environment) and 'control of investment risks' (contract stability, expropriation, non-discrimination, etc).

[71] That is to say, according to the domestic constitutional context, whether such treaty obligations are part of domestic law directly (incorporation) or require some act of domestic implementation (transformation).

[72] For example, the construction of domestic law so far as possible so as to conform to the international obligations of the state.

[73] Simma above n 1 at 594. [74] Ibid.

[75] These non-binding principles were adopted in 2003 and may be found at <http://www.equator-principles.com> (accessed 2 October 2011). They are based on the International Finance Corporation (IFC) performance standards on social and environmental sustainability (available at <http://www.ifc.org/ifcext/sustainability.nsf/Content/PerformanceStandards> (accessed 2 October 2011) and on the World Bank Group's Environmental, Health and Safety general guidelines (available at <http://www.ifc.org/ifcext/sustainability.nsf/Content/EHSGuidelines>, accessed 2 October 2011) and apply to all new Project Finance arrangements above US$10million. In common with many corporate social responsibility voluntary initiatives, it is up to financial institutions to adopt the Equator Principles framework, and some 72 institutions have done so, most recently (1 May 2011) the Ahli United Bank (Bahrain).

[76] I use 'adopt' here consistent with the Equator Principles' own terminology, while clearly acknowledging their non-legally binding status (as indeed are the performance standards on which the Principles are based). Of course, such principles/standards may be 'hardened' through loan conditionalities (contractual provisions) for example. For general assessment see: B. Richardson, 'The Equator Principles: The Voluntary Approach to Environmentally Sustainable Finance' (2005) *European Environmental Law Review* 280; ibid., *Socially Responsible Investment Law: Regulating the Unseen Polluters* (Oxford: Oxford University Press, 2008); and D. Ong, 'From "International" to "Transnational" Environmental Law? A Legal Assessment of the Contribution of the "Equator Principles" to International Environmental Law' (2010) 79 *Nordic Journal of International Law* 35.

[77] There is scope for mutual reinforcement in the project agreements: see, for example, the BTC Pipeline Code of Practice, above n 43. Indeed, Simma also observes that such an audit 'should not be seen as a radical departure from industry benchmarks in the international investment regime': n 1 at 594.

international human rights treaties'.[78] Again, the BTC case study is apposite, given the role of the Implementation Committee thereunder, and the Deed Poll by the project investor concluded subsequent to the initial project agreements following sustained criticism from NGOs in particular.[79] There are the distinct advantages of the flexibility, and speed, of voluntary contractual undertakings over revising the IGA/HGAs for example (as is the 2003 Joint Statement interpretative approach). Moreover, the BTC case is striking in the degree of disclosure (ie transparency) of concession documentation for public scrutiny.[80] Simma's purpose is clear: to integrate human rights considerations into the pre-establishment phase and as a complement to 'the post-establishment entry of human rights treaty norms through investment treaty interpretation', and to reduce the instances where investor protection and host state compliance with human rights norms are determined by the 'uncertain vagaries of mere "compensability" in the dispute settlement phase between host states and foreign investors'.[81]

The Nord Stream project is also ample illustration of the regulatory challenges of transboundary pipelines in the absence of overarching holistic regulation, in particular here of the environmental consequences of pipeline siting and construction. Rather, surveying, siting, and construction of submarine gas pipelines are regulated by wider, and diverse, environmental instruments addressed to marine pollution and conservation, habitat, and species conservation. The importance of the harmonizing role of an overarching project environmental assessment, and ongoing cooperation between the *lex specialis* for the Baltic marine region, the Helsinki Convention, and the ex ante procedural requirements of Espoo, are evident. There is clearly a strong, and necessary, role here for domestic and regional regulation, with international standards performing a more ephemeral, but nonetheless essential, 'back stop' function. Indeed, the unique circumstances of both case studies considered here underscores the doubt expressed at the outset regarding the utility of negotiating a generic pipelines treaty, and hence the inevitability of bespoke legal arrangements for such 'complex megaprojects'. Yet, in the face of diverse and potentially conflicting domestic laws, both case studies demonstrate the important harmonizing and benchmarking role of a wide range of existing international rules and standards through incorporation by reference in flexible contractual and treaty arrangements.

[78] Ibid at 596, relying *inter alia* on L. Cotula, 'Reconciling Regulatory Stability and Evolution of Environmental Standards in Investment Contracts: Towards a Rethink of Stabilization Clauses' (2008) 1 *Journal of World Energy Law & Business* 158–179.

[79] Cameron also cites the example of the benchmarking of evolving standards in the BTC case as an illustration of joint investor–state initiatives: above n 2 at 408.

[80] Transparency plays a more prominent role in the new generation of bilateral investment treaties (BITs): see, for example, article 19 of the 2004 Canadian Model BIT.

[81] Ibid. At this stage the capacity of, eg, NGOs directly to influence the process is limited by the extent to which 'participation' is permitted such as through the submission of amicus briefs: for recent discussion see Ishikawa above n 36.

7

Protecting Energy Infrastructure in the EU: The Impact of External Damages on Supply Security

Martha M. Roggenkamp

I. Introduction

The energy sector and the regular supply of energy resources involve different types of infrastructure such as onshore and offshore oil and gas production installations, electricity generators, and renewable energy installations like wind farms, but also above-ground and subsoil electricity and gas transmission and distribution lines and all kinds of ancillary installations like compressor stations and monitoring and controlling installations. Each type of infrastructure plays a specific role in securing energy supply.

Cables and pipelines are an essential element in achieving a reliable and secure energy supply, as these networks connect energy producers on the one hand and energy consumers on the other hand. The latter include large industrial consumers, small consumers, and household consumers. Generally the networks are divided into several categories. The first category includes the pipelines and cables connecting production facilities with the main grid. Whereas cables usually are considered as part of the production installation, a separate category applies to these pipelines as they generally are referred to as upstream pipelines. The next category involves the main grid, ie high-pressure and high-voltage networks. These so-called transmission (pipe)lines usually cover the entire territory of a state and are considered as the 'national energy highways'. Transmission lines are again connected to another—third—category of infrastructure, ie distribution lines, which operate under lower pressure and/or voltage and have a direct connection to consumers. Each category of networks can be governed by a different set of legislation.[1]

[1] See also the introduction to this book. If you have not read the introduction to get this background, you should!

When discussing the need to secure long-term energy supply, the focus is usually on the availability of primary resources like oil and gas and the need to have access to these resources. The further away these resources are located, the more complex such access will be from a technical but also from a legal point of view. Remote resources have another important drawback, which is that they need to be transported to consumer areas and this involves the need to construct an extensive system of energy networks.[2] Networks need to be maintained and be safe in order to avoid any supply interruptions. This chapter will concentrate on the latter aspect of supply security: the reliability of the networks, which again involves several different aspects. First of all, it implies the need to secure sufficient investments in the networks for reasons of maintenance and upkeep and any necessary network expansion. Second, it entails the need to protect networks from damages. Such damages can be unintentional as, for example, a result of construction activities, but also intentional as a result of an explicit plan to damage networks and society as a whole, ie a terrorist attack. This research will centre on the latter aspect: the possibility of damages to energy infrastructure and the laws and regulations available to avoid such damages. The focus will be on the situation in the EU. First, the chapter discusses the extent to which EU law provides instruments to protect energy networks from external disruptions and to reinstall energy supply in case of a major disruption. Thereafter it will discuss the regime in two member states, ie the Netherlands and Belgium.

II. EU Legislation

A. Introduction

Since the establishment of the European Economic Community (EEC) in 1957,[3] the process of European integration has gradually developed. The number of EU member states (MS) has increased from six to 27 and economic goals have been extended to include other policy objectives like environmental and consumer protection and external policy. The Community's main aim is still the need to create an internal market, ie an area without internal frontiers in which the free movements of goods, services, persons, and capital is ensured and without hindrance to competition. These principles also apply to the energy sector, as Article 194 of the Treaty on the Functioning of the European Union (TFEU)[4] now explicitly provides that the EU policy on energy shall aim, in a spirit of solidarity, to ensure the functioning of the internal market, security of supply, and the interconnection of national energy network systems. The latter is, amongst others, achieved

[2] The legal issues relating to the development of such large-scale grids are assessed elsewhere in this book. See, for example, chapter 6 by Catherine Redgwell and chapter 5 by Mohammed Bekhechi.

[3] Treaty of Rome, 25 March 1957.

[4] Following the Treaty of Lisbon of 2009 the Treaty establishing the European Community was amended and renamed as the Treaty on the Functioning of the European Union. See Official Journal of the European Union (OJ) 30 March 2010, C 83/01.

by a separate EU policy on the establishment and development of trans-European networks.[5]

Below we will discuss the way in which these goals have been transposed into secondary EU law and, in particular, how these measures address the reliability of energy networks. We will therefore focus on the impact of the liberalization process on the rules governing operation and maintenance of the grids and the extent to which EU laws provide for the protection of energy networks from external damages.

B. The operation of networks in the internal energy market

1. Directives governing the establishment of an internal electricity and gas market

The most important measures in the sphere of energy result from the policy to establish an internal energy market and the legislation deriving from that policy. Since the publication of the 1988 working document on the establishment of an internal energy market,[6] the energy sector has been confronted with radical changes. The most important and far-reaching changes involved the downstream electricity and gas sector and resulted from the implementation of directives on the internal electricity and gas markets issued in 1996, 1998, 2003, and 2009, respectively.[7]

In the Internal Energy Market (IEM) all consumers, suppliers, and producers of energy should have freedom of choice. Primarily, the IEM aims at providing all consumers the freedom to choose an energy supplier. As of 2007 all consumers have been given such choice and a complete market opening has been achieved. In addition, the Directives envisage freedom to generate electricity and gas (subject to some authorization and/or tendering procedures) and therefore exclude the possibility of establishing or maintaining production monopolies. As energy networks are considered natural monopolies, all market parties should have non-discriminatory access to the grid. In order to create such non-discriminatory access to the grid, network operators need to be independent from supply and production activities. The Directives therefore require unbundling of production, supply, and network activities.[8] As a result, MS are required to appoint independent transmission system operators (TSOs) and distribution system operators (DSOs) who are responsible for operating an adequate and reliable network system. Only at the upstream level is such unbundling not required.

The IEM Directives aim to regulate the networks, especially the use and access to the networks. The Directives generally provide that the TSOs/DSOs must operate,

[5] TFEU Art 170–172. [6] COM(88) 238 final.

[7] Directive 96/92/EC on the internal electricity market and 98/30/EC on the internal gas market were amended by directives 2003/54/EC and 2003/55/EC and more recently by directives 2009/72/EC and 2009/73/EC.

[8] See I. del Guayo, G. Kühne, and M.M. Roggenkamp, 'Ownership Unbundling and Property Rights in the EU Energy Sector' in A. McHarg, B. Barton, A. Bradbrook, and L Godden (eds) *Property and the Law in Energy and Natural Resources* (Oxford: Oxford University Press, 2010) 326–59.

maintain and develop, under economic conditions, secure, reliable, and efficient transmission/distribution systems.[9] It follows from this that the network operators are responsible for the grid and its maintenance and repair. The first Directives did not establish any conditions or prerequisites for construction and maintenance of the grids. Increasingly, however, provisions have been included regulating the quality and the standards of the network itself. As of 2009 the Directives explicitly state that MS shall ensure the application of technical safety criteria as well as technical rules establishing the minimum technical design and operation standards for the system.[10] Since the Directives merely refer to the need to apply minimum technical standards and do not elaborate on the requirements constituting adequate and reliable networks, it is clear that each MS is free to decide whether or not to issue further rules for the network operators to apply. In practice, the technical rules applying to the construction of networks are based on self-regulation and involve equipment standards (such as, for example, the type of steel used in gas pipelines), the depth at which pipelines need to be buried, and zoning requirements. Such standards are usually included in technical norms issued by organizations like the European Committee for Standardization CEN (Comité Européen de Normalisation), CENELEC (Comité Européen de Normalisation Electronique) or ISO (International Organization for Standardization). These technical standards are therefore also relevant for safeguarding supply security.

The 2009 Directives unequivocally require MS to monitor the quality and maintenance of the networks if necessary for security of supply.[11] Given the fact that the term security of supply is usually applied in the broadest sense, it would include the requirement to monitor the security and reliability of the grid itself. This is illustrated by the fact that MS may deviate from the general access regime by applying public service obligations involving, for example, maintenance and construction of energy networks.[12] In case of a sudden crisis in the energy market and where the physical safety or security of persons, apparatus, or installations or system integrity is threatened, MS may also take the necessary safeguard measures such as prioritizing certain consumers in grid access. Such measures must cause the least possible disturbance in the functioning of the internal market and must not be wider in scope than is strictly necessary to remedy the crisis.[13]

2. *Safeguarding electricity and gas supply*

As the EU liberalization policy does not necessarily run parallel with the need to secure energy supply, separate legal measures were taken in 2004/2005 and 2010 to stimulate MS to take the necessary steps to safeguard an adequate level of electricity and gas supply.[14]

[9] See, for example, arts 12 and 25 of Directive 2009/72/EC.
[10] Art 5 Directive 2009/72/EC and Directive 2009/73/EC.
[11] Art 4 Directive 2009/72/EC and Directive 2009/73/EC.
[12] Art 3 Directive 2009/72/EC and Directive 2009/73/EC.
[13] Art 42 Directive 2009/72/EC and art 46 Directive 2009/73/EC.
[14] Directive 2004/67/EC as replaced by Regulation EU 994/2010 and Directive 2005/89/EC.

Additional measures to safeguard an adequate level of security of gas supply were provided by Directive 2004/67/EC and were meant to apply in case of a 'major supply disruption', ie a situation where the Community would risk to lose more than 20 per cent of its gas supply from third countries and the Community—within the framework of solidarity—is unlikely to be able to manage adequately the disruption through national supply measures. Although the Directive provided a range of measures to ensure that MS had sufficient levels of gas supply, it did not contain any reference to the reliability of the networks and their role in supplying consumers with sufficient quantities of gas. By contrast, Regulation (EU) 994/2010 repealing Directive 2004/67/EC clearly recognizes the role of a reliable gas infrastructure to secure gas supply. MS must now ensure that all necessary measures are taken so that in the event of disruption of the single largest infrastructure, the capacity of the remaining infrastructure is able to satisfy total gas demand. In order to assess whether the remaining capacity is sufficient, the Regulation introduces the N-1 formula. According to the N-1 formula MS need to ensure that '[i]n the event of a disruption of the single largest gas infrastructure, the capacity of the remaining infrastructure [...] is able [...] to satisfy total gas demand of the calculated area during a day of exceptionally high gas demand occurring within a statistical probability of once in 20 years'.[15] As MS without sufficient infrastructure connection—so-called 'gas islands'—are the most vulnerable in case of a major gas supply disruption, the Regulation specifically requires member states to take all necessary measures, such as the construction of adequate interconnections. Other measures involve the establishment of storage facilities and/or demand-side measures, such as fuel switching at power plants.[16] The system operators are required to make the necessary investments by 3 December 2014.[17]

Security of supply and infrastructure investment in the electricity sector are governed by Directive 2005/89/EC of 18 January 2006, which aims at providing an adequate level of generation capacity, an adequate balance between supply and demand, and an appropriate level of interconnection between member states. In addition, it refers to 'operational network security' as the continuous operation of the transmission and, where appropriate, the distribution networks under foreseeable circumstances. Consequently, MS or the competent authorities must ensure (i) that TSOs set the minimum operational rules and obligations on network security; and (ii) that TSOs and, where appropriate, DSOs comply with the minimum operational rules and obligations on network security. In order to maintain an appropriate level of operational network security, TSOs shall maintain an appropriate level of technical transmission reserve capacity for operational network security and cooperate with the TSOs to which they are interconnected. Operational network security rules define the level of foreseeable circumstances in

[15] Art 6(1) Regulation 994/2010. Annex I of the Regulation provides for a calculation model and also defines what the largest infrastructure is, depending on the specific situation in each MS.

[16] Recitals 2 and 14 Regulation No. 994/2010.

[17] Art 6 Regulation 994/2010. See also S. Goldberg, 'Regulation 994/2010: A Measure to Improve the Security of Gas Supply in the EU?' in M.M. Roggenkamp and Ulf Hammer (eds), *European Energy Law Report VIII* (Cambridge: Intersentia, 2011) 59–90.

which security shall be maintained. MS shall, in particular, ensure that interconnected network system operators exchange information relating to the operation of networks in a timely and effective fashion in line with the minimum operational requirements. The same requirements shall, where appropriate, apply to operators of systems that are interconnected with networks outside the Community.

3. *Developing trans-European energy networks*

Interconnectors are crucial for both developing an internal energy market and securing energy supply. It is therefore not surprising that special provisions governing such infrastructure have been included in the EU Treaty by means of the introduction of a EU policy on the development of trans-European networks, *inter alia*, in the area of energy infrastructures.[18] Accordingly, within the framework of a system of open and competitive markets, the Community shall promote the interconnection and interoperability of national networks, as well as access to such networks. In order to achieve these objectives, the EU Treaty provides that the Community shall establish a series of guidelines covering the objectives, priorities and broad lines of measures envisaged in the sphere of trans-European networks. These guidelines shall identify projects of common interest, which may be supported by the Community. MS are also required to coordinate, in liaison with the Commission, their national policies having a significant impact on the achievement of the establishment of trans-European networks.

Consequently, the Community issued separate guidelines on the development of trans-European energy networks. The most recent guidelines were presented on 6 September 2006 following the entry into force of the 2003 IEM Directives and the need to integrate fully the new MS and the accession and candidate countries into the guidelines.[19] On the basis of those guidelines, the Community identifies projects of common interest. Such projects may include electricity and gas transmission grids and ancillary installations such as monitoring and control systems and storage facilities and, in addition, play an important role in creating an internal market and securing energy supplies, for example by strengthening relations with energy-producing countries outside the EU. The Community can even prioritize specific projects of common interest, if these projects solve problems of bottlenecks and missing links. Such priority projects can be declared to be of European interest. As a result, a European coordinator will be appointed who, amongst others, will promote the European dimension of the project, contribute to the coordination of the national procedures, and report annually to the Commission on the progress of the project(s) and on any difficulties and obstacles which are likely to result in a significant delay.

The policy on the development of trans-European energy networks aims at the development of (cross-border) infrastructure necessary to develop the internal

[18] The policy on trans-European networks was included in the Treaty of Maastricht in 1992 and provisions were included in article 3n and title XII of the Treaty (XVI TFEU).

[19] Decision no. 1364/2006/EC repealing decision 96/391/EC and decision no. 1229/2003/EC, OJ 2006, L 262/1.

energy market and supply security and does not provide for additional safeguards to protect these networks. Following the rules on subsidiarity, such safeguards would probably be considered as being part of the individual member states' competences.[20]

4. *The reliability of energy networks in the EU*

The above shows that the EU recognizes the need for good functioning energy infrastructure in order to establish an IEM and achieve supply security. A variety of instruments exist to develop and maintain such infrastructure, but little mention is made of the need to protect the existing infrastructure. Although the EU recognizes the need to safeguard supply security, the term 'supply security' is not clearly defined but could include measures preventing any activity threatening the supply of energy through gas and electricity grids. The reasons and background for such threat and/or disruption are indifferent and not an issue in the EU laws analyzed above, which basically charge the network operators with the task to secure the reliability of the networks. The situation changed in 2009 when the EU Commission issued a Directive providing for specific protection of some parts of the energy infrastructure from threats from terrorist attacks, as discussed below.

C. Regulating the protection of energy networks against major disruptions

1. *Background*

Following the terrorist attacks in the USA (2001) and Spain (2004), the Commission launched the idea of identifying and protecting critical infrastructure. Critical infrastructure was defined as:

> those physical and information technology facilities, networks, services and assets which, if disrupted or destroyed, would have a serious impact on the health, safety, security or economic well-being of citizens or the effective functioning of governments in the member states. Critical infrastructures extend across many sectors of the economy, including banking and finance, transport and distribution, energy, utilities, health, food supply and communications, as well as key government services.[21]

This idea and consequent EU policy resulted in Directive 2008/114/EC on the identification and designation of European critical infrastructures and the assessment of the need to improve their protection.[22] In contrast to the original idea, the Directive only applies to two specific sectors—transport and energy—and only as far as the infrastructure has an impact on the EU as a whole. Hence, the

[20] The principle of subsidiarity entails that the EU may act in areas which do not fall within its exclusive competence only if and insofar as the objectives of the proposed action cannot be sufficiently achieved by the member states, but can rather, by reason of the scale or effects of the proposed action, be better achieved at EU level (art 5 Treaty on the European Union).

[21] 'Critical Infrastructure Protection in the Fight Against Terrorism', Communication of 20 October 2004.

[22] OJ 23 December 2008, L 345/75.

introduction of a policy governing the identification and protection of 'European critical infrastructure' (ECI).[23]

2. *The concept and identification of European critical infrastructure*

In order to identify which parts of the energy infrastructure have a pan-national significance and can be considered as ECI, the Directive applies a two-step approach. The first step is to identify within each member state all critical infrastructures, being:

> an asset, system or part thereof located in member states which is essential for the maintenance of vital societal functions, health, safety, security, economic or social well-being of people, and the disruption or destruction of which would have a significant impact in a member state as a result of the failure to maintain those functions.[24]

The next step is to assess whether a critical infrastructure also can be a European critical infrastructure. For this purpose the critical infrastructure should be located in one member state but its disruption or destruction needs to have a significant impact on at least two MS. The Directive presents some specific criteria for assessing whether or not a critical infrastructure also qualifies as a European critical infrastructure.[25]

The first and easiest criterion to be applied is an assessment of the sector involved, ie transport or energy. As far as the energy sector is concerned, the Directive concentrates on the facilities necessary for the production and transmission of energy[26] and, by contrast to the 2005 Green Paper,[27] excludes distribution lines. This can be explained by the fact that distribution lines by nature do not have a cross-border effect.[28] In order to subsequently identify potential European critical energy infrastructure, the Directive then requires member states to apply three cross-cutting criteria:

- the number of casualties;
- the economic impact (including potential environmental effects); and
- the public impact (psychological and political impact).

The precise threshold of each criterion will be determined on a case-by-case basis. The Commission, together with the member states, shall develop guidelines for the application of the criteria and approximate thresholds to be used to identify ECIs. The criteria shall be classified and their use shall be optional for MS.[29]

[23] See also J.A. Hoyos Pérez, 'European Commission Initiatives for a Better Protected Energy System', in M.M. Roggenkamp and U. Hammer (eds) *European Energy Law Report V* (Cambridge: Intersentia, 2008) 129–39.

[24] Art 2 under a Directive 2009/114/EC.

[25] Art 3 and 4 Directive 2009/114/EC.

[26] See Annex 1 Directive 2008/114/EC.

[27] 'Green paper on a European Programme for Critical Infrastructure Protection' of 17 November 2005, COM(2005) 576 final.

[28] Offshore upstream pipelines can cross several national borders.

[29] Each member state shall inform the Commission on an annual basis of the number of infrastructures per sector for which discussions were held concerning the cross-cutting criteria thresholds.

In applying the sectoral and cross-cutting criteria, MS should be able to identify potential ECIs. The Commission can be involved in this process either as a result of a request from a MS for assistance or by drawing the attention of the relevant MS to the existence of potential critical infrastructure that may be deemed to satisfy the requirements for designation as an ECI.[30]

MS were required to implement the Directive and identify potential ECIs by 12 January 2011.[31] Within one year following the designation of an ECI each member state shall conduct a threat assessment.[32] In addition, each member state shall provide every two years to the Commission a brief report presenting generic data on the types of risks, threats, and vulnerabilities encountered per ECI sector.[33] Reports shall be classified as deemed necessary by the member state. Based on these reports, the Commission and the member states shall assess whether further protection measures at the Community level should be considered for ECIs.

3. Cooperation between member states

Given the fact that infrastructure can only be designated as an ECI if disruption or destruction has a significant impact on at least two member states, any designation by one MS will have an effect on another MS. The Directive therefore includes a procedure of information sharing and mutual recognition.

Each MS shall start with informing potentially affected MS that an ECI may be designated and provide the reasons for designating it as such. All MS involved shall then engage in bilateral and/or multilateral discussions. The Commission may participate in these discussions but shall not have access to detailed information which would allow for the unequivocal identification of a particular infrastructure. A MS that has reason to believe that it may be significantly affected by a potential ECI designation, but has not been approached by the MS on whose territory the infrastructure is located, may inform the Commission about its wish to be engaged in bilateral and/or multilateral discussions. The Commission shall without delay communicate this wish to the MS on whose territory the potential ECI is located. These discussions need to result in an agreement between the MS involved. Without such an agreement, ie the acceptance of all MS involved, an ECI cannot be designated as such. The MS on whose territory a designated ECI is located shall inform the Commission annually on the number of designated ECIs and on the number of MS involved in each designated ECI.

The identification of an infrastructure as ECI implies that the infrastructure is essential for supply security and therefore needs some special protection. Consequently, the designation of ECIs is considered as classified information. The Commission may therefore participate in the discussions between MS but is not

[30] Article 3 Directive 2009/28/EC. Each member state and the Commission shall continue on an ongoing basis the process of identifying potential ECIs.

[31] Thereafter they will review regularly the list of identified objects. See below section III A.

[32] Article 7 Directive 2009/114/EC.

[33] Common methodological guidelines for carrying out risk analyses in respect of ECIs may be developed by the Commission in cooperation with the member states. The use of such guidelines shall be optional for the member states.

entitled to be informed of the identity of an ECI. Similarly, only those MS that may be significantly affected by an ECI shall know its identity and are allowed to inform the owner/operator of the infrastructure concerning its designation as an ECI.[34]

4. Impact on infrastructure owners and operators

The designation of an ECI may have some consequences for the owner/operator of the infrastructure. The Directive defines owners/operators of ECIs as

> those entities responsible for investments in, and/or day-to-day operation of, a particular asset, system or part thereof designated as an ECI.

In case of energy networks, the focus will mainly be on TSOs. Depending on the situation in each MS, the TSO can either be completely unbundled or legally unbundled, as a result of which the investment decisions can be separate from the day-to-day operation. By referring to both operators and owners, the Directive seems to acknowledge such a distinction.

TSOs operating an ECI have to establish an 'operator security plan' (OSP), which needs to be reviewed by the MS within one year following designation of the critical infrastructure as an ECI.[35] Each MS shall assess whether the designated ECIs located on its territory possesses an OSP or has in place equivalent measures. If a MS finds that such OSP or equivalent exists, no further implementation action is required. The aim of these OSPs is to identify the critical infrastructure assets of the ECI and which security solutions exist or are being implemented for their protection, ie all activities aimed at ensuring the functionality, continuity, and integrity of critical infrastructures in order to deter, mitigate, and neutralize a threat, risk, or vulnerability.[36] The Commission may support the owners/operators of designated ECIs by providing access to available best practices and methodologies, support training, and the exchange of information on new technical developments relating to critical infrastructure protection.[37]

In order to liaise with the operators/owners of an ECI, each MS will be required to appoint a Security Liaison Officer (SLO) unless such an officer already is in place or an equivalent exists. The main task of the SLO is to act as the point of contact for security-related issues between the owner/operator of the ECI and the relevant MS authority.[38] For this purpose each MS shall implement an appropriate

[34] Art 4 Directive 2009/114/EC.

[35] See art 5 Directive 2009/28/EC. This period may be extended under exceptional circumstances by agreement with the member state authority and with notification to the Commission. Where supervisory or oversight arrangements already exist in relation to an ECI such arrangements are not affected by this article and the relevant MS authority referred to in this article shall be the supervisor under those arrangements. Compliance with measures including Community measures which in a particular sector require or refer to a need to have a plan similar or equivalent to an OSP and oversight by the relevant authority of such a plan, is deemed to satisfy all the requirements of MS under, or adopted pursuant to, that article. The guidelines for application referred to in article 3(2) shall contain an indicative list of such measures.

[36] The minimum content to be addressed by an ECI OSP procedure is set out in Annex II.

[37] Art 8 Directive 2009/114/EC.

[38] Art 6 Directive 2009/114/EC.

means of communication between the relevant MS authority and the SLO with the objective of exchanging information concerning identified risks and threats involving an ECI. In addition to a SLO, each MS shall appoint a European critical infrastructure protection contact point (ECIP contact point). The ECIP contact points shall coordinate European critical infrastructure protection issues within the MS, with other MS, and with the Commission.[39]

5. *Confidentiality of information*

The designation of infrastructure as ECI implies that all information about such critical infrastructure, by its nature, is sensitive. Information about a critical infrastructure is sensitive if when disclosed it could be used to plan and act with a view to causing disruption or destruction of critical infrastructure installations. The designation of infrastructure as being critical therefore has some consequences.

First, the designation of ECIs is considered as classified information. Hence, any person handling classified information pursuant to this Directive on behalf of a MS or the Commission shall have an appropriate level of security vetting.[40] MS as well as the Commission and relevant supervisory bodies shall ensure that written and non-written sensitive information concerning the protection of ECI submitted to the MS or to the Commission is not used for any purpose other than the protection of critical infrastructures.

Second, not all information on ECIs will be made known. MS's reports on the types of risks, threats, and vulnerabilities encountered per ECI sector are classified at the level of confidentiality deemed necessary by the originating MS. The Commission may, for example, participate in the discussions between MS on the designation of ECIs but is not entitled to be informed of the identity of an ECI. Similarly, only those MS that may be significantly affected by an ECI shall know its identity and are allowed to inform the owner/operator of the infrastructure concerning its designation as an ECI.

D. Protecting the networks and securing energy supply in the EU

The EU acknowledges that energy networks are essential for creating an internal energy market and securing long-term energy supply. The existing legal regime governing the establishment of an IEM, trans-European networks, and security of energy supply focuses on the need to develop energy networks in order to achieve these goals. Once energy networks have been developed, the TSOs and DSOs are charged with the task to provide consumers with regular and reliable energy supplies. In case of a sudden crisis, however, a MS may take the necessary safeguard measures resulting in voluntary and mandatory consumption reductions and/or taking control of generating resources. The need for a proper back-up system is acknowledged by Regulation 994/2010, *inter alia*, introducing the N-1 rule as a

[39] Art 10 Directive 2009/114/EC.
[40] Art 9 Directive 2009/114/EC.

result of the application of which the disruption of one national pipeline can no longer jeopardize national gas supply.

Although relevant, these EU laws do not include any provisions regarding the need to protect the grid against external damages. The EU legislator probably considers this as a general task of the TSOs and DSOs and part of their obligation to maintain the grid and secure energy supply. Due to the subsidiarity principle any rules governing infrastructure protection and external safety will be issued at the national level. However, the number of technical and safety standards drafted at the EU level may imply that safety issues concerning the construction of safe networks are further harmonized than may be expected at first glance. Additional EU and national safety measures may apply as a result of the entry into force of Directive ECIP, albeit this is limited to some specific designated energy infrastructure, ie transmission grids and interconnectors which, if damaged, would have a cross-border impact.

So far, major accidents and disruptions affecting cross-border energy supply have been more or less zero. This does not mean, however, that accidents do not take place. Since the EU gas and electricity transmission systems have a length of approximately 215,000km and 305,000km respectively, and the distribution grids are far more extensive,[41] there is a good chance for damages to take place. It is also acknowledged that the main threat to underground gas pipelines is caused by third-party damages.[42] Such damages can be caused by digging and construction activities and can be avoided if the precise location of the infrastructure is publicly known. In this case, transparency may be more important than confidentiality. Be that as it may, in case of such damages the TSOs and DSOs that are responsible for maintaining the grid are also responsible for repairing the grid and reinstalling supply.

As important parts of infrastructure protection are dealt with at the national level, the next part of this chapter will present some examples of the ways in which MS regulate the protection of energy networks.

III. Network Protection at the National Level

A. Implementation of EU law

The above-mentioned Directives on the IEM and supply security need to be implemented in national law as, for example, an Electricity Law, Gas Law, or Energy Law. The ECIP Directive had to be implemented before January 2011 and the picture so far is diverse. Some MS have enacted a special ECIP law and others have argued that sufficient equivalent measures exist and a special implementation process is not necessary. MS have ample room to design and apply laws if deemed necessary to protect energy networks.

[41] In 2005 the gas distribution grid in the EU was estimated to have a total length of 1.444.400km. See presentation by Daniel Hec, Marcogaz, Technical Association of the European Natural Gas Industry, 'European Pipeline Safety Regulations and Standards', Geneva, 22–23 January 2008.

[42] Approximately one accident per 5000km pipeline takes place each year. Ibid.

By contrast to the ECIP Directive, these national approaches often aim at protecting the final and household consumers from supply interruptions. Such interruptions may often be the result of unintended damages due to construction works, storms, etc. The effects can be quite severe and resulting in blackouts/brownouts lasting several hours or even days.

Below we will present two case studies involving the Netherlands and Belgium. The reason for choosing these countries is, amongst others, that they are linked through several cross-border gas and electricity networks and as result damages to the Dutch grid may have an effect on energy supply in Belgium and vice versa. Moreover, in 2004 Belgium experienced a major accident as a gas pipeline near the town of Ghislenghien (Gellingen) was damaged due to construction activities nearby.[43]

B. The Netherlands

1. Integrity and reliability of energy networks

Obviously, the integrity and quality of the grid is an important aspect of supply security. Network operators are basically responsible for the operation and the reliability of the grid.[44] These grid management tasks are carried out on the basis of Technical and Tariff Codes issued by the energy regulator (the office of energy and transportation of the competition authority).[45]

The Tariff Code determines the way in which grid access tariffs are calculated. These tariffs are based on the CPI-X and Q formula. Within this formula the X-factor refers to a discount, which intends to promote efficient operation by the grid operators.[46] The Q-factor was added to the tariff formula in 2005 and represents the need that network operators also guarantee the quality and reliability of the grid. Each network operator is required to issue regularly a quality plan indicating the level of reliability they intend to achieve.[47] The tariff revenues are meant to maintain the grid. If network operators use the revenues for other purposes and/or do not invest in the upkeep of the networks, the Minister may decide to withdraw approval, with the appointment of another company as network operator.[48]

The Technical Codes regulate a range of issues and instructions governing the cooperation between grid operators, balancing of the grid, metering, grid connections, and a range of safety requirements. These codes and regulations do not fully guarantee the reliability of the grid, as external damages are usually the main reason for supply disruption.

[43] The incident took place on 30 July 2004 and resulted in 24 deaths.
[44] Art 16 Electricity Act and art 10 Gas Act.
[45] See chapter 3 Electricity Act and chapter 2 Gas Act.
[46] Art 41 Electricity Act 1998 and art 80 Gas Act. See for a more detailed explanation of the CPI-X formula also chapter 17 of this book.
[47] *Tweede Kamer* 2002–2003, 29 023, nr 1, Annex 2, p. 10.
[48] Art 13 Electricity Act. A similar provision is not included in the Gas Act.

2. *Damages to the grid*

Supply interruptions are often the result of damage to energy infrastructure like transformer stations and subsoil cables and pipelines. In the Netherlands some 1.25 million kilometres of cables and pipelines are buried in the subsoil. Considering the amount of subsoil infrastructure and the fact that the Netherlands is a densely populated country, it is not surprising that subsoil infrastructure is easily damaged. Per annum some 40,000 incidents are registered and the repair costs only are estimated at approximately 40 to 75 million Euros.[49] One important reason for these damages is that the precise location of the subsoil infrastructure is not clear. Construction activities may therefore result in damages to subsoil cables and pipelines and thus supply interruptions. Such situations could be avoided if the exact position of the infrastructure were known.

Since the 1960s attempts have been made to organize some sort of registration of subsoil cables and pipelines. The discussion concentrated on the questions (i) what to register; and (ii) how such registration should be organized. The end result was a self-regulation regime providing for network operators to be registered instead of the networks. For this purpose several provincial cable and pipeline registration offices were established and later merged into one organization: *Kabels en Leidingen Informatie Centrum* or KLIC. All contractors and builders were supposed to approach these offices for information about the network companies operating in the area. Subsequently these companies could provide the contractors with detailed information on the subsoil infrastructure before the start of activities. In case they had not made such inquiries and damages arose, the contractors or builders were usually found liable by the courts.[50]

In practice this registration system was not considered satisfactory as neither network companies nor the contractors were obliged to involve the register. In order to decrease the number of incidents and thus improve the reliability of supply, the Act on Information Exchange Concerning Subsoil Networks (WION) entered into force on 1 January 2008.[51] This new regime is to a large extent similar to the previous one but now has a legal basis and is no longer optional. In addition, the KLIC is now part of the official land register (*Kadaster*). As before, any person intending to undertake digging or excavation activities is required to approach the register/KLIC to get information on the network operators in the area. They then need to approach the network operator(s) for additional information on the possible location of the subsoil grid. They also have to carry out tests before actually starting the activities. If they do not follow these instructions/guidelines and damage occurs, they are likely to be held liable. This liability may not only include damages to the network itself but also other resulting damages, including the compensation the

[49] See, amongst others, information on the new Act published by the Ministry on Economic Affairs at <http://www.igl.nl/downloads/grondroerdersregeling.pdf> (accessed 2 October 2011).

[50] See also F.J. van Velsen, 'Grondroerdersregeling: Chaos in de ondergrond werkt door in wetsvoorstel' (2006), 5 *Bouwrecht* 424–31

[51] *Wet informatieuitwisseling ondergrondse netten*, 1 January 2008, *Staatsblad* 120.

network operators are required to pay consumers if a supply disruption lasts longer than four hours.[52]

3. *European Critical Infrastructure Protection (ECIP)*

a. The concept of vital infrastructure

The Netherlands recognized in 2001 the need to protect important infrastructure against terrorism. In a subsequent 'Action plan on safety protection and against terrorism' it determined that infrastructure is vital if a disruption would result in economic or social unbalance, many casualties (directly or indirectly) or long-term impact due to time-consuming repairs and lack of alternatives during the period of repair. On the basis of these criteria the government identified 12 vital sectors, including energy (electricity, gas, oil, and nuclear), telecommunications/ICT, water supply, and transportation. Thereupon a more detailed risk-analysis was completed for each sector.

It appeared from the risk-analysis that the level of security and reliability of the energy sector was relatively high. Some safety issues were nevertheless identified such as third party access to information due to outsourcing, the need to protect specific objects, and the extent to which free access should be possible to vital parts/installations of the energy infrastructure.[53] This resulted in some recommendations involving a reassessment and further harmonization of the existing security and safety policies, the possibility of closing the air space above certain vital locations, and the improvement of the emergency communication system.

In addition some changes were made to the organizational structure. The Minister of Internal Affairs was appointed as the core department responsible for a) the coordination of all measures concerning vital infrastructure; and b) public order and safety together with the Ministry of Defense. As a separate agency of the Ministry of Internal Affairs, the General Intelligence and Security Service (AIVD) is charged with the task to gather intelligence and to promote security measures. In addition, new organizations have been established to adequately protect vital infrastructure.[54] These include the National Advice Centre on Vital Infrastructures (NAVI)[55] and the Strategic Consultation on Vital Infrastructure (SOVI), which is charged with improving communications between the various vital sectors.[56] Besides, the National Coordinator for Counterterrorism (NCTb)

[52] The Supreme Court of the Netherlands ruled on 29 April 2011 that a constructor needs to compensate a DSO for the payments made by the DSO to household consumers after a power failure of more than four hours. Although the DSO was legally obliged to compensate the households, the constructor was to be blamed for the damages to the grid and thus the power failure. See Raad 29 April 2011, 10/02729, *LJN* BQ2935.

[53] Ministerie van Binnenlandse zaken en Koninkrijksrelaties, *Bescherming Vitale Infrastructuur*, 2005, p 14.

[54] *Tweede Kamer* 2006–2007, 26 643, nr 83.

[55] Letter of the Minister of Internal Affairs to the Parliament, Pilot *Nationaal Adviescentrum Vitale Infrastructuren* (NAVI), 8 November 2006.

[56] Instellingsbesluit Strategisch Overleg Vitale Infrastructuur (SOVI), *Staatscourant* 2006, nr 81.

is responsible for coordinating anti-terrorist security measures.[57] Last but not least, the Counter Terrorism Alert System came into effect in December 2005 as an instrument to warn governmental services and economic sectors of a heightened threat so that measures can be taken quickly in order to minimize the risk of terrorist attacks.[58]

b. Implementation of ECIP Directive

On 20 December 2010 the Dutch government published in the Official Gazette an announcement concerning the implementation of the ECIP Directive.[59] Taking into account the above developments regarding the protection of vital infrastructure, it was considered unnecessary to enact special legislation as equivalent solutions are in place. Instead, the implementation is based on a policy plan identifying the steps to designate ECI as described in the Directive.[60] A crucial element in the plan is the establishment of an additional organization—the CIP contact—under the Ministry of Internal Affairs and consisting of all relevant ministries involved.

Electricity and gas grids were already designated in 2001 as vital infrastructure. In 2005 the General Intelligence and Security Service (AIVD) identified two major threats: terrorism and espionage. The more recent 2010 risk assessment provided little information on risks to the energy sector and limited its analysis to risks to electricity supply. In the meantime the ECIP Directive has entered into force and consequently an assessment has to be made on the possibility that vital infrastructure may be classified as European critical infrastructure

4. Protection of infrastructure versus protection of information

Electricity and gas infrastructure are vital for providing regular and reliable energy supply and therefore need to be protected from external damages, intentional or unintentional. Unintentional damages are usually the result of construction and building activities and can be addressed by providing information on the exact routes and locations of the subsoil infrastructure. An important dilemma in making such information public is, however, that it would also increase the risk of terrorism or sabotage, especially when such installations are at the core of the economy. The solution chosen in WION to supply information on network operators rather than networks may therefore be a suitable compromise between the need for confidentiality and transparency.

[57] 'Instellingsregeling Nationaal Coördinator Terrorismebestrijding (NCTb)', *Staatscourant* 2005, nr 127. The NCTb falls under the joint responsibility of the Ministers of Justice and Interior Affairs. The Minister of Justice has so-called 'Extended Authority' and may in the event of acute threats of terrorism take measures involving other ministers.

[58] *Tweede Kamer* 2005–2006, 29754, nr 73.

[59] *Staatscourant* 24 December 2010, no. 20996.

[60] *Implementatieplan voor EPCIP-richtlijn* of 16 December 2008 published as an Annex to a letter from the Minister of Internal Affairs to the Chairman of the Lower House of Parliaments of 4 February 2009, *Tweede Kamer* 20008–2009, 22112, no. 793.

C. Belgium

1. Introduction

The Belgian and Dutch energy infrastructures are coupled by several cross-border interconnections. These interconnections were made as early as the 1960s following, amongst others, the gas export contracts concluded after the discovery of the Groningen gas field. In Belgium the gas and electricity transmission systems are operated by national companies—Fluxys and Elia—and governed by federal legislation. In addition, there are several regional distribution grids governed by regional energy laws such as the Flemish energy decrees. TSOs and DSOs are responsible for the operation and the reliability of the grid. The regulatory regime is to a large extent similar to the Dutch one, ie based on regulated TPA, detailed tariff regulation, and technical codes supervised by energy regulators (at national and/or regional level).

2. Damages to the grid

As in the Netherlands, construction activities may cause damages to subsoil energy infrastructure. In Flanders, for example, the total length of subsoil infrastructure is at least 500,000km and each day some 90 incidents take place.[61] The above-mentioned incident near the Walloon town of Gellingen (*Ghislenghien*) in 2004 is a clear example of the devastating effects such construction activities can have. In this case an excavator damaged a gas pipeline operated by Fluxys. A pressure increase some weeks later resulted in a rupture of the pipeline followed by a major gas explosion and fire. The disaster resulted in major damages, 132 casualties, and 24 deaths. How could this happen? The contractor was, after all, obliged to investigate the possible presence of gas pipelines in the area of excavation. As a general rule, if the existence of a pipeline becomes known, the pipeline operator has to be informed and excavation near the pipeline is prohibited. It appears, however, that in this particular case *proper* plans of the subsoil gas grid were missing. In the following court procedure, the Belgian Court of first instance decided on 22 February 2010 that the constructor was to be held liable, as previous works in the area should have clarified that it was better not to use an excavator. The Court of Appeal in June 2011 arrived at completely the opposite conclusion and held Fluxys liable because of negligence and violation of safety measures.[62]

This accident led to a review of the applicable legislation. It was found that the complex and non-transparent procedures for getting information on the location of networks and the lack of reliability of the information itself were the main reasons for the frequent damages. Therefore a new approach was introduced based on the one-stop-shop principle of the Dutch regime described above. In 2008 the government introduced an electronic system providing (access to) information

[61] See brochure 'graafschade' available at <http://www.blankenberge.be/01/MyDocuments/KLIP_20090309.pdf> (accessed 2 October 2011). Exact figures about the total length of cables and pipelines in the Walloon region do not seem to be available.

[62] Fluxys decided on 11 July 2011 to put the case before the highest appeal court.

on subsoil infrastructure.[63] In fact, there are two systems: one at the federal level (KLIM) and one at the regional Flemish level (KLIP).[64] As of 1 June 2009 any person who wishes to carry out construction activities can register at the KLIP website and locate the network company operating in a specific area. An announcement is sent electronically to the network company, which then checks the presence of any subsoil networks. If some exist, the network company sends a copy of the network construction plan to the interested person. Excavation may start upon receipt of these plans.

An important element in establishing this online information system was the initiative taken by the utility companies as early as 1998 to develop a digital topographical map (*Grootschalig referentie bestand*) indicating the exact location of subsoil infrastructure. A legal basis was created by Decree in 2004, including the establishment of a separate Agency on Geographical Information in Flanders.[65] The aim is that for the entire region of Flanders online topographical information will also be available in 2015. The current registration procedure via KLIP takes into account the fact that not all information is available online yet.

3. The ECIP Directive

The Belgian government decided to implement the ECIP Directive in a separate Act on Safety and Protection of Critical Infrastructure, which entered into force on 15 July 2011.[66] The Act goes beyond the Directive as it covers two additional areas: finance and ICT.

The Act defines three types of infrastructure: critical infrastructure (national infrastructure which is essential and if damaged has an impact on vital parts of society), national critical infrastructure (critical infrastructure which is essential for the country as a whole), and European critical infrastructure (national critical infrastructure which if damaged has an impact on at least two member states of the EU). The national government is required, on the basis of preset selecting criteria, to designate these three types of infrastructures. The procedure for designating national and European critical infrastructures is included in a separate annex to the Act. The operators of the designated critical infrastructures are required to appoint a safety coordinator and draft internal safety plans describing the degree of protection needed.

Two national organizations are charged with tasks involving the selection and protection of critical infrastructures: the General National Crisis Centre of the federal Ministry of Internal Affairs (*Algemene Directie Crisiscentrum* or ADCC) and the Coordinating Body on Threat Analysis (*Coordinatieorgaan voor de dreigingsanalyse* or OCAD). The ADCC is also appointed as the Belgian ECIP contact

[63] Decree of 14 March 2008, *Belgian Official Journal* of 6 May 2008 as amended on 10 December 2010, *Belgian Official Journal* of 24 December 2010.

[64] Whereas KLIM is an abbreviation for 'Kabels en Leidingen Informatie Meldpunt' and involves information on transmission lines, KLIP is short for 'Kabels en Leidingen Informatie Portaal' (contact points for information on cables and lines) and provides information on distribution networks.

[65] *Decreet houdende het Grootschalig Referentie Bestand* of 16 April 2004, BS 5 July 2004.

[66] Belgian Official Journal of 15 July 2011, no. 1799, pp 42320–8.

for liaising with the European Commission and other EU member states. The ADCC and the OCAD must operate in close cooperation with the police and gather all information necessary to protect critical infrastructure. On the basis of threat analysis the ADCC may decide that additional external safety measures are required. For each designated sector a separate inspection will be undertaken to supervise compliance of the Act.

D. Comparison

The Netherlands and Belgium can be compared with each other for several reasons. The neighbouring countries have both densely populated areas near the coast, major harbours (Rotterdam and Antwerp), and a comparable terrain—subsoil—as a result of which the energy transmission and distribution networks are below ground. In addition, both countries are connected through several cross-border electricity and gas interconnections, of which the latter are of specific importance for large imports of natural gas from the Netherlands and for securing national energy supply.

Both countries are faced with similar problems in relation to protecting the infrastructure from external damages, most of all damages following construction works. Both countries apply a similar solution in the sense that one point of contact is established which all persons who wish to start construction activities affecting the subsoil need to approach. The way in which the Dutch KLIC and Belgium's KLIP (or KLIM) operate are largely similar. Two major differences can be noted. First, in the Netherlands there is a legal obligation to approach KLIC on the basis of the WION Act. That is not the case in Belgium. Second, in Belgium the parties acquiring information on subsoil infrastructure may also have access to electronic topographic maps showing the exact location of the infrastructure, as a result of which the information is more transparent but could in principle also be used to intentionally damage the infrastructure instead of protecting it.

Both countries had to implement the ECIP Directive but applied two opposite implementation processes. Whereas the Netherlands decided to make use of the existing organization and implement it through a policy plan, Belgium decided to implement it on the basis of a separate Act. Although the outcome of both procedures may be the same, the Dutch procedure is not very transparent. Apart from this, both regimes are based on the concept that national vital or critical infrastructure should be designated first and be used as a basis to designate European critical infrastructure. Whether or not such infrastructure has been designated is not known because such procedure and designation is considered confidential.

Given that the Netherlands is a major gas supplier in the EU it may well be that serious damages to some of its major gas supply systems will have an impact on gas supply in neighbouring countries like Belgium, Germany, and even France. However, as the gas supply is not limited to one gas transmission line and one cross-border connection, the complete system needs to be damaged in order to have such an effect. The accident in Ghislenghien seems to illustrate that a major disruption of a gas transmission grid results not necessarily in disruptions of gas supply. It is therefore not very likely that the electricity and gas grids in the Netherlands and Belgium, although being considered as national vital infrastructures, will also be labeled as European critical infrastructure. The requirement to apply the N-1 formula will further guarantee that parallel supply grids will be in place and thus that there most likely will be few European critical energy infrastructures designated.

IV. Conclusion

A regular and reliable supply of electricity and gas is crucial in our current society. Such supply depends on reliable infrastructure. The reliability of the energy infrastructure depends on the use of proper material, upkeep, and investments in the grid. It also depends on guaranteeing that the infrastructure is not damaged. In order to avoid such damages and protect citizens from the consequences of possible damages, the grid operators are required to apply rules on external safety. These rules are often based on European or international standards (CEN, CENELEC, or ISO standards).

It is particularly necessary to protect networks from damages caused by construction activities. In the EU rules protecting subsoil infrastructure are a matter of national law. Due to the large number of accidents there is a tendency to create more transparency so that any person who starts activities affecting the subsoil knows beforehand where pipelines and cables are situated. This tendency towards more transparency may, however, interfere with the need to keep information on European critical infrastructure confidential. Nonetheless, this may be countered by the fact that the policy to establish more cross-border interconnections and supply routes may result in a limited need for a policy on European critical infrastructure.

PART II

NEW ENERGY SOURCES AND INNOVATIVE NETWORK MANAGEMENT

8

Smart Grids and Intelligent Energy Systems: A European Perspective

Anita Rønne

I. Introduction

To achieve the goals set up by IPCC[1] of holding the increase in global temperature below two degrees Celsius,[2] to meet the challenges of an ageing infrastructure and increasing demand for electricity, and to achieve the CO_2 and green energy targets set by the European Union (EU)[3] and many national governments,[4] there is a need for developing energy systems that are based on more flexible, innovative, and interactive solutions.

The EU has as part of its action on climate change agreed on a reduction of CO_2 emissions by 20 per cent in 2020 (and may go beyond that target up to 30 per cent) combined with a target for renewable energy at 20 per cent in 2020, including a specific provision for a minimum of 10 per cent renewable energy in transport to ensure that targets are not only addressed in electricity and heating, and finally an indicative target for improvement of energy efficiency at 20 per cent also by 2020.[5]

[1] Intergovernmental Panel on Climate Change. See <http://www.ipcc.ch> (accessed 28 September 2011).

[2] Latest recognized in the United Nations Climate Convention's Copenhagen Accord and Cancun Agreements. See further <http://www.unfccc.int> (accessed 28 September 2011).

[3] The 20–20–20 targets as put forward in the EU Climate and Energy Package from March 2007 and implemented by law in 2009 and the 'Energy 2020—A strategy for competitive, sustainable and secure energy', COM(2010) 639.

[4] Department of Energy & Climate Change, 'UK Renewable Energy Roadmap', July 2011; The Energy Agency. 'Denmark: Energy Strategy 2050—From Coal, Oil and Gas to Green Energy', 5 March 2011; 'Target of 67.5 per cent for Norway's renewable energy share by 2020', Press release of 21 July 2011, available at <http://www.regjeringen.no/en/dep/oed> (accessed 28 September 2011).

[5] See above n 3 and A McHarg & A Rønne, 'Reducing Carbon-based Electricity Generation: Is the Answer Blowing in the Wind?', in D. Zillman, C. Redgwell, Y. Omorogbe, & L.K. Barrera-Hernández (eds), *Beyond the Carbon Economy—Energy Law in Transition*, (Oxford: Oxford University Press, 2008) and H.T. Anker, B.E. Olsen, and A. Rønne (eds) *Legal Systems and Wind Energy*, (Copenhagen: DJØF Publishing, 2009).

Many other regions and countries—eg the US and China—have adopted equivalent ambitious targets and are investing heavily in clean technologies and reinventing their energy system on a low-carbon model.[6] It is beyond doubt that renewable energy resources once insignificant are gradually expanding their role in global energy supply.

The design of the future grids must thus enable the integration of substantially higher amounts of fluctuating energy that are depending upon climatic conditions while ensuring energy security. Moreover, energy systems must be able to accommodate the use of new production and end-use technologies. An increasing number of smaller production units and decentralized production[7] at the consumer level, plus the linking of new sectors like transport in the form of plug-in and electric cars, present other challenges to the existing network systems. By improving information and communication technologies and ensuring smart metering and automatic devices, it is envisioned that electricity supply is capable of matching more effectively consumer demand while stabilizing the electricity system and reducing peak demand. Consequently, the energy systems of the future will have to undergo major structural changes and to develop more intelligence into the grid, or to use a more popular term—to become 'smart'.

The objective of this chapter is partly to set the scene for energy systems of today and focus on the most recent developments as well as the need for changes, and partly to analyse the challenges they pose to the organization of the existing network structures and the legal regime governing the grid. First, however, it is important to clarify what we are talking about.

II. What are Smart Grids and Intelligent Energy Systems?

There seems to be a mix-up between various terms like intelligent energy systems, smart grids, and smart metering. Although there is no standard global definition, consensus may be identified that whereas smart metering is just one technology option, intelligent energy systems or smart grids include the full supply circle from production over transmission and distribution to end-use and all the stakeholders

6 The Obama–Biden, 'New Energy for America Plan', 2008, cf <http://www.whitehouse.gov/issues/energy-and-environment/new-foundation/learn-clean-energy-economy> (accessed 28 September 2011) and 'Blueprint for a Secure Energy Future', 30 March 2011, available at <http://www.whitehouse.gov/sites/default/files/blueprint_secure_energy_future.pdf> (accessed 28 September 2011); 'China Launches Green Power Revolution to Catch Up on West—Plan to Hit 20% Renewable Target by 2020', cf <http://www.guardian.co.uk/world/2009/jun/09/china-green-energy-solar-wind?intcmp=239> (accessed 28 September 2011); *China Daily*, 'The Draft 12th Five-Year Plan (2011–2015)', 7 March 2011, see <http://www.china.org.cn/environment/2011-03/07/content_22073772.htm> (accessed 28 September 2011) and the National Development and Reform Commission, 'China's Energy Conditions and Policies', see <http://en.ndrc.gov.cn/policyrelease/P020071227502260511798.pdf> (accessed 28 September 2011).

7 Often described as 'distributed production'—a terminology that is also used in the EU Electricity Market Directive 2009/73/EC (see under V.B. of this chapter) and defined as 'generation plants connected to the distribution system'.

involved. In that sense, and illustrated below, intelligent energy systems and smart grids seem to cover the same and be synonymous.

A good starting point for the definition of smart grids is provided by the European Technology Platform for Electricity Networks of the Future, also called SmartGrids ETP: 'electricity networks that can intelligently integrate the behaviour and actions of all users connected to it—generators, consumers and those that do both—in order to efficiently deliver sustainable, economic and secure electricity supplies'.[8] The definition is further elaborated by adding: 'SmartGrids do not only supply power but also information and intelligence. The "smartness" is manifested in making better use of technologies and solutions to better plan and run existing electricity grids, to intelligently control generation and to enable new energy services and energy efficiency improvements.'

In the EU Commission's Communication from April 2011, *Smart Grids: From Innovation to Deployment*, these grids are described as 'an upgraded electricity network to which two-way digital communication between supplier and consumer, intelligent metering and monitoring systems have been added. Intelligent metering is usually an inherent part of Smart Grids.'[9]

The International Energy Agency (IEA) also employs the smart grid terminology and describes it as 'an electricity network that uses digital and other advanced technologies to monitor and manage the transport of electricity from all generation sources to meet the varying electricity demands of end-users. Smart grids coordinate the needs and capabilities of all generators, grid operators, end-users and electricity market stakeholders to operate all parts of the system as efficiently as possible, minimising costs and environmental impacts while maximising system reliability, resilience and stability.'[10]

Another definition takes its starting point from 'intelligent energy system', which is defined as:

> An intelligent grid is the infrastructure connecting energy demand and supply using the latest developments in digital technology and communication technology in order to increase efficiency, reliability and security of the system. In a wider context, an intelligent system facilitates a high degree of interaction between electricity, heat and gas sectors. The energy system is being transformed from the former centralised producer-controlled system to a system much more decentralised and consumer-interactive. It involves all stakeholders in the sector.[11]

So in conclusion there seems to be no difference between smart grids and intelligent energy systems. Their key features are thus:

- innovative digital and advanced technologies;
- increased information among and between consumers, energy suppliers, and other grid users;

[8] <http://www.smartgrids.eu> (accessed 28 September 2011).

[9] COM(2011) 202 of 12 April 2011, p 2.

[10] International Energy Agency (IEA), 'Technology Roadmap—Smart Grid', (2011), p 6.

[11] Danish Industries Federation (DI), 'Intelligent Energy Systems—A White Paper with Danish Perspectives', May 2010.

- integrated and holistic planning and management of both supply and demand;
- two-way control and influence on production and end-use;
- efficiency improvement of the entire electricity system;
- new investment needs.

III. The Role of the Power System

Generally, in the energy sector of today there may be three different grid systems: one for power, one for district heating, and one for natural gas. In Northern Europe it is in particular solutions based on electricity that will play the biggest role for the transition to a low carbon society. A precondition in electricity systems is that production and consumption must always be in balance. When integrating a much higher share of renewable energy sources—many of which have variable output—it will be necessary to have reserve production capacity available so that a stable electricity supply may be ensured at all times. Moreover, the network must better facilitate the connection and operation of generators of all sizes and technologies. As pointed out by IEA, 'If a power system is sufficiently flexible, in terms of power production, load management, interconnection and storage, the importance of the variability aspect is reduced.'[12]

As demand will increase and existing grids in Europe are ageing, considerable investments in networks are also required, and increasingly, the infrastructure must cross national borders with effective interconnections to ensure continuity of supply at all times. However, should renewable energy be integrated solely in the conventional manner, it would demand huge investments in grid and electricity production capacity.[13] It will also be essential to develop well-functioning international energy markets to ensure that surplus electricity can be sold.

Concurrently, it will be essential to influence demand so that it is matching the production profile. In other words when the wind is blowing and electricity prices are getting lower, it will be an advantage if demand can be controlled more efficiently than today. Other technologies like heat pumps and electric vehicles will be able to contribute to balancing production with demand. The consumers must be allowed to play a bigger part and participate in optimizing the operation of the system, to get better information and options for choice of supply. As production at the same time will be decentralized, the consumer will get a double role—as a consumer and as a producer and both take and deliver electricity to the network—the term 'prosumer' is sometimes used in this connection.

The future energy systems must therefore be able to handle complex interactions between inputs and output and between grid interconnections. This is in fact the very core of intelligent energy systems and smart grids that the networks are able to

[12] IEA Report, 2008, 'Empowering Variable Renewables—Options for Flexible Electricity Systems', p 5.

[13] DI Report 2010, 'Intelligent Energy Systems—A White Paper with Danish Perspectives'.

control, regulate, and monitor themselves to a greater extent than is the case today, allowing the energy to be used effectively by way of metering, communications, market frameworks, and regulatory frameworks for generation, transportation, and consumption. This would at the same time imply the benefits that maintenance of the grids, losses of energy in the transmission and distribution process, congestion, and reaction to unexpected events could be improved, eliminated, and optimized.

However, it is crucial to recall that major investments will be needed for variable technologies at all levels—generation, transmission, distribution, and consumption—and that a proper balance between costs and benefits must be sought and carried out in a transparent way. In other words, these new investments must be justified as there is only one to pay in the end—the consumer!

IV. Policy Developments in a Region and a Country that Want to Take Leadership

In Europe, developments are driven by both national and EU policies. Greening energy is at the very forefront on the political agenda and a wide range of policy action plans and legal initiatives have been taken at both the regional and national levels. Increasingly, focus is put on the needed energy systems to support a development that is more climate-friendly and resource-efficient while ensuring competitiveness, energy at fair prices, and a safe and reliable energy supply at all times.

The EU and the case study of Denmark provides illustrative examples of a region and a country that in parallel wants to change energy patterns in a more sustainable way but with a clear focus on the economic alleys that will be opened for new jobs and technology exports by taking the lead in promoting new solutions and advancing technological developments.

A. The European Union

As a landmark decision, on 8 and 9 March 2007, the European Council[14] committed itself for the first time to a binding Europe-wide environmental target: a 20 per cent reduction in greenhouse gas emissions by 2020 compared with 1990. As mentioned above, the Council also agreed on a binding commitment that renewable energy will comprise 20 per cent of the EU's total energy consumption by 2020.

[14] The Presidency Conclusions of the Brussels European Council, 8–9 March 2007—7224/07. See also the Commission reports: 'An Energy Policy for Europe', COM(2007) 1; 'Renewable Energy Road Map—Renewable Energies in the 21st Century: Building a More Sustainable Future', COM(2006) 848; 'Limiting Global Climate Change to 2 Degrees Celsius—The Way Ahead for 2020 and Beyond', COM(2007) 2; 'Towards a Future Maritime Policy for the Union: A European Vision for the Oceans and Seas', COM(2006) 275; and 'A European Strategy for Sustainable, Competitive and Secure Energy', COM(2006) 105. For more details on renewable energy within EU see the Climate Action Network Europe website at <http://www.climnet.org/EUenergy/renewables.html> (accessed 28 September 2011); the Commission's Sustainable Energy Europe Campaign at <http://www.sustenergy.org/tpl/page.cfm?pageName=home>; and The European Renewable Energy Council (EREC) at <http://www.erec-renewables.org> (accessed 28 September 2011).

This is a follow-up of the European energy strategy published by the Commission as a Green Paper in March 2006.[15] The energy strategy seeks to provide solutions for the three core objectives of EU energy policy: sustainable development, competitiveness, and security of supply. Renewables play a significant role in this context and as will be illustrated below the energy systems as such attract more and more attention in this respect.[16]

On 19 September 2007, the Commission published a third package of legislative proposals to improve liberalization, market transparency, and cross-border trade, followed four months later by a major legislative package of 23 January 2008 on climate and energy. It included proposals for new directives on the promotion of renewables and emissions trading and new guidelines for state aid. The legislative initiatives also covered harmonization of the allocation of national targets for increased use of renewables including for the transport sector.

The proposals to improve liberalization, market transparency, and cross-border trade in the electricity sector have now been adopted as Directive 2009/72/EC supported by several regulations, cf further below under section V B.[17] Likewise, the proposal on renewables has now become Directive 2009/28/EC and maintains the main objective of establishing overall binding national renewable energy targets of a 20 per cent share of renewable energy of total energy consumption in 2020, and a binding 10 per cent minimum target for the use of renewables in transport (originally for bio-fuels but subsequently changed to cover renewables as such) to be achieved by each member state.[18] The directive is intended to remove unnecessary barriers to the growth of renewable energy—for example, by simplifying the administrative procedures for new renewable energy developments.[19]

An increasing focus on development of infrastructure is illustrated by the issuance of the European Strategic Energy Technology Plan (SET-Plan). The SET-Plan supports European energy and climate policies through technology innovation,[20] and aims to coordinate efforts at national and EU level through joint strategic planning and effective implementation mechanisms. As far as the electricity grid is concerned, it is highlighted that creating a real internal market is still a main target but also the

[15] 'Green Paper on a European Strategy for Sustainable, Competitive and Secure Energy', COM(2006) 105.

[16] For earlier policy documents and reports see also 'Energy for the Future: Renewable Sources of Energy', COM(1997) 599; 'The White Paper for a Community Strategy and Action Plan'; 'The Share of Renewable Energy in the EU', COM(2004) 366; and 'The Support of Electricity from Renewable Sources', COM(2005) 627l.

[17] Directive 2009/72/EC concerning common rules for the internal market in electricity, Regulation No. 713/2009 establishing an Agency for the Cooperation of Energy Regulators and Regulation No. 714/2009 on conditions for access to the network for cross-border exchanges in electricity—all of 13 July 2009.

[18] The emissions trading proposal—now Directive 2009/29/EC calls for a reduction in EU emissions of at least 20 per cent by 2020 compared with 1990 levels, and by 30 per cent, provided that other industrialized countries commit to comparable efforts within the framework of a global agreement to combat climate change post-2012. A major change is that there will be one EU-wide cap on the amount of emissions allowances instead of 27 national caps.

[19] See also Ulf Hammer's chapter (chapter 11) of this book.

[20] 'Investing in the Development of Low Carbon Technologies—(SET-Plan)', COM(2009) 519 and 'A Technology Roadmap', Commission Staff Working Document SEC (2009) 1295.

increase of the share of intermittent energy sources in total energy production, and to manage complex interactions between suppliers and customers. The goal is to connect 50 per cent of traditional electricity networks to plants generating renewable energy by 2020. Moreover, the creation of market opportunities for energy efficiency technologies is put forward as a priority, including smart networks. It is estimated that €3–8 billion will be needed to finance the various proposed initiatives and that the market alone will not deliver but must be supported by public investments.

Another important initiative is the publication of a new energy strategy in year 2010,[21] followed by a comprehensive range of other initiatives.[22] This Communication sets out the European Commission's energy strategy in the period to 2020 and outlines five priorities: limiting energy use in Europe; building a pan-European integrated energy market; empowering consumers and achieving the highest level of safety and security; extending Europe's leadership in the development of energy technology and innovation; and finally, strengthening the external dimension of the EU energy market. It calls for changes in the way energy infrastructures and networks are planned, constructed, and operated and among other initiatives to launch new, large-scale European projects concerning smart grids linking the whole electricity grid system and on electricity storage.

Energy infrastructures are also at the forefront of the flagship initiative 'Resource-efficient Europe'. This Communication provides a vision of what is needed for making networks within the EU efficient. It puts forward a new method of strategic planning to map out necessary infrastructures, qualify which ones are of European interest on the basis of a clear and transparent methodology, and provide a toolbox to ensure their timely implementation, including ways to speed up authorizations, improve cost allocation, and target finance to leverage private investment. More details have been included in a Proposal for a regulation on 'Guidelines for trans-European Energy Infrastructure' presented on 11 October 2011.[23]

The focus on the grid system has accelerated more recently. A 'Roadmap for Moving to a Competitive Low-carbon Economy in 2050'—the EU Commission's Communication from March 2011[24]—defines pathways as to how to reach EU's objective of cutting greenhouse gas emissions by 80–95 per cent of 1990 levels by 2050. It identifies smart grids as a key enabler for a future low-carbon electricity system, facilitating demand-side efficiency, increasing the shares of renewables and distributed generation, and enabling electrification of transport.

The European Commission adopted the Communication *Smart Grids: From Innovation to Deployment* one month later to drive forward the deployment of

[21] 'Energy 2020—A Strategy for Competitive, Sustainable and Secure Energy', COM(2010) 639 of 10 November 2010.

[22] 'EUROPE 2020—A strategy for smart, sustainable and inclusive growth', COM(2010) 2020; 'The Future Role of Regional Initiatives', COM(2010) 721; 'North Seas Countries' Offshore Grid Initiative, Memorandum of Understanding', ENTSO-E (European Network of Transmission System Operators for Electricity), 3/12, 2010.

[23] 'Energy Infrastructure Priorities for 2020 and Beyond—A Blueprint for an Integrated European Energy Network', COM(2010) 677. See also COM (2011) 658 of 19 Oct 2011 identifying 'projects of common interest', eligible for funding from the EU under the 'Connecting Europe Facility'.

[24] COM(2011) 112 of 8 March 2011.

future European electricity networks. The key political message is that bringing together latest progress in information and communication technologies and network development will allow electricity current to flow exactly where and when it is needed at the cheapest cost. Smart grids will give in particular to consumers the ability to follow their actual electricity consumption in real time: smart meters will give consumers strong incentives to save energy and money. Estimates show that smart electricity grids should reduce CO_2 emissions in the EU by 9 per cent and the annual household energy consumption by 10 per cent. They also help to ensure secure functioning of the electricity system and are a key enabler of both the internal energy market and integration of vast amounts of renewables. However, even if the publication is devoted to deployment of smart grids, it is weak on concrete actions and on identification of regulatory needs.

B. Denmark

Denmark has been a pioneer and maintains its position as one of the leading countries in the development and use of renewable energy and especially of wind energy. Since the energy crisis of the 1970s there has been consensus on giving green energy priority and in 2003 the goal of providing 20 per cent of Danish electricity consumption from renewable resources was achieved, compared to an average of 2.4 per cent for Europe as a whole.[25] In 2009 renewables accounted for 19.7 per cent of total Danish energy consumption and 27.4 per cent of electricity was supplied from renewable sources. Wind power contributes the largest share (around 18.3 per cent) compared with other renewable energy sources.[26] An illustrative example in this connection is that already in November 1997, the island of Samsø was designated as Renewable Energy Island.[27] The goal was for the island to be self-sufficient in renewable energy in 10 years, and today the island's land-based wind turbines produce 100 per cent of the island's electricity. On a calm day, the island can take electricity from the mainland grid, and as soon as the wind blows again and energy is produced it may 'return' the electricity to the grid.

In the late 1980s and early 1990s, the political focus gradually shifted from supply security considerations, minimization of costs, and local environmental effects to wider environmental considerations, notably the goal of achieving long-term sustainable development.[28] The development of the integration of energy markets within the European Union brought a need for a new foundation for energy policy to ensure that overall environmental objectives could be maintained under the new open market conditions.[29] The strategy 'Energy 21', from 1996, states that

[25] Danish Wind Industry Association, <http://www.windpower.org> (accessed 28 September 2011).

[26] The Energy Agency, 'Energy Statistics 2009', cf <http://www.ens.dk> (accessed 28 September 2011).

[27] <http://www.ens.dk/en-US/Info/news/Factsheet/Documents/samsoe170709.pdf%20engelsk.pdf> (accessed 28 September 2011).

[28] This change in priorities is reflected in a plan of action called 'Energy 2000'; see Danish Ministry of Energy, 'Energy 2000: A Plan of Action for Sustainable Development', 1990.

[29] These challenges were reflected in the fourth of the energy strategies—Danish Ministry of Environment and Energy, 'Energy 21: The Danish Government's Action Plan for Energy', 1996.

the Danish Government should aim for an average annual increase in the use of renewable energy of one per cent per annum until 2030, constituting approximately 35 per cent of the gross energy consumption and 50 per cent of electricity production.[30]

The climate issue has been closely related to the development of renewables. Denmark has some of the highest per capita carbon dioxide emissions in the world, and the prospect of climate change is the main reason for the current focus on renewables. As part of the allocation of obligations within the EU, Denmark must reduce its emissions by 21 per cent compared to 1990.[31] The more recent long-term Danish strategy on energy policy, Energy Strategy 2025 from 2005, confirms the Danish Government's intention to use the market as a basis for the continued increase in the use of renewable energy. The strategy was followed up in January 2007 in the Danish Government's draft energy policy vision which set the target of doubling the share of renewable energy to 30 per cent of energy consumption by 2025.[32] In fact the target goes further and includes a long-term vision of total independence from fossil fuels, and the replacement of all coal, oil, and natural gas by the renewable energy.

On 21 February 2008, the Danish Government entered into a comprehensive new energy agreement with the parliamentary parties.[33] The parties agreed that renewable energy should cover 20 per cent of total Danish energy consumption in 2011 and includes better terms for wind turbines. Aside from significantly raising the transfer rate for electricity from land wind turbines, the parties agreed to install 400 MW from new offshore wind turbines by 2012. Moreover, a compensation scheme for neighbours of wind turbines was proposed and was implemented in a new Act on Promotion of Renewable Energy[34] that gather the most relevant legislation (from among other acts—the Electricity Act) to create greater transparency about the legislative framework for this sector.

In February 2011, the then Danish Government published its proposal for how the country can achieve independence from fossil fuels by 2050 in its *Energy Strategy 2050*—a follow-up of the report and suggestions of the scientific Commission on Climate Change.[35] By 2020, the strategy calls for the energy industry to reduce its consumption of fossil fuels by 33 per cent, compared with 2009. In addition, the strategy will increase renewable energy's share of gross energy consumption to 33

[30] Cf n 29, p 39.

[31] Denmark ratified the Climate Convention on 21 December 1993 and the Kyoto Protocol on 31 May 2002 together with the EU. See also A Rønne, 'The Danish Way of Combating Greenhouse Gas Emissions', in P.D. Cameron & D. Zillman (eds), *Kyoto: From Principles to Practice* (The Hague: Kluwer Law International, 2001).

[32] 'A Visionary Danish Energy Policy', which describes Danish energy policy objectives up until 2025, 19 January 2007.

[33] See <http://www.ens.dk/en-US/policy/danish-climate-and-energy-policy/Documents/Energy%20Policy%20Agreement%2021%20Feb%2008_final.pdf> (accessed 28 September 2011).

[34] Act no 1392 of 27 December 2008.

[35] The Danish Government, 'Energy Strategy 2050—From Coal, Oil and Gas to Green Energy', February 2011 and the Danish Commission on Climate Change, 'Green Energy—the Road to a Danish Energy System Without Fossil Fuels', 28 September 2010.

per cent in 2020, and reduce energy consumption by 6 per cent in 2020, compared with 2006, through a major focus on energy efficiency.

The increase in renewable energy will be obtained from wind, biomass, and biogas, including the construction of new 600 MW offshore wind farm at the Kriegers Flak to be tendered and come online by 2018–2020. Studies are to be carried out of coastal areas that would identify locations suitable for a further 400 MW of small offshore wind turbines and also placement of new wind turbines on land is to be promoted. Wind power alone is expected to cover more than 40 per cent of overall electricity consumption by 2020, compared with about 20 per cent today. Building requirements are likewise to be tightened, and from 2012, new buildings may not be built with oil or gas furnaces and from 2017, oil furnaces may not be installed in existing buildings. Finally, there will be support for the development of minor renewable energy technologies, including solar and wave power, and demonstration projects for large heat pumps intended for use by district heating plants, as well as funding for studies of geothermal energy. By 2020 more than 60 per cent of electricity consumption will be covered by renewable energy.

The strategy is also indicating a raise of the energy savings target that energy companies have to implement amongst their consumers by 50 per cent starting in 2013 and by 75 per cent in the period 2017–2020. The development of intelligent energy systems is emphasized and includes the establishment of new international electricity-sharing capacity in connection with Kriegers Flak offshore wind farm, and to require that all electric meters installed after 2015 be intelligent electric meters. In economic terms, the proposed initiatives are not to burden the state finances or the competitiveness of the industry. It is thus stated that homeowners will experience moderate increases in the costs of heat and electricity, but that they will also be given opportunities to lower their energy expenses through greater efficiency. Also companies can expect added expenses amounting to 0.1 per cent of the rise in their gross revenue growth by 2020. It is expected that an agreement with the majority of the political parties can be entered into later in 2011 and new legislation will follow subsequently. The consequences of the new government of 3 October 2011—from the political parties of the Social Democratic, the Socialist People's Party, and the Social-Liberal Party—will imply an even higher focus on green technologies and the ambitious goal that the entire electricity and heat supply must come from renewable energy already in 2035. Moreover, the new government has published on 25 November 2011, its new plan 'Our Future Energy'. It is based on the previous government's 'Energy Strategy 2050'. It pushes, however, the pace of developments a bit higher. The Strategy will form the basis for negotiating with the opposition to reach a new political agreement that will cover the period till 2020.[36]

Flexible production and monitoring a decentralized supply system is not something new in Denmark. In 1980, there were only 15 central power plants but today

[36] Opening speech by the Prime Minister of 4 October 2011, see <http://www.stm.dk/_p_13523.html> (accessed 15 October 2011) and the statement is further elaborated as: 'We must speed up the pace of the green restructuring. We need to have more wind turbines in the seas around Denmark. More biomass from our agriculture. And we must be even better at saving energy.' See 'Our Future Energy', The Danish Government, November 2011.

there are several thousands of electricity producers, including not only central power plants, but wind turbines and de-centralized CHP (combined heat and power) units. In contrast to this, flexible consumers are something new, and very much in need due to the fact that already today, a lot of excess wind power is sold at the Scandinavian market at very low prices.

A new EcoGrid project[37] has just received 10 million euros in funding from the EU and the Danish island of Bornholm will be the world's biggest and a full-scale laboratory for the intelligent power system of the future. It is a real cooperative effort, with EU partners from Denmark, Norway, the Netherlands, Belgium, Germany, Spain, Portugal, Estonia, Austria, and Switzerland. Over half of the island's electricity consumption will come from renewable energy and Bornholm has been selected as a test area because of the island's isolated grid, which is connected only by a cable to Sweden. By disconnecting the cable it is possible to make Bornholm energy-independent. As it is explained, the vision for the intelligent energy systems is to have electric cars set to charge at night and possibly let them function as storage capacity and return power to the system if needed. Moreover, heaters may turn themselves on in a strong wind, and ice machines may start when the sun is highest in the sky. The idea is that a central unit will collect data on the production and consumption and utilize this information to regulate the electricity production and use in the most efficient way. The consumers will be able to follow the price of electricity every five minutes so that they use the electricity when the electronic devices inform that electricity is cheap. Consumers will be able to choose between two models depending upon how much control they want to maintain or hand over to the electricity company that will be able to turn electricity on and off by a remote control system.

The following will focus on the legal framework within the European Union and thus within which the 27 member states must act and adopt legislation.

V. The New EU Legal Framework for Energy Systems of Today

Success in achieving the political objectives does not only depend on technology fixes, but also on the legal and regulatory regimes. This section will analyse the current legal framework within the EU to identify possible constraints and gaps for further developments.

A. The EU Treaty Framework on energy and infrastructure

The Treaty on the Functioning of the European Union (TFEU),[38] the consolidated version of the Treaty establishing the European Community as amended by the Lisbon Treaty as per 1 December 2009, confirms the increasing priority of

[37] See <http://energinet.dk/en/forskning/EcoGrid-EU/sider/EU-EcoGrid-net.aspx > (accessed 15 October 2011). See also the Danish Newspaper *Berlingske*, 20 August 2011, 1 section, p 8.

[38] OJ C115/47, 9.5.2008.

the energy sector by including a special title (XXI) devoted to 'Energy'. It follows up on a course that was initially set by the Treaty on European Union (the 1992 Maastricht Treaty) that added energy measures to the list of the Community's activities[39] and the promotion of trans-European networks in the areas of energy infrastructures to the text of the EC Treaty.[40] In addition to the statement that 'The [European] Union shall contribute to the establishment and development of trans-European networks in the areas of transport, telecommunications and energy infrastructures' (article 170), the European Union should also aim at 'promoting the interconnection and interoperability of national networks as well as access to such networks'. For this purpose a series of guidelines should be established identifying projects of common interest, and implement any measures that may prove necessary to ensure the interoperability of the network (article 171). In this connection member states are required to (and shall) coordinate among themselves the policies pursued at national level. TFEU's article 194 includes the objectives of the EU energy policy:

> In the context of the establishment and functioning of the internal market and with regard for the need to preserve and improve the environment, Union policy on energy shall aim, in a spirit of solidarity between member states, to:
>
> a) ensure the functioning of the energy market
> b) ensure security of energy supply in the [European] Union
> c) promote energy efficiency and energy saving and the development of new and renewable forms of energy, and
> d) promote the interconnection of energy networks.

It is added that: 'Such measures shall not affect a member state's right to determine the conditions for exploiting its energy resources, its choice between different energy sources and the general structure of its energy supply' but at the same time there is included authority to adopt measures significantly affecting a member state's choice between different energy sources and the general structure of its energy supply if the Council is acting unanimously in accordance with a special legislative procedure and after consulting.[41]

The legal basis for Community action on the environment was introduced in 1987, with the provision of the basis for legislation and programmes on energy efficiency and the requirement that general environmental protection issues must be integrated into the elaboration and implementation of all Community policies and activities.[42] Much more vigorous enforcement of the competition and free movement rules has indicated significant changes in the organization of energy activities and the balance of regulatory initiatives between the Commission and the member states.

[39] EC Treaty, article 3(u): 'the activities of the Community shall include... measures in the spheres of energy, civil protection and tourism'. TFEU set out the principle of shared competence in article 4(2)(1).

[40] EC Treaty, article 154 (TFEU article 170).

[41] Article 192(2)(c). See also chapters 7 and 20 of this book for a discussion on trans-European networks.

[42] TFEU, article 11, previously EC Treaty article 6 (earlier article 130r).

Although the member states have retained their sovereignty over primary energy sources[43] and choice of energy mix, Community rules require notification of certain investment projects in the petroleum, natural gas and electricity sectors, including major power stations and transmission lines.[44] It emphasizes the need for sustainable development and for promoting measures at international level to deal with regional or global environmental problems, and in particular, combating climate change.[45]

B. The electricity market and energy services directives

The overall aim of the three subsequent electricity market directives[46] is establishing common rules for generation, transmission, and distribution of electricity, and setting out rules for the functioning and organization of the sector, access to markets, operation of the system, tender procedures, and granting of authorizations. The goal has been to create an integrated, liberalized, and single-energy market within the European Union that should lead to a better functioning market with better prices, greater supply security, and environmental sustainability. The third liberalization package as outlined above had to be implemented by the member states by 3 March 2011 so new national electricity legislation may be found in all member states.

Briefly described, the Electricity Market Directive 2009/72/EC[47] includes general rules for the organization of the sector (articles 3 to 6 and Annex I); public service obligations relating to security of supply, regularity, quality and price, environmental protection, etc (article 3(2)); freedom of all customers to choose a supplier from any EU member state (article 3(4)), and promotion of regional cooperation between member states and regulatory authorities (article 6). Moreover, the Directive regulates authorization and tendering procedures for generation (articles 7 to 8) and more effective unbundling for the Transmission System Operation (articles 9 to 16) and of Distribution System Operation (articles 24 to 29). Finally, the Directive ensures third party access (TPA) to the grid (article 32) based on published tariffs that are applicable to all customers and applied objectively without discrimination between system users. However, access may be refused by the system operator if there is a lack of capacity and when giving duly substantiated reasons. National Regulatory Authorities are regulated (in articles 35–40), but recognizing that national regulators are insufficient, an Agency for the Cooperation of Energy Regulators (ACER) took up operations, 3 March 2011. The prime aim

43 See the Preamble to Directive 94/22/EC of the European Parliament and of the Council of 30 May 1994 on the conditions for granting and using authorizations for the prospection, exploration, and production of hydrocarbons; and TFEU, article 194.

44 Council Regulation (EC) No 736/96 of 22 April 1996 on notifying the Commission of investment projects of interest to the Community in the petroleum, natural gas, and electricity sectors. This includes power stations with a capacity of 200 MW or more, and transmission lines of 345 kV if in the air and 100 kV if underground and constituting an essential link.

45 Articles 11 and 191.

46 Directive 1996/92/EC, Directive 2003/54/EC and Directive 2009/72/EC.

47 Directive 2009/72/EC of 13 July 2009 concerning common rules for the internal market in electricity and repealing Directive 2003/54/EC.

is to complete the internal market in the EU for electricity and gas by improving the regulatory framework at Community level. ACER shall develop non-binding framework guidelines and provide opinions and recommendations regarding issues such as necessary binding rules, regional cooperation, and network development plans. ACER may take individual decisions in specific cases on terms and conditions for access to and operational security of cross-border infrastructure.[48]

Connected to the third liberalization package are also the issuance of regulations on access to the network for cross-border exchanges in electricity and enhanced power for regulators, eg to monitor the sector and carry out unannounced inspections. The aims are setting fair rules for cross-border exchanges, thus enhancing competition within the internal market. It establishes a new cooperation between all transmission system operators through a new entity—ENTSO: European Network of Transmission System Operators (articles 4 to 5)—and regulates that Network Codes (articles 6, 7 and 8(6)) shall provide rules for, *inter alia*, network security, connection, TPA, data exchange, interoperability, capacity allocation and congestion management, transparency, and harmonized transmission tariff structures.

The promotion of developing smart grids is, however, also set forward directly in the Electricity Market Directive recital 27 that states that 'member states should encourage the modernisation of distribution networks, such as through the introduction of smart grids, which should be built in a way that encourages decentralised generation and energy efficiency'. To this should be added that it is not an explicit obligation but merely encouragement and that the concept of smart grids involves the full supply cycle and not only distribution networks. Different technologies may be added to both generation and transmission, although the distribution activity is the most complex section with the highest number of actors.

In article 3(10), demand-side management and energy efficiency measures shall be implemented by member states to achieve the objectives of among others environmental protection, combating climate change and energy security. The requirement is not strict, but on the contrary softened by adding the wording 'where appropriate'. Smart grids and intelligent metering systems are directly encouraged in article 3(11) as examples of energy efficiency measures. Member states or regulatory authorities shall thus strongly recommend that electricity undertakings optimize the use of electricity, for example by providing energy management services, innovative pricing formulas, or introducing intelligent metering systems or smart grids, where appropriate. It is, however, not a legal obligation, only an option, and smart grids are not defined in more detail.[49]

Demand-side management is further elaborated in article 25(7): 'When planning the development of the distribution network, energy efficiency/demand-side management measures or distributed generation that might supplant the need to upgrade or replace electricity capacity shall be considered by the distribution system operator.' As can be observed, it is not a strict legal obligation either to

[48] See for an analysis of this Directive also chapters 7, 19, and 20 of this book.
[49] An equivalent provision is found in the Gas Market Directive 2009/73/EC, article 8.

implement demand-side management, but only a requirement to the distribution system operator to 'consider' the possibility.

To improve the information to consumers it is also a requirement that they are properly informed of actual electricity consumption and costs frequently enough to enable them to regulate their own electricity consumption. That information shall be given by using a sufficient timeframe, which takes account of the capability of customer's metering equipment and the electricity product in question. Due account shall be taken of the cost-efficiency of such measures. No additional costs shall be charged to the consumer for that service, cf the Directive's Annex I(i).

To facilitate the implementation of intelligent metering systems and smart grids as such, member states shall assist the active participation of consumers in the electricity supply market, cf Annex I(2). The implementation of such metering systems may be subject to an economic assessment of all the long-term costs and benefits to the market and the individual consumer or which form of intelligent metering is economically reasonable and cost-effective and which timeframe is feasible for their distribution.

In this connection, member states are obliged to assess the roll-out of intelligent metering systems as a key step towards the implementation of smart grids. Moreover, 80 per cent of those that have been positively assessed must be equipped with intelligent metering systems by 2020 and an implementation plan and timetable must be defined before 3 September 2012. Reflecting the fact that the implementation of intelligent metering systems varies between countries, member states shall ensure the interoperability of those metering systems to be implemented within their territories and shall have due regard to the use of appropriate standards and best practice and the importance of the development of the internal market in electricity.[50]

Smart grids are thus identified as a way for member states to meet their obligations to promote energy efficiency and related hereto is the requirement that there must be consumer access to consumption and billing information. Smart grid deployment should proceed at a similar pace between member states.

The purpose of the Energy Services Directive 2006/32/EC[51] is to make the end use of energy more economic and efficient by establishing indicative targets and the legal frameworks needed to eliminate market barriers which prevent efficient end use of energy. Moreover, it aims to create the conditions for the development of a market for energy services and other measures aimed at improving end-use energy efficiency. The Directive applies to the distribution and retail sale of energy and calls for metering that accurately reflects the final customer's actual energy consumption and provides information on actual time of use (article 13(1)). Moreover, it sets an obligation to remove volume-based incentives (to allow efficiency gains).

[50] More on the negotiation process and background for the result may be found in C. Jones (ed), *EU Energy Law, Volume I, The Internal Energy Market, The Third Liberalization Package* (Leuven: Clayes & Casteels Publishing, 2010) p 436ff.

[51] Directive 2006/32/EC of 5 April 2006 on energy end-use efficiency and energy services and repealing Council Directive 93/76/EEC.

VI. The Role of Law for Smart Grids and Intelligent Energy Systems of Tomorrow

The implementation of smart grids raises a lot of questions that need to be solved. Among those IEA rightly asks, how should smart grid investment costs be recovered? If shortfalls in benefits occur, how should they be shared between utilities and consumers? How can additional services (such as balancing, demand response, energy retailing) be enabled by new regulations and smart grid technologies? Should electricity rate options be compulsory or voluntary? Should vulnerable customers be protected from the possibility of higher bills? If so, how? Should advanced technology investments such as smart grids, which carry the extra risk of technology obsolescence, be treated differently from other utility investments? Should some customer groups less able to participate in dynamic pricing be excused from bearing the extra costs of smart grids or being subject to new service conditions? If so, what can or should be done for these customers? What is the impact of differing tariff structures between interconnected regions?[52]

Below some of these issues are addressed in more detail: new standards, regulatory incentives, increased co-ordination, new market models, and protection of data, but the author is aware of the fact that this does not cover all the relevant matters or provide the exhaustive analysis that needs to be considered to justify the huge investments that are needed to smarten existing and new infrastructure projects with advanced technology.

A. New standards

In order to implement smart grid technologies there is a need to develop new standards for equipment, data transport, interoperability, and cyber security in order to be able to exchange information. As pointed out by IEA, 'a broad range of product and service providers who have not worked together in the past will have to collaborate in smart grid deployment'.[53]

First of all, technical standards for smart grids and meters must be adopted to ensure interoperability of smart utility meters (electricity, gas, water, and heat). Second, technical standards for electric vehicle charging systems must be adopted, as harmonization is needed to allow users to use the same charger for all electric vehicles and to be connected and operated all over Europe. The relevant EU standardization bodies were requested in March 2009 to agree to common EU technical standards for smart meters by the end of 2011. Because of the lack of progress a new mandate was issued in March 2010 for smart grids based on industry consensus, and a new 2012 deadline has been set.

[52] IEA, 'Technology Roadmap—Smart Grids' (2011), p 34.
[53] Cf n 52, p 32.

B. Regulatory incentives

Today there is a regulatory gap to meet uncertainties for the application of smart grid technologies. The Electricity Market Directive only refers explicitly to meters but is silent with respect to other segments of smart grids apart from referring to the modernization of the distribution networks. However, the obligation for member states to roll out smart grids creates the potential for developing differences country-by-country. Moreover, the Directive does not regulate how the investments should be carried out and by whom and whether it may be financially supported by the state.

There is also a need for a better demand response to consume less at times of high wholesale market prices or when system reliability is jeopardized. Today consumers are not billed according to real production costs and transportation capacities. By making greater use of real time-differentiated electricity prices, consumers may be motivated to change their demand pattern and change the way they use electricity.

Also the price-setting of the future will need to reflect external costs to the society. The real environmental costs such as greenhouse gas emissions must thus be reflected in rates and operating costs. Today, electricity companies build and expand the electricity networks and the costs are covered over a certain period of time based on regulated rates of return which do not include proper incentives. To meet the need for cost savings, eg in peak generation, a network operator should at the same time be allowed to earn revenues that are not linked to additional sales but based on efficiency gains. The rate design, risk allocation, and price regulation thus need to be reconsidered. In this process protection of certain groups of consumers must be assured, as some may not be able to change behaviour and to take advantage of new pricing structures. Likewise, the option of making the rate regime optional or mandatory should be considered.

C. Definition of roles, responsibilities, and need for co-ordination

A power system with a high penetration of fluctuating energy sources requires special consideration for control and coordination of the overall system monitoring, protection, and operation. Liberalization has included the unbundling of the previous vertical structure of the electricity industry to increase competition and prevent cross-subsidization. The organizational and ownership structure of the networks has also changed via requirements of unbundling. The electricity sector in the EU is today divided into market-based sectors—production and trade/marketing—and regulated sectors such as the grid-based transmission and distribution sectors under the auspices of separate system operators. In many ways this development is implying a more complex system for operation and institutional set-up. Today decisions are taken at different levels and by different entities.

The adoption of smart grids, however, will play a role in all parts of the electricity infrastructure and raises a need for overall planning of the whole electricity system. Costs and required investments need to be considered on a broader system basis,[54] and in this respect the separation of activities in different entities may create a barrier. Consequently, increased co-operation between the distribution system operators (DSOs) and the transmission system operators (TSOs) and between the consumer/prosumer and the generators must be ensured, together with the most efficient implementation of the necessary investments. Since an increasing part of the generation units will be located at the distribution level, a more active role should be allocated to the DSOs. Such obligations may need to be adopted and a possible fourth electricity market directive may be anticipated to ensure developments.

D. Market models and contractual arrangements

Smart grids imply that supply may be interrupted by direct control of either the system operator or the consumer. New contract models may therefore provide conditions to price different services also at the residential consumer level. Previously, interruptive consumers have only been defined among big industrial consumers. In the future, however, the increased information on prices and consumption makes it possible for consumers to consider whether they are in need of the same quality of power at all times or whether they only need a guaranteed base load, whereas grades of power may vary at certain hours or days.

In the contractual arrangements it has to be regulated how the responsibilities will be divided between the supplier and the grid company and the relationship to the consumer. Who will be entitled to act and under which conditions? The protection of consumer from remote and abrupt service disruption that smart grids will make possible should also be considered, although this new capability may also result in economic saving cuts.

E. Access to and protection of data

The construction of an intelligent system hinges on data access in multiple directions and levels in order to operate effectively and efficiently. In this connection consumer access to energy data must be facilitated but at the same time the exchange of private data of businesses and households must be protected. Smart grids and smart metering creates large amounts of detailed information of the individual consumer. Questions like ownership, access, storage, use, and sale of this information must be addressed and regulated adequately to protect the consumer from adverse impact and to ensure equal competition between energy providers.[55]

Increased reliance on computer technology also involves a greater risk for cyber attacks. Consequently, effective mechanisms must be developed to cope with such

[54] See also n 52 p 23.

[55] In EU, the general data protection directive would be applicable, cf Directive 1995/46/EC of 24 October 1995, on the protection of individuals with regard to the processing of personal data and on the free movement of such data. The Directive sets strict limits on the collection and use of personal data and demands that each member state set up an independent national body responsible for the protection of these data.

challenges and lessons may be learned from other infrastructure systems, such as the banking and telephone sectors.

VII. Conclusions

The decarbonization of the energy sector is one of the critical challenges of the twenty-first century. The EU, Denmark, and other major world economies are investing heavily in clean technologies. There are different pathways to decarbonization and no single measure or technology will suffice. In each country the choice of means to be deployed depends on a combination of policy decisions, economic conditions and resources, public acceptance, and extant legal and regulatory frameworks. Market forces may be an important driver of change in this connection. However, experience with voluntary targets for renewables, energy efficiency, and CO_2 emission reductions, confirm that public regulation is also necessary.

As demonstrated by the development of the EU energy market, regulation may stimulate competition between private players. The third EU Energy Package includes a balance between market and non-market interventions, as illustrated by Directive 2009/72/EC concerning common rules for the internal market in electricity, and Directive 2009/28/EC on the promotion of renewable energy. In this regard, traditional top-down regulation instruments, such as orders, taxes, state aid, licensing, and permit systems, may be streamlined and supplemented with instruments like private–public and producer–consumer cooperation and trade, which contribute to ensure energy security, investments in new infrastructure, price and supply stability, and balanced environmental decisions. Such objectives, however, may not be achieved at the national or at the EU level alone, and require coordinated international action promoting low carbon technologies across borders. In this regard, regulation should be seen holistically at all levels—national, regional, and international, as well as in between the producers, transmitters, distributors, and consumers.

Today there are no clear rules on basic questions such as who is entitled to regulate what and how. Procedures for the construction and renewal of energy grids need to be streamlined and optimized, and regional regulatory barriers and obstacles must be identified and tackled. In this connection, decisions on the division of roles and responsibilities on ownership, possession, and access to data must comply with international standards and EU law.

Smart grid technologies offer much potential for developing a greener energy supply that is more energy efficient and more sustainable and at the same time, as President Obama declared recently, 'the countries that lead the twenty-first century clean-energy economy will be the countries that lead the twenty-first century global economy'.[56] It is an evolving set of technologies that we may not know the

[56] Remarks by the President on America's Energy Security, 30 March 2011, see <http://www.whitehouse.gov/the-press-office/2011/03/30/remarks-president-americas-energy-security>(accessed 28 September 2011).

full scope of today, and that may be deployed with different pace and in steps, and may spread beyond the energy sector as it may be applied to other kinds of infrastructure systems, such as water and gas. Comparisons have been made to the development of the internet and the mobile phone and as with respect to these advanced technologies one may envision first-, second-, and third-generation smart grids.

However, it is also a technology (or technolog*ies*) that will imply huge investments that will be needed in some way or the other to be covered by the consumers. Proper planning and coordination is another element that must increase—covering all segments of the supply chain and not only the particular unbundled activity. There is therefore a fundamental need to consider the costs and benefits very carefully and at the same time to ensure that the liberalization process will not create a barrier for the application and development of smart grids, ie the requirement of unbundling the production/supply and network activities versus the close cooperation between supplier and network company that are needed to apply smart grid solutions. In this context, there is a need to balance competing priorities—market against intervention—and address the tension between policy stability and flexibility.

Moreover, consumer awareness of potential benefits must be accentuated. Experience within energy-saving initiatives emphasizes that the need for analysis of consumer behaviour and the value of education and consumer campaigns should not be underestimated.

Finally, as pointed out by IEA, 'The intelligent network concept represents a paradigm shift in the generation and use of electricity, and this is in itself likely to be something of a barrier as it must evolve from the existing system, which cannot be simply turned off while the necessary upgrades are installed.'[57]

[57] IEA Report, 'Empowering Variable Renewables—Options for Flexible Electricity Systems', July 2008, p 24.

9

Demand Response and Infrastructure Development in the United States

*LeRoy Paddock and Charlotte Youngblood**

I. Introduction

Energy networks in the United States are becoming increasingly complex. These networks must be capable of responding to the challenges posed by climate change, a growing desire to encourage energy efficiency, and the introduction of new technologies that require interactive networks. In this ever-diversifying landscape, demand-side resources have a key role to play in managing electricity networks.

Demand response (DR) describes approaches to reducing electricity demand by end-users rather than adding generation to increase the 'supply' of electricity. DR programmes attempt to encourage consumers to either change their electricity usage at certain time periods ('load shifting' or 'peak load shaving'), or to reduce overall electricity use permanently (energy efficiency). Under United States law, 'demand response' is defined as 'a reduction in the consumption of electric energy by customers from their expected consumption in response to an increase in the price of electric energy or to incentive payments designed to induce lower consumption of electric energy'.[1] In the coming years, demand response may have a significant impact on generation, transmission, and distribution infrastructure development: deferring the need for new generation and transmission lines, relieving congestion in existing transmission lines, providing some of the balancing capacity needed to enable increased use of intermittent generation capacity such as wind and solar power, and facilitating the creation of two-way, interactive transmission networks. DR programmes in the United States have evolved since the late 1970s as interest in energy conservation has waxed and waned, and as a result of market liberalization. DR programmes are currently the subject of renewed interest by both policy-makers concerned about climate change and electricity professionals interested in tapping demand response resources to help keep the grid

* Some of the research associated with this article was conducted in connection with Ms Youngblood's thesis entitled 'Smart Policy in a Climate of Change: the Role of Demand Response in Securing the Future of the U.S. Electricity System'.

[1] 18 CFR s 35.28.

functioning properly and efficiently. These programmes, however, are complex and subject to both state and federal regulation. This chapter will explore the evolution of DR programmes in the United States and their impact on electricity networks.

II. Overview of Market Developments

A. Market liberalization

Beginning in the 1970s, regulatory liberalization permitted companies that are not public utilities to generate electricity that is fed into the electric grid. Liberalization at both the state and federal level also resulted in the break-up of vertically integrated, investor-owned utilities in several states, often separating generation, transmission, and distribution functions. In 1996 the Federal Energy Regulatory Commission (FERC) issued Order 888, requiring utilities that owned transmission to provide 'open' and non-discriminatory access to their transmission, and requiring the 'functional unbundling' of generation and transmission[2] to prevent undue discrimination against non-utility generators.[3] Functional unbundling did not require divestment of transmission facilities but, together with state deregulation, did lead a number of companies to become either generation companies or distribution companies. Now, over 3,100 entities play roles in the electricity market,[4] ranging from traditional regulated public utilities that still generate, transmit, and distribute electricity, to independent power producers who simply generate electricity to be fed into the transmission and distribution network, to largely unregulated organizations that aggregate individual consumer demand response efforts and bid the demand response commitments into organized wholesale markets.

B. The increasing focus on reliability

The federal government has become more involved in assuring the reliability of the national electric grid. This role became especially important after a widespread 2003 blackout throughout the north-east part of the United States and in southern Canada. The North American Electric Reliability Corporation (NERC) was formally approved by the United States Government in 2006 as the certified 'electric reliability organization', tasked with 'ensuring reliability' in 'bulk power' in North America.[5] NERC develops reliability standards for both the United States and

[2] Federal Energy Regulatory Commission (FERC), Order 888, Promoting Wholesale Competition Through Open Access Non-discriminatory Transmission Services by Public Utilities; Recovery of Stranded Costs by Public Utilities and Transmitting Utilities, 75 FERC 61,080 (1996).

[3] Ibid.

[4] M. Willrich, Massachusetts Institute of Technology (MIT) Energy Innovation Project, 'Electricity Transmission Policy for America: Enabling a Smart Grid, End to End' (2009) 2, available at <http://web.mit.edu/ipc/research/energy/pdf/EIP_09-003.pdf> (accessed 28 September 2011).

[5] United States Government Accountability Office, Electricity Grid Modernization (2011) 11, available at <http://www.gao.gov/new.items/d11117.pdf>; see also <http://www.nerc.com> and <http://www.nerc.com/page.php?cid=1|7|11> (all accessed 28 September 2011).

Canada. In the United States, those standards are then reviewed and approved by the FERC for application in the US. Since June 2007, NERC has had the authority to enforce its reliability standards in the United States.[6] FERC has also granted regional reliability organizations the authority to enforce federally-approved reliability standards.[7] Demand response is one tool that NERC and regional reliability organizations can use to help maintain a reliable transmission network.

According to NERC in 1999, transmission infrastructure development was not keeping pace with generation, as vertically integrated utilities divested their generation portfolios.[8] To address some of the complexities associated with this disconnect between generation and transmission, the FERC began to encourage 'system operators' to play a role in coordinating electricity transmission in the United States. Two types of system operators emerged: Regional Transmission Organizations (RTOs) and Independent System Operators (ISOs). Although they have different names, RTOs and ISOs are in most ways functional equivalents. The FERC did not order the creation of RTOs and ISOs, but encouraged their creation in Order 2000, issued 20 December 1999.[9] Today they coordinate electric supplies for two-thirds of the United States. RTOs and ISOs now play a significant role in organizing markets that allow demand response to be bought and sold in a manner similar to generation resources. This capacity to meet energy needs through energy reduction is sometimes referred to as a 'demand resource'. In this intricate system, coordinating the grid to keep it balanced is extremely challenging.[10] Increasingly RTOs and ISOs are looking to demand response to aid in this network-balancing process.

Reaching the full potential for demand response will depend on an advanced metering infrastructure that can support programmes that interrupt power during

[6] *Mandatory Reliability Standards for the Bulk Power System*, Order No. 693, 72 Fed. Reg. 16,416 (April 4, 2007), FERC Stats. & Regs. ¶ 31,242 (2007), *order on reh'g*, Order No. 693-A, 120 FERC ¶ 61,053 (2007), available at <http://www.ferc.gov/whats-new/comm-meet/2007/071907/E-6.pdf> (accessed 28 September 2011).

[7] See, for example, 133 FERC ¶ 61,066, available at <http://www.ferc.gov/whats-new/comm-meet/2010/102110/E-8.pdf> (accessed 28 September 2011).

[8] FERC, Order 2000, Regional Transmission Organizations (1999) at 18. ('In the period from 1988 to 1998, electricity demand in the US grew by 30 per cent, yet only 15 per cent of new transmission capacity was added.') Clark W. Gellings, *The Smart Grid: Enabling Energy Efficiency and Demand Response* (Lilburn, Georgia: The Fairmont Press, Inc, 2009) at 7.

[9] Ibid.

[10] Grid balancing has been described as follows: 'The electricity grid is a dynamic entity in a constant state of flux as the levels of supply from generators and demand from all consumers constantly change. Electricity cannot yet be stored economically: excess supply one day cannot be used to meet a surge in demand on another day. So the grid's operators must ensure that the amount of available electricity matches the amount consumed at any time, and vice versa. The grid must be able to respond to predictable movements in demand patterns as well as random, second-by-second changes. The grid's operators must also have plans in place to manage a major interruption to supply, such as a power station failing. To keep the grid balanced, traditional generators have a throttle control that enables them to vary their output instantaneously—this is known as response. The grid also maintains a buffer of spare capacity from 'balancing stations' which can be called on at extremely short notice. If a power station goes down, or there is a problem in transmission, these 'spare' stations swing into action. The problem the grid operators have is that this insurance policy is expensive. It adds to the cost of power supply and, because it must be kept running at a low level at all times, it produces CO_2 emissions even when power is not being used.' See <http://www.rltec.com/gridbalancing> (accessed 28 September 2011).

periods of peak demand, take advantage of energy efficiency gains, and help integrate new infrastructure into the energy system such as electric vehicles, advanced storage technologies, dynamic buildings, and 'smart' appliances. While the federal government is poised to play a significant leadership role in reducing the regulatory uncertainties associated with demand response, the majority of decisions that impact demand response are made at the local and state level. As a result, the footprint of demand response on infrastructure is likely to vary substantially across the country.

III. Overview of Electricity Regulation in the US

A. Federal versus state authority

Electricity regulation in the United States is divided between the federal and state governments. The federal government regulates 'wholesale' sales of electricity (and related activities including interstate transmission) because these transactions are primarily inter-state in character, while the states regulate retail markets and other intra-state activities. The authority for state government regulation of retail electricity markets is found in state 'police powers' that are reserved to the states under the Tenth Amendment to the United States Constitution.[11] When the US Constitution established a federal government, those powers not specifically delegated to the federal government were reserved to the states.[12] Among the state powers preserved were those necessary 'for the protection of the health, safety, morals, and general welfare' of the public.[13]

In contrast, the federal government in the US is a government of limited powers. There must be a constitutional basis for federal regulation. The United States Congress has authority to regulate wholesale sales of electricity (and other activities affecting them) under the Commerce Clause, which authorizes the federal government to regulate commerce among the states.[14] Congress has delegated the authority to regulate wholesale sales of electricity (and other related practices) to the FERC. However, the division of authority between the states and the FERC in areas such as demand response is not entirely clear.

[11] 'The powers not delegated to the United States by the Constitution, nor prohibited by it to the States, are reserved to the States respectively, or to the people.' United States Constitution, Amendment X. For a discussion of state authority in electricity regulation, see Richard F. Hirsh, *Power Loss: The Origins of Deregulation and Restructuring in the American Electric Utility System* (Mass: The MIT Press, 1999) 15.

[12] Ibid.

[13] Richard F. Hirsh, *Power Loss: The Origins of Deregulation and Restructuring in the American Electric Utility System* (Mass: The MIT Press, 1999) 15.

[14] The power 'To regulate Commerce with foreign Nations, and among the several States, and with the Indian Tribes' was vested in Congress, United States Constitution, art 1, s 8, cl 3.

B. Legal framework

1. Early regulatory history

Early in the development of the electricity sector, concerns about monopolistic power predominated. States treated the industry as a natural monopoly affecting the public interest. This monopoly required close oversight, especially to prevent price discrimination. In *Munn v Illinois*, the Supreme Court held that state governments may regulate rates charged by otherwise private industry when that industry is a 'necessity' for society, and when it is a 'virtual monopoly'.[15] These public necessity 'virtual monopolies' became known as 'public utilities'.[16] Because of the impact electric utilities have on the public interest and their tendency toward price discrimination, states created public utility commissions to review the rates charged by utilities to ensure that the rates were 'reasonable'.[17]

The federal government entered the utility regulatory picture in 1935 when the Roosevelt administration pushed Congress to enact the Public Utility Holding Company Act (PUHCA) in an effort to break up holding companies.[18] The Act subjected holding companies to oversight by the Securities and Exchange Commission and broke up many of the conglomerates.[19] The Federal Power Act of 1935 (FPA) gave a federal agency, the Federal Power Commission (FPC), authority over the interstate sale of electricity.[20]

2. PURPA

In 1978, another fundamental change in federal authority over electric utilities occurred with the passage of the Public Utility Regulatory Policies Act (PURPA). PURPA amended the FPA by opening the US wholesale electricity generation market to non-utility generators.[21] This marked the beginning of competition in wholesale generation and a shift in the infrastructure planning process that required consideration of resources supplied by, among other things, alternative energy production facilities, and ultimately demand response.

PURPA preserved the delicate balance between state and federal jurisdiction.[22] The federal government retained the authority to regulate '...the transmission of electric energy in interstate commerce and...the sale of electric energy at wholesale in

[15] See above n 13 at 16–17 (citing to *Munn v Illinois* (1877) 94 US 113).

[16] Ibid at 17 (citing to *The Compact Edition of the Oxford English Dictionary*, 2nd edn, 1987).

[17] Ibid at 26–28. The theory of calculating 'reasonable' rates evolved over the years, with the Supreme Court in 1898 in *Smyth v Ames* finding that a regulated company was entitled to 'a fair return upon the value of that which it employs for the public convenience'. 169 US 466, 547 (1898).

[18] For a discussion of the Public Utility Holding Company Act, see ibid at 40–1 (referring to the Public Utility Holding Company Act of 1935, since repealed by the Public Utility Holding Company Act of 2005, s 1263 of the Energy Policy Act of 2005).

[19] Ibid.

[20] The Federal Power Act of 1935. The Federal Power Commission (FPC) later became the Federal Energy Regulatory Commission (FERC).

[21] The Public Utility Regulatory Policies Act of 1978.

[22] It extended Federal regulation 'only to those matters which are not subject to regulation by the States'. Ibid.

interstate commerce'.[23] PURPA also left intact the requirement that rates be 'just and reasonable', and mandated that public utilities file 'schedules showing all rates and charges for any transmission or sale subject to the jurisdiction of the Commission, and the classifications, practices, and regulations affecting such rates and charges, together with all contracts which in any manner affect or relate to such rates, charges, classifications, and services'.[24]

PURPA sought to diversify electricity generation by opening the market to small generators that relied on traditional sources of power, as well as by encouraging increased use of renewable energy.[25] The Act required utilities to purchase power offered for sale by 'qualifying facilities' (QFs) (including 'small power production facilities' and 'cogeneration facilities')[26] so long as the price of their power was less than the 'avoided cost'. The avoided cost described the cost of power purchases from other utilities, or for the production of electricity by the utility from existing capacity or the construction of new generation capacity.[27] This meant that the utilities could not discriminate against the electricity offered by the new generators if it was price-competitive. This non-discrimination concept is today emerging in FERC orders that address requirements to treat demand response in certain circumstances in the same manner as they would treat conventional generation sources.

3. 1992 Energy Policy Act

The 1992 Energy Policy Act (1992 Act) began the process of liberalization by giving FERC the authority to mandate that utilities allow 'open access' to their transmission lines. Pursuant to the 1992 Act, FERC issued Order No. 888 in 1996 in an effort to have utilities 'unbundle' their generation and transmission activities, and allow other entities 'open access' to their transmission lines.[28] With the advent of this open access to transmission, wholesale electricity could more easily be traded among a wider range of entities. While PURPA allowed non-utility generators to emerge, the 1992 Act opened up the opportunity for markets in wholesale power and related services to emerge.

[23] 16 USC, s 824(a). By 1978, this Federal authority was vested in the Federal Energy Regulatory Commission (FERC) rather than the earlier Federal Power Commission.

[24] 16 USC, s 824d(c). PURPA also gave FERC the authority to establish a rate in the event that it found that a rate (or other related practice) submitted by a utility was 'unjust, unreasonable, unduly discriminatory or preferential'. 16 USC, s 824e.

[25] Above n 13 at 82.

[26] Edison Electric Institute, *PURPA: Making the Sequel Better than the Original* (2006).

[27] 'Avoided costs means the incremental costs to an electric utility of electric energy or capacity or both which, but for the purchase from the qualifying facility or qualifying facilities, such utility would generate itself or purchase from another source.' 18 CFR, s 292.101(b)(6). Since 'qualifying facilities' are not considered 'utilities', they are exempt from most requirements of the Federal Power Act and the Public Utility Holding Company Act. See also 'FERC: Industries—What are the Benefits of QF Status?' (FERC 2010), available at <http://www.ferc.gov/industries/electric/gen-info/qual-fac/benefits.asp> (accessed 28 September 2011).

[28] Federal Energy Regulatory Commission (FERC), Order 888, Promoting Wholesale Competition Through Open Access Non-discriminatory Transmission Services by Public Utilities; Recovery of Stranded Costs by Public Utilities and Transmitting Utilities, 75 FERC 61,080 (1996).

In addition to increasing competition at the wholesale level that was prompted by federal law, states enacted liberalization legislation. By the late 1990s, deregulation and other economic forces led to the dismantling of many vertically integrated utilities, significantly changing the way the utilities planned for energy production. However, not all states followed the same trajectory. While wholesale markets were being developed at the inter-state level, states still had control over the degree of deregulation they would allow at the 'intra'-state level. According to the US Energy Information Administration, in September of 2010, 28 states had little or no deregulation in place; the District of Columbia and 15 states had deregulated electricity markets to some degree; and seven states had started the process of deregulation, but subsequently suspended it.[29]

4. *The shift to wholesale markets*

The traditional model for ratemaking, known as 'cost-of-service' ratemaking, involved utilities that owned their own generation and filed their tariffs for approval with the FERC. The tariffs generally allowed utilities to recover their average costs, and to make a 'reasonable return on equity'.[30] With the development of wholesale markets operated by RTOs and ISOs, however, there is no longer the same opportunity for controlled regulatory oversight to match the costs of generation with rate recovery because not all generators are utilities that are required to submit tariffs for review. Instead, RTOs and ISOs oversee 'wholesale capacity markets' where generators of all kinds (both 'utilities' and 'non-utility generators') can bid their generation capacity into the market where it can be purchased by 'retailers'—or business entities that buy the bulk power for resale at the retail, or individual consumer, level. Roughly two-thirds of US electricity consumers are served by an RTO or ISO operating in an organized wholesale market.[31] These markets include trading in 'real time' as well as trading 'day-ahead' electricity sales.

Although these markets are the product of 'deregulation' or 'restructuring', they are not wholly unregulated. The FERC still has the obligation to ensure that 'wholesale electricity rates, including rates in these organized markets, are just and reasonable and are not unduly discriminatory or preferential'.[32] This includes rates paid to organizations that meet capacity needs through DR.[33]

[29] See <http://www.eia.gov/cneaf/electricity/page/restructuring/restructure_elect.html> (accessed 28 September 2011).

[30] See M. Wittenstein and E. Hausman, 'Incenting the Old, Preventing the New: Flaws in Capacity Market Design, and Recommendations for Improvement' (2011), available at <http://www.publicpower.org/files/PDFs/2011APPACapacityMarketsReport.pdf> (accessed 28 September 2011).

[31] Hon. J. Wellinghoff, D. L. Morenoff, J. Pederson, M.E. Tighe, Federal Energy Regulatory Commission, 'Creating Regulatory Structures for Robust Demand Response Participation in Organized Wholesale Electric Markets 1', available at <http://www.ferc.gov/about/com-mem/wellinghoff/ACEEE_article.pdf> (accessed 28 September 2011). The following ISOs and RTOs were active in the United States as of 2009: ISO New England, New York ISO, PJM Interconnection, MISO ISO, SPP RTO, ERCOT RTO, and California ISO.

[32] Ibid at 2.

[33] There is a hotly contested debate among participants in the electricity sector about whether wholesale markets are resulting in just and reasonable rates and whether/how these markets should

IV. Law of Demand-side Management

Although the rate of increasing demand for electricity in the US has slowed in recent years due to the economic downturn, the Energy Information Administration (EIA) projects an approximate increase in total electricity demand of approximately 30 per cent from 2008–2035,[34] and that approximately 250 gigawatts (GW) of new capacity will need to come on line between 2009 and 2035 to meet that demand.[35] Projections suggest that the technology choices to meet this need for new capacity will depend on a wide range of factors, including federal and state climate change policies. Some technology choices are likely to replicate the current central power station model, requiring potential upgrades in transmission, but not necessarily new transmission lines. Other more 'distributed' technology choices such as wind and solar power facilities will require additional transmission lines to transport the electricity from more remote locations to high-demand areas. Still other choices may result in new and expanded demand response programmes, displacing the need for new generating and transmission capacity.

Demand response may be elicited through prices that encourage reductions in usage, or through programme-based incentives provided by the utility or an energy service company.[36] The National Energy Conservation Policy Act (NECPA),

be reformed. See <http://www.publicpower.org/files/PDFs/EMRICompetitiveMarket.pdf> for The American Public Power Association's position on market reform (accessed 28 September 2011).

[34] The Annual Energy Outlook, Energy Information Administration (2010). See <http://www.eia.gov/oiaf/aeo/pdf/trend_3.pdf> (accessed 28 September 2011). This scenario is referred to as the 'reference case'—which means that projections are based on current law. The full report also includes 38 'sensitivity' scenarios with different assumptions. The full report can be accessed at <http://www.eia.gov/oiaf/archive/aeo10/index.html> (accessed 28 September 2011).

[35] See <http://www.eia.gov/oiaf/archive/aeo10/electricity.html> (accessed 28 September 2011). The precise amount of new capacity in generation required will depend on the rate of actual economic growth.

[36] The FERC generally divides Demand Response programmes into 'dispatchable' and 'non-dispatchable'. The Dispatchable programmes include 'curtailable or interruptible rates', meaning that customers get reduced rates for agreeing to lower or discontinue use when they are called upon by the retail energy provider. Dispatchable programmes also include 'direct load control', or agreements between customers and retail energy providers that the provider will have some degree of access to control appliances in the customer's home or business. Finally, dispatchable programmes include incentive programmes at the wholesale level where RTOs/ISOs/Aggregators will pay customers to change usage behaviour according to system needs. This final dispatchable category can involve agreements between certain customers and RTOs/ISOs/Aggregators, or can involve signals from wholesale markets. If a customer bids demand response into a market and the bid is accepted, the 'signal' to the customer comes from this market interaction but is still triggered by a decision on the part of the system operator or service provider that a change in behaviour is desirable to meet demand. Non-dispatchable programmes describe the suite of programmes that use price to encourage consumers to change their behaviour. In this set of programmes, dynamic pricing is used to capture the cost of additional energy use in times of peak demand. These programmes are 'non-dispatchable' because they rely on consumer behaviour. The only 'signal' from the energy provider is in the form of higher or lower prices. Examples of dynamic pricing schemes include: 'real-time pricing', where prices are not set ahead of time and literally fluctuate every hour; 'critical peak prices', prices set higher for peak energy-use periods; and 'static time-varying prices', or prices that vary depending on the time of day, week, or year. The Federal Energy Regulatory Commission (FERC), *National Action Plan on Demand Response* (2010) 3–4.

passed in 1978,[37] has been cited by some as the advent of 'modern' DR programmes in the United States.[38] One of the stated purposes of NECPA was to 'reduce the growth in demand for energy in the United States'.[39] Among other conservation programmes, NECPA required that utilities offer energy audit opportunities to their customers, the beginning of institutionalized energy efficiency education.

The 1992 Energy Policy Act (1992 Act) amended PURPA to require that electric utilities consider adopting an 'integrated resource planning' (IRP) model, meaning that planning had to address the full range of resource options, including both demand- and supply-side alternatives.[40] The 1992 Act did not demand that utilities engage in IRP, but rather required *consideration* of IRP and outlined the requirements that utilities would need to follow if they chose to engage in the IRP process. If a utility chose to adopt IRP, then the utility was required to file the plan with the state public utility commission (or other regulatory authority) and had to engage in public comment.[41] Some states, such as Minnesota, mandated integrated resource planning.[42] As a result of both federal and state efforts, demand resources were integrated into utility plans and encouraged by an array of programmes.

Other DR initiatives were, of course, market driven. Utilities sometimes turned to DR programmes to help defer the cost of building expensive new generating plants. They also used peak load management approaches to reduce the high cost of purchased power during times of high demand. A number of states enacted Conservation Improvement Programmes (CIPs) that required utilities to spend a designated amount of the utility's annual gross operating revenues on projects that reduce the consumption of electricity. Utilities recover these expenses in their rate base. For example, Minnesota set the CIP investment rate at 1.5 per cent of gross operating revenues for conventional power production facilities and 2 per cent for nuclear facilities.[43]

[37] 42 USC, s 8201 et seq. (1978).

[38] J. Eto, *The Past, Present and Future of U.S. Utility Demand-Side Management Programs* (1996) 5. See <http://eetd.lbl.gov/ea/emp/reports/39931.pdf> (accessed 15 October 2011).

[39] 42 USC, s 8201(b).

[40] 'The term "integrated resource planning" means, in the case of an electric utility, a planning and selection process for new energy resources that evaluates the full range of alternatives, including new generating capacity, power purchases, energy conservation and efficiency, cogeneration and district heating and cooling applications, and renewable energy resources, in order to provide adequate and reliable service to its electric customers at the lowest system cost. The process shall . . . treat demand and supply resources on a consistent and integrated basis'. 16 USC, s 2602(19).

[41] The 1992 Energy Policy Act amended 16 USC s 2601 et seq. See 16 USC s 2621 for requirements that utilities consider adopting certain standards, including integrated resource planning.

[42] For example, public utilities in Minnesota must develop 'a set of resource options that a utility could use to meet the service needs of its customers over a forecast period, including an explanation of the supply and demand circumstances under which, and the extent to which, each resource option would be used to meet those service needs. These resource options include using, refurbishing, and constructing utility plant and equipment, buying power generated by other entities, controlling customer loads, and implementing customer energy conservation.' Minn. Stat., s 216B.2422, subd. 1(d) (2010).

[43] Office of the Legislative Auditor, State of Minnesota, The Conservation Improvement Program (2005) at 3, available at <http://www.auditor.leg.state.mn.us/ped/pedrep/0504ch1.pdf> (accessed 28 September 2011).

These early DR programmes delivered mixed results[44] and gradually declined in prominence as states deregulated electric utilities. According to the US EIA, in 1996, electric utilities attributed 29,893 megawatts of reduced peak load to energy efficiency and 'load management' programmes.[45] In 2000 and 2003, only 22,901 and 22,904 megawatt reductions resulted from these demand-side management programmes.[46] The EIA attributed this decline, in part, to deregulation causing utilities to scale back demand-side management programming.[47] However, after 2003, investment in demand-side management programs began to steadily rise again.[48] In addition, 19 states retained conservation improvement programmes or enacted new 'public goods charges'[49] that created 'public benefit' programmes that often support DR initiatives.[50]

In 2005, the Energy Policy Act of 2005 (2005 Act) mandated that the FERC gather relevant data on demand response and advanced metering programmes and publish an annual report on the country's progress.[51] The Energy Independence and Security Act of 2007 (EISA) required that the FERC complete a National Assessment of Demand Response and an Action Plan for Demand Response.[52] Congress also required that the FERC in cooperation with the Secretary of Energy submit a proposal to implement the National Action Plan to Congress.[53] The National Assessment of Demand Response Potential has identified a significant amount of untapped potential in energy savings through DR programmes.

Also in 2007, the FERC issued orders 890 and 890-A. Order 890 authorized the use of demand response in transmission planning and ancillary service provision, recognizing that some wholesale markets were already including demand response in both.[54] Order 890-A clarified that demand response should be considered equivalent to other resources in transmission planning.[55] In 2008, Order 719 required

[44] 'Some utilities, notably those with large DSM programs, had saved energy at cost of less than $.02/kWh, while others had saved energy at a cost in excess of $.10/kWh.' J. Eto, *The Past, Present and Future of U.S. Utility Demand Side Management Programs* (1996) at 12. See <http://eetd.lbl.gov/ea/emp/reports/39931.pdf> (accessed October 15, 2011).

[45] US Energy Information Administration, Annual Energy Review 2009, Table 8.13 Electric Utility Demand-Side Management Programs, 1989–2008, August 2010. Available at <http://www.eia.gov/totalenergy/data/annual/pdf/aer.pdf> (accessed 28 September 2011).

[46] Ibid.

[47] Ibid, note 4.

[48] Ibid.

[49] For example, in California, 'a public goods charge was established that ensured minimum funding levels for "cost effective conservation and energy efficiency" for the 1998–2002 period, and then (in 2000) extended through the year 2011'. The World Bank, 'Primer on Demand Side Management' (2005) 40, available at <http://siteresources.worldbank.org/INTENERGY/Resources/PrimeronDemand-SideManagement.pdf> (accessed 28 September 2011).

[50] See The Pew Center on Global Climate Change, Public Benefits Funds: <http://www.pewclimate.org/what_s_being_done/in_the_states/public_benefit_funds.cfm> (accessed 28 September 2011).

[51] See <http://www.ferc.gov/legal/staff-reports/2010-dr-report.pdf> (accessed 28 September 2011).

[52] Energy Independence and Security Act of 2007, Public Law 110–140, s 529.

[53] Ibid.

[54] FERC defines ancillary services as 'services necessary to support the transmission of electric power from seller to purchaser given the obligations of control areas and transmitting utilities within those control areas to maintain reliable operations of the interconnected transmission system'. See <http://www.ferc.gov/market-oversight/guide/glossary.asp> (accessed 28 September 2011).

[55] FERC, Order 890-A, Preventing Undue Discrimination and Preference in Transmission Service (2007).

ISOs and RTOs to allow demand response to be bid into markets on equal footing with other resources for provision of some ancillary services.[56] The FERC also required that RTOs and ISOs review their markets to ensure that the resources were reflecting accurate values in times of 'operating reserve shortages'.[57]

V. The Future of Demand Response

A. Demand response as a resource

Although the United States has for perhaps two decades relied on demand-side resources to shave peak demand and respond to temporary crises, demand resources are increasingly viewed as longer-term 'resources' that are part of the basic electricity portfolio. These demand resources can be used in planning for how to meet energy demand, reducing load on transmission lines, and assisting in load balancing as new, intermittent sources come on line. In addition, while participation in demand response programmes has predominantly been by large consumers of electricity in the commercial and industrial sectors, the future of fully integrated demand response programmes envisions far broader participation at the individual household level. The question of how demand response will impact energy networks and how these networks will impact it depends largely on how these planning processes are structured to integrate demand response plans in the coming years at the national, regional, state, and local levels.

B. Potential for demand response to impact infrastructure development

1. Deferring the need for building new generation and associated transmission lines

The National Assessment of Demand Response Potential looked at four different scenarios to project the potential for demand response participation to impact peak demand.[58] Under the 'business as usual' scenario, if current demand response programmes were implemented as planned, FERC forecasts that demand response programmes could be responsible for a four per cent reduction in peak demand by 2019.[59] In the second scenario, if all states adopted demand response programmes, some advanced meters were installed, and five per cent of customers participated in dynamic pricing programmes, the US could see peak demand reduced by nine per cent by 2019.[60] In the third scenario, or the 'achievable scenario', if there were

[56] FERC, Demand Response Compensation in Organized Wholesale Energy Markets (2011) 10.

[57] Ibid.

[58] FERC, A National Assessment of Demand Response Potential (2009).

[59] Ibid at x. This percentage reduction was calculated by comparing the potential forecast with a baseline scenario with no demand response programmes.

[60] Ibid at xi. This reduction is again compared to a baseline scenario with no demand response programmes.

'universal' deployment of advanced meters, dynamic pricing as the 'default', and high participation rates in both dynamic pricing and other demand response programmes, there could be a peak demand reduction of 14 per cent by 2019.[61]

Finally, in the 'full participation' scenario, if there were universal deployment of advanced metering infrastructure and all customers participated in dynamic pricing, there could be a 20 per cent peak demand reduction by 2019.[62] According to FERC, the difference between the 'full participation' scenario and the 'business as usual' scenario could be the equivalent of 2000 'peaking' power plants or a 150 GW peak load reduction.[63] These scenarios paint a clear picture that the degree of demand response participation in the future will depend on a number of variables, including the deployment of advanced metering technology and consumer participation, among others.

While the report focused on reduction in demand for new generation, the fact that the 'achievable scenario' in the National Assessment would reduce peak demand by 14 per cent also indicates that more pervasive DR programmes are likely to have significant effects on the need for new transmission and on congestion within existing transmission lines. Further, the fact that the achievable scenario envisions 'universal' adoption of smart metering indicates that the scenario relies on a two-way network in contrast to the traditional use of transmission lines as a one-way source of supply to consumers.

2. *Role of demand response resources in the RTOs and ISOs*

In areas where an ISO or RTO has been established, the ISO or RTO plays a role in coordinating multiple plans of various entities to ensure there are adequate resources and services for a balanced grid.[64] Generally, in the RTO/ISO context, demand resources may participate in multiple markets, including capacity markets, ancillary service markets, and energy markets. Capacity markets are the organized wholesale markets where purchasers of bulk power look to buy enough 'capacity' to meet whatever obligations are imposed by law. Energy markets are commodities markets that involve trading actual energy supply, as opposed to the capacity to generate energy. Exactly how DR is treated varies from one RTO or ISO to another.

3. *Demand resources in the Pennsylvania–New Jersey–Maryland RTO*

The Pennsylvania–New Jersey–Maryland Interconnection (PJM) is an RTO that coordinates wholesale electricity in 13 states and the District of Columbia.[65] Generation planning happens at both the state level and the PJM level. NERC

[61] Ibid at xii. This reduction is again compared to a baseline scenario with no demand response programmes.

[62] Ibid.

[63] Ibid at x.

[64] Deutsche Bank Group, *Natural Gas and Renewables: A Secure Low Carbon Future Energy Plan for the United States* (2010) 104.

[65] See <http://www.pjm.com/about-pjm.aspx> (accessed 28 September 2011) for background information on PJM.

established a series of 'generation planning reserve margin requirements' to ensure reliability. To meet these requirements, either the 'load-serving entities' provide the supply or demand resources to meet those requirements, or the ISO/RTO obtains the supply or demand resources needed through a capacity auction.[66] PJM is also responsible for transmission planning and develops Regional Transmission Expansion Plans that project transmission needs to 15-year time horizons.[67] Because RTOs and ISOs are also responsible for 'interconnections' (literally linking generation to transmission at the wholesale level), PJM is also responsible for 'scheduling' which generation resources will be used to maintain reliability in the RTO's territory.

PJM represents a particularly sophisticated example of the range of demand response programmes and the various ways demand resources may factor into short- and long-term planning for generation and transmission. In other words, when planners are considering whether new-generation capacity or transmission lines will be needed to meet future demand, PJM has devised a way to incorporate these demand response programmes as one type of 'resource' that may be used in place of new generation, or as a means of reducing demand on transmission in congested areas.

PJM manages demand response programmes in a number of different wholesale markets. In capacity markets, PJM allows demand response to participate in auctions as any other source of capacity for generation would participate as a 'forward capacity resource'. This means that consumers that are willing to provide reductions in energy usage are viewed, by the market, as resources similar to a new generating plant. Both provide (or 'free up') capacity.[68] Since it is unusual that a single consumer could provide sufficient capacity to become relevant in the market, PJM does not usually deal directly with consumers, but rather deals with companies that 'aggregate' the demand resources from a number of different end-users into a package that the market recognizes as relevant. One type of company that provides these aggregation services is known as a Curtailment Service Provider (CSP). These CSPs include utilities, energy service companies, and other firms that focus on demand response services.[69]

In addition to providing the opportunity for demand resources to participate in capacity markets, PJM also allows demand resources to participate in ancillary services markets[70] and energy markets.[71] How much money is made by the retail

[66] 2010 ISO/RTO Metrics Report (2010) 17, available at <http://www.isorto.org/atf/cf/%7B5B4E85C6-7EAC-40A0-8DC3-003829518EBD%7D/2010%20ISO-RTO%20Metrics%20Report.pdf> (accessed 28 September 2011).

[67] Ibid at 261.

[68] See Retail Electricity Consumer Opportunities for Demand Response in PJM's Wholesale Markets (undated) 2, available at <http://www.pjm.com/markets-and-operations/~/media/markets-ops/dsr/20101203-end-use-customer-fact-sheet.ashx> (accessed 28 September 2011).

[69] Ibid at 1.

[70] Ibid at 3. 'There are three Ancillary Services markets in which economic demand response resources may participate: Synchronized Reserves (the ability to reduce electricity consumption within 10 minutes of PJM dispatch), Day Ahead Scheduling Reserves (the ability to reduce electricity consumption within 30 minutes of PJM dispatch) and Regulation (the ability to follow PJM's regulation and frequency response signal).'

[71] 'PJM's Economic Load Response program enables demand resources to voluntarily respond to PJM locational marginal prices (LMP) by reducing consumption and receiving a payment for the

customers that participate in these programmes depends on how much value is placed on this 'service' or 'resource'. Lessons learned from the experience of ISOs and RTOs such as PJM will be invaluable as the US moves forward to further integrate demand response.

C. The potential impacts of technology and infrastructure development on demand response

1. *Smart grid technology*

Smart grid technology is considered essential to enabling demand response and energy efficiency to reach their 'full potential'.[72] The 'smart grid' has been defined in many different ways. As explained by Clark Gellings in his recent book, *The Smart Grid: Enabling Energy Efficiency and Demand Response*, the 'smart grid' broadly refers to 'the use of sensors, communications, computational ability and control in some form to enhance the overall functionality of the electric power delivery system'.[73] The smart grid offers opportunities to build a new kind of network that better integrates distributed sources of generation, improves grid reliability, and allows consumers and utilities to improve efficiency by offering real-time and detailed information about electricity use. By enabling consumers to have real-time information about the cost of electricity at a given time, the smart grid provides the necessary information to vastly improve demand response participation rates.[74] As outlined in the National Assessment, smart grid technologies, like advanced metering infrastructure, are considered key to successful demand response participation.

The speed with which the infrastructure is dispatched will undoubtedly play a role in how quickly participation in demand response programmes increases, and how much of an impact demand response is able to have on infrastructure planning in the short term. Moreover, although the federal government has played a significant role in encouraging smart grid deployment, the decisions to allow rate recovery for the portion of the costs of the smart grid paid by utilities will largely be made at the local and state levels.[75]

reduction. Using the day-ahead alternative, qualified market participants may offer to reduce the load they draw from the PJM system in advance of real-time operations and receive payments based on day-ahead LMP for the reductions.' See <http://pjm.com/markets-and-operations/demand-response/dr-energy-market.aspx> (accessed 28 September 2011).

[72] See The Brattle Group, 'Transforming America's Power Industry: The Investment Challenge 2010–2030' (2008) vi, available at <http://www.eei.org/ourissues/finance/Documents?Transforming_Americas_Power_Industry.pdf> (accessed 28 September 2011).

[73] Clark W. Gellings, *The Smart Grid: Enabling Energy Efficiency and Demand Response* (Lilburn, Georgia: The Fairmont Press, Inc., 2009) at 1.

[74] The smart grid and demand response are so linked that a collaborative between FERC and the National Association of Regulatory Utility Commissioners (NARUC) merged collaborative efforts on demand response and the smart grid into one collaborative on 'Smart Response'. The National Association of Regulatory Utility Commissioners. See <http://www.naruc.org/Ferc/default.cfm?c=3> (accessed 28 September 2011).

[75] The National Association of Regulatory Utility Commissioners has put together a website with state information on smart grid initiatives. See <http://www.naruc.org/SmartGrid/> (accessed 28 September 2011).

2. *Plug-in electric vehicles (PEVs)*

The deployment of plug-in electric vehicles (PEVs) in the US may also impact on demand response. PEVs, especially when combined with improved storage, have the potential to provide critical ancillary grid-balancing services that may help accommodate the continued expansion of intermittent power sources. However, PEVs may also place new pressures on the electricity network.[76] Although the federal government has supported the development of PEVs and electric vehicle infrastructure,[77] the development of enabling infrastructure is still uncertain and will fall predominantly on states and local governments to sort out.[78]

3. *Variable and renewable generation sources*

Variable and renewable generation sources make the need for demand response more acute and may enable higher participation rates. Wind, in particular, may pose intermittency challenges to the grid that could be somewhat alleviated by improved demand response programmes.[79] Wind tends to provide most energy at night, when demand for electricity is at its lowest. If, for example, electric vehicles were coupled with effective demand response programmes through an interactive electricity network, consumers could charge their car batteries over night, drawing down the excess power on the grid. While the current grid lacks the information required to enable real-time coordination of intermittent small-scale generation sources, a smart grid coupled with increased demand response participation could help enable more streamlined use of distributed generation sources and transmission networks.

4. *Smart buildings and appliances*

Finally, 'smart buildings' and 'smart appliances' hold tremendous potential for enabling higher participation rates in demand response programmes, and may have even greater impacts on energy use when combined with distributed generation. 'Smart buildings' are those that have a centralized communication system for controlling systems such as air conditioning. With consumer consent, these 'smart buildings' can communicate with utilities or other service providers to respond in real time to changes in energy usage or pricing.

[76] D.B. Sandalow (ed), *Plug-In Electric Vehicles: What Role for Washington?* (Washington, DC: Brookings Institution Press, 2009) 51.

[77] See <http://avt.inel.gov/pdf/phev/phevInfrastructureReport08.pdf> (accessed 28 September 2011).

[78] Local governments are, in many ways, already taking the lead. For example, the 'C40 cities' have developed an 'Electric Vehicle Network' to support one another's efforts in deploying electric vehicles as part of efforts to reduce greenhouse gas emissions. Fourteen of the 'C40 cities', including Chicago and Los Angeles in the United States, have committed to 'make their cities more electric vehicle friendly' by focusing on efforts such as streamlining, permitting, and facilitating electric vehicle infrastructure development. See <http://www.c40cities.org/news/news-20091216.jsp> (accessed 28 September 2011).

[79] R. Malme, P. Davis, and J. Strömbäck, 'Demand Response 2.0', *Spark Fortnightly* (2010) 10. See <http://www.fortnightly.com/uploads/August10_SPARK.pdf> (accessed 15 October, 2011).

In addition, 'smart appliances' are becoming increasingly common. Smart appliances describe the suite of appliances that have sensor and control technologies enabling them to 'communicate' with consumers and utilities. In some pilot areas, smart appliances are being used in combination with smart meter pilots implemented by utilities.

Technology and infrastructure decisions related to electricity networks promise to impact the efficacy of demand response programmes and their ability to, in turn, impact the use and reliability of the networks. Decisions about how demand resources are integrated into resource planning to ensure adequate generation and transmission capacity will impact the degree of demand response engagement. Decisions about enabling infrastructure such as the smart grid, electric vehicles, storage technology, and dynamic buildings and appliances will also impact the degree of demand response engagement. While full participation of demand response holds tremendous potential to defer new network investments and to maximize the value of existing networks, investments in enabling technologies will be required to fully realize these benefits.

VI. Legal Issues in Realizing Fully Integrated Demand Response

A. What *is* demand response?

One of the major legal issues related to full engagement of demand response is the uncertainty associated with *what* demand response is, how it should be compensated, and by whom. According to the definition of 'demand response' in the FERC regulations, demand response is generally considered only a reduction in energy use. As FERC found in its recent *EnergyConnect* decision, 'demand response', as currently defined by regulations, is not a wholesale sale of electricity because there is no 'feed-in' of electricity back into the grid for resale.[80]

However, the future 'vision' of demand response involves allowing 'demand response resources' to provide a range of reliability and balancing services.[81] This broader role for demand response has been encouraged by the federal government and, under current law, may bring activities of 'aggregators of retail consumers', such as the Curtailment Service Providers, potentially under the jurisdiction of FERC for some activities. Sales of DR in capacity markets increasingly look like wholesale sales of electricity. If these aggregators are subject to FERC authority, they may be required to obtain 'market-based rate authorization' for transactions that involve resale of energy.[82] Subjecting these aggregation activities to federal

[80] FERC, Energy Connect Inc, 130 FERC P 61031 (2010) at 4–6. The decision also notes, however, that the purchase and subsequent resale of ancillary services may be considered a wholesale sale of electricity, subjecting the selling entity to regulation as a public utility under subsections 201(b) and (e) of the Federal Power Act.

[81] Ibid. [82] Ibid.

authority is controversial given the historic role of states in managing (almost exclusively) the retail side of electricity provision.

B. How should demand response be compensated?

There has also been a heated debate about how demand resources should be compensated. In March 2011, FERC issued a final rule on compensation for demand response in organized wholesale markets.[83] Prior to the March 2011 rule, the FERC had allowed RTOs and ISOs to develop their own compensation regimes, leading to wide variation in compensation for demand resources.[84] The March 2011 rule sets out a standardized compensation format based on a 'net benefits test'. When 'a demand resource' in an organized wholesale market is available to balance the grid and is 'cost-effective', the rule requires that the demand resources be compensated at the 'locational market price'.[85] The rule also addresses how the costs of demand response should be allocated among benefitting consumers.[86]

The comments FERC received on this rule illustrated the deep divisions about what demand resources are and what they are worth. Some were in favour of the outcome of the rule, arguing that demand resources should be identical to supply in terms of compensation.[87] Others argued that the compensation for demand resources should not include payment for generation and transmission, since a demand response 'reduction' does not involve either. Others suggested that, since demand resources already save the consumer money (by lowering the consumer's bill), paying the consumer for that service is tantamount to 'double' compensation.[88] Moreover, some stakeholders were also concerned that this 'double' compensation of demand resources would encourage inefficiencies in demand response—that consumers would choose to lower electricity resources even when supply resources would otherwise offer greater benefits without the 'double' counting.[89] The impact of the rule remains to be seen. It is, at the very least, an attempt at regulatory clarity to encourage demand response participation.

C. How far does the federal government's constitutional authority to regulate demand response extend?

Although the federal government continues to more closely regulate demand response, there may be challenges to the FERC's constitutional authority to do so—especially since some demand resources are originating through aggregation of consumer decisions at the retail level. As explained above, regulatory jurisdiction in the United States is allocated between the federal government and the states. Grounded in the commerce clause of the US Constitution and established

[83] FERC, Demand Response Compensation in Organized Wholesale Energy Markets (2011).
[84] Ibid at 12–13. [85] Ibid at Summary. This price is the 'locational marginal price' (LMP).
[86] Ibid at 15. [87] Ibid. [88] Ibid.
[89] For a full discussion of these pricing debates, see FERC, Demand Response Compensation in Organized Wholesale Energy Markets (2011) 17 *et seq*,

in the Federal Power Act, the federal government has the authority to regulate the 'transmission of electric energy in interstate commerce and... the sale of electric energy at wholesale in interstate commerce'.[90] FERC also has the authority to ensure that rates, charges, and classifications, are just, reasonable, and not unduly discriminatory or preferential, and to ensure that any '... rule, regulation, practice, or contract affecting such rate, charge, or classification' is also just, reasonable, and not unduly discriminatory or preferential.[91]

While FERC has ruled that demand response is not a 'wholesale sale of electricity',[92] FERC may retain jurisdiction to regulate demand response activities based on arguments that demand response involves 'practices' that affect rates, as outlined in Order 719-A. In the compensation order, FERC explained that it has 'jurisdiction over demand response in organized wholesale energy markets, because it directly affects wholesale rates'.[93] FERC may also retain jurisdiction over other wholesale transactions that are related to demand response, but that fall outside of the Commission's technical definition of demand response. As FERC noted in the *EnergyConnect* decision, the purchase and resale of certain ancillary services may constitute a wholesale sale of electricity.[94]

Even if federal authority extends to regulation of demand response in wholesale markets, whether states may prevent individual consumers from participating in demand response that is bid into wholesale markets remains unclear. At the moment, some states allow demand response programmes implemented by utilities but have stopped short of allowing non-utility participation in demand response efforts. The jurisdictional questions surrounding demand response are likely to continue and may shape the level of participation allowed and encouraged by states. The federal government is likely to continue playing a key leadership role in identifying nationwide standards and facilitating the participation of demand response. States, however, may carry a significant amount of responsibility in determining the future of their retail consumers' participation.

VII. Conclusion

Demand response holds tremendous potential to defer the need for new generation capacity and new transmission networks, to relieve congestion on existing transmission networks, and to enable a more reliable, efficient, interactive, and climate-friendly grid. The patchwork of approaches in the United States provides a unique opportunity for the development of locally relevant and innovative approaches. It also raises the potential for conflict and confusion. Looking ahead, full participation of demand response will depend on closely coordinated federal regulatory leadership and state implementation. Federal efforts will be necessary to create a

[90] 16 USC, s 824(b)(1).
[91] 16 USC, s 824e.
[92] FERC, Energy Connect Inc, 130 FERC P 61031 (2010).
[93] FERC, Demand Response Compensation in Organized Wholesale Energy Markets (2011) 86.
[94] FERC, Energy Connect Inc, 130 FERC P 61031 (2010) at 4–6.

stable regulatory environment and encourage investment in enabling infrastructure and technology. State efforts will be needed to provide the necessary regulatory framework and rate incentives to support expansion of demand response programmes, appropriate for local geography. The diversity and flexibility of the US system may present unique challenges, but it also affords a unique opportunity to build an effective decentralized, modern electricity network in which demand response plays a pivotal role.

10

Establishing an Offshore Electricity Grid: A Legal Analysis of Grid Developments in the North Sea and in US Waters

Olivia Woolley, Peter J. Schaumberg, and Graham St. Michel

I. Introduction

Offshore wind energy generation could make a significant contribution to meeting European electricity demand in the coming decades, and, in doing so, assist the compliance of Member States of the European Union (EU) with their respective commitments under the Renewable Energy Directive[1] to increase the proportion of energy from renewable sources in total national energy consumption.[2] The exploitation of this resource could also contribute to reducing the EU's carbon emissions whilst enhancing its energy security. Forty-nine offshore wind farms have already been constructed and are grid-connected in the coastal waters of European states, with the construction of many more planned by EU states during the remainder of this decade.[3] In contrast, no offshore wind farms have been constructed as yet in the US (although one has been authorized).[4] However, the significant potential for wind energy generation in the shallow waters off the Atlantic coastline has been recognized.[5] The Obama administration has recently announced that US$50m

[1] Council Directive 2009/28/EC on the promotion of the use of energy from renewable sources and amending and subsequently repealing Directives 2001/77/EC and 2003/30/EC [2009] OJ L140/16, Article 3 (Renewable Energy Directive).

[2] European Environment Agency, 'Europe's Onshore and Offshore Wind Energy Potential: An Assessment of Environmental and Economic Constraints' (Technical report No. 6/2009, 2009) Executive Summary.

[3] European Wind Energy Association, 'Offshore Wind', available at <http://www.ewea.org/index.php?id=203> (accessed 17 October 2011).

[4] US Department of the Interior, 'Secretary Salazar Announces Approval of Cape Wind Energy Project on Outer Continental Shelf of Massachusetts' (DOI News, 28 April 2010), available at <http://www.doi.gov/news/doinews/Secretary-Salazar-Announces-Approval-of-Cape-Wind-Energy-Project-on-Outer-Continental-Shelf-off-Massachusetts.cfm> (accessed 24 March 2011).

[5] Wendy Koch, 'Obama Fast-tracks Mid-Atlantic Offshore Wind Energy' (USA Today, 7 February 2011), available at <http://content.usatoday.com/communities/greenhouse/post/2011/02/us-mid-atlantic-offshore-wind-energy-/1> (accessed 24 March 2011).

funding will be made available to promote the offshore wind sector's growth in this coastal region.[6]

A foundational consideration for the expansion of this sector is how offshore wind farms are to be linked to onshore electricity transmission systems. Standard practice to date has been to connect them through dedicated cables as and when they are constructed. However, it has been widely argued that a more strategic approach in which plans for offshore energy and associated infrastructure development are coordinated should be adopted if best use is to be made of the offshore wind energy resource.[7] Factors that favour coordinated development include: the costs savings if several wind farms use the same cable connection;[8] the benefits of a likely reduction both in the environmental stresses that multiple separate cable connections may generate and of permitting requirements for offshore construction;[9] the stimulus that firm plans for infrastructure development would provide for offshore wind energy—a problem to date having been a reluctance by developers to commit funds until the availability of a grid connection is certain;[10] and the possibility of using new interconnector capacity to overcome congestion in onshore transmission systems as well as for connecting offshore generation.[11] In addition, strategic planning for the sector could alleviate some of the difficulties that attaching many separate sources of intermittent energy would create for onshore grid operation by aiming to maximize the 'smoothing' effect of linking geographically disparate wind farms that can take advantage of moving (or different) weather fronts to a single offshore transmission system.[12]

These amongst other considerations have prompted recent proposals in the EU and the US for major offshore electricity transmission projects. In the EU, the states bordering the North Sea have jointly committed to explore the technical feasibility of developing a North Sea offshore grid, and, as part of this, to examine how possible legal and regulatory barriers to such a development could be overcome.[13] In the US, investors announced their intention in October 2010 to construct the

[6] Ibid.

[7] Commission, 'Energy Infrastructure priorities for 2020 and Beyond—A Blueprint for an Integrated European Energy Network', COM(2010) 677/4, 26–9; Commission, 'Offshore Wind Energy: Action needed to deliver on the Energy Policy Objectives for 2020 and Beyond', COM(2008) 768 final, 7–9; G.W. Adamowitsch, 'European Coordinator's First Annual Report' (Report to Commission, September 2008) 8.

[8] COM(2010) 677/4 (above n 7) at 27; M.M. Roggenkamp et al, 'Market and Regulatory Aspects of Transnational Offshore Electricity Networks for Wind Power Interconnection' (2010) 13 *Wind Energy* 483–4.

[9] A. Woyte et al, 'European Concerted Action on Offshore Wind Energy Deployment: Inventory and Analysis of Power Transmission Barriers in Eight Member States' (2007), 10 *Wind Energy* 367.

[10] Ibid 364. [11] COM(2008) 768 (above n 7) at 5.

[12] A. Woyte et al (2007) (above n 9) at 365.

[13] The North Sea Countries Offshore Grid Initiative, 'Memorandum of Understanding', 3 December 2010 (Memorandum of Understanding). Possibilities for connecting offshore wind farms to interconnectors are also being explored in the Baltic Sea (for example, see the Kriegers Flak project between Germany, Denmark, and Sweden). Our focus is on the North Sea as it is only in this marine area that all of the coastal states have expressed an interest in a possible collaboration on the development of an interstate offshore grid. However, the legal issues and questions associated with proposals for a North Sea offshore grid that we discuss in this chapter would also be applicable to equivalent development in the Baltic Sea.

Atlantic Wind Connection, a 350-mile long offshore interconnector between the grids of New Jersey and Virginia (with additional connections at two points) to which wind farms can be linked as they are constructed.[14] Whilst the nature of the North Sea grid is to be determined, it seems likely that this too would be built up from cables serving a dual role of interconnecting national transmission systems and conveying offshore wind energy (although the resulting grid, because of the many interconnections involved, would be of much greater complexity than the Atlantic Connection).

Whilst an obvious difference between the European and US schemes is that the former would involve transboundary development whilst the latter would lie solely within US waters, there are legal issues that are common to both projects. First, they will both be constructed in areas in which the legal certainty that territorial sovereignty provides for onshore transmission construction is replaced by a set of rights under international law for states to exploit the seas and to exercise jurisdiction over marine activities. For both projects, there is a need to establish whether international law would allow the envisaged development and, of equal importance, control over its construction and operation by the states involved. Second, both projects require that the adequacy of existing legal frameworks for regulating transmission systems be assessed. Are the necessary laws in place to regulate the siting, construction, and operation of offshore transmission infrastructure? If not, how can appropriate arrangements for governing offshore grid developments be established? A related question is whether current regulatory practices for onshore grid construction and operation could have a chilling effect on offshore development. If so, would it be preferable to remove potential barriers through legal reform or to overcome them by creating new regulatory regimes that are purposefully designed for offshore transmission?

The legal issues associated with proposals for offshore grid development in the EU and the US, and a comparison of the means by which these might be addressed provide the subject matter of this chapter. In section II, we examine the international legal framework for offshore development and consider the key question of whether it provides sufficient support for offshore grid projects of the type that are now being proposed. In section III, we consider how suitable regimes could be put in place to regulate the development and use of transmission infrastructure in the North Sea. In section IV, we provide an overview of the gradual progress that is being made towards creating a regulatory framework for offshore electricity generation and transmission in the US. We conclude in section V by identifying key legal considerations that analysis of the North Sea and Atlantic Connection projects reveals should be taken into account in strategic planning for combined offshore wind energy generation and transmission projects in general.

[14] Matthew Wald, 'Offshore Wind Power Line Wins Backing' (*New York Times*, 12 October 2010), available at <http://www.nytimes.com/2010/10/12/science/earth/12wind.html?_r=1&th&emc=th>, accessed 24 March 2011).

II. Offshore Grid Development under International Law

A. The United Nations Convention on the Law of the Sea

The key statement of the rights of states to undertake and regulate offshore activities and the duties they should observe in exercising them is contained in the 1982 United Nations Convention on the Law of the Sea (UNCLOS).[15] The Convention separates the seas into zones to which different rights and duties for states party to the Convention attach.[16] The zones of particular relevance for the offshore infrastructure projects with which we are concerned are the territorial sea (which extends to 12 nautical miles from states' coastlines) and the exclusive economic zone (EEZ), an area established by coastal state declaration that encompasses all of the waters and seabed beyond the territorial sea to an outside limit of 200 nautical miles. The relevant coastal state may, subject to certain limitations, exercise full jurisdiction over the territorial sea and extend its legislation (including for regulating electricity generation and transmission) to it if it so chooses.[17] By contrast, coastal states possess only a limited set of rights, as laid down in Part V of UNCLOS, to undertake and exercise jurisdiction over marine activities in the EEZ. These include 'sovereign rights... for the economic exploitation and exploration of the zone, such as the production of energy... from the winds', and an exclusive right to authorize and regulate the construction and operation of 'installations and structures' that are built in order to make use of the rights conferred.[18]

B. UNCLOS and offshore electricity cables

1. Options for offshore transmission infrastructure

Agreement was reached on the contents of UNCLOS nine years before the world's first offshore wind farm began operation,[19] and at a time when there was limited experience with laying transboundary submarine cables for electricity transmission.[20] It is therefore unsurprising that the express provision it makes for cable use and regulation in connection with energy generation is limited. To the extent that UNCLOS does make provision for cable laying and use, its treatment of cables would also differ according to whether they are used: to transmit electricity

[15] UNCLOS's text is at <http://www.un.org/Depts/los/convention_agreements/texts/unclos/closindx.htm> (accessed 28 September 2011).

[16] It is important to note that the US has not ratified UNCLOS. However, the position described in this section is also likely to be the same for the US as (i) it declared an EEZ in 1983 and has exercised EEZ rights within it thereafter; and (ii) the full package of EEZ rights and duties set out in Part V UNCLOS may now have become part of customary international law. See D. Attard, *The EEZ in International Law* (Oxford: Clarendon Press, 1987) 287–95.

[17] M.M. Roggenkamp, 'Petroleum Pipelines in the North Sea: Questions of Jurisdiction and Practical Solutions' (1998), 16 *Journal of Energy and Natural Resources Law*, 93.

[18] Articles 56(1)(a) and 60(1)(b) UNCLOS.

[19] The Vindby wind farm was built in Danish waters in 1991.

[20] Rainer Lagoni, 'Legal Aspects of Submarine High Voltage Direct Current (HVDC) Cables' (Hamburg: LIT Verlag, 1998) 2.

generated by offshore wind farms in a state's EEZ to its own onshore transmission system; to link offshore wind farms on one state's EEZ to another state's national transmission system through cables that may cross that other state's EEZ and/or territorial sea; and to transmit electricity for other purposes, including through interconnectors linking onshore transmission systems. In view of this, we consider in the following sections four different possibilities for offshore transmission infrastructure (ranging from comparatively straightforward wind farm to shore connections to a complex multilateral grid) and the legal position applying to each of them under international law.

a. Option 1: Offshore wind power plant connected to the onshore transmission network of the same state (State A)

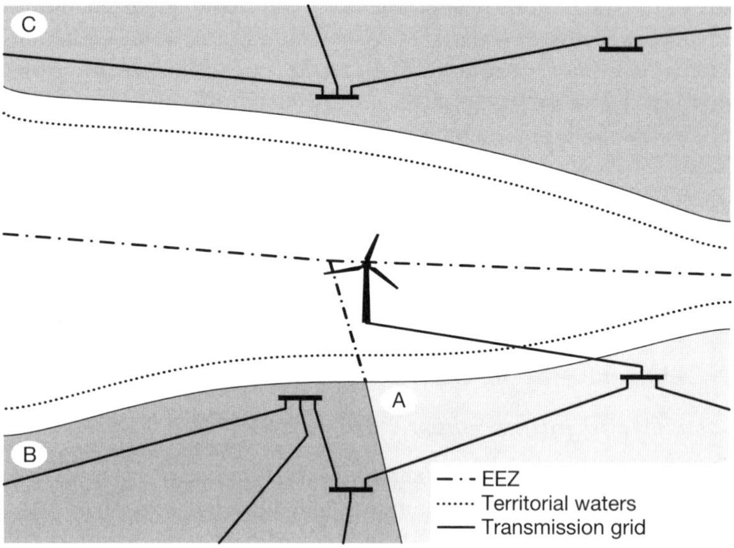

Figure 10.1

This, as depicted in Figure 10.1, is the only one of the four options for which UNCLOS makes reasonably clear provision. States have a right to exploit offshore wind energy resources in their own EEZs.[21] Although it is not expressly provided for in the Convention, it is strongly arguable that this right also empowers the state concerned to take actions necessary for exploiting offshore wind energy, including laying cables on the state's EEZ to transmit the energy generated and regulating their construction and operation.[22] It is confirmed in UNCLOS that each coastal state has 'jurisdiction over cables . . . constructed or used in connection with . . . the

[21] Article 56(1)(a) UNCLOS.

[22] See the discussion of the expression 'sovereign rights' at M.H. Nordquist, S. Rosenne, and L.B. Sohn (eds), *United Nations Convention on the Law of the Sea 1982: A Commentary* (Vol. 2, The Hague: Martinus Nijhoff, 1993) 541–2.

operations of… installations and structures under its jurisdiction'.[23] These 'installations' would include offshore wind farms situated on its own EEZ.

b. Option 2: Offshore wind power plant on EEZ of State A but connected to the onshore transmission network of another state (State B)

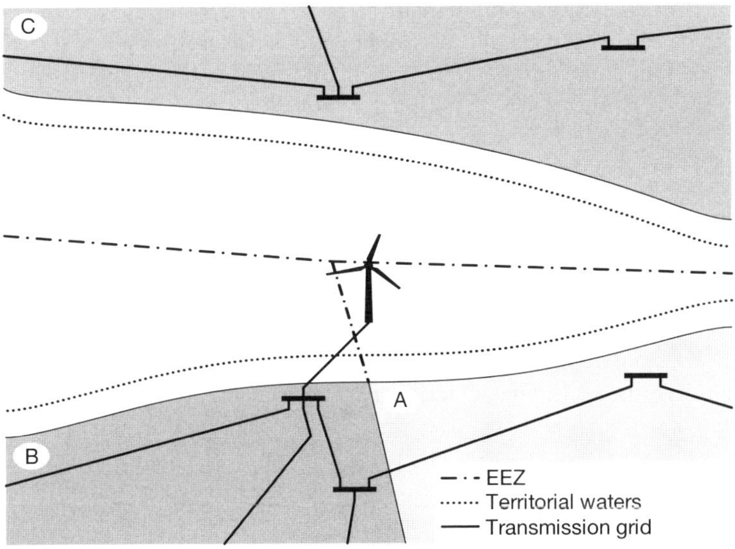

Figure 10.2

The legal position would be more complicated if it were thought necessary or desirable (whether for economic, environmental, or geophysical reasons) to connect the offshore wind farm to the transmission system of another state using a cable that crosses that other state's different maritime zones (as shown in Figure 10.2). As mentioned at IIA above, states have sovereignty over the territorial sea. Accordingly, the permission of State B to lay the cable in these waters would be required. State B would also have the authority to regulate the construction and operation of that part of the cable situated in its territorial sea. The practical effect of this partial control over a single cable is that State B would need to be involved with determining the legal basis for its construction and operation as a whole. In addition, State B would presumably require that it has regulatory control as necessary over the wind farm to which its transmission system is being linked for reasons including its ability to comply with the grid code for State B's transmission system and other requirements for access to the network. Grid codes lay down the technical specifications that electricity-generating plants must meet to ensure the proper functioning of the relevant transmission system. This would create a potential conflict with the exclusive jurisdiction that states possess over wind farms on their own

[23] Article 79(4) UNCLOS.

EEZs.[24] Again, therefore, it would necessitate discussion and agreement on how the wind farm and cable are to be regulated by the states concerned.

If the cable were also to cross over State B's EEZ, an element of uncertainty over who has the right to regulate that part of the cable would be added to the picture. All states have the freedom to lay electricity cables in any part of the seas, including on the seabeds of other states' EEZs.[25] However, there is no corresponding provision in UNCLOS for the state undertaking the cable project to regulate the construction and operation of cables laid in pursuance of this right. All states have an express 'functional' jurisdiction in their EEZs which allows them to enact and enforce laws in connection with national security, laws that allow for the exercise of their own rights in the EEZ, and for the protection of the marine environment.[26] In practice, North Sea states including Denmark and Germany ensure that their EEZ laws are complied with by requiring that parties laying a cable that crosses their EEZ (a 'transit cable') should apply to the transit state for a permit.[27] However, this limited law-making authority would not include, for instance, a right for State B to dictate how that part of the cable lying on its EEZ should be operated.

In view of the above, questions arise over which state should have the right to regulate the construction, operation, and other aspects of transit cables. These questions may be of limited practical relevance in the given example as agreement between State A and State B over the cable's construction and operation and who should have the authority to regulate it in different circumstances would be required in any event because they both have clear claims to exercise full jurisdiction over parts of the cable. Even so, uncertainty over who has jurisdiction over a cable transiting states' EEZs (save for the functional jurisdiction of transit states) creates a scope for disagreement and conflict that is unwelcome for major infrastructure projects. This potential is of much greater significance for, and is more likely to impact, multi-state transmission grid projects of the type that are now being contemplated.

c. Option 3: Offshore wind power plant on State A's EEZ but connected to an interconnector between State A and State C

For this option, we look separately at (a) the legal position for an interconnector which is constructed without wind farms attached to it; and (b) how that position would be affected if wind farms are attached subsequently.

A standalone interconnector (ie without a wind farm attached)

An offshore interconnector linking two states and connecting their transmission systems will, as shown in Figure 10.3 necessarily pass through the territorial seas and exclusive economic zones of both states. Accordingly, each of them would have some degree of clearly established jurisdiction over the cable in view of their exclusive

[24] Article 60(1)(b) UNCLOS. [25] Article 79(1) UNCLOS.

[26] Article 56 UNCLOS.

[27] M.M. Roggenkamp, 'Submarine Electricity and Gas Interconnectors: A Treaty Perspective' in *European Energy Law Report IV* (M.M. Roggenkamp and U. Hammer, (eds), Antwerp: Intersentia, 2007) 252–3, 264–5.

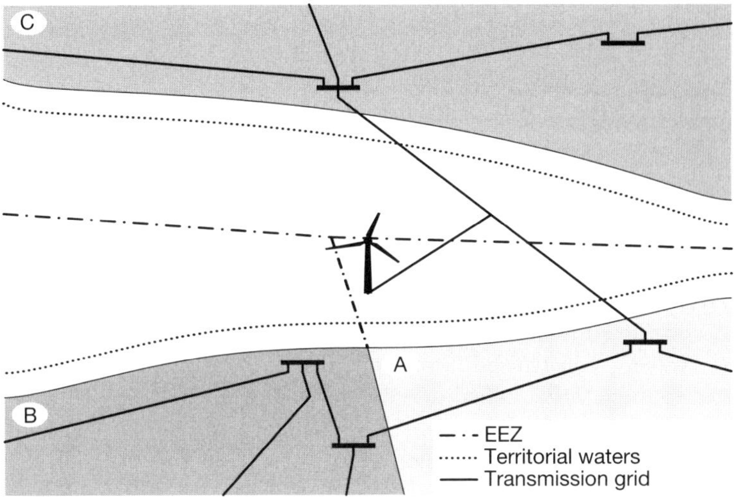

Figure 10.3

rights to permit and regulate cable laying and operation in the territorial sea. It will be necessary because of the shared territorial sea jurisdiction for both states to take part in authorizing the cable's construction and operation and for them to reach agreement on how it should be operated unless they are mutually content to leave such matters to their respective transmission system operators (TSOs) acting under European and national regulatory rules for the use of interconnectors.[28]

The position with transit cables described for Option 2 (including the freedom for all states to lay cables and the functional jurisdiction of each state over the cable laid) applies equally to those parts of the interconnector lying on the EEZ. The uncertainty over who has the right to regulate matters not covered by the functional jurisdiction is also present although, to date, this has not been viewed as problematic where offshore interconnectors have been developed only to link two transmission systems.[29] States have allowed their construction and operation by national TSOs (or, in some cases, non-TSO investors) without determining which state would have jurisdiction over the interconnector if questions concerning its construction or use arise that have not been addressed by the developing parties.

An interconnector to which wind farms are subsequently linked

The possibility of attaching wind farms to interconnectors would introduce legal complications to the situation with a standalone interconnector. An important consideration is who would decide whether a wind farm can be linked to the interconnector

[28] The relevant rules are contained in Council Regulation (EC) 714/2009 of 13 July 2009 on conditions for access to the network for cross-border exchanges in electricity and repealing Regulation (EC) No. 1228/2003 [2009] OJ L211/15 (Regulation on Cross-Border Exchanges).

[29] M.M. Roggenkamp, 'Submarine Electricity and Gas Interconnectors' (above n 27) at 252–3, 264–5.

in view of the uncertain regulatory authority over transit cables discussed above. As noted above, states may have been content to date to allow the operation of offshore interconnectors within the framework provided by existing European and national legislation for electricity market regulation, and without reaching any additional agreement on who would have the right to regulate the interconnector in particular circumstances. However, if it is proposed that wind farms should be attached to the interconnector then agreement between the states will be necessary on who should authorize, provide, and bear the costs for providing that connection.

In practice, the states concerned (or TSOs and/or regulators acting on their behalf) will need to agree that the wind farm can be attached, by whom this is to be effected, and the terms on which this will be allowed. The possibility that interconnectors now being constructed in the North Sea will ultimately have wind farms connected to them (or be linked to other developments including wind farms) is widely anticipated. It would therefore be prudent for states involved with interconnector development to reach a formal agreement on how this situation should be addressed at the time the interconnector is developed, rather than when the connection is proposed, with all the attendant potential for conflict that would create.

An additional concern, as with Option 2, arises over who would have jurisdiction over the attached wind farms and the transmission of electricity generated by them. If the wind farm depicted above transmits electricity only to State A's transmission system, then the right to regulate the wind farm and the transmission of electricity from it could be exercised as for Option 1. However, if the energy from the wind farm could be transmitted to either state, both States A and C may wish to have control over the operation of the wind farm. In practice, it would be necessary in this situation to determine who should regulate the wind farm and related electricity transmission and in what circumstances. Grid code requirements for the access of generating stations to the respective transmission systems and arrangements for allocating interconnector capacity would also need to be harmonized.

d. Option 4: Offshore wind power plant connected to a transnational offshore transmission network

Under this final option, offshore wind farms are connected to an offshore grid system that would allow the transmission of energy generated to several states (Figure 10.4). In this case, the legal difficulties raised by Option 3 concerning operational and regulatory authority over different parts of the transmission network and connected wind farms are applicable. However, they are likely to be magnified as more states become involved with developing the infrastructure (or with hosting it as transit states). In reality, agreement between the states concerned would be necessary over who has jurisdiction over such a grid (or different sections of it). A practical response may be for the states concerned to appoint a single operator for the grid as a whole, as well as harmonizing their regulations for cable construction and operation and establishing combined regulatory arrangements.

One additional aspect of Option 4 that should be considered is how cables that interconnect disparate parts of a transnational grid, but which are not themselves

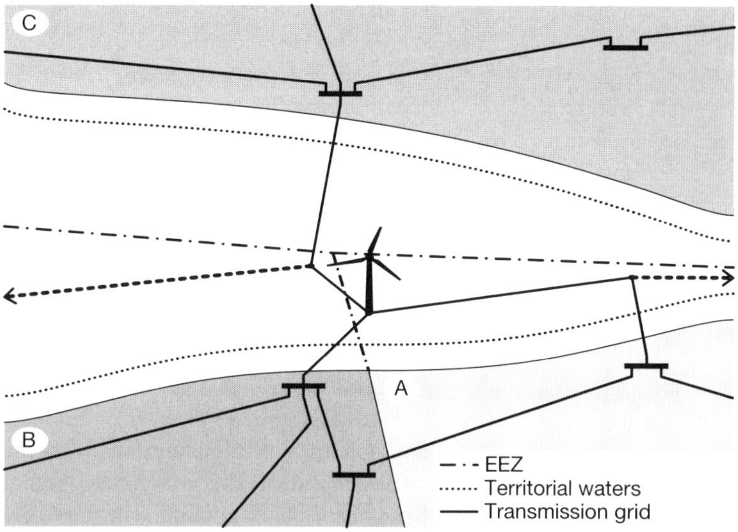

Figure 10.4

connected directly to offshore wind farms or to onshore transmission systems, would be viewed under international law. On a basic level, the position would be similar to that with other 'transit cables' in that there is a right for states to construct them, but no corresponding provision under UNCLOS concerning jurisdiction. However, the question of who should have regulatory authority over such cables will be less easy to establish in practice than for interconnectors or other transit cables. There would be no clear basis for a state to claim jurisdiction based on the cable's connection with its national transmission system. They may also be used to transmit energy from wind farms lying on the EEZs of several different states. The likely difficulty with attributing jurisdiction to these 'orphan' cables provides a good illustration of the practical need for states involved with development of complex offshore grid arrangements to agree on how their construction and operation should be regulated and by whom.

C. Conclusion

It is clear from the above that the legal framework provided by UNCLOS is lacking in some respects both for EU and US offshore grid development. The lack of clarity concerning jurisdiction over interconnector cables is unlikely to present a practical problem for the Atlantic Wind Connection as this lies entirely in US waters. However, for the more complex transboundary solutions that may be pursued in the North Sea, some serious problems could arise with determining who would have the right to regulate the cable infrastructure and attached wind farms, and to what extent. It is, of course, important that there should be clarity over jurisdictional rights to ensure that grid infrastructure: is constructed satisfactorily, to common specifications that allow for future grid expansion, and in ways which respect the North Sea states' international and European environmental commitments;

and that grid operation is secure, and allows investments to be recouped from its use. The bigger underlying concern, however, is that without legal certainty over who holds the rights to build, operate, and make money from a North Sea grid and what those rights entail, it will not be possible to secure the massive investment required for an infrastructure project on this scale. We consider in the following section how regimes for governing offshore transmission infrastructure in the North Sea that provide this certainty could be established.

III. Governing Electricity Transmission in the North Sea

A. Developing appropriate regulatory regimes

Whilst international law confers on states the right to authorize and to exercise jurisdiction over the construction and operation of offshore grids, it is for states themselves to choose whether to make use of those rights, and, if they decide to do so, how the conduct of related activities should be regulated. The picture is complicated somewhat for the North Sea states (excluding Norway) by their membership of the European Union.[30] The EU has legislative competence, to the extent laid down in the Treaty on the Functioning of the European Union,[31] over matters of relevance for offshore grid development. These include the question of how European electricity markets should be structured, and their regulation by national authorities.[32] As a result, the freedom of Member States to legislate in certain areas may be constrained by parameters already established in European law.

In this section, we consider the key issues associated with creating regulatory regimes for the development and operation of electricity transmission systems in the North Sea. We discuss the extreme difficulty with formulating an appropriate system of governance for some aspects of offshore transmission (particularly the siting and the subsequent operation and regulation of transboundary infrastructure) at a stage when the North Sea states have not decided how, or indeed whether, to proceed with offshore grid development. However, we argue that consideration can and should be given now to how an offshore electricity transmission sector could be provided for in overarching European and national legal frameworks for the electricity sector. In particular, questions arise over whether the regulatory regime for electricity transmission established under the European laws that prescribe rules for the internal electricity market, and which Member States are

[30] Although Norway is not a Member State of the EU, it may still be required to implement EU legislation, including for the energy sector, because it is a party to the European Economic Area Agreement. See F. Arnesen et al, 'Energy Law in Norway' in M.M. Roggenkamp, C. Redgwell, I. del Guayo and A. Ronne (eds) *Energy Law in Europe: National, EU and International Regulation* (2nd edn, Oxford: Oxford University Press, 2007) Section 12.17, 885.

[31] Consolidated Version of the Treaty on the Functioning of the European Union [2010] OJ L83/47.

[32] Ibid article 194.

required either to give effect to in national legislation or to observe directly, should be applied without modification to new offshore grid development.[33]

B. Governing grid development and operation

It is not possible, at present, to develop fully detailed proposals for the governance of offshore transmission infrastructure in the North Sea. The principal reason for this is that the appropriate form and content of governance arrangements will very much depend on prior agreement over how (if at all) the creation of a North Sea grid is to be progressed. For example, if the North Sea states decide not to collaborate on this, but to concentrate on developing their own offshore wind energy resources, then the extension of national laws for regulating transmission to marine areas and the passage of national legislation for permitting offshore cable construction and operation may suffice. Similarly, if states are to concentrate in the short term on constructing offshore interconnectors between transmission systems (with the establishment of an offshore grid as a long-term goal), then they may take the view, as they have done to date, that existing European and national regulation for the development and operation of interconnectors is adequate for this purpose.

However, as we contend in section II, more complex configurations than this in which interconnectors are linked to each other and to offshore wind farm clusters with a view to creating a North Sea grid would raise questions of jurisdiction that states involved with relevant projects may need to settle, as well as necessitating the development of new interstate arrangements for governing the siting, construction, and operation of transmission infrastructure projects. Legal structures that are tailored to address the complexities associated with individual projects are likely to be required, although it is difficult to say at this juncture whether these would best be provided by framework agreements between states, by more detailed operational agreements between states and developers, by the adoption of a harmonized legal system by the North Sea states for offshore electricity transmission, or by a combination of these approaches.

This need for prior agreement before detailed regulatory regimes can be formulated affects two key areas of governance for North Sea grid development. The first is decision-making on the siting of new transboundary infrastructure, particularly the identification of cable routes that would be compatible with the obligations of North Sea states for marine environmental protection under international and European law. This would need to be approached by the states concerned on an ad hoc basis as there is, as yet, no framework in place for governing offshore transboundary development in the North Sea region. This situation may change as all of the North Sea states are, with differing degrees of proactivity, currently developing marine spatial planning and permitting regimes that will provide mechanisms

[33] See Council Directive 2009/72/EC of 13 July 2009 concerning common rules for the internal market in electricity and repealing Directive 2003/54/EC [2009] OJ 211/55 (2009 Electricity Market Directive); and Regulation on Cross-Border Exchanges (above n 28).

for controlling sea uses in waters under their own sovereignty and jurisdiction.[34] However, these national approaches have not yet translated into broader interstate cooperation over planning uses of the North Sea as a whole, rather than as an entity divided into separate national spaces. Indeed, the North Sea states make clear in their Memorandum of Understanding that their exploration of how marine spatial planning and permitting procedures can be coordinated does not indicate any intention to develop 'a spatial plan for the North Seas' or to establish 'centralized administrative structures'.[35]

If a regional planning framework is not in place by the time decisions have been reached on configurations for an offshore grid, it may be that the states involved with individual projects will need to establish separate arrangements for agreeing on where cables are to be situated and for assessing the environmental effects of different possibilities for grid development. If this proves to be the case, the framework agreed for environmental assessment and permitting of the Nord Stream pipeline by the five Baltic Sea states in whose waters it is being constructed may provide some guidance as to how the potential environmental consequences of transboundary development in the North Sea can be managed in a way consistent with state duties for environmental protection.[36]

The second key area of governance concerns the arrangements that would be required for governing the operation of transmission infrastructure. A fundamental question for establishing suitable controls is by whom transboundary offshore infrastructure would be operated. The answer to this will depend on what approach is best suited to managing the options chosen for grid development. For example, it may be desirable, where it would be appropriate to operate a transboundary development as a standalone grid, for the states involved to appoint a single operator rather than have separate operators for those parts of the infrastructure lying in their EEZs and territorial seas. In such circumstances, it would also be sensible for the parties to the development to agree on a single set of rules for operating the offshore transmission system. Conversely, states may be content to allow interconnectors, even where they have wind farms linked to them, to be operated by those responsible for the interconnected onshore transmission systems with cooperation between them as necessary on matters such as capacity allocation and controlling attached generating stations.

A related consideration is by whom transboundary developments would be regulated. Possible options include a single authority agreed upon by the states concerned to regulate the grid system on their behalf, a committee comprised of persons appointed by the different states to represent their interests in decision-making over the operation and development of the infrastructure, or control by existing national regulatory authorities of those parts of the grid recognized as

[34] See the reports on national marine spatial planning regimes produced under the SeaEnergy research project at <http://www.seanergy2020.eu/publications-and-results/national-maritime-spatial-planning-regimes-wp2/> (accessed 28 June 2011).

[35] Memorandum of Understanding (above n 13) at Annex 3.

[36] T. Koivurova and I. Polonen, 'Transboundary Environmental Impact Assessment in the Case of the Baltic Sea Gas Pipeline' [2010] *The International Journal of Marine and Coastal Law* 151.

being subject to the national jurisdiction of the different states. The appointment of a single regulatory authority acting on behalf of all parties concerned with the development may be preferable to splitting responsibility between national bodies for overseeing grid operation where proposed grid developments would lie in the marine areas or be connected to the transmission systems of several states. However, it is again the case that the suitability of different possibilities for grid operation and regulation will depend on the nature of the infrastructure to be constructed. Inevitably, political considerations will also be a major determinant of what operational and regulatory structures are adopted.

The potential benefit of concluding a framework treaty to provide a legal platform for subsequent grid development in the region is something that the North Sea states should examine at this early stage of their collaboration. In this regard, the UK and Norway entered into a treaty in 2005 that is intended to create a framework within which detailed arrangements for subsequent cross-boundary petroleum cooperation, including over the construction, laying, and operation of pipelines, can be agreed.[37] The treaty records broad commitments by the states to collaborate over and to agree on the terms for pipeline projects falling within its ambit. As with the UK–Norway agreement, a legal framework for transmission infrastructure in the North Sea could be used by the coastal states to provide a clear statement of their respective positions in relation to transboundary development, and thus to facilitate the planning and execution of subsequent infrastructure projects.[38]

A particular purpose that such a treaty could serve in this context would be to ensure that the possibility of creating a larger North Sea grid is kept open if the states decide on a gradual approach to grid development in which, initially, states authorize separate projects, but with the intention that they would ultimately be linked together. A framework agreement could require that the North Sea states consult with each other, and that they should agree on technical specifications, grid capacity, and other matters that may affect the development of a shared grid before undertaking separate infrastructure projects.

C. Offshore electricity transmission and European rules for the internal electricity market

All aspects of the electricity sector in the EU are subject to an overarching regime laid down in European law.[39] The Member States are required to promulgate national laws that restructure their own electricity markets so that these

[37] Framework Agreement Between the Government of the United Kingdom of Great Britain and Northern Ireland and the Government of the Kingdom of Norway Concerning Cross-Border Boundary Petroleum Cooperation (signed on 4 April 2005).

[38] A.L. Brautaset, 'The Ormen Lange Field, the Langeled Pipeline and the New UK Norway Framework Agreement Concerning Cross Boundary Petroleum Cooperation' in *European Energy Law Report II* (M.M. Roggenkamp and U. Hammer, eds, Antwerp: Intersentia, 2005) 201–2.

[39] 2009 Electricity Market Directive (above n 33); and the Regulation on Cross-Border Exchanges (above n 28).

correspond with the European model. The dominant concerns of European legislative activity in this area have been with liberalizing a sector that, historically, was under public ownership in order to create a competitive market for electricity production and supply, and with removing barriers between the national markets in Member States with a view to creating a single European internal electricity market.[40] Accordingly, the relevant laws address matters that may impact on competition and integration such as the ownership of electricity undertakings, and the relationship between different market actors.

The separation of the operation and, save where specific circumstances apply, the ownership of the transmission and production of electricity is required by the European rules to minimize risks of anti-competitive behaviour.[41] Transmission, which is regarded as a monopoly activity that is unsuited to competition, is also heavily regulated to ensure that owners and operators of grids do not abuse their dominant position, including by setting unjustifiably high charges for grid use.[42] Similarly, the development of grid networks is made subject to national regulatory review both to ensure that grids *are* upgraded, whilst preventing owners from making grid improvements (and, more importantly, including the costs within transmission tariffs) whose public benefit cannot be demonstrated.[43]

The European Commission strongly advocates grid development in the North Sea. It has identified this as one of its four priorities for enhancing European energy infrastructure by 2020.[44] However, no guidance has been given by the Commission (or other European institutions) as to whether existing European legislation for the electricity sector will be applied without revision to the development and operation of a North Sea offshore transmission system. In this regard, the laws adopted under the third package of reforms for the European electricity market of 2009 make no reference even to the possibility of offshore grid development in the North Sea (or elsewhere).[45] Directive 2009/28/EC (the Renewable Energy Directive), which places general obligations on Member States to ensure that existing grid systems are able to accommodate energy from renewable sources, is also silent on the possibility of creating new grids for transmitting offshore renewable energy.[46]

The question of whether the expansion of an offshore transmission sector is to proceed within bounds set by the current European legal framework and its regulatory provision for electricity transmission is a significant one for future grid development in the North Sea and other European marine areas. The reason for this is that the framework's strictures, taken together with its lack of specific provision for offshore transmission, could operate to limit the type and scale of offshore development that Member States are able to undertake in a number of respects.

[40] See the accounts of European legislative activity concerning the electricity sector in M.M. Roggenkamp, C. Redgwell, I. del Guayo and A. Ronne (eds), *Energy Law in Europe: National, EU and International Regulation* (2nd edn, Oxford: Oxford University Press, 2007) Chapter 5, Section I; and J. de Jong, 'The 2007 Energy Package: The Start of a New Era?' in *European Energy Law Report V* (M.M. Roggenkamp and U. Hammer, eds, Antwerp: Intersentia, 2008) 95.

[41] 2009 Electricity Market Directive (above n 33) at article 9.

[42] Ibid chapters IV and V.

[43] Ibid article 37 subparas (a) and (g).

[44] COM(2010) 677/4 (above n 7) at 10.

[45] 2009 Electricity Market Directive (above n 33); Regulation of Cross-Border Exchanges (above n 28).

[46] Renewable Energy Directive (above n 1) at article 16.

First, its application may result in constraints that were designed to address concerns with existing onshore transmission being imposed unnecessarily on offshore development. For example, it must be questioned whether the separation of energy production and transmission is essential for grid systems that are being constructed in previously undeveloped areas, and whose primary and perhaps exclusive purpose may be to convey electricity from clusters of offshore wind energy-generating stations. It is also feared that the strict regulation of cost recovery for grid development may make it difficult to secure the investment required for new offshore infrastructure.[47] Second, some of the new uses of transmission infrastructure that may be employed in offshore grid development, such as combining interconnectors with wind farms, are not covered by the current laws or by definitions of infrastructure types contained in them. There is therefore no legal certainty over how the new types of infrastructure (or new uses of existing infrastructure) would be regulated.

Finally, the current European legislative framework clearly envisages that ongoing grid management and development (including the seawards expansion of onshore grids) will be conducted by the owners and operators of national transmission systems. Although there is no express bar on new entrants to the transmission sector, there is also little encouragement of this. The dominance of existing TSOs in the market structures under European law is reinforced by the central role given to ENTSO-E, an organization that consists of representatives from Europe's national system operators, in governance for the future development of an integrated European grid.[48]

A development programme led solely by established system operators may be desirable in some respects. It would benefit from their experience as well as from the strategic coordination that making one body responsible both for offshore development and upgrades to the onshore grid would bring. At the same time, huge investment in onshore transmission systems is already needed to overcome problems with congestion and aging infrastructure.[49] The risk therefore is that offshore grid development may be viewed by onshore TSOs and national regulators alike, notwithstanding European support for this, as being a lower priority when there are already more pressing concerns to be addressed with the capacity and security of existing networks. With this in mind, the possible involvement with offshore transmission by parties that are not encumbered with the baggage of onshore systems and the means by which this could be promoted in the relevant legislation should at least be explored.

Again, the extent to which European rules for the internal electricity market could constrain offshore grid development will depend on the approach to this that the North Sea states adopt. However, it is reasonably clear from the above that European legislation for the electricity sector does not make appropriate provision for combined offshore wind energy and grid development. It is therefore

[47] COM(2010) 677/4 (above n 7) at 8.
[48] Regulation on Cross-border Exchanges (above n 28) at articles 4 to 8.
[49] COM(2008) 768 (above n 7) at 6.

incumbent on the Commission, if it is to avoid hampering the development of the infrastructure that it calls for, to see to it that the EU's own legislative house is put in order. Either new legislation is required at the European level to cover the new offshore infrastructure, or alternatively, derogation should be permitted from current laws that would allow Member States involved with particular projects to develop suitable regional regulatory regimes for them.

IV. Offshore Wind Energy Development in the United States and the Atlantic Wind Connection Proposal

A. Developing a regulatory framework

US offshore wind development is promising but nascent. The US leads the world in installed, land-based wind energy capacity, but is not yet home to a single offshore wind farm (although several projects have been proposed and one project recently completed its permitting process).[50] The federal law expressly authorizing offshore wind energy development within US territorial waters was not enacted until 2005, and the detailed regulatory framework for obtaining leases and other permissions for offshore renewable energy development has only been in effect since June 2009. Yet despite its early stages, offshore wind energy in the US is garnering increasing interest. As noted in section I, one recent proposal known as the 'Atlantic Wind Connection' may provide key infrastructure to boost the financial and practical viability of offshore wind farms in the Atlantic.[51] If completed, the Atlantic Wind Connection will be able to connect up to 7,000 megawatts of offshore wind.

This section provides an overview of the regulatory framework, planning considerations, and other environmental laws governing wind farm and transmission line development in the US, as well as a brief discussion of the relevant power pricing issues related to the concept of connecting multiple wind farms to a backbone transmission line with a limited number of landfall points.

B. US regulatory authority to permit offshore electricity generation and transmission

In the US, the states have jurisdiction over the first three nautical miles from shore (except on the Gulf Coast of Texas and Florida, where state waters extend three leagues, or approximately nine miles). The federal government has jurisdiction over the Outer Continental Shelf (OCS), which is the submerged land lying between the seaward extent of the states' jurisdiction and the seaward extent of federal juris-

[50] Cape Wind News Release (7 January 2011), available at <http://www.capewind.org/news1174.htm> (accessed 28 September 2011).

[51] Information about the Atlantic Wind Connection can be found at <http://atlanticwindconnection.com/> (accessed 28 September 2011).

diction (approximately 200 nautical miles offshore).[52] In 1953, Congress charged the US Department of the Interior (DOI) with the administration of mineral exploration and development in the OCS.[53] Pursuant to that authority, DOI has issued some 8,000 oil and gas leases on approximately 43 million OCS acres.[54]

When the first US offshore wind farm was proposed in 2001, uncertainty arose over which federal agency, if any, had authority to permit the project. Congress resolved this uncertainty with the Energy Policy Act of 2005 (EPAct), which gave DOI new authority to regulate renewable energy projects on the OCS.[55] Under EPAct, a sub-agency within DOI known as the Bureau of Ocean Energy Management, Regulation and Enforcement (BOEMRE), may issue a lease, easement, or right of way on the OCS for activities that 'produce or support production, transportation, or transmission of energy from sources other than oil and gas'.[56]

Perhaps the greatest boost to offshore wind development in the US arrived in April 2009, when BOEMRE released regulations pursuant to EPAct establishing a detailed regulatory regime for development of wind and other renewable energy resources on the OCS.[57] Similar to BOEMRE's longstanding regulatory program for OCS oil and gas development, the renewable energy regulations create a competitive leasing framework and financial obligations for lessees and grantees to ensure a fair return to the US government.

The regulations provide for commercial leases which allow electricity generation with a 25-year operating term. Leases must be issued through a competitive process. The agency may issue a non-competitive lease only if it determines there is no competition for a particular area. Limited leases with a term of up to five years can be used for testing and site assessment, but allow only limited electricity generation. Commercial leases for wind energy production include the right to a project easement across the OCS for transmission lines or other delivery media, and limited leases may also include the right to a project easement (including authorization for off-lease transmission lines), if necessary. In addition to leases, BOEMRE has the authority to grant separate and independent rights of way (ROWs) and rights of use and easements (RUEs) on the OCS to support renewable energy development. These grants are separate from the leases authorizing the wind farm, and would be sought, for example, by a developer interested in constructing a transmission line or other support installation independent from a specific wind farm lease. ROW grants allow construction and use of a cable across the OCS for the purpose of transmitting electricity generated from renewable energy sources (either

52 43 USC s 1331(a). While the US has not ratified UNCLOS, interest in offshore wind energy may provide arguments in favor of ratification. See, eg, Dwyer, Kieran, 'UNCLOS: Securing the United States' Future in Offshore Wind Energy', (2009) 18 *Minn. J. Int'l L.* 265.

53 The Outer Continental Shelf Lands Act ('OCSLA'), 43 USC ss 1331 *et. seq.*

54 US Dept of the Interior, Minerals Mgmt. Serv., *Leasing Oil and Natural Gas Resources*, available at <http://www.boemre.gov/ld/pdfs/greenbook-leasingdocument.pdf> (accessed 28 September 2011).

55 43 USC s 1337(p).

56 Id. s 1337(p)(1).

57 Renewable Energy and Alternate Uses of Existing Facilities on the Outer Continental Shelf, 74 Fed. Reg. 19638 (29 April 2009) (codified at 30 CFR pts 250, 285, 290). Those regulations became effective on 29 June 2009.

on the OCS or from projects not located on the OCS, such as a project under state jurisdiction which traverses the OCS before landfall). RUE grants enable placement and operation on the OCS of a facility or other installation that supports the production, transportation, or transmission of energy from renewable resources. ROW and RUE grants are issued through the same competitive process as for leases. The grant term generally lasts as long as the right of way or easement is properly maintained and continues to support the activities for which it was granted.[58]

C. Marine spatial planning and other applicable siting and permitting considerations

Overlaying the new regulatory regime for offshore energy development is a concept known as marine spatial planning (MSP), which is defined by the United Nations Educational, Scientific, and Cultural Organization (UNESCO) as 'a public process of analyzing and allocating the spatial and temporal distribution of human activities in marine areas to achieve ecological, economic and social objectives that are usually specified through a political process'.[59] More simply, it is land use planning for the marine environment, intended to identify areas most suitable for various classes of activities and balance those uses with other competing interests.[60] Its relevance to US offshore wind development is that it requires analyses to ensure renewable energy projects do not conflict with other marine use priorities such as critical habitat for sensitive species, commercial fishing, navigation, aesthetics, and oil and gas development, to name but a few.

MSP has received increased attention in the US with the dawning of offshore renewable energy and the reminders of the potential impacts of human activity in the OCS such as the Deepwater Horizon oil spill. Indeed, MSP is now an express part of the US government's newly proclaimed National Policy for the Stewardship of the Ocean, Coasts, and Great Lakes.[61] With an emphasis on MSP and intergovernmental engagement, state governments look to play an increasing role in offshore wind development, especially with respect to projects that will need state-issued permits such as wind farms or power hubs located within three nautical miles from shore (i.e. located within state jurisdiction) and the transmission lines from federal OCS wind farms which must cross state waters to make landfall at some given point. Already, the federal Coastal Zone Management Act (CZMA)

[58] See 30 CFR ss 285.300–16.

[59] *Marine Spatial Planning*, A Report from UNESCO Intergovernmental Oceanographic Commission (2009), available at <http://www.unesco-ioc-marinesp.be/> (accessed 28 September 2011).

[60] See website of the National Oceanic Atmospheric Administration regarding coastal and marine spatial planning: <http://www.cmsp.noaa.gov/> (accessed 28 September 2011).

[61] See 'Executive Order—Stewardship of the Ocean, Our Coasts, and the Great Lakes' (19 July 2010), available at <http://www.whitehouse.gov/the-press-office/executive-order-stewardship-ocean-our-coasts-and-great-lakes> (accessed 28 September 2011). The Executive Order adopts the final recommendations from the Administration's Interagency Ocean Policy Task Force, available at <http://www.whitehouse.gov/files/documents/OPTF_FinalRecs.pdf> (accessed 28 September 2011).

encourages states to enact coastal zone management plans and requires federal activities (including permit actions in federal waters) to be consistent 'to the maximum extent practicable' with the enforceable policies of approved state management plans.[62] The CZMA gives a state an opportunity to review proposed projects and object if the state determines the reasonably foreseeable effects of the project would be inconsistent with the state's programme.[63] Now, several Atlantic states have begun drawing on MSP principles to develop comprehensive ocean management plans, which could be folded into the consistency analyses required by the CZMA. For example, the states of Rhode Island and Massachusetts have undertaken efforts to zone waters off their coasts for uses that include renewable energy.[64] While MSP represents additional levels of bureaucratic review, it may also benefit offshore wind infrastructure by enhancing state and federal coordination and by green-lighting certain areas for project development.

In addition to the spatial planning considerations of state and federal government, numerous US environmental laws can impact the siting and permitting of wind farms and their transmission lines. Foremost among them is the National Environmental Policy Act (NEPA), which requires that federal agencies evaluate and disclose the environmental consequences of their actions.[65] BOEMRE must comply with NEPA in its decisions to issue leases and grants. If the proposed activity is likely to have a 'significant impact' on the environment, BOEMRE must prepare an Environmental Impact Statement (EIS) setting forth the impacts and the agency's analysis of all reasonable project alternatives.[66] While NEPA is a procedural statute, federal agencies have been dealing with it for over 40 years, and it should not present a substantial barrier to wind energy infrastructure. On the other hand, NEPA can provide opponents of federally permitted projects a basis to delay such projects by challenging the NEPA analyses in court.

Construction of renewable energy installations on the OCS such as wind turbines, submarine transmission systems, and landfall transition structures also require a Section 10 permit under the Rivers and Harbors Appropriations Act of 1899.[67] This permitting programme is administered by the US Army Corps of Engineers, and with respect to projects on lands under lease from BOEMRE, is limited to an evaluation of the impact on navigation and national security.[68] The Section 10 permit is important given the potential for offshore infrastructure to interfere with navigation and shipping lanes, but otherwise should not stymie wind energy development. Moreover, the Section 10 permit is unlikely to present

[62] 16 USC s 1456(a). Summaries of the various state ocean and coastal management plans are available from the US Department of Commerce at <http://www.cmsp.noaa.gov/> (accessed 28 September 2011).

[63] 16 USC s 1456(c).

[64] Massachusetts's Ocean Management Plan is available at <http://www.mass.gov/eea/ocean-coastal-management/mass-ocean-plan/final-massachusetts-ocean-management-plan.html> (accessed 13 October 2011). Rhode Island's Plan is available at <http://seagrant.gso.uri.edu/oceansamp/> (accessed 28 September 2011).

[65] 43 USC ss 4321, *et seq.*

[66] 40 CFR s 1502.14(a), (d).

[67] 33 USC s 403.

[68] 33 CFR s 322.5(f).

significant challenges to the transmission lines, which, in most cases, are unlikely to interfere with shipping lanes.

Finally, depending on the circumstances of a given proposal, wind energy projects may be subject to a host of federal laws that protect endangered species (Endangered Species Act),[69] marine species (Marine Mammal Protection Act),[70] migratory birds (Migratory Bird Treaty Act),[71] fish habitats (Magnuson-Stevens Fishery Conservation and Management Act),[72] estuaries (Estuary Protection Act),[73] marine sanctuaries (National Marine Sanctuaries Act),[74] and archaeological resources (National Historic Preservation Act of 1966 and Archeological and Historical Preservation Act of 1974).[75] Each of these laws has the potential to impede energy infrastructure on the OCS, but they do not present challenges particularly unique to wind farms or the backbone transmission line. After the first US offshore wind projects are sited and become operational, government agencies, developers, and investors will better understand the technology and its impacts in practice, leading to more efficient review and permitting of subsequent projects.

At this time, it is uncertain to what extent federal and state authorities will implement MSP principles in offshore wind development, particularly for transmission lines. Siting and planning for the first wind farms remains the focus in the US, with somewhat less attention being given to how spatial planning and traditional US environmental laws may affect a backbone transmission line. Increased understanding of the implications for transmission lines is likely to follow the permitting and development of the first wind farms, many of which may obtain approval for their own transmission lines. For now, MSP and the environmental laws discussed above have the potential to affect siting of both wind farm and independent transmission line projects.

D. Energy transmission and sale[76]

Wind power generated offshore must be transmitted to the national energy grid either by individual radial lines (ie wind farm to shore cables) or a proposed backbone transmission line that would aggregate energy output from multiple offshore facilities and make landfall at a few designated locations. In the US, the wholesale transmission and sale of power from offshore wind facilities is regulated by the Federal Energy Regulatory Commission (FERC) under the Federal Power Act 1935 (FPA).[77] Typically, generators with a production capacity of more than 20 megawatts obtain market-based rate authorization from FERC, which allows them to charge negotiated rates for wholesale power sales, subject to FERC's authority to revisit the rates charged. Upon obtaining market-based rate authorization, a

[69] 16 USC ss 1531–44; 50 CFR pt 402.
[70] 16 USC ss 1361–407.
[71] 16 USC ss 702–3.
[72] 16 USC s 1802(10); 50 CFR s 600.920(a)(1).
[73] 16 USC ss 1221–1226.
[74] 16 USC ss 1431–45; 15 CFR pt 922.
[75] 16 USC ss 470–470t; id. at ss 469–469c.
[76] The authors wish to thank Margaret A. Moore, Julia Scarpino Wood, and Jessica C. Friedman, attorneys at the law firm of Van Ness Feldman, for contributing to this discussion.
[77] 16 USC ss 824–825r.

generator may sell power into the wholesale market on a merchant basis (ie in the day ahead or real-time markets) or make sales pursuant to one or more power purchase agreements for a specified term. As an alternative to obtaining market-based rate authorization, a generator may make wholesale power sales at FERC-approved cost-based rates.

State public utility commissions also may play a role in power purchase arrangements between offshore wind developers and electric utility purchasers. Given the relatively high projected cost of offshore wind power as compared to onshore alternatives, most, if not all, offshore wind power in the US initially will be sold pursuant to long-term power purchase agreements (as opposed to on a merchant basis). The primary purchasers of the power will be electric utilities with franchised service territories, which must satisfy state-mandated renewable portfolio standards or other legislative mandates intended to facilitate the development of offshore wind power.[78] Power purchase agreements may be subject to approval for pass-through purposes by the state public utility commission that regulates the retail rates and franchise territory of the purchasing electric utility. If the power purchase agreement is contingent on state approval of the utility's prudence of purchase,[79] disapproval by the state commission could nullify the agreement.[80]

One major consideration relevant to the pricing of offshore wind power is the cost of interconnection and transmission service. The interconnection of an offshore wind facility with the integrated onshore transmission grid can substantially increase the development costs associated with a project. For a project seeking to interconnect directly with the transmission grid via a radial transmission line that serves only the offshore wind facility, the private developer is responsible for the costs of building all of the interconnection facilities from the offshore wind project to the point of interconnection with the transmission grid plus the costs of any necessary network upgrades on the transmission provider's system. To the extent that interconnection facilities can be designed to benefit other transmission customers, a developer may be able to share the cost of the facilities with other users of the transmission system. The Atlantic Wind Connection is an example of a transmission line designed to have regional benefits, the costs of which would be shared by onshore users of the transmission system in addition to the offshore wind facilities. Such cost sharing, however, is conditioned on the inclusion of the transmission facility in the appropriate regional transmission plan, which involves

[78] For example, several states along the Atlantic coastline have implemented legislation designed to promote offshore wind development by designating preferred developers and, in certain instances, requiring utilities to negotiate power purchase agreements for power produced by offshore wind facilities.

[79] For example, legislation by the State of Rhode Island requiring the local electric utility to negotiate a power purchase agreement with the State's preferred offshore wind developer required state public utility commission approval as a condition to the effectiveness of the power purchase agreement.

[80] In addition, a state may have siting authority over an offshore wind project's facilities located in state waters or on state lands.

an extensive stakeholder review process. To date, it has proven difficult to obtain stakeholder approval for regional sharing of the costs of such transmission lines.[81]

Service on transmission facilities such as a backbone transmission line that accesses offshore wind facilities (ie a transmission facility other than sole-use radial interconnection lines funded by the generators), may be offered at negotiated or cost-based rates, depending on the preference of the transmission owner.[82] Offshore wind developers may be able to negotiate what are known as 'anchor tenancy arrangements' to reserve a portion of the capacity on an unaffiliated transmission line to the extent that the transmission owner has sought FERC approval for such arrangements. Anchor tenancy arrangements generally require the generator to assume a portion of the development costs and risks. These issues regarding the sale and transmission of power from offshore wind farms present additional complications for developers, especially at this early stage in advance of any grid connection by a single offshore facility.

E. Backbone transmission lines: stimulating offshore wind energy generation

While offshore wind development in the US is still in its infancy, BOEMRE's new renewable energy regulations provide a strong framework and stable regulatory environment for authorizing the development of offshore wind farms and associated cable infrastructure. Clearly, the regulations provide BOEMRE authority to issue leases for wind farms and independent ROW and RUE grants for the backbone transmission line and power hubs that will run up and down the Atlantic OCS, and also with authority to issue any project easements needed for wind farm leaseholders to connect to the backbone line. Lease applicants are now navigating through the process, and state and federal agencies are better coordinating their efforts to balance development opportunities and environmental protection.

Still, more reasonable and predictable leasing approval and permitting deadlines are needed to facilitate the financial investments for getting wind harnessing technologies in the water.[83] One encouraging example of such an effort is BOEMRE's

[81] Given the distance of most proposed offshore wind facilities from their point of interconnection with the terrestrial power grid and their utility-scale size, direct current (DC) transmission lines are most likely to be used to interconnect offshore wind facilities. FERC Chairman Jon Wellinghoff has recently stated that FERC may adopt a different regulatory approach for DC lines because they fundamentally differ from the alternating current (AC) lines that make up a majority of the US power grid. An initiative to encourage development of DC lines could also facilitate development of transmission to access offshore wind farms and, as a result, facilitate development of offshore wind in the US.

[82] FERC has established rules providing incentives in order to attract investment in new transmission facilities, including transmission facilities that would serve offshore wind developments. See 18 CFR s 35.35. For example, under section 219 of the FPA, 16 USC s 824s, FERC may grant incentive-based rate treatment for the transmission of electric energy in interstate commerce when it benefits consumers by increasing reliability and reducing the cost of delivered power by reducing transmission congestion.

[83] The current leasing and permitting process typically takes around seven to ten years; too long for many developers and investors.

recently announced 'Smart from the Start' initiative intended to reduce the length of the leasing process by two to five years and otherwise streamline offshore wind activity.[84] The initiative invokes MSP principles by calling for the zoning of certain portions of the OCS for wind farm and transmission line development. The zoned 'Wind Energy Areas' will benefit from large-scale environmental analysis to expedite approvals.[85] The initiative also announces BOEMRE's plans to 'move forward aggressively' processing applications to build offshore transmission lines such as the Atlantic Wind Connection. At this early stage, the speed and success of US offshore wind development will likely depend on these and additional government efforts to ensure that projects are developed in a timely and cost-effective manner.

US Atlantic offshore wind is currently far behind offshore wind development in Europe. Nevertheless, there is great promise for Atlantic wind, and important infrastructure such as a backbone transmission line may benefit under the jurisdiction of a single nation-state as compared to Europe, where multiple sovereigns may give rise to increased complexity. In any event, a backbone transmission line such as the Atlantic Wind Connection project may provide long-awaited stimulus for more rapid offshore wind development in the US.

V. Conclusion

Our review of proposals for electricity transmission infrastructure in the North Sea and off the US Atlantic seaboard reveals some concerns that are common to these projects, and which are also likely to affect similar schemes wherever they are to be constructed. First, some degree of jurisdictional uncertainty is present: in the North Sea Grid's case as a product of the lack of clear guidance in UNCLOS concerning state jurisdiction over the development and operation of transboundary marine infrastructure; for the Atlantic Wind Connection, as a consequence of compound jurisdiction and multiple layers of regulatory approvals between federal and state governments. In both cases, possibilities for addressing the resulting difficulties and thereby providing the legal certainty required for major infrastructure development are being explored. The North Sea states are looking at how barriers to developing a shared grid may be removed, including by the harmonization of relevant national laws. Marine spatial planning and zoning are being used in the US as initiatives to enhance coordination between separate federal and state planning regimes.

Second, a fundamental consideration is how sufficient regulatory structures can be created for offshore transmission infrastructure. In the US, the federal government has taken steps in recent years to put a legal framework in place for offshore wind energy generation and transmission. However, questions remain about

[84] See <http://www.doi.gov/news/pressreleases/Salazar-Launches-Smart-from-the-Start-Initiative-to-Speed-Offshore-Wind-Energy-Development-off-the-Atlantic-Coast.cfm> (accessed 28 September 2011).

[85] BOEMRE has so far identified four WEAs offshore of Delaware, Maryland, New Jersey, and Virginia. 76 Fed. Reg. 7226 (9 February 2011).

potential barriers such as stakeholder approval for regional cost-sharing and state approval of power purchase agreements that may hinder the sale and transmission of offshore renewable power. As for the North Sea states, some preliminary decisions must be made about how an offshore grid is to be developed before matters concerning the regulation of this enterprise can be settled. Even so, it is clear now that legislative action at the European level is necessary to accommodate offshore transmission infrastructure within the overarching legal framework for the internal electricity market.

Third, it is necessary to examine whether existing legal and regulatory regimes that were not designed either with offshore electricity transmission in mind or with other ends in view than promoting renewable energy generation and development that contributes to reducing carbon emissions have the potential to retard grid construction. Under both the EU and US regulatory systems, one concern is that mechanisms for determining how costs associated with offshore transmission infrastructure would be recouped may place more emphasis on low energy prices than on providing financial security for projects that advance competing public goods such as energy security, mitigating climate change, and environmental protection.

Finally, these concerns are linked by a common theme: the absence of suitable legal frameworks at international, regional, and national levels for projects of a nature and on a scale that have only become possible through recent advances in technology for offshore electricity generation and transmission. Institutions with the authority to make or reform law must create the legal foundation that is required if the potential benefits of coordinating the expansion of offshore renewable energy generation with offshore grid development are to be realized.

11

The Role of Energy Networks in Facilitating the Production and Use of Renewable Energy Sources in Norway

Ulf Hammer

I. Introduction

The purpose of this chapter is to describe the role of energy networks in facilitating the production and use of renewable energy sources, and in particular the regulation of the Directive 2009/28/EC of the European Parliament and the Council of 23 April 2009 on the promotion of the use of energy from renewable sources (the renewable energy directive).[1] There is a close functional connection between generating installations, networks, and consumer installations. However, the main focus here will be on the networks.[2] They are the physical frame of the electricity market.

Norway will be taken as an example when discussing the above. In the following, I will therefore first discuss the Norwegian market system for renewable energy. The focus here will be on hydropower. Almost 100 per cent of Norwegian electricity production is based on hydropower. Other forms of renewable energy, eg wind power, will not be discussed. They play a marginal role for the time being. Then, I will present the renewable energies directive. This will be followed by a brief conclusion.

[1] Special thanks go to former research assistant Ingerid Tungen who, under my guidance, wrote a thesis on the renewable energies directive. Her work has been a considerable inspiration for this chapter. Her thesis is yet to be published.

[2] The term 'networks' is wide. When I talk about networks in this chapter, the term includes grids and associated facilities, eg transformers.

II. The Norwegian Market System

A. Short-term balance

In a hydropower system there has to be an instantaneous balance between production and consumption. In the short term, this balance is handled by the electricity market. In the long term, the balance must be handled by the building of new energy infrastructures so that electricity can be directed from surplus areas to shortage areas.

In the winter of 2011 there was a shortage of water in the Norwegian hydropower system. The system operator, Statnett, has the authority to split the market.[3] The Norwegian part of the Nordic electricity market was split into five price areas. Within each price area there has to be a balance between supply and demand located in the area. Naturally, this leads to price differences between areas. But price differences also indicate where new infrastructures have to be built.[4]

B. Long-term balance

A long-term balance between supply and demand necessitates the building of new infrastructures to supply shortage areas. In addition, electricity is imported from other countries to improve supplies. Import and export of electricity requires a licence pursuant to the Energy Act (EA) section 4-2.[5]

An electricity line on Norwegian territory requires a building licence pursuant to the EA (section 3-1).[6] The licence is granted by the Ministry of Petroleum and Energy (MPE), which in these cases has delegated its authority to the Norwegian Water Resources and Energy Directorate (NWED).[7]

The EA (section 2-1). regulates the licence procedure. It implies an extensive procedure. The starting point here is the objective of the Act: EA section 1-2. It says that energy activities shall be conducted in a rational manner, taking into account public and private interests that are affected. A distinction can be made between internal and external factors. By internal factors I mean technical and economic factors directly associated with the development, such as security of supply and project economy. On the other hand, the external factors relate to various kinds of interests that are affected by the project. The most important are different kinds

[3] System operation regulations section 5.

[4] Statnett uses two additional markets in its system operation: the balancing market and the capacity reserve market. See U. Hammer 'From Self-regulation to Public Regulation' in B. Barton, Lila K. Barrera Hernandez, Alastair R. Lucas, and Anita Rønne (eds) *Regulating Energy and Natural Resources* (Oxford: Oxford University Press, 2006) 179–81.

[5] This licence is formally a quantitative restriction, which is forbidden pursuant to the EEA agreement articles 11 (imports) and 12 (exports). But in practice this licence is granted to Statnett, the system operator, and Nord Pool Spot, the manager of the spot market. Both are neutral entities in relation to competition in the market.

[6] The EA does not apply on Norwegian sea territory: EA section 1-1 second paragraph. However, Statnett needs a building licence pursuant to the EA section 3-1 when new export cables on the sea bed together with transformers on land shall be built.

[7] EA section 2-2.

of environmental interest. The classical environmental interests are well known, but recently the concern to protect the biodiversity of the environment has gained increased attention. I will soon revert to that.[8]

The wide aims of the Act affect the licence procedure. Impact analyses pursuant to the Planning and Building Act (PBA) chapter 14 shall be attached to the application. The PBA does not apply to energy infrastructures, except the PBA chapter 2 and 14: PBA section 1-3, second paragraph. This implies that local and regional municipalities have no decision-making power in energy infrastructure cases, but their views in the hearing process have to be taken into account, which is illustrated by the Sima–Samnanger case.[9]

Public organs and directly affected private interests shall receive the application. Other interests can see the application in the NWED and the affected municipalities. All parties can submit comments within a certain deadline.

C. The Nature Diversity Act

The Nature Diversity Act (NDA) has recently been enacted.[10] The purpose of the NDA is to protect biological, geological, and ecological processes and landscape diversity. The means are conservation and sustainable use.[11] The principles for official decision-making in sections 8 to 12 shall serve as guidelines when public permits are granted: NDA section 7. This means that the principles have to be taken into account when a licence is granted pursuant to the EA. Moreover, the decisions shall state that the principles have been applied.

What are these principles? I will here focus on the main issues. First, NDA section 8 requires a certain knowledge base prior to decision-making. Decisions shall be based on scientific knowledge of the population status of species, the range and ecological status of habitat types, and the impacts of environmental pressures. Second, a precautionary principle applies. According to NDA section 9 the aim shall be to avoid possible significant damage to biological, geological, or landscape diversity. Third, an ecosystem approach applies. Any pressure on the ecosystem shall be assessed on the basis of cumulative environmental effects on the ecosystem now or in the future: NDA section 10. This implies that the effects of other projects, in operation or planned, have to be taken into account. In other words, there is a wide scope of concerns, both in space and time.

Needless to say, the above requirements on public decision-making suggest an increased workload. As to the energy sector, this is illustrated by the Sima–Samnanger case, which I will soon present.

The NDA also has regulations regarding protected areas (national parks, nature reserves, etc). This is a common feature in European law. According to Council Directive 92/43/EEC of 21 May 1992 on the conservation of natural habitats

[8] See Section II.C of this chapter. [9] See Section II.D of this chapter.

[10] Act of 19 June 2009. It entered into force on 1 July 2009 as regards the parts of the act, that are subject matter here. [11] NDA section 1.

and of wild fauna and flora, several sites of Community importance have been protected. I will not deal further with protected areas in this article.

D. The Sima–Samnanger case

The BKK area, that is, the area round Bergen, Norway's second largest city, is strained as regards long-term security of supply. New energy networks have to be built to transport electricity into the area. Statnett has practised a separate price area as to the BKK area in spring 2011, but this is only a short-term solution.[12]

Statnett filed an application on 1 June 2006 for a licence to build and operate a new 410 kV transmission line from Sima power station to Samnanger transformation station, a distance of approximately 90 kilometres. The NWED granted the licence to build an air transmission line on 30 May 2008. The licence met a lot of resistance from various entities, including local and regional public and private organizations. They, 49 in total, complained to the MPE. The MPE made its decision on 2 July 2010. It confirmed the decision of the NWED.[13] But the case was not over yet: in a subsequent decision of 1 March 2011, the MPE considered whether submarine cables should be used as an alternative to an air transmission line. The MPE confirmed its previous decision. The whole decision-making process took almost five years.

The transmission line does not cross a long distance. However, the landscape is on the west coast with a lot of biological diversity, including populations of wild reindeer. In its decision of 2 July 2010 the MPE considered all the different concerns, both according to the EA and the NDA. A lot of documentation had to be assessed, and this is described in the decision. It is comprehensive, 57 pages in total.

According to the EA section 7-1, grid owners on all levels—ie the central grid, regional grids, and distribution grids—have to take part in energy planning. Statnett, the owner of the central grid, has comprehensive plans to develop new energy grids along the coast.[14] The aim is long-term security of supply. Resistance has already been announced. This includes demands for sub-sea cables. It also includes demands from the Sami population, whose interests are a legitimate concern according to the NDA section 1.

Consequently, Sima–Samnanger will not be the last difficult case.

III. The Directive on Renewable Energy

A. Legal basis, purpose, and scope

This is an important Directive. It raises a new investment perspective as regards renewable energy sources, including grids to transport this energy to the market.

[12] See Section II.A, of this chapter, as to short-term balance.

[13] The MPE also granted an expropriation permit pursuant to the Expropriation Act section 2.

[14] According to Statnett's grid development plan 2010, Statnett plans to invest 40 billion Norwegian Krone in new or updated infrastructure towards 2020. See <http://www.statnett.no/no/Kraftsystemet/> (accessed 29 September 2011).

Norway is not a member of the EU. However, Norway is part of the European Economic Area (EEA) and has signed the EEA agreement.[15] In Norway, Directive 2009/28/EC of the European Parliament and the Council of 23 April 2009 on the promotion of the use of energy from renewable sources (the renewable energy directive) is in the process of being included in the EEA agreement. The next step will be to transpose the directive into Norwegian law.

The directive has a dual legal basis: article 175(1) of the EC Treaty (now article 192 TFEU) and article 95 thereof (now article 114 TFEU) in relation to articles 17, 18, and 19 of the Directive. The latter articles deal with biofuels and bioliquids, and will not be discussed here. In the following I will refer to the TFEU (which has replaced the EC Treaty).

Article 192 TFEU refers to environmental objectives of article 191 TFEU, including the reduction of climate change. Energy production from renewable sources is beneficial in the latter regard. Therefore, the EU has decided to increase its production of energy from renewable sources to 20 per cent of internal consumption by 2020. This aim forms the background of the renewable energy directive and is also formulated directly there: articles 1 and 3(1) of the directive. Member states are obliged to reach this target; they 'shall ensure' that the share of energy from renewable sources reaches the national target within 2020. In practice, this implies a duty to invest. Such a duty represents a novelty. It also represents a very short time horizon for investments. This creates problems, which I will discuss.[16]

It should be noted that the aim of the directive refers to final consumption of energy. Norway`s production of electricity is almost 100 per cent based on hydropower, but its final consumption is based on other sources as well, including fossil fuels.[17]

The directive establishes a common framework for the promotion of energy from renewable sources. The latter term is defined in article 2(a). It comprises energy from renewable non-fossil sources, such as wind, solar, and hydropower. This means that fossil fuel and nuclear energy are outside the scope of the directive.[18]

B. Parts of the planning system revisited—the green certificates market between Norway and Sweden

The aim of the Directive represents a common target for the EU. But within this common target, member states must develop national targets pursuant to Annex I of the directive. The scope of the national targets is a negotiating matter between the member state and the EU. This also applies to Norway when the directive is

[15] This is an agreement between the EFTA states (except Switzerland) and the EU, giving the EFTA states access to the EU internal market.

[16] See Section III.C of this chapter.

[17] Norway's consumption of electricity (based on hydropower) constitutes 46 per cent of national consumption. Fossil fuels constitute 39 per cent. See *Aftenposten* 24 June 2011. On 20 July 2011 the EFTA states adopted a proposal to increase the Norwegian share of renewable energy sources consumption to 67.5 per cent in 2020. This proposal is part of the negotiations to include the renewable energies directive in the EEA agreement.

[18] Nuclear energy is not included in the above definition of the renewable energies directive.

included in the EEA agreement. Consequently, national targets may differ, but combined they shall achieve the common target. In order to control the national targets, the directive introduces a planning system. This is not a novelty, seen from a historic Norwegian perspective.

Prior to the Norwegian market reform, which was introduced by the EA of 1990, the Norwegian Parliament stipulated Statskraftverkene's (now Statkraft's) prices and other contract terms. The pricing was an important element in the long-term planning of the energy sector. Future demand patterns were calculated, which formed the basis for the planning of new production and transmission capacities.[19]

Parts of the old system concerning long-term planning of capacities have been introduced in the directive on renewable energy. But the basic prices are determined by the market according to the electricity directive.[20] Consequently, the electricity directive applies in addition to the renewable energies directive.

The planning system of the renewable energies directive is as follows: each member state shall adopt a national renewable energy action plan: article 5(1) of the Directive. This plan shall set out national targets for the share of energy from renewable sources consumed in transport, electricity, and heating and cooling in 2020. The member states should have notified the Commission of these plans by 30 June 2010.[21] If the member states' share of energy from renewable sources falls below national plans, revised plans must be forwarded to the Commission.[22] The directive does not stipulate special sanctions in such cases. Ordinary procedures pursuant to the TFEU apply.

According to the renewable energies directive, member states must finance the costs of renewable energy. It is up to member states to develop the appropriate price instruments, which have to be compatible with EU law. This matter is not fully harmonized. Accordingly, price instruments vary between member states and will not be further discussed here.[23]

However, Norway and Sweden have organized a common green certificates market, which will be in effect from 1 January 2012. The two states have entered into an agreement (treaty) as from 29 June 2011 which establishes a general framework for the respective national regulations. The common purpose of the agreement is the goal to produce 26.4 TWh renewable energy by 2020.[24] For Norway this constitutes 13.2 TWh and approximately half of today's electricity production to consumers. It should be noted that according to the recital of the agreement, a well-functioning grid is necessary for a common green certificates market. In the recital the parties are also open for an enlargement of the green certificates market to other states.

[19] See U. Hammer, 'Norway's Experience of Liberalisation' (1988), *J. Energy & Natural Resources L.*, 30–1. See also U. Hammer, *Tilrettelegging av kraftmarkedet* (Oslo: Universitetsforlaget, 1999) 432–5 (English Summary).

[20] Directive 2009/72/EC.

[21] Article 5(2) of the directive.

[22] Article 4(4) of the directive.

[23] This implies that if certificates are traded between member states, the rules on the four freedoms (goods, services, capital, and people) will apply.

[24] Article 2(1) of the agreement.

Pursuant to the Norwegian regulation of this market, which was enacted on 1 July 2011, producers of renewable electricity have a right to green certificates, while suppliers of electricity are obligated to buy green certificates. The cost of green certificates is ultimately paid by consumers. For consumers this means an extra component to the electricity price. This component will finance the extra costs of developing renewable energy. The green certificates are registered in a green certificates registry, operated by Statnett, and can be subject to trade.

The renewable energy directive also has rules on priority access for generating installations using renewable energy sources, but these rules will not be dealt with here, since access issues are not central in this project.[25] Anyhow, it is hard to see what effect these rules will have for the Norwegian part of the Nordic electricity market, which is based on open access for electricity based on hydropower sources, which constitute almost 100 per cent of Norwegian electricity production. But the access rules may be important for other countries in this market, which have considerable sources of energy based on fossil (Denmark) and nuclear energy (Sweden and Finland).

C. The development of new grids—legal conflicts

Fulfilling the targets of the renewable energy directive requires the building of new grids.[26] This is especially important for Norway, where hydropower sources are situated in remote districts. The renewable energy directive article 16 deals with grids.

According to article 16(1) of the Directive, member states shall take 'appropriate steps' to develop transmission and distribution grid infrastructure. But here arise problems due to the short time horizon of the directive, ie 2020. According to recital 41 of the Renewable Energies Directive, administrative approval procedures should be 'streamlined' with transparent timetables for installations using energy from renewable sources. To a certain extent this has already been implemented in Norway. Regional and local municipalities no longer have decision-making power in energy cases pursuant to the Planning and Building Act, but the procedure is still heavy and time-consuming, as illustrated by the Sima–Samnanger case.[27] The right to complain could be restricted, but that requires legislative amendments.

The relationship between the renewable energy directive and other directives and regulations protecting wild fauna and flora is clear. The various directives and regulations will apply combined. The environment is a wide concept and climate change is only a part of it: article 191(1) TFEU. Therefore, the renewable energies directive itself is not in conflict with other EU directives and regulations protecting the environment.

When the directive is transposed into Norwegian law, it shall have priority over other acts, regulating the same topic: the EEA act section 2. Maybe the procedures

[25] Article 16(2) of the renewable energies directive.

[26] In addition, there are the requirements of the agreement between Norway and Sweden.

[27] See Section II.D of this chapter.

pursuant to the Nature Diversity Act will have to be adapted. But this will be a result of the efficiency requirements of the renewable energies directive. The Nature Diversity Act causes a lot of work and is time-consuming.[28] But it is not in direct conflict with other acts, eg the EA.

IV. Conclusion

The balance between production and consumption, which is essential for security of supply in the electricity system, necessitates in Norway the building of new grids to maintain this balance in the long term. In addition, the Renewable Energies Directive, which will soon be included in the EEA agreement, requires the construction of new generating capacity and necessitates the building of new grids to transport electricity from production to consumption. In addition, the agreement between Norway and Sweden presupposes the building of new grids.

The renewable energy directive has a short time horizon, ie 2020. Consequently, licence and application procedures need to be streamlined to reach the ambitious target of the Directive within time. This means challenges for all countries involved, including Norway.

[28] See Section II.C of this chapter.

12

The Role of Energy Networks in Developing Renewable Energy Sources in China

Wang Mingyuan

I. Introduction

China is in the transformation phase from a rural society to an industrial society, from a planned economy to a market economy. The major challenges which China's power industry is facing are: how to establish and improve an effective power market mechanism and the supporting governmental administrative system; how to improve energy efficiency; how to safeguard security of energy supply; how to improve the coal-oriented energy mix; and how to promote the healthy and orderly development of renewable energy. The purpose of this chapter is to provide an introduction to the market reforms in China's power industry, to analyse the influence of renewable energy sources on energy network enterprises such as grid companies, and to discuss the basic duties, difficulties, and the possible solution of China's network companies with regard to the promotion of renewable energy sources.

II. Basic Background of the Development of Renewable Energy in China

A. The market-oriented reform of China's power industry

The electricity sector was one of the major fundamental industries and one of the final fortresses in China's planned economy. Since the end of the 1970s, while China carried out the reform and opening up policy, the power industry began a difficult road to market reform.

Under a planned economy the electricity companies were governmental instruments and the state uniformly made plans, approved projects, and appropriated funds for the development of the electricity sector, which was solely administrated by the authority of the central government responsible for the electricity sector.[1]

China's power industry reform started without a clear division line between the functions of the government and those of the companies. In 1997, the State Council decided to incorporate the national power company. This company was established by the State Council, as a solely state-owned enterprise, and managed as an enterprise group.[2] In this way, the national power sector began to 'separate government functions from enterprise management', and the power industry swiftly developed.

However, it was evident that there were some systematic shortcomings in the state monopolistic operation, since the inter-provincial market barriers hindered both the establishment of a trans-provincial power market and the optimum distribution of power resources, and the previous power system failed to adapt to the requirements of the market economic system. In order to improve the development of the power industry and enhance the national economic competitiveness, China must accelerate the progress of the power system reform. For this purpose, the National Development and Reform Commission (NDRC) began to lead the drafting of a *Power System Reform Plan* (hereinafter referred to as 'Reform Plan') in 2000. The State Council adopted and issued the plan, so as to as determine the guiding ideas, general targets, main tasks, and basic measures for China's power system reform,[3] in order to lead China's power industry towards a new period of market reform.

In accordance with the Reform Plan, the guiding ideas are that the market should play its fundamental role in allocating resources efficiently, the accelerated improvement of modern enterprise system, and the establishment of the power system adapting to a system of socialist market economy. General targets are to break existing monopolies, introduce competition, improve the power pricing system, optimize distribution of resources, promote the grid cover of the whole country, develop an open power market system with a clear line between the functions of the government and companies/industries, fair competition, and orderly development. The main measures being adopted so far are the following: a separation between production activities and network (transport) activities on the basis of competitive pricing; formulating the power market operation rules and governmental supervision system; establishing a new mechanism to stimulate the development of clean power plants; and promoting the direct supply of power to large users.[4]

[1] Wangxiang Chen: 'Reform and Opening-up Pushing Power Industry Approaching to New Period of Scientific Development—Reform and Opening-up of China's Power Industry for Three Decades', available at <http://www.chinapower.com.cn/article/1145/art1145084.asp>, (accessed 10 April 2011).

[2] The State Council of the People's Republic of China, the Circular on Incorporation of State Power Company (Guofa [1996] No. 48, 7 December 1996), available at <http://www.china.com.cn/law/flfg/txt/2006-08/08/content_7060160.htm> (accessed 10 April 2011).

[3] The State Council of the People's Republic of China, Power System Reform Plan (Guofa [2002] No. 5, 10 February 2002), available at <http://policy.mofcom.gov.cn/section/claw!fetch.html?id=g000013888> (accessed 10 April 2011).

[4] See supra n 3.

The Reform Plan 2000 required that assets managed by the national power company should be divided according to two kinds of activities: generation and transport. Following this separation two network companies were established, the State Grid Corporation and the China Southern Power Grid Corporation. The Reform Plan also required the establishment of a State Electricity Regulatory Commission (enjoying the rank of a ministry), which should perform the following electricity regulatory functions and responsibilities according to the authorization of the State Council: formulating the rules for power market operation, regulating the market operation, and maintaining fair competition. Depending on the market situation, the Commission can issue suggestions on electricity price adjustment to the governmental price authority, supervise and check electricity production quality standards, issue and administer electricity business licenses, resolve market disputes, and supervise the implementation of social public service policies.[5]

The evaluation of these reforms shows that the Chinese electricity sector has achieved prominent results: (i) China has successfully separated both government functions from enterprise management and generation plants from the grid; (ii) competition has been applied to power generation, strengthening the vigour of the power companies; (iii) the reform of electricity pricing has been increased; and (iv) the regional power market has started to come into being. All in all, the power industry has rapidly developed, powerfully supporting national economic and social development.[6]

With regard to the development of a regulatory regime, in 2008 the Chinese government established a new special energy industry authority, the National Energy Administration, which is a state bureau with the rank of a deputy minister under the administration of the NDRC. The National Energy Administration operates in addition to the administrative functions and responsibilities on electricity issues held by the NDRC and the State Electricity Regulatory Commission respectively, and is responsible for the industrial administration of energy sources such as coal, oil, natural gas, power (including nuclear power), and renewable energy sources. It organizes and formulates energy industrial standards, monitors the development of the energy sector and the balance between supply and demand, and directs and coordinates the supply of energy to rural areas.

With regard to electricity prices, the National Energy Administration would make the proposal for the adjustment of the prices, and report it to the NDRC for ratification. If the suggestion is approved by the NDRC, then it is passed on to the State Council for examination. If the NDRC adjusts the energy prices, it shall ask the National Energy Administration for an opinion.[7]

[5] See supra n 3.

[6] Power System Reform Working Group: The Opinions for Implementation of Deepening Power System Reform during 'Eleventh Five-year', available at <http://www.gov.cn/zwgk/2007-04/12/content_580305.htm> (accessed 10 April 2011).

[7] The Central Government of the People's Republic of China: The Approved Plan for 'Three Determinations' of National Energy Administration, Establishing Nine Departments and 112 Establishment Ceiling, available at <http://www.gov.cn/gzdt/2008-07/29/content_1058473.htm> (accessed 10 April 2011).

However, there is still a long way to go. The general reform is far from completed; at the same time, new circumstances and problems arise. These problems include an unreasonable power supply structure, a relatively slow development in power grid construction, an inadequate market for the distribution of electricity resources, and an increasing pressure on the protection of the environment and combating climate change.[8] Therefore, China should continue with the process of market reform, further accelerate power market development, promote a reform in electricity pricing, and establish an open and orderly power market with fair competition and sound development under the government macroeconomic control and effective regulation.

B. Characteristics and structure of energy resources

In general, China's energy resources are characterized by the following points.

There are abundant fossil fuels as well as renewable energy resources. Coal is dominant within fossil fuels, but there are relatively few proven oil and natural gas reserves. There is great potential for unconventional energy resources, such as shale oil and coal bed methane. In 2006, China had 1,034.5 billion tons of coal reserves, and the remaining proven reserves accounted for 13 per cent of world coal reserves, ranking third place in the world. In the renewable energy sector, the reserves of hydropower resources is equal to 6,190 billion kilowatt hours (kWh); theoretically, the annual power generation on the basis of economic development may be about 1,760 billion kWh, accounting for 12 per cent of hydropower resources in the world and ranking first place in the world.

Per capita possession of energy resources is low. As China has a large population, per capita possession of coal and hydropower resources is equal to 50 per cent of the average level of the world, and the per capita oil and natural gas only account for one-fifteenth of the average level in the world. The cultivated land is less than 30 per cent of the average level of the world, so as to restrain the development of biomass energy.

The distribution of energy resources is not balanced either. Coal resources are mainly situated in north and north-west areas, while hydropower resources are distributed in south-west areas, and oil and natural gas are available in the eastern, middle, and western parts of China, as well as offshore. The main energy consumption areas are located on the south-east coast with the highest economic development in China. It is obvious that there are significant differences between the areas where the energy resources are located and where energy is consumed. It also shows how energy flows and the basic routes of energy transportation in China. Transportation projects have enormous economies of scale and long distances such as transportation of coal from northern to southern areas, carriage of oil from northern to southern areas, transmission of natural gas from western to eastern areas, and transmission of power from western to eastern areas.

[8] See supra n 6.

It is difficult to develop energy resources. The exploitation conditions of coal resources are worse in China, since mineral workers must mine most of the coal reserves, and very few of the coal resources may be mined in the open air. Also with regard to oil and natural gas, China faces complicated geological conditions, as reservoirs are located in deep underground structures requiring sophisticated technological expertise to explore and develop. The undeveloped hydropower resources are concentrated in the high mountains and deep valleys in the south-west, far away from load centres, so the difficulty in their development is huge and development costs are enormous. Unconventional energy resources are explored to a lower extent, deliver worse economic benefits, and create an inadequate competitive force in China.[9]

The design of the Chinese energy market is based on the current situation of available resources and has been based for a long time on a consumption structure where 'coal is dominant'. However, in comparison with 1952, in 2008, the proportion of coal consumption had decreased from 95 per cent to 68.7 per cent and the consumption of oil increased from 3.37 per cent to 18 per cent, natural gas increased from 0.2 per cent to 3.8 per cent, hydropower, nuclear power, and wind power increased from 1.61 per cent to 9.5 per cent. Within the energy sector, fossil fuels with low heat value are dominant but impose significant effects on the ecological environment.[10] China is facing climate change problems, which illustrates that there is an obvious and significant contradiction between the current development and utilization of energy resources and the need for low-carbon energies development in China.

II. The Promotion of Renewable Energy in China: Problems and Challenges

A. Relevance of renewable energy

The development of renewable energy in China takes place against the abovementioned background.

Renewable energies include hydropower, biomass energy, wind energy, solar energy, geothermal energy, ocean energy, etc. These energy sources are considered as being beneficial to the harmonious development of mankind and nature, by virtue of the huge potential of such resources, low environmental pollution, and perpetual utilization.

[9] The Information Office of the State Council of the People's Republic of China, China's Energy Situations and Policies (December 2007), available at <http://news.xinhuanet.com/english/2007-12/26/content_7315441.htm> (accessed 16 October 2011).

[10] Minxuan Cui: 'Transformation of China's Energy Strategy in Low-carbon Times', available at <http://www.ndrc.gov.cn/xxfw/hyyw/t20100224_331907.htm>, (accessed 16 October 2011).

Moreover, the development and utilization of renewable energy is beneficial to meet energy demand, improve energy mix, reduce environmental pollution, and boost economic development. It is significant in the following aspects:[11]

i) It meets the scientific development goals whilst realizing sustainable growth. A sufficient, safe, and clean energy supply is the fundamental safeguard for economic development and social progress. With the pressure of increasing energy demand, it is a major strategic choice for China to accelerate the development and utilization of renewable energies, so that the contradiction between energy supply and economic development is overcome.
ii) It protects the environment and addresses climate change objectives. Since China is facing a severe environmental pollution problem and has a fragile ecological system, the rapid development and utilization of fossil energy sources like coal impose adverse effects on the environment. Therefore, the development and utilization of renewable energy play a very important role in the optimization of energy mix, environmental protection, greenhouse gas emission reduction, and provide a response to climate change issues.
iii) It helps to construct new socialist rural areas. Rural areas have the weakest economic and social development in China, and energy infrastructure is usually underdeveloped. There are about 11,500 thousand people without a power supply in the whole country, and energy supply in many rural areas depends on the traditional utilization of inefficient and direct combustion of biomass such as straw and fuel wood. By virtue of abundant renewable energy resources in rural areas, the acceleration of utilization of renewable energy, on the one hand, may take advantage of the local resources and adopt the measures suitable for the local situation to resolve the problems of power supply in remote areas and energy supply for rural residents; on the other hand, that trend may also transform the biomass resource in rural areas into commercial energy resource so that renewable energy becomes the characteristic industry in rural areas, and thus effectively extends the agricultural industry chain, enhances the agricultural benefits, increases the farmers' income, improves rural environment, and promotes the economic and social sustainable development in rural areas.
iv) It provides for new economic growth, pushing economic transformation and expanding employment. By virtue of extensive supply of renewable energy sources, every area has a certain number of conditions for the development and utilization of renewable energy. The development and utilization of renewable energy mainly make use of the local natural resources as well as local human resources, and it is significant to promote the local economic development. At the same time, as renewable energy is also a high and new technology and burgeoning industry, the rapid development of renewable

[11] National Development and Reform Commission of the People's Republic of China: Mid- and Long-term Development Planning for Renewable Energy, available at <http://www.ndrc.gov.cn/zcfb/zcfbtz/2007tongzhi/W020070904607346044110.pdf>, last accessed on 10 April 2011.

energy has become a new source of economic growth, and may effectively push the development of related industries such as equipment manufacturing industry. It will have an impact in adjusting industrial structures, promoting the transformation of economic growth, expanding employment, and pushing economic and social sustainable development.

B. Problems being faced by China in developing renewable energy

The renewable energy sources with largest potential for development in China are hydro, biomass, wind, and solar. After a continuous development over the years, hydropower has become an important part of the electricity industry. The use of household methane is extended to a large extent, and there are obvious achievements in the utilization of wind power, photovoltaic power generation, solar energy for heating purposes, and an efficient utilization of biomass energy. Together these resources can make a great contribution to the adjustment of the energy supply structure, the need for environmental protection, and the promotion of economic and social development.[12] However, the proportion of renewable energy consumption in total energy consumption is very low, and the slow technological progress and weak industrial base stand in the way of the requirement of sustainable development. China is confronted with the following major problems regarding the development of renewable energy:[13]

i) The policies and incentive measures are not sufficient. Except for hydropower and solar energy for heating purposes, which are able to participate in market competition, most renewable energy sources are of a small scale and have an irregular production and are under the current technological and policy conditions not competitive due to high costs in the development and utilization, therefore it is necessary to give them some economic support. At present, support policies are insufficient in relation to wind power, biomass energy, and solar energy, etc. With weak economic incentives, a lack of harmony among relevant policies, and unstable policies, China fails to have any long-term effective mechanism to support the sustainable development of renewable energies.

ii) The market safeguard mechanism is not perfect. For a long period, the development of renewable energy in China lacked a clear target. Although the state gradually gives more support to the development of renewable energy, due to the imperfect compulsory market safeguard policies, China fails to constitute a stable market demand, and the development of renewable energy lacks the impetus from the market, slowing down the new technologic development of renewable energies.

iii) The technological development and the industrial system are weak. Except for hydropower generation, solar energy for heating purposes and methane, the technology of other renewable energy is at a low level and lacks the capa-

[12] See supra n 11.

[13] See supra n 11.

> bility to foster technological research and development. Due to the weak capability of the manufactures, the technology and necessary equipment depend on imports, and there is a clear gap between the technological level and manufacturing capability in China and the advanced level in foreign countries. At the same time, resource prospecting, technological standards, product testing and certification, etc, are not perfect, and the training of personnel cannot satisfy the requirement of rapid market development. As a result, China fails to constitute the technological services system which may support the development of renewable energy industry.

The Standing Committee of the National People's Congress formulated the Renewable Energy Law of the People's Republic of China (hereinafter referred to 'Renewable Energy Law') in 2005, and amended and improved the law in 2009.[14] Its aims are to accelerate the development of renewable energy, enhance the proportion of renewable energy in energy consumption, solve the problem of population without power in remote areas, solve the problem of lack of energy supply in rural areas, push the energy-oriented use of organic waste, promote the industrialization of renewable energy technological development, promote energy conservation and emission reduction, actively address climate change, and better satisfy the demand of economic and social developments. The National Development and Reform Commission analysed the status quo of the renewable energy resources, technologies, and industrial development in China, used renewable energy development experience in other countries for references, and formulated the *Medium & Long-term Development Plan for Renewable Energy* in 2007. The plan includes the guiding ideas, main tasks, development targets, key fields, and safeguard measures for renewable energy development by 2020.The Plan should lead China to fully utilize the renewable energies with mature technologies and outstanding economic benefits, such as hydropower, methane, solar thermal application, and geothermal energy, to accelerate the industrial development of wind power generation, biomass power generation, solar energy power generation, to gradually increase the proportion of superior and clean renewable energy in the energy mix so that the consumption of renewable energy accounts for 15 per cent of total energy consumption by 2020.[15]

C. Challenges which the development of renewable energy imposes on the Chinese grid

Since the development and utilization of renewable energy sources are affected to such a great extent by their intermittent and random characteristics, it may have some unfavourable influences on the safe operation of the power grid system.[16] The

[14] Standing Committee of the National People's Congress: Renewable Energy Law of the People's Republic of China, available at <http://www.gov.cn/flfg/2009-12/26/content_1497462.htm>, last accessed on 10 April 2011.

[15] See supra n 11.

[16] For example, wind energy resources are sometimes nothing, larger or smaller; solar energy resources are available in the day, but nothing in the night.

current power grid system is designed to cater for the large power stations, and the direction of the power flow is fixed, mainly transmitting power from power plants to power consumers. Due to the small number of power stations in comparison with the large number of consumers, the grid administrators/operators arrange for power generation based on demand requirements, thus maintaining the balance between supply and demand and ensuring the safe operation of the power grid system. With the development of renewable energy sources, the power grid will be transformed from the current domination of centralized power supply to the combination between centralized power supply and distributed power supply. For example, each consumer (including households) may instal a photovoltaic (PV) power station and consequently receive power from the grid as well as supply power into the grid. Such a power grid system will work like the internet. The characteristics of renewable energy sources require the power grid system to be as large as possible so as to facilitate more renewable energy power generation facilities, and to be adapted to the requirements of the transform from centralized power supply to distributed power supply so that the grid should be able to accept an increased level of renewable electricity and at the same time still accept electricity from the main central generators.

For a long period, China's technological research and development of renewable energy sources mainly focused on the development of generation units and the manufacture of relevant products, thus ignoring the importance of the issue of renewable energy power grid connection. As for the development of wind power, since it is more difficult to schedule wind power development and thus balancing the power grid system, the problem of a grid-connected operation has become a key factor in the restraint of wind power development. Power generation technologies of other renewable energy sources are facing similar problems. Generally speaking, China has not adapted the mode of power grid system operation technologies and management to the characteristics of renewable energy. One of the major challenges of the power grid system is to adapt to the characteristics of renewable energy to promote the development and utilization of renewable energy.

III. Legal Duties of Grid Enterprises in the Development of Renewable Energy Power

A. Main interested parties in the development of renewable energy—government, power generation enterprises, and grid enterprises

The Renewable Energy Law[17] addresses the above challenges as it establishes targets for renewable energy development and utilization (articles 4 and 7), the prospecting, development, and utilization planning of renewable energy sources (articles 6, 8, and 9), a directive catalogue for the industrial development of renewable energy sources (article 10), a system of administrative licenses or registration

[17] See supra n 14.

of renewable energy grid connection generation projects (article 13), the guarantee of purchase of qualified renewable energy power (article 14), grid connection tariffs for renewable energy generation projects (article 19), expenses sharing (articles 20 and 21), and renewable energy development funds (article 24). These are the major elements of the legal framework providing to ensure and promote the development of renewable energy.

Specific rules for the implementation of the Renewable Energy Law have been issued by a number of government authorities falling under the responsibility of the State Council. The National Development and Reform Commission, for example, formulated the Provisions on the Administration of Renewable Energy Power Generation in 2006[18] and the State Electricity Regulatory Commission formulated the 'Measures for the Supervision and Administration of Grid Enterprises Purchasing Qualified Renewable Energy Power'[19] in 2007. These laws and regulations specify the positions and connections of the interested parties such as government, power generation enterprises, grid enterprises, and power users in the development of renewable energy. In general, they are very beneficial for power generation companies.

Since the government is the planner and the regulator of the renewable energy market, as well as the pusher and leader of the renewable energy development, its major functions include:

- organizing and harmonizing the prospecting of renewable energy resources;
- formulating the mid- and long-term targets for total volume of development and utilization of renewable energy;
- formulating, planning, and organizing the development and utilization of renewable energy;
- determining the proportion of renewable energy power generation accounting for total power generation volume, which should be reached during the period, according to the national development and utilization plan for renewable energy;
- formulating and publishing the directive catalogue of renewable energy industrial development;
- granting preferential tax treatments to the projects listed in the directive catalogue of renewable energy industrial development;
- determining and adjusting in a timely manner the grid connection tariffs for the power generation projects of renewable energy; and
- setting up the renewable energy development funds.

[18] National Development and Reform Commission: The Provisions on the Administration of Renewable Energy Power Generation, available at <http://www.ndrc.gov.cn/zcfb/zcfbtz/tz2006/t20060206_58735.htm> (accessed 12 April 2011).

[19] State Electricity Regulatory Commission: the Measures for the Supervision and Administration of Grid Enterprises Purchasing Renewable Energy Power Quantity in Full, available at <http://www.gov.cn/flfg/2007-08/01/content_702636.htm> (accessed 12 April 2011).

When a power generator wishes to connect a renewable energy production facility to the grid, it is under an obligation to apply for an administrative licence or a registration from the relevant authorities in accordance with the laws or the regulations of the State Council, and has the obligation to cooperate with grid enterprises to ensure the safety of the grid.

Grid companies—the State Grid Corporation of China and China Southern Power Grid Corporation—are obliged to operate and maintain the grid which should be friendly to the development of renewable energy. This includes the obligation to expand and reinforce the grid, to improve the grid operation and management, and to give priority to schedule and purchase qualified power generated by the renewable energy generation plants connected to the grid in conformity with the technological standards applying to the grid. So to speak, grid companies have the obligation to resolve the transmission and schedule issues of renewable energy.

In addition, grid companies may collect additional premium for power price on the basis of sold power quantities in the whole country. The difference originates when the expense inccurred in the purchase of renewable energy power at the grid-connected power tariffs determined by the government is higher than the expense inccurred if the power price is calculated on the basis of the average grid-connected power price of the conventional energy; the reasonable grid-connected expenses and other reasonable expenses paid for purchasing renewable energy power may be accounted into the costs of transmission of the grid enterprises, and recovered from the selling power price. In other words, the above expenses and costs will be borne by final power consumers in the whole country.

B. Provisions on the legal duties of grid enterprises in the laws and regulations

Taking into account that electricity grids are considered as natural monopolies, the extreme importance of the development and utilization of renewable energy, the social responsibilities of the grid companies—the Renewable Energy Law, the Provisions on the Administration of Renewable Energy Power Generation, and the Measures for the Supervision and Administration of Grid Enterprises Purchasing Qualified Renewable Energy Power—determine the grid companies' legal duties in the development of renewable energy, and the liabilities for any breach of the relevant provisions. According to these provisions, grid companies are required to bear more and larger duties and responsibilities in the initial phase of the development of renewable energy, in addition to the governments playing the leading role.

1. The relevant provisions in the Renewable Energy Law

In accordance with the Renewable Energy Law, grid companies shall enter into grid connection agreement with those renewable power generators that have legally obtained an administrative licence or for which registration has been made, giving priority to dispatching, and the purchase of renewable energy produced by a facility connected to its grid, in conformity with the grid-connected technological standards. Therefore, grid companies are also required to reinforce the grid if necessary and expand the scope of renewable energy power grid, develop and apply the technologies of smart grids and energy storage, improve the management of grid operation, enhance the grid's capability of transmitting renewable energy power, provide grid-connection services for renewable energy generators, and to cooperate with power generators to ensure the safety of the grid (article 14).

If grid companies breach the above provision and fail to purchase renewable power offered, which results in economic loss of the renewable power generator, such companies shall be liable for compensation, and the state electricity regulatory authority shall order them to make a correction within a stipulated period of time. In case they refuse to make a correction, a fine of less than the economic loss of the renewable power generator shall be imposed (article 29).

2. The relevant articles in the Provisions on the Administration of Renewable Energy Power Generation

Chapter 3 of the Provisions on the Administration of Renewable Energy Power Generation further specifies the duties of the grid companies and operators.

Grid companies at the provincial or higher level shall formulate the construction plans for the power grids facilitating renewable energy production, according to the medium- and long-term renewable energy power generation plan of the people's government at the provincial or higher level, which is included in the national and provincial power grid development plans. The construction plan shall be implemented after approval by the provincial people's government and the National Development and Reform Commission (article 10).

Moreover, grid companies shall vigorously investigate and design power grids according to the planning requirements, and construct and renovate power grids based on the progresses and needs of renewable energy generation projects so as to ensure the supply of all the electricity produced to the grids (article 11).

In addition, grid companies are responsible for the construction and management of the connection of renewable energy generation projects. For the medium- and large-scale hydropower, wind, and biomass power projects directly connected to transmission grids, the connection system should be built by the grid companies at their own costs. In principle, the grid companies are responsible for building the connection to the grid of small-scale solar power and biogas power generation projects. However, power generators (or individuals) can also invest and construct the connection themselves but this requires consultation with grid companies (article 12).

Grid companies are also responsible for measuring and registering the amounts of renewable energy power purchased. The provincial grid companies are required to summarize and report this information to the energy regulatory department of provincial government prior to 20 January each year, with a copy filed to the National Development and Reform Commission (article 13).

Furthermore, chapter 5 on supplementary provisions stipulates that any dispute between the power grid and the power generator can be submitted to the National Development and Reform Commission or the State Electricity Regulatory

Commission for mediation, or be subject to civil proceedings in the court in the event of mediation failure.

3. *Main provisions in the Measures for the Supervision and Administration of Grid Enterprises Purchasing Qualified Renewable Energy Power*

The *Measures for the Supervision and Administration of Grid Enterprises Purchasing Qualified Renewable Energy Power* ('the Measures') include five chapters and 24 articles, specifying the applicable scope, supervisory functions and responsibilities, regulatory measures, and liabilities.

In accordance with chapter 2 of the Measures, the following activities of the grid companies shall be regulated by the State Electricity Regulatory Commission and its agencies (hereinafter referred to 'electricity regulatory authorities'):

(i) The grid companies at the provincial or higher level shall formulate the construction plans for the power grids facilitating renewable energy generation, which shall be reported to the electricity regulatory authorities for record after the approval of the provincial people's government and the competent department of the State Council. Grid companies shall construct or renovate the power grids facilitating renewable energy generation according to the planning requirements, and put it into use as scheduled, to ensure the transmission of the power produced by grid-connected renewable energy generators (article 5).

(ii) Grid companies shall sign the power purchase contract and the grid-connected dispatching agreement with renewable energy power generators. The grid-connected renewable energy generators shall conform to the grid-connected technological standards for renewable energy power, stipulated by the state, and pass the appraisal of grid-connected safety organized by electricity regulatory authorities (article 6).

(iii) Grid companies shall provide grid-connection services for the renewable energy power generators (article 7).

(iv) The power dispatching institutions shall compile the power generation dispatching plan in accordance with the relevant state provisions and the requirements to ensure grid-connected renewable energy power generation, and organize the implementation. The power dispatching institution shall formulate the detailed operational rules in such a way that they are in conformity with the characteristics of renewable energy generators and to ensure that the renewable energy power plant is connected to the grid and report such rules to electricity regulatory authorities for its records (article 8).

(v) Grid companies shall improve the maintenance of the transmission equipments and the technological supporting system, improve the reliability of the grid, ensure the safety of the equipment, and prevent power failures caused by the equipment of renewable energy power projects connected to the grid (article 9).

(vi) Grid companies shall purchase all qualified power produced by renewable energy generation plants connected to the grid. In case a renewable energy power plant will be disconnected from the grid due to *force majeure* or due to circumstances endangering the safety and stability of the grid, the grid company shall send a timely written notice to the renewable energy power generator on the lasting time, the anticipated power quantity, and the specific reason for the disconnection. The grid company shall also report the problem of the disconnection, and the corresponding rectification measures to the electricity regulatory authorities, and the power regulatory authorities shall supervise whether the grid company fulfils such rectification (article 10).

(vii) Grid companies shall settle all electricity charges and subsidies in a timely manner strictly in accordance with the electricity tariff and subsidy standards for renewable energy power generation approved by the state, and the electricity purchase contract (article 11).

If one of the following behaviours of a grid company or an electricity dispatching institution causes an economic loss of the renewable energy power generator, the grid company shall be liable for compensation, and the state electricity regulatory authority shall order them to make correction within a stipulated period of time; in case of refusal to make a correction, a fine of less than the economic loss of the renewable power generation enterprises shall be imposed. Such a measure applies when:

(i) failing to construct or delaying the construction of the connection to the grid for the renewable energy generation projects;

(ii) refusing or hindering the conclusion of the electricity purchase contract and the grid-connected dispatching agreement with renewable energy generators;

(iii) failing to provide or delaying grid-connection services to renewable energy power generators;

(iv) failing to give priority to dispatching of renewable energy power;

(v) failing to purchase qualified renewable energy power due to the grid company or power dispatcher.

Grid companies shall pay such compensation within 15 days as of the date that power regulatory authorities confirm the economic losses of the renewable energy power generator (article 20).

C. Wind power generation: achievement, predicament, and outlet

Wind energy resources are rich in China. The total wind energy capacity is estimated at about one billion kilowatts, of which some 0.3 billion kilowatts can be produced on land and the remaining part offshore. Wind energy resources are divided into two wind bands: one consists of 'three north areas' (ie the north-east area, the northern region of north China, and the north-west area) and the other of the eastern coastal land, islands, and offshore areas. In addition there are some areas with abundant wind energy resources in further inland areas.[20]

The technological and economic performance of the wind power industry is quite mature. During the 'eleventh five-year plan' period (2006–2010), the grid-connected wind power industry developed rapidly, and it has shown a tendency to explosive growth since 2007. The wind power industry has developed rapidly in the areas with rich wind energy resources such as Inner Mongolia, Xinjiang, Liaoning, Shandong, and Guangdong, to name a few. The wind power target of 30,000 megawatts installed capacity for the year of 2020 was already achieved by China in 2010.[21]

In 2010, the newly added installed capacity of wind power reached 18,928 megawatts in China, accounting for 48 per cent of newly added installed capacity globally. China surpasses the United States and becomes the first largest wind power market in the world. The accumulative installed capacity in the whole country exceeds 40,000 megawatts, and wind power offshore was initially developed to a large extent. During the 'twelfth five-year plan' period, China's wind power industry will continue to develop at the speed that each year more than 10,000 megawatts will be installed, and as a consequence wind energy will become one of the most important fields of investment, which includes the construction of wind farms as well as the manufacturing of wind power equipment.[22]

However, wind energy resources are usually supplied in the reverse direction of the load centre. It is very difficult to develop wind power at a large scale and transmit power over long distance, and the mechanism of harmonization and cooperation between wind power generators and grid companies is very poor. Due to the limited load capacity of the regional grid, the local grid can only absorb limited amounts of renewable energy. Although large quantities of electricity can be produced by wind turbines, it can often not be transmitted and supplied to the end consumers. For example, the installed wind power capacity in Jiuquan, Gansu is expected to be 15 million kilowatts in 2015, and will exceed 20 million kilowatts in 2020, while the maximum load of northwest grid is 20 million kilowatts (excluding independent Xinjiang grid). Even if the 750-kilovolt transmission lines are completed for the purpose of transmitting the power produced by the Jiuquan wind power base, the current load capacity in the north-west area is only able to resolve the capacity problem in 2015.

Furthermore, the approvals of wind power generation projects and grid projects are separated from each other. Many power generation enterprises and local

[20] National Development and Reform Commission of the People's Republic of China: Mid- and Long-term Development Planning for Renewable Energy, available at <http://www.ndrc.gov.cn/zcfb/zcfbtz/2007tongzhi/W020070904607346044110.pdf>, last accessed on 2 October 2011.

[21] Xiujie Zhou: (2011–2015) 'China's Wind Power Generation Industry Analysis and Prospect Prediction Report', available at <http://www.ocn.com.cn/reports/2006005fenglifadian.htm> (accessed on 16 October 2011).

[22] See supra n 21.

governments expand the wind power generation scale blindly, ignoring that the conditions of being grid-connected are not satisfied. Because the management of the wind power generation project is not in agreement with the management of connection projects, the difficulty in being grid-connected will be further aggravated.[23]

Wind power in areas with rich wind energy resources such as the north-west, Inner Mongolia, and the north-east, and which is not used locally, must be transmitted to areas such as Beijing, Tianjin, and Tanggu, East China, and Middle China, etc. If power can be transmitted by means of a national grid of nearly 900 million kilowatts, China will establish a grid system aiming at balancing wind power and hydropower, and balancing wind power and photo-electricity in south-west area and three-north areas where renewable energy generation is still in the preliminary development phase at present. China may then be able to develop more renewable energy at a lower price.[24]

However, due to factors such as load, transmission capacity, and the absence of electricity storage options, each local grid is operating relatively independent from the other grids in China. At present, the mechanism of trans-provincial transmission is not completely set up, and it lacks effective means to balance the interests of all parties involved. As a result, wind power in Inner Mongolia can scarcely be transmitted to the area of Beijing, Tianjin, and Tanggu, and the wind power in the north-west area cannot be transmitted to the Middle China and East China areas. It is the conflict of 'regional saturation' and 'overall hunger'. Moreover, it is an evident problem that the development and utilization plan for renewable energy is separated from energy planning, power planning, and grid planning. The quantity and speed of investment of grid companies in the connection system are always unable to satisfy the need for wind power development, and some power generation companies have to make investments in the construction of connection systems for their own wind power plants.[25]

In order to solve the abovementioned problems, it will be necessary to harmonize the various plans, improve the development of renewable energy as part of the national overall strategy of energy development, specify the formulation and implementation of local plans based on national plans, and to maintain the obligation that grid companies shall purchase renewable energy power. For this purpose, the Standing Committee of the National People's Congress issued the Decision on the Amendment to the Renewable Energy Law of the People's Republic of China on 26 December 2009. This decision affects mainly the development and utilization plan of renewable energy and the system safeguarding the purchase of renewable energy power. The energy authority of the State Council shall, on the basis of the mid- and long-term total volume targets of renewable energy throughout the country, prepare a national renewable energy development and utilization plan, which is to be implemented after being approved by the State

[23] Huifeng Xue, Zhongying Wang: 'Amendment and Implementation of Renewable Energy Law', *Energy Review* (2010, Issue 2).

[24] See supra n 23.

[25] Supra n 23.

Council. The energy authorities of the people's governments at the level of province, autonomous region and municipality directly under the State Council shall, on the basis of the national development and utilization plan for renewable energy and the local mid- and long-term target for the development and utilization of renewable energy, cooperate with relevant authorities of the people's governments at their own level in preparing local renewable energy development and utilization plans for their own administrative regions, which shall be implemented after being approved by people's governments at their own level, and reported to the energy authority of the State Council and the state electricity regulatory authority for its records. Apart from this, the energy authority of the State Council shall, together with the state electricity regulatory authority and the financial authority of the State Council, determine the share of renewable energy generation in the total amount of power generation which should be reached during the planning period in accordance with the national development and utilization plan for renewable energy, and formulate the specific measures that grid companies will apply to give priority to scheduling and purchasing qualified renewable energy power. The energy authority of the State Council shall supervise this process and urge its annual implementation, together with the state electricity regulatory authority. Finally, grid companies are obliged to purchase qualified renewable energy power produced by plants connected to the grid in conformity with the technological standards applying to the grid, reinforce the grid, improve the operation and management of the grid, enhance the capability of transmitting renewable energy power, and provide appropriate connection services to renewable energy power generators.

These measures are beneficial to overcome the 'bottlenecks' in the development of wind energy, but there are doubts as to whether it is possible to implement the necessary provisions effectively.

IV. Conclusion

With the reform of the entire energy sector, China has established a legal and administrative system for grid companies to ensure and promote the development of renewable energy, and has achieved sound results. However, the construction of power grids and the reform of the power tariff system are lagging behind, the reform of the whole power system and of the market is far from completed, and it is necessary to further improve the corresponding legal system and governmental administrative system on electricity. At the same time, grid companies are required by law to ensure and to promote the development of renewable energy, but they are not fully ready to comply with the technological and administrative exigencies imposed on the system operators, and by the need of adapting it to the characteristics of renewable energy. It is evident that these challenges, which the power industry is facing in relation to the operation and the administration of the grid, seriously restrict the implementation of the applicable laws and policies and the sound development of renewable energy. While China is intensifying the reform of the electricity market and improving the relevant legal and governmental administrative systems, it is also trying to improve the construction and development of the grid and the application of technologies with regard to smart grids and energy storage, and to powerfully promote the development of renewable energy.

13

Economic Regulation and the Design of a Carbon Infrastructure for Alberta

*Nigel Bankes and Rick Nilson**

I. Introduction

This chapter deals with the development and potential regulation of a new form of infrastructure, a carbon infrastructure, to accommodate the adoption of industrial-scale carbon capture and storage (CCS)—a greenhouse gas mitigation technology. The chapter focuses on the development of this infrastructure in Alberta, Canada and in particular the carbon dioxide (CO_2) pipeline and storage elements of the infrastructure. The chapter poses two main questions: (1) is there a case for economic regulation of either or both of these elements of the infrastructure? and (2) if so, what form might that regulation take?

The province of Alberta is the centre of Canada's oil and gas industry. It has a very carbon-intensive economy. Most of the electric power generated in the province is from coal,[1] crude oil represents a significant portion of Alberta's exports,[2] and production of bitumen and synthetic crude oil from the oil sands is a major source of economic activity.[3] All of this results in significant levels of greenhouse gas (GHG) emissions,[4] which threaten not only the global atmosphere but also

* Nigel Bankes' work on the legal issues associated with carbon capture and storage is supported by the Institute for Sustainable Energy, Economy, and Environment (through a grant from Natural Resources Canada) and by Carbon Management Canada. Nilson's contribution to this chapter is based on his major paper in his course-based LLM degree, which he completed at The University of Calgary in 2011.

[1] Alberta Energy, 'Our Business: Coal' (Alberta Energy, 8 June 2011), available at <http://www.energy.alberta.ca/OurBusiness/coal.asp> (accessed 2 October 2011).

[2] Government of Alberta, 'Alberta's International Merchandise Exports' (Government of Alberta, September 2010), available at: <http://www.albertacanada.com/documents/SP-EI_ABMerchandiseExports.pdf> (accessed 2 October 2011).

[3] Government of Alberta, 'Oil Sands Economic Benefits' (Government of Alberta, March 2011), available at <http://www.oilsands.alberta.ca/FactSheets/fs_Economic_Benefits_online.pdf> ('[F]rom 1999–2008, $90.71 billion has been invested in the oil sands'. Accessed 2 October 2011).

[4] Alberta Environment, 'Regulating Greenhouse Gas Emissions' (Alberta Environment, 2011) available at <http://environment.alberta.ca/0915.html> (accessed 2 October 2011. Alberta's GHG emissions in 2008 were 244 megatonnes (Mt)).

Alberta's access to export markets.[5] One wedge in Alberta's emissions reduction strategy is CCS.[6] Indeed, the government of Alberta is counting on CCS for 70 per cent of targeted reductions.[7]

CCS is a process by which carbon dioxide (CO_2) is captured from industrial emitters, transported to a suitable disposal site and sequestered underground in geological formations.[8] There are three components of the CCS value chain: (1) the capture and compression of CO_2 emissions; (2) the transportation of CO_2 to a suitable disposal site; and (3) the permanent[9] sequestration of CO_2 in a depleted oil or gas reservoir, a deep saline aquifer, or some other target such as an unmineable coal seam. These are all very costly activities[10] and there is a financial gap[11] between the cost of CCS and the current cost of alternative compliance mechanisms.[12] To address this gap the Government of Alberta has decided to subsidize CCS and has selected four industrial-scale projects to receive grants totalling two billion Canadian dollars.[13]

In addition to the economic challenges to the widespread adoption of CCS, the literature and government task forces also recognize the need to provide an appropriate legal and regulatory framework for CCS projects. In previous work we have

[5] Alberta faces the risk that other states might impose restrictions on imports of oil products derived from bitumen based upon the carbon intensity of those products, see Robert K. Strumber, 'NAFTA Services and Climate Change' (November 2009), Pardee Center Task Force Report 11, 15–18, available at < http://www.bu.edu/pardee/files/2009/11/Pardee-Report-NAFTA.pdf> (accessed 2 October 2011).

[6] Robert Socolow and Steve Pacala, 'Stabilisation Wedges: Solving the Climate Problem for the Next Fifty Years with Current Technology' (2004) 305 *Science* 968.

[7] Alberta Government, 'Alberta's 2008 Climate Change Strategy: Responsibility/Leadership/Action' (Alberta Government, 2008) 1 at 24, available at <http://environment.gov.ab.ca/info/library/7894.pdf> (accessed 2 October 2011. The Government of Alberta is targeting a 50-Mt reduction in GHG emissions by 2020 and a 200-Mt reduction by 2050 as compared to a business-as-usual case. CCS is expected to reduce GHG emissions by 139 Mt).

[8] International Panel on Climate Change Working Group III, IPCC Special Report on Carbon Dioxide Capture and Storage in B. Metz, et al (eds), Cambridge: Cambridge University Press, 2005, available at <http://www.ipcc.ch/pdf/special-reports/srccs/srccs_wholereport.pdf> (accessed 10 October 2011)

[9] CO_2 can be used temporarily for EOR in maturing reservoirs, which enhances the economics of CCS. Ultimately these reservoirs will be depleted and the CO_2 will be permanently sequestered.

[10] Alberta Carbon Capture and Storage Development Council, 'Accelerating Carbon Capture and Storage Implementation in Alberta' (Alberta Energy, March 2009), available at <http://www.energy.alberta.ca/Org/pdfs/CCS_Implementation.pdf> (accessed 10 October 2011) (CCS Council) (the cost of CCS 'range[s] from $70 to more than $150/tonne' at 10).

[11] Ibid at 37.

[12] Alberta has adopted intensity-based targets for large final emitters (emissions >100,000 tons pa). A covered entity can stay in compliance by making the required reductions, purchasing emissions offsets, purchasing emissions performance credits (from a covered entity that has beaten its target) or by contributing $15 tonne to the Alberta Climate Change and Emissions Management Fund, see Specified Gas Emitters Regulation, Alta Reg 139/2007 (SGER).

[13] The four projects are: (1) Alberta Carbon Trunk Line, a proposed 240km CO_2 pipeline between Redwater and Lacombe, including a gathering system at the north end and a distribution system at the south end; (2) Pioneer, TransAlta Corporation's project to capture 1 Mt of CO_2 annually from its Keephills coal-fired power plant west of Edmonton; (3) Shell's Quest project to capture 1.2 million tonnes (Mt) of CO_2 annually from its Scotford Upgrader and pipeline it to a saline acquifer storage site; and (4) Swan Hills Synfuels in-situ coal gasification project. For summaries see Alberta Government, 'CCS Major Initiatives' (Alberta Government, 8 June 2011), available at <http://www.energy.gov.ab.ca/Initiatives/1897.asp> (accessed 2 October 2011).

emphasized that this framework needs to address four main groups of issues:[14] property issues (including pore space ownership), regulatory and permitting issues (eg the choice of regulator and adaptation of existing rules for analogous projects, especially acid gas disposal projects[15]), liability issues (eg the possible transfer of long-term liability), and the interface between CCS projects and carbon trading schemes.[16] The province of Alberta made considerable progress in addressing the first three of these issues with a series of amendments to the Mines and Minerals Act[17] and the Oil and Gas Conservation Act (OGCA)[18] adopted in the autumn of 2010.[19] The province has yet to deal with the interface between CCS projects and carbon trading schemes.

This chapter deals with a question that fits within the second, regulatory group of issues: is it necessary to provide for a degree of economic regulation in order to encourage investment in CCS technologies and projects and to protect the public interest in ensuring that CCS technology is implemented at the lowest possible cost? The chapter addresses the second and third links in the CCS chain, ie the transportation link and the storage link. We assume that the facility owner (the emitter) will always be responsible for the capture part of the chain and will be responsible for delivering captured CO_2 that meets the specifications of the transportation entity. In some cases it is possible that the transportation entity or an aggregator may assume responsibility for some of the processing and compression functions. While the literature discusses other potential transportation options (ships, rail, road, etc), we assume that industrial-scale plants in Alberta will use pipelines. In some cases these will be existing pipelines converted to CO_2 use and in some cases they will be new pipelines.

The second part of this chapter describes the three elements of the CCS chain in Alberta. Part three examines the case for economic regulation of the transportation and storage elements of the chain. Part four discusses how these issues were dealt with in the European Union's (EU's) CCS Directive[20] and in the proposals

[14] N. Bankes, J. Poschwatta, and E. M. Shier, 'The Legal Framework for Carbon Capture and Storage in Alberta' (2008) 45(3) *Alberta Law Review* 585 and N. Bankes and M. Roggenkamp, 'Legal Aspects of Carbon Capture and Storage' in D.N. Zillman, C. Redgewell, Y.O. Omorogbe, L.K. Barrera-Hernández (eds), *Beyond the Carbon Economy: Energy Law in Transition* (Oxford: Oxford University Press, 2008) 339.

[15] Acid gas disposal projects entail the capture of an emissions stream from a natural gas processing plant and its geological injection. The emissions stream will vary from plant to plant but will likely include SO_2 and H_2S. For a review, see S. Bachu and K. Haug, 'In Situ Characteristics of Acid-Gas Injection Operations in the Alberta Basin, Western Canada: Demonstration of CO_2 Geological Storage' in S. Benson (ed), *Carbon Dioxide Capture for Storage in Deep Geologic Formations—Results from the CO_2 Capture Project* (vol. 2, Amsterdam: Elsevier, 2005) 867. N. Bankes and J. Poschwatta, 'Carbon Capture and Storage in Alberta: Learning from the Acid Gas Disposal Analogy' (2007) 97 *Resources* 1.

[16] See S. Russell, 'The Treatment of Carbon Capture and Storage Projects Within Emissions Trading Schemes' (LLM thesis, University of Calgary, 2011).

[17] Mines and Minerals Act, RSA 2000, c M-17.

[18] OGCA, RSA 2000, c O-6.

[19] Carbon Capture and Storage Statutes Amendment Act, SA 2010, c 14. For commentary see N. Bankes, 'Alberta's New Carbon Capture and Storage Legislation' (2011) 1(2) Greenhouse Gases: Science and Technology 134.

[20] Council Directive 2009/31/EC of 23 April 2009 on the geological storage of carbon dioxide and amending Council Directive 85/337/EEC, OJ L140/1, available at <http://eur-lex.europa.eu/

of the United Kingdom (UK) to implement the Directive. Part five considers the options for Alberta.

II. The Elements of the CCS Chain in Alberta

A. The capture element: the large emitters

Most of the large emitters in Alberta are concentrated in a few geographical locations. See Table 13.1 identifying the main sources of CO_2 in Alberta.

The concentration of sources shown in the table provides Alberta with some opportunities to combine CO_2 streams from different CCS projects, especially from coal-fired generating plants and upgraders.[21]

B. The storage element: storage sites in Alberta and the legal right to store

Storage sites in Alberta are more broadly distributed but their size and quality vary, as do the targets. In general, Alberta is well endowed with potential storage opportunities in all three of the major classes, depleted oil and gas reservoirs, saline aquifers, and unmineable coal seams. Dahowski and Bachu offer the following

Table 13.1 Main sources of Co_2 in Alberta[22]

Source	Region	2008 GHG Emissions	% of Alberta's Emissions
Oil sands producers	Fort McMurray region (North-East Alberta)	78 M tonnes	36%
Coal-fired electric power generators	Wabamun region (West of Edmonton)	46 M tonnes	21%
Upgraders and refiners	Heartland region (East of Edmonton)	29 M tonnes	13%
Petrochemical facilities	Red Deer region (Central Alberta)	25 M tonnes	12%
Total Alberta emissions		215 M tonnes	100%

LexUriServ/LexUriServ.do?uri=OJ:L:2009:140:0114:0135:EN:PDF> (accessed 2 October 2011. EU CCS Directive).

21 The upstream oil sands sector presents more challenges given the diversity of point sources (eg heavy trucks for mining projects and individual well head facilities for in-situ projects). See generally J.W. Bergerson and D.W. Keith, 'The Truth About Dirty Oil: Is CCS the Answer?' (2010), 44 (16) *Environmental Science and Technology* 6010 (questioning whether it makes sense to invest in CCS to reduce GHG emissions in oil sands, rather than alternative mitigation measures).

22 Table created from information in CCS Council, above n 10 at 24. Identified sources do not total 100 per cent as there are also smaller, less concentrated sources of CO_2 emissions.

summative assessment of the availability of storage in depleted reservoirs (which likely represent the first targets given the opportunities for enhanced hydrocarbon recovery):[23]

> Most... pools with very large capacity for CO_2 storage are located in a band approximately 150 to 200km wide parallel to the Rocky Mountains.... There are many potential storage sites southwest of Edmonton and along the Edmonton–Calgary corridor, which provides lots of CO_2 storage opportunities for the large CO_2 emitters in these regions.... Forty-seven of the 227 pools with a capacity greater than 5 Mt CO_2 each are located in the Thrust and Fold belt of the Rocky Mountains where access is more difficult, and these potential sinks most likely will be used at a later stage when infrastructure will be in place. The other 180 pools with very large CO_2 storage capacity are located in the undeformed part of the Alberta Basin where access is easier. There are good CO_2 storage opportunities in south-eastern Alberta and northeastern British Columbia as well, but much fewer in northeastern Alberta, and none within a 100km distance of the oil-sands plants in the Fort McMurray area. Carbon dioxide emissions from the oil-sands plants will have to be pipelined to the south and southwest.

The *Carbon Sequestration Atlas of the United States and Canada* estimates total storage capacity in depleted reservoirs at 10,090 million metric tons. Saline aquifer sites, well distributed through the province, have an estimated capacity of between 35,000 and 39,000 million metric tons and coal seam storage has an estimated capacity of 800 million metric tons. Annual emissions from large final emitters are estimated at 208 million metric tons per year.[24]

The right to dispose of CO_2 into the target geological formation in Alberta will be governed by provincial law and in particular by recent amendments (2010) to the province's Mines and Minerals Act.[25] The Act vests pore space ownership in the Province. It also creates a two-stage tenure disposition scheme for sequestration rights comprising evaluation agreements and injection agreements.[26] The Act allows such agreements to be issued by competitive bid or on application.[27]

C. The transportation element of the chain: pipelines

There is an extensive natural gas pipeline system in Alberta but only a small amount of CO_2 pipeline infrastructure associated with a number of existing EOR (enhanced oil recovery) projects.[28]

[23] R.T. Dahowski and S. Bachu, 'Assessing the Effect of Timing of Availability for Carbon Dioxide Storage in the Largest Oil and Gas Pools' (U.S. Department of Energy, 2007) at 159, available at <http://www.pnl.gov/main/publications/external/technical_reports/PNNL-16137.pdf> (accessed 2 October 2011).

[24] US Department of Energy, '2010 Carbon Sequestration Atlas of the United States and Canada' (US Department of Energy, 2010) at 159, available at <http://www.netl.doe.gov/technologies/carbon_seq/refshelf/atlas/> (accessed 2 October 2011).

[25] Mines and Minerals Act, RSA 2000, c M-17.

[26] Ibid ss 116, 117. CO_2 injection as part of an EOR activity would be premised on the existence of another tenure, ie a petroleum or natural gas production right.

[27] Ibid s 16. See also the Carbon Sequestration Tenure Regulation, Alta Reg 68/2011.

[28] The longest CO_2 pipeline in Canada is the 325km international Souris pipeline bringing CO_2 from North Dakota to the Weyburn and Midale EOR operation in Saskatchewan. That pipeline is

III. Carbon Infrastructure: The Case for Economic Regulation

The literature and government policy reports discuss several options for the development of a carbon infrastructure (pipelines, etc) for the second and third links in the CCS chain. Much of the literature focuses on the pipeline element of the chain. Chrysostomidis et al, for example, describe two basic options: (1) a series of point-to-point pipelines matching a specific source with a specific storage location funded by the project developers; and (2) the development of pipeline networks, including a backbone pipeline system which allows for common carriage of CO_2 from multiple sources to multiple sinks.[29] A consultation paper issued by the UK's Department of Energy and Climate Change (DECC) describes similar options under the headings of a 'Decentralised Model' and a 'Centralised Model'. A key part of the centralized model is a single organization with responsibility for network design, investment, and operation.[30] In such a case 'the owner of the network also has an obligation to develop the infrastructure to meet the demand of users and is usually required to provide access to the system on a non-discriminatory and transparent basis'.[31] Under this centralized model, the owner or operator of the facility would be regulated (including the rate of return) as a monopoly supplier. The DECC report also discusses a variant on the centralized option in which the infrastructure might be built or operated by a public authority.[32]

The centralized model emphasizes the need for central planning to facilitate the building of an adequate infrastructure, while the decentralized model relies more on market forces to construct the infrastructure. In either case it is important to ask whether there is a need for economic regulation. This is an obvious question in

regulated by the federal National Energy Board. See National Energy Board, Reasons for Decision, Souris Valley Pipeline Limited, MH-1-98, October 1998. One such Alberta project is Glencoe Resources' EOR project at Ponoka/Chigwell, which takes CO_2 from petrochemical facilities owned by MEGlobal/Dow and Nova Chemicals at Prentiss and Joffre through an 80km pipeline system using some new pipe but also some existing pipeline. See Alberta Economic Development Authority, 'Enhanced Oil Recovery Through Carbon Capture and Storage: An Opportunity for Alberta' (January 2009) at 15–16, available at <http://www.assembly.ab.ca/lao/library/egovdocs/2009/aleda/173910.pdf> (accessed 2 October 2011).

[29] I. Chrysostomidis et al, 'Assessing Issues of Financing a CO_2 Transportation Pipeline Infrastructure' (2009) 1 *Energy Procedia* 1625. See also K. Bliss et al, 'A Policy, Legal, and Regulatory Evaluation of the Feasibility of National Pipeline Infrastructure for the Transport and Storage of Carbon Dioxide' (Interstate Oil and Gas Compact Commission, 2010), available at <http://www.iogcc.state.ok.us/Websites/iogcc/Images/PTTF%20Final%20Report%202011.pdf> (describing intrastate and interstate models in the United States, with different regulators; accessed 2 October 2011).

[30] Department of Energy and Climate Change, 'Developing Carbon Capture and Storage (CCS) Infrastructure: Consultation on Implementing the Third Party Access Provisions of the CCS Directive and Call for Evidence on Long Term Development of CCS Infrastructure' (URN: 10D/989, 2010), available at <http://www.decc.gov.uk/assets/decc/Consultations/ccs-third-party-access/1008-cons-ccs-third-party.pdf>, hereafter DECC Consultation Paper.

[31] Ibid at pt 3.21.

[32] Ibid at pt 3.32, referring to a 'thinkpiece' by Oxburgh et al, available at <http://www.ucl.ac.uk/cclp/pdf/Oxburgh_thinkpiece-1.pdf> (accessed 2 October 2011).

the case of the centralized model but it is equally relevant to the decentralized for the reasons discussed below.

A number of arguments support some degree of economic regulation of the transportation and storage elements of the chain. Pipelines, whether natural gas pipelines or CO_2 pipelines, exhibit the typical characteristics of a natural monopoly in which unit prices decline with scale (up to certain pipeline dimensions).[33] In such a case competition is destructive and leads to the building of unnecessary and duplicative infrastructure. Better then, so the argument goes, to authorize a single pipeline serving particular locations and facilities but provide for some form of regulation to simulate the returns that would be available in a competitive market.[34]

There is a second distinctive argument in favour of regulation in an emerging infrastructure market characterized by considerable uncertainties. These uncertainties include uncertainties as to future carbon prices and related government policies. Uncertainties create a risk of underinvestment in the development of carbon infrastructure, with the result that the public will not receive the benefits associated with economies of scale. Individual project developers will size the pipeline to the needs of the particular project, while a longer-term public perspective would size the pipe based on projections of future needs.[35] On the other hand, overbuilding poses the risk that assets will be stranded or incompletely utilized. This suggests the importance of both some degree of central planning and also the need for governmental financial support, at least during the early years, while demand builds. Not all are convinced, however. Some commentators suggest that private parties may have just as good access to accurate information about future possible demand as governments and that it is possible to harness this information through techniques such as pipeline open seasons, which provide opportunities for parties to reveal future needs and procure capacity on pipelines that are proposed for construction.[36] Others argue that the prospect of regulation will act as a serious disincentive to investment.[37] There is some middle ground. For example, a

[33] NERA Economic Consulting, 'Developing a Regulatory Framework for CCS Transportation Infrastructure' (vol. 1, 2009), available at <http://www.decc.gov.uk/assets/decc/what%20we%20do/uk%20energy%20supply/energy%20mix/carbon%20capture%20and%20storage/1_20090617131338_e_@@_ccsreg1.pdf> (accessed 2 October 2011), hereafter NERA Report. The authors point out that in the offshore gas pipeline context there are economies of scale associated with pipelines up to 30 or 36 inches in diameter but not beyond because of the limited availability of the necessary specialized equipment.

[34] For a general review Stephen Breyer, *Regulation and its Reform* (Cambridge, Mass: Harvard University Press, 1982), especially chapter 1.

[35] The possible scale benefits may be considerable. For example, the DECC Consultation paper (above n 30 at pt 3.7) refers to internal work which suggests that a CO_2 pipeline sized to serve multiple sources with an aggregate demand of 25M tonnes per year would cost double that of a pipeline designed to serve a single source of 2.5M tonnes. See also Tom Mikunda et al, 'Towards a CO_2 Infrastructure in North-Western Europe: Legalities, Costs and Organizational Aspects' (2011) *Energy Procedia* 2409.

[36] See NERA Report (above n 33) at 12.

[37] See eg Adam Andrews, 'Picking Up on What's Going Underground: Australia Should Exempt Carbon Capture and Geo-sequestration from Part IIIA of the Trade Practices Act' (2008) 17 *Pacific Rim Law & Policy Journal* 407 (arguing strongly that the uncertainties associated with potential third party access under the Commonwealth legislation would be a barrier to investment, and accordingly

regulator might require operators proposing to construct new facilities to launch an open season according to prescribed rules.[38]

The case for regulating the storage link in the chain is perhaps not as obvious. There is no suggestion in the literature that CO_2 storage and disposal facilities exhibit natural monopoly characteristics and while natural gas storage may have been subject to full economic regulation historically this is perhaps because it was developed in most cases in conjunction with, if not by, regulated distribution utilities. In more recent times there have been moves to deregulate natural gas storage where competitive conditions can be established.[39] A number of arguments have been developed to extend some degree of economic regulation to this link in the chain. First, there are a limited number of suitable sites for geological storage of CO_2. As the demand for CO_2 sequestration increases, the concentration of these sites in the hands of a few could pose barriers to entry by others who wish to pursue CCS and thus 'Carbon capturers [become the] captive customers'[40] of those with this market power. In addition, there is much to be said on environmental grounds for the careful selection of a smaller number of excellent large sites, rather than the proliferation of a larger number of smaller sites.

Second, because of the high cost of CCS, many jurisdictions are considering various state subsidies. These subsidies may provide some companies with an unfair cost advantage and lead to unfair competition. However, the subsidies tend to be targeted at the capture end of the chain, rather than at either of the downstream elements.[41] Third, vertical integration may emerge along the different elements of the CCS chain.[42] This may foreclose competing energy firms from accessing these facilities too. It may therefore be desirable to consider providing some form of regulated access to the required infrastructure.

Fourth, there may be horizontal integration in one particular component of the CCS chain as firms seek the economies of scale and expertise that come from specialization. While this is desirable in order to reduce the overall costs of CCS, it

the governments should make it clear at the outset that the legislation would not be applied to CCS projects).

[38] Open seasons allow an operator to assess the need for new infrastructure and allocate capacity on a transparent and non-discriminatory basis. For examples of open-season rules see European Regulators' Group for Electricity and Gas, 'ERGEG Guidelines for Good Practice on Open Season Procedures (GGPOS)' (C06-GWG-29-05c, 2007), available at <http://www.energy-regulators.eu/portal/page/portal/EER_HOME/EER_PUBLICATIONS/CEER_PAPERS/Gas/2007/C06-GWG-29-05c_GGPOS.pdf> (accessed 10 October 2011), and Federal Energy Regulatory Commission (FERC), 'Regulations Governing the Conduct of Open Seasons for Alaska Natural Gas Transportation Projects' (revised 1 June 2005), available at <http://www.ferc.gov/industries/gas/indus-act/angtp.asp> (accessed 2 October 2011).

[39] This is the case in, eg, Ontario. For discussion, see Bill Rupert, 'Deregulation: The Ontario Natural Gas Storage Case' in Gordon Kaiser and Bob Heggie (eds), *Energy Law and Policy* (Toronto: Carswell, 2011) 291.

[40] Hans Vedder, 'EC Competition Law and the Organization of CCS' in M.M.Roggenkamp and E. Woerdman (eds), *Legal Design of Carbon Capture and Storage: Developments in the Netherlands from an International and EU Perspective* (Antwerp: Intersentia, 2009) 150, hereafter Roggenkamp, *Legal Design*. The arguments developed in this and the following paragraph apply to pure disposal projects. For a variety of reasons they would not apply to an EOR injection project.

[41] But not always. The ACTL project in Alberta includes a backbone pipeline system (above n 13).

[42] Eg in Alberta the vertically integrated Shell Quest project (above n 13).

can also result in market power. This horizontal integration could occur at the transportation stage of CCS, or at the sequestration stage.[43]

Some or all of these arguments were clearly convincing to the EU, since it chose to address the subject of access to carbon infrastructure in its CCS Directive. The next section examines both the relevant articles of the Directive and the UK's proposed implementation of these articles of the Directive.

IV. The European Response: The Access Provisions of the EU's CCS Directive and the UK's Implementation Proposals

A. The EU Directive

The EU adopted its CCS Directive[44] in 2009. Member states (MS) were required to transpose it into national law by 25 June 2011. The Directive addresses the issue of access in its preamble and in articles 20 and 21. Others have provided a detailed commentary on articles 20 and 21 and it is therefore only necessary to emphasize some key elements.[45] First, and for the reasons given above, article 20 requires MS to take the necessary measures to provide user access to *both* transportation networks and storage sites. Access shall be provided in 'a transparent and non-discriminatory manner' applying the objectives of 'fair and open access' and shall take into account available capacity and capacity that might reasonably be made available, domestic requirements, technical specifications, and the 'need to respect' the needs of the owner and the interests of other users. Second, an operator may refuse to provide access (with reason) where there is no available capacity. And third, MS must take measures to ensure that where access is denied for lack of capacity the operator is required to take the necessary enhancements to provide access where economic to do so and where doing so would not negatively impact environmental security. Finally, article 21 requires MS to put in place a dispute settlement mechanism to resolve any disputes relating to access that may arise.

In her review of the Directive, Professor Roggenkamp concludes that the access regime provided for in the Directive is a 'weak regime', although in the final analysis much will depend on 'the exact wording of the national implementing laws and the extent to which the national TPA [third party access] regimes provide third

[43] There are other possible arguments. For example, the NERA Report (above n 33) emphasizes the problem of externalities noting (at 3) that parties will under-invest in carbon infrastructure if they are not able to capture the benefits of those investments (eg where there is no price or too low a price on carbon emissions). The EU also relies on the essential facilities doctrine developed in the United States. '[R]efusals to provide access to potential competitors by monopolists owning or controlling a facility…essential to the activities of the potential competitor…could constitute an abuse of a dominant position.' Allowing third party access to key CCS infrastructure mitigates this monopoly power. See Martha M. Roggenkamp, 'The Concept of Third Party Access Applied to CCS' in Roggenkamp, *Legal Design* above n 40 at 278 [hereafter, Roggenkamp, 'Concept']. Whether parts of CCS infrastructure are seen to be essential facilities will depend on whether national regulations require CCS as a means of reducing GHG emissions and the degree of market power, in relation to scarce CO_2 sequestration sites.

[44] See EU CCS Directive (above n 20).

[45] See Roggenkamp, 'Concept' (above n 43) at 273–99; and Mikunda et al (above n 35).

parties sufficient certainty...'[46] It is possible that a variety of TPA regimes will emerge in the different jurisdictions.[47]

B. The UK proposals for implementing articles 21 and 22

In December 2010, DECC released both a consultation memorandum[48] and a set of draft regulations[49] following this up in April 2011 with the government response to the consultation.[50] The UK TPA proposals are based on existing provisions in the Petroleum Act 1998 (offshore pipelines) and the Pipeline Act 1962 (onshore pipelines). These provisions put the onus on the parties to reach agreement but provide for resort to the regulator if the parties are unable to agree. The draft regulations contain three main ideas. The first is that a party wishing to construct new carbon infrastructure may be required to incorporate additional capacity, incorporate design modifications or, in the case of a pipeline, adopt a different route.[51] In so doing the regulator shall provide an opportunity for the relevant parties to be heard and may make orders as to the costs to be borne by the party seeking access. Second, the regulations deal with access to existing infrastructure. If the parties are unable to agree, the regulator may make an appropriate order, fixing the charges that may be payable.[52] In so doing the regulator must take into account:[53]

(a) capacity which is or can reasonably be made available in the infrastructure;
(b) that the composition of the CO_2 stream to be conveyed by, injected into or stored in the relevant infrastructure is compatible with the use of the relevant infrastructure for the purpose for which it has been designed;
(c) any incompatibilities of technical specification which cannot reasonably be overcome;
(d) [any other] difficulties which cannot reasonably be overcome and which could prejudice the efficient, current and planned future transport or storage of carbon dioxide;
(e) the reasonable needs of the owner and any associate of the owner for the conveyance and storage of carbon dioxide;
(f) the interests of all users and operators of the relevant infrastructure;

46 Ibid 'Concept' at 279–80. 47 Ibid 298. 48 Above n 30.

49 DECC, 'Statutory Investments: Environmental Protection—The Storage of Carbon Dioxide (Access to Infrastructure) Regulations' (hereafter Draft regulations, 2011), available at <http://www.decc.gov.uk/en/content/cms/consultations/ccs_3rd_party/ccs_3rd_party.aspx>.

50 DECC, 'Government Response to the Consultation on Implementing the Third Party Access Provisions of the European Union Carbon Capture and Storage Directive' (URN 11D/0041, 2011), available at <http://www.decc.gov.uk/assets/decc/consultations/ccs-third-party-access/1666-gov-response-cons-third-party-eu-ccs.pdf>. The final regulations were promulgated 15 September 2011 as The Storage of Carbon Dioxide (Access to Infrastructure) Regulations 2011, available at <http://www.bailii.org/uk/legis/num_reg/2011/uksi_20112305_en_1.pdf> (accessed 10 October 2011).

51 Draft regulations above n 49 ss 5–9 ('variation conditions'). 52 Ibid s 10(9).

53 Ibid s 10(5).

(g) the proportion of the United Kingdom's carbon dioxide reduction obligations pursuant to international legal instruments and to European Union legislation that will be met through capture and geological storage of carbon dioxide; and

(h) the number of parties involved in the dispute.

The regulator may only make the relevant order if it is satisfied that any person suffering a loss can be made whole or that the order will not prejudice:[54]

(a) the efficient operation of the relevant infrastructure concerned;

(b) the safety or environmental security of the conveyance or storage of carbon dioxide by or in the relevant infrastructure;

(c) the conveying by or storage in the relevant infrastructure of the quantities of carbon dioxide which the owner or an associate of the owner requires or may reasonably be expected to require; or

(d) the conveying by or storage in the relevant infrastructure of the quantities of carbon dioxide which another person with a right to have carbon dioxide so conveyed or stored requires to be conveyed or stored in the exercise of that right.

Where appropriate, the regulator may issue an order to expand the capacity of the relevant infrastructure or make an interconnection—subject to appropriate notice and cost recovery obligations. In doing so the regulator will award costs in a manner that will not reduce the incentive for parties to invest in the development of infrastructure and will take account of how the parties propose to allocate risks, including contingent liabilities, as between the storage site owner and the supplier of the CO_2.[55]

Third, in the interests of transparency and to facilitate access, owners of the relevant infrastructure must publish, at least once a year, information relating to available capacity and any relevant technical and operating requirements.

Those who responded to the draft regulations highlighted a number of concerns. Many emphasized the differences between pipelines and storage sites. Pipelines are engineered while storage sites are natural, and thus while it is relatively easy to define pipeline capacity there is more uncertainty associated with storage sites. Furthermore, in the case of a storage site the parties and the regulator need to take account of both injection capacity and accumulated reservoir capacity. A further concern was that the regulations did not provide enough guidance as to how the regulator would establish the terms of third party access. The government responded to this concern by requesting the Carbon Capture and Storage Association to propose topics that might be covered by more detailed guidance to be issued in the future.[56]

[54] Ibid s 7(7). [55] Consultation memorandum above, n 30 at 2.23–2.24.
[56] Government Response, above n 50 at paras 10, 26.

In sum, the UK implementation proposals place the initial onus on the parties to negotiate access. The regulator does have the authority to step in and impose access terms and conditions where necessary. Such terms and conditions may require additional facilities such as additional injection wells. The regulations will also impose some transparency requirements with respect to reporting on available capacity. In developing the regulations, the regulator was clearly sensitive to concerns that regulation that was too heavy-handed and which failed to take account of the risks facing operators would act as a disincentive to investors.

V. The Position in Alberta

Alberta's current regulatory scheme does not provide for the economic regulation of the transportation and storage links in the CCS chain.[57] This gap exists because existing economic regulation provisions simply do not apply to CO_2 pipelines or storage facilities. For example, the Gas Utilities Act does not apply to CCS infrastructure since CO_2 does not fall within the definition of natural gas as used in that Act.[58] Similarly, the common order provisions of the OGCA (discussed below) do not apply to all CO_2 pipelines and there is no relevant form of common order dealing with CO_2 injection and storage. This does not mean that the provincial government has turned its head against economic regulation and potential TPA rules. Rather it is simply a case of the government not yet having addressed the issue. There are signs that it will do so as part of its ongoing CCS Regulatory Framework Assessment.[59]

When the government does address the issue it seems likely that, just as in Europe, it will draw upon its regulatory experience in the natural gas sector.[60] Consequently this part of the chapter begins with an assessment of that experience before turning to carbon infrastructure. In assessing the regulation of the natural gas industry in Alberta[61] it is useful to distinguish between: (1) the upstream exploration and production sector including the upstream gathering and processing

[57] Non-economic regulation including pipeline permitting and the permitting of storage sites taking account of environmental, safety, and public-interest concerns are dealt with elsewhere. For reviews see Stefan Bachu, 'Legal and Regulatory Challenges in the Implementation of CO_2 Geological Storage: An Alberta and Canadian Perspective' (2008), 2 *International Journal of Greenhouse Gas Control* 259.

[58] Gas Utilities Act, RSA 2000, c G-5.

[59] See CCS Regulatory Framework Assessment, Alberta Energy, available at <http://www.energy.alberta.ca/Initiatives/2840.asp>. N. Bankes is a member of the Steering Committee for the RFA.

[60] Most discussion focuses on the natural gas sector and we will follow that example here. However, it is possible that the electricity sector may also offer some useful suggestions, including the more proactive approach to system-wide planning for backbone infrastructure which characterizes that industry. In Alberta this is the responsibility of the Independent System Operator who manages the interconnected electric system [IES] and must, *inter alia*, 'forecast the needs of Alberta and develop plans for the [IES] to provide efficient, reliable and non-discriminatory system access service and the timely implementation of required transmission system expansions and enhancements', Electric Utilities Act, SA 2003, c E-5.1, s 33(1).

[61] The focus is on provincial rules as applied to facilities under provincial jurisdiction. For the federal rules which apply to interprovincial and international pipelines see National Energy Board Act, RSC 1985, c N-7, s 71; and for discussion of the Board's jurisprudence and practice see J. David Brett

facilities; (2) the main transmission pipeline system downstream of the field-based processing facilities; and (3) the distribution network. For present purposes we will ignore the distribution network other than to observe that historically this sector was occupied by incumbent franchised monopoly utilities that were subject to full economic regulation based on the North American regulated public utility model. As elsewhere in Europe and the United States, some unbundling and deregulation has occurred.[62]

Each of the following sections describes the current position in relation to a different model of economic regulation in the upstream and transmission sectors of the natural gas industry and then asks whether the model can be applied to the transportation and storage links in the CCS chain. Three models are discussed: (1) common carrier status (the common orders); (2) designation as a gas utility; and (3) the Alberta Gas Trunk Line model.

In general, access to upstream gathering and processing facilities and producer access to markets—ie a gas purchase contract—(and the economic terms and conditions of such access) are not subject to any form of economic regulation (other than general competition law[63]). Consequently, parties construct their own facilities or obtain access to the facilities of others on market terms.[64] Upstream gas storage is generally subject to the same market-based regime.[65]

There are, however, a number of exceptions which might inform the options for possible regulation of CO_2 infrastructure. One group of exceptions, the so-called common orders, is found in the OGCA[66] administered by the Energy Resources

and Nadine E. Berge, 'Oil and Gas Transportation: Is Contract a Viable Alternative to Traditional Regulation' (2007), 44 *Alberta Law Review* 93.

[62] For reviews see Keith Miller, 'Energy Regulation and the Role of the Market' (1999), 37 *Alberta Law Review* 419; and Nikol Schultz, 'Light-Handed Regulation' (1999), 37 *Alberta Law Review* 387.

[63] For a review of the application of competition rules in the upstream oil and gas sector see John Lowman et al, 'A Practical Guide to Competition Act Compliance in the Oil and Gas Industry' (2008), 45 *Alberta Law Review* 691. There are some similarities between the debate about the role of economic regulation in the CCS industry and the debate over the regulation of gathering facilities in frontier upstream areas. For this discussion see, National Energy Board, Mackenzie Gas Project, GH-1-2004, Reasons for Decision, Volume 2, December 2010, available at <http://www.neb-one.gc.ca/clf-nsi/rthnb/pplctnsbfrthnb/mcknzgsprjct/rfd/rfd-eng.html> (accessed 2 October 2011) section 8.3 at 191–5.

[64] Brett and Berge (above n 61) at 95: 'Hundreds of gas lines operate within Alberta moving gas from field to pipeline. They are producer-owned and many carry third party gas. While in some cases access was ordered by a regulator, most of these arrangements (including charges) were entered into through negotiation between the parties.'

[65] Most natural gas storage in Alberta is used to help manage contract deliverability and to hedge the commodity market. Some storage was originally acquired by distribution utilities in the province but by and large is no longer 'used and useful' for utility purposes and has thus been removed from the regulated rate base: see *ATCO Gas & Pipelines Ltd v Alberta* (Energy & Utilities Board) 2006 SCC 4, [2006] 1 SCR 140; for a discussion of natural gas storage issues see N. Bankes and J. Gaunce 'Natural Gas Storage Regimes in Canada: A Survey', (2009) ISEEE, University of Calgary, available at <http://www.iseee.ca/media/uploads/documents/pdfs/researchreports/Nat_gas_storage_cdn_survey.pdf> (accessed 10 October 2011).

[66] OGCA, RSA 2000, c O-6 (OGCA).

Conservation Board (ERCB)[67] of the province. A second exception falls within the Gas Utilities Act[68] administered by the Alberta Utilities Commission (AUC).[69]

A. The common orders

The 'common orders' comprise the common purchaser, common carrier, and common processor orders.[70] They were originally developed (and especially the common purchaser order) to protect a producer from drainage. The orders were based on the premise that in an oil and gas jurisdiction based on the rule of capture[71] a producer should be able to obtain access to the market for its production (ie obtain a gas sales contract—the common purchaser order), a pipeline (the common carrier order), and a processing plant (common processor), otherwise such a producer might be subject to drainage.[72] Currently, the pre-conditions for obtaining a common carrier order are as follows:[73]

- producible reserves available for transportation through an existing pipeline;
- a reasonable expectation of a market for the substance that is proposed to be transported by the common carrier operation;
- the applicant could not make reasonable arrangements;
- the proposed common carrier operation is the only economically feasible way, the most practical way to transport the substance in question, or clearly superior environmentally.

In sum, in addition to establishing the need for pipeline [processing] capacity, the applicant must show both that the pipeline [processor] is acting unreasonably and that there is some good reason beyond this why the applicant should have access to the pipeline [processor] facilities based on economics, 'practicality'

[67] The ERCB is a classical oil and gas conservation regulator in the North American tradition. See D. Breen, *Alberta's Petroleum Industry and the Conservation Board* (Edmonton: University of Alberta Press, 1993).

[68] Gas Utilities Act, RSA 2000, c G-5.

[69] The AUC is a classical public utility regulator.

[70] For a more detailed account of the ERCB's practice in relation to the common orders, see Robert M. Perrin, 'Declaratory Relief Under Oil and Gas Legislation' (1980) 18 *Alberta Law Review* 26; and Robert M. Perrin, 'Declaratory Relief Under Oil and Gas Legislation—Update' (1980) 19 *Alberta Law Review* 34. For the common-order statutory provisions see the OGCA (above n 66 ss 48–54). An additional analogous order is the rateable take order, s 36 OGCA.

[71] Under the rule of capture, A commits no actionable wrong when a well completed entirely on and under A's land drains production from under B's land. For the Alberta position see Cecilia Low, 'The Rule of Capture: Its Current Status and Some Issues to Consider' (2009) 46 *Alberta Law Review* 799; See generally Terence Daintith, *Finders Keepers?: How the Law of Capture Shaped the World Oil Industry* (Washington, DC: Resources for the Future Press, 2010).

[72] See Brett and Berge (above n 61) at 96: 'For parties seeking transportation of oil and gas, a declaration of common carrier (or enforcement of that statutory obligation) has frequently been the method of choice for access.'

[73] ERCB, 'Directive 065' (ERCB, 2010) at s. 1.3.4, available at <http://www.ercb.ca/docs/documents/directives/Directive065.pdf> (accessed 2 October 2011). The text applies mutatis mutandis to the common processor order. The ERCB continues to require evidence of drainage for the common purchaser order.

or potential environmental impact.[74] In making an order the Board may prorate capacity.[75]

The common order application is potentially a two-stage process. Step 1 is the ERCB designation of a party as a common carrier (or processor/purchaser) as a consequence of which the owner must provide non-discriminatory access to the pipeline or facility and may apportion capacity. Step 2, which may or may not be necessary, is taken when the parties cannot agree on the rates to be charged, in which case the matter may be referred to the AUC for it to establish what will effectively be just and reasonable rates for the service provided, based on general public utility principles.[76] Thus, step 1 provides access and some of the terms and conditions; step 2 triggers additional regulation economic elements of the terms and conditions of access. The OGCA contemplates that the ERCB may relieve an owner from fulfilling all of the duties of a common purchaser, carrier, or processor in appropriate circumstances.[77]

The key characteristics of the common order scheme are the initial reliance on the market and then a two-stage complaint-driven procedure that allows for regulatory intervention if the parties cannot agree on market conditions for access or agree upon resulting tolls and tariffs.

Is this model a useful model in the context of CCS? The current language of the OGCA does not deal expressly with a carbon dioxide pipeline. While some argument could be made that an upstream CO_2 pipeline carrying CO_2 produced from a gas-processing plant from a stream of CO_2-rich raw gas might be a pipeline for the purposes of the OGCA, the definitions could not reasonably be extended to a CO_2 pipeline that was moving CO_2 captured from a thermal generating plant. In sum, the relevant section of the Act would need to be amended. There is also no direct analogue in the OGCA that would provide a common order with respect to the injection link in the chain. It would therefore be necessary to create a new form of common order which we could think of as a common injection order. A crucial part of drafting such a change would be to ensure that existing parties were adequately protected. Thus an applicant might have to show that there is surplus injection capacity and that existing injectors would still be able to inject all of their captured volumes.

We think that it would be fairly straightforward to draft these changes; the real question is whether this approach makes sense. There are several flaws in the model. First, the common order approach assumes that there is no need for an integrated planning approach to pipeline/processing infrastructure. It assumes that the private contractual model is appropriate and that it is only necessary to interfere in

[74] In recent years the ERCB has become increasingly concerned with the so-called proliferation issue, ie underutilized capacity at existing facilities the use of which should be optimized before building new facilities.

[75] See OGCA (above n 66) at ss 48(4)(b) (common carrier), 51(4)(b) (common purchaser of gas) and 53(5)(b) (common processor).

[76] See OGCA (above n 66) at s 55.

[77] Section 49 relief from the duty of carrying any oil, gas, or synthetic crude oil of inferior or different quality or composition or from any other duties that in its opinion are unreasonable. And to the same effect see ss 52 and 54 dealing with the other orders.

extraordinary circumstances (historically driven by conditions of drainage). In sum, the common order approach does not address one of the key economic issues associated with planning a carbon infrastructure, which is to gain economies of scale so as to reduce the long-term unit costs associated with transportation. Second, and related, the common order approach as currently conceived is both ad hoc and reactive, meaning that a party can only make an application once it is actually in a position to take advantage of the order (ie it has actually drilled and completed the well and is ready to tie in). This hardly seems to meet the needs of a capture entity which will want to know with certainty that it will have access on reasonable terms to both transportation service and an injection site if it goes to the capital expense of developing a capture facility. And third, if the starting premise is that of private contract, the various players along the chain will view the possibility of a common order declaration as a serious business risk, if, for example, it requires them to share or prorate a pipeline or injection facility that was sized to meet the interests of their particular project. While it may be possible to deal with these latter concerns by setting appropriate pre-conditions to making the order (such as those discussed in the UK consultations on implementing the EU Directive), the first two objections remain.

B. Gas utility status

There is a second model of economic regulation in Alberta which arises under the terms of the province's Gas Utilities Act, where a facility and the owner of a facility (eg a pipeline) is declared to be a gas utility\owner of a gas utility and therefore subject to full prospective rate-making, including the duty to file tariffs for all services provided. This may occur where it can be shown that the facility is providing service to 'any member of the public' and should not otherwise be exempted from the application of the Act. As a matter of practice this does not happen very often but in an appropriate case (eg reasons of market power and absence of alternatives) the Utilities Commission may claim jurisdiction.

We have seen one recent example of this in the case of the Ventures Pipeline[78] originally constructed as a contract carrier to provide processed natural gas supplies to Suncor's oil sands upgrading facilities. Ventures was a subsidiary of Nova Gas Transmission Ltd (NGTL)—the supplier of regulated gas transmission service throughout Alberta.[79] Subsequently, NGTL decided to provide service to oth-

[78] See Suncor Energy Inc (24 October, 2006), 2006-105, online: <http://www.auc.ab.ca/applications/decisions/Decisions/2006/2006-105.pdf> (accessed 2 October 2011). See also *TransCanada Pipeline Ventures Ltd v Alberta* (Energy and Utilities Board) 2008 ABCA 55; TransCanada Pipeline Ventures Ltd & Suncor Energy Inc (20 May 2009), 2009-065, online: AUC <http://www.auc.ab.ca/applications/decisions/Decisions/2009/2009-065.pdf> (accessed 2 October 2011); and TransCanada Pipeline Ventures Ltd v Alberta (Utilities Commission) 2010 ABCA 96. This was not an upstream pipeline; it was a fairly large capacity transmission pipeline.

[79] NOVA Pipeline Ventures Ltd, AEUB Decision 98–20, online: <http://www.ercb.ca/docs/documents/decisions/1998/d98-20+addendum.pdf> (accessed 2 October 2011).

ers by purchasing capacity on the Ventures pipeline through Transportation by Others (TBO) contracts with Ventures.[80]

Suncor concluded that this development radically changed the circumstances underlying the original contract and applied to the AUC to have it inquire into the affairs of Ventures with a view to possible rate regulation.[81] Ventures objected, claiming that it was not a gas utility within the meaning of the Gas Utilities Act since it did not provide service to the public. Furthermore, the Commission should not exercise its jurisdiction to make the inquiry since Suncor was a sophisticated party and well able to protect itself through contractual negotiations.

The Commission rejected Ventures' arguments. It held that Ventures was providing service to the public. Suncor was a member of the public, but even beyond that, Ventures was indirectly providing service to the public through the TBO contracts. At the subsequent inquiry the Commission concluded that Suncor had established that the contractual rates charged to Ventures were unjust or unreasonable or unjustly discriminatory. It also concluded that there was no real competition for gas transportation in the Fort McMurray area and that NGTL, Ventures' parent corporation, had created a dysfunctional market by first of all refusing to provide regulated service in the Fort McMurray area and then reversing itself and doing so through the TBO contracts.

This was no doubt an extreme case but it does illustrate a number of points. First, that the definition of 'gas utility' is extremely broad, and second, that in deciding whether or not to exercise its jurisdiction to make further inquiries the Commission will examine whether there are competitive conditions or whether one party is able to exercise market power. The result of regulating a carrier as a gas utility, however, is drastic since it incorporates complex principles of prospective rate-making, requiring the Commission *inter alia* to establish a rate base and just and reasonable tolls.

Is this model appropriate for transportation and storage elements of the CCS chain? At the present time a CO_2 pipeline could not be brought within the operation of the Act simply because the definitions of 'gas' and 'gas pipeline' as used in the Act contemplate a methane-based gas stream. These terms would have to be amended to include carbon dioxide. The Act would require further amendment if this mode of regulation were to apply to the injection phase as well as the transportation phase. But more generally there are reasons for thinking that this model is inappropriate. First, this model represents the most stringent form of economic regulation of the methods discussed here because it is full prospective regulation (*ex ante*) in which the regulator approves just and reasonable tolls before the operator can provide service. It is simply not clear that *ex ante* regulation is necessary in

[80] In effect, NGTL was reversing itself; having originally declined to provide a regulated service it was now proposing to offer a regulated service to others through the TBO contracts.

[81] This was an application under s 24 of the GUA. The Court (following the Board) described the application process as a three-step process to regulate the tolls that Ventures could charge: (1) establish the jurisdiction of the Board\Commission, (2) investigate to determine if there is a market problem; and (3) if there is, assess the appropriate remedy which might require an Order in Council: *TransCanada Pipeline Ventures Ltd v Alberta (Energy and Utilities Board)* 2008 ABCA 55 at para 53.

the case of carbon infrastructure and if it were implemented would be a significant disincentive to investors. Second, and related, it seems advisable to address the issue of utility status up front before parties invest in infrastructure and negotiate contracts perhaps believing that the operators of the transportation and storage links will not be subject to utility-style regulation and prorating. The procedure (and the risk) of making declarations of utility status after the fact creates additional uncertainty for investors in this area.

C. The main natural gas transmission pipeline system: the AGTL model of economic regulation

The construction of the main natural gas transmission system intra-Alberta began in the mid-1950s under the auspices of Alberta Gas Trunk Line Limited (AGTL). AGTL was a unique collaborative joint venture structured by a special Act of the legislature.[82] AGTL was not a Crown corporation (ie a state pipeline company) but the province reserved the power to appoint two of the seven members of the board of the corporation with the balance being appointed by gas utilities, gas exporters, and gas producers. The company was to act as a common carrier (s 13) of natural gas. Pipeline construction was subject to provincial permitting by the ERCB under the terms of the Pipeline Act[83] but the economic regulation of AGTL in terms of the tolls, tariffs, and terms and conditions of service fell to what was then the Public Utilities Board (PUB, now the AUC) under the terms of the AGTL statute and the Gas Utilities Act.[84] A key policy decision was to establish a postage stamp toll for AGTL pipeline services.[85]

One crucial difference between PUB regulation of AGTL and PUB regulation of all other utilities in the province was that the PUB's regulation of AGTL was complaint-driven and retrospective,[86] whereas in the case of distribution utilities the PUB applied full prospective rate-making in which the regulated utility cannot charge customers a rate until the PUB had approved those rates as just and reasonable rates. But in the case of AGTL, AGTL established its rates, leaving it to the shippers to initiate a complaint to the PUB if they thought that the rates were discriminatory or not just and reasonable.[87]

AGTL subsequently used its pipeline base to expand into the petrochemicals business, changing its name to NOVA. The province later relinquished its

[82] Alberta Gas Trunk Line Company Act, SA 1954, c 37, s 13(1).

[83] See now Pipeline Act, RSA 2000, c. P-15.

[84] Gas Utilities Act, RSA 2000, c G-5.

[85] Under a postage stamp toll all gas pays essentially the same tariff no matter how far it travels on the system. This effectively provided a subsidy for those exploration areas that were distant from markets.

[86] Alberta Gas Trunk Line Company Act, SA 1954, c 37, s 30 provided that the Directors should fix just and reasonable rates but that on application to the then Board of the Public Utility Commissioners the Board might vary or fix the rates originally established by AGTL. The provision was updated in 1970; see Alberta Gas Trunk Line Company Act, SA 1954, c 37, as amended by SA 1970, c 5.

[87] In practice very few complaints were launched, but see the following example that led to litigation: *NOVA v Amoco Petroleum Company* [1981] 2 SCR 437.

controlling position in NOVA. Since then, the entire NOVA pipeline system[88] has come under the jurisdiction of the federal National Energy Board under the terms of the National Energy Board Act[89] on the grounds that the Alberta transmission system is part of a broader interprovincial system operated by TransCanada PipeLines Ltd.[90]

In sum, the history and evolution of AGTL provides some useful lessons when thinking about the options for regulating an emerging carbon infrastructure. First, it provides an interesting example of a collaborative approach between government and the different sectors of the industry. This approach allowed the government to provide strategic direction in the emergence of the natural gas pipeline infrastructure. It also provided a forum within which centralized planning of the infrastructure could occur. Second, it provides an example of a light-handed form of economic regulation in the form of an *ex post* complaints-based scheme, rather than a scheme of review and prior approval of proposed rates.

The AGTL model speaks directly to the situation in which a jurisdiction is developing a new infrastructure for a new industry, characterized by a significant public interest. It therefore seems useful in the context of the CCS chain, especially if government is providing financial support for this part of the chain. The characteristics of the model include common carrier status, and public influence on key strategic decisions made by the pipeline entity through government-appointed directors. The different sectors of the CCS industry might also be represented on the board of such an entity, as was the case with AGTL when it was originally structured by Act of the provincial legislature. The entity would be responsible for broad system planning taking account of both capture volumes and injection and storage capacity. The actual regulation of tolls and tariff on the pipeline system and injection and storage facilities might be relatively light-handed, with full regulation only being imposed in the event of a complaint. The entity might have a legal or de facto franchise monopoly on the provision of CO_2 pipeline service.[91] New legislation would be required to implement this concept. The entity could be confined to the pipeline element of the chain but it might also offer an integrated injection and storage service.

One of the four projects financially supported by the province seems to draw explicitly on the AGTL model for its inspiration. This is the Alberta Carbon Trunk Line (ACTL). ACTL is a proposed 240km CO_2 pipeline between Redwater and Lacombe, including a gathering system at the north end and a distribution system at the south end. One of the proponents of ACTL is North West Upgrading, which is planning a new upgrader with CO_2 capture capability. ACTL intends to

88 The system includes in excess of 23,500km of pipe, varying in sizes and number of aboveground facilities, including 49 compressor stations and approximately 1,200 meter stations.

89 National Energy Board of Canada Act, RSC 1985, c N-7.

90 TransCanada Pipelines Ltd, Reasons for Decision (February 2009), GH-5-2008, online: NEB <http://dsp-psd.pwgsc.gc.ca/collections/collection_2010/one-neb/NE22-1-2009-1-eng.pdf> (accessed 2 October 2011).

91 So far as we are aware, AGTL never had a legal exclusive franchise on the provision of gas transmission service, but a postage stamp rate tends to enforce a de facto exclusive franchise.

transport CO_2 from the North West Upgrader, and other sources in the Heartland area to EOR operations in the Lacombe area. The Alberta Government is supporting both projects.[92] The publicly available information about the terms of support does not discuss questions of access and economic regulation.

Thus, while the ACTL project shares some elements of the AGTL concept, others appear to be lacking. Indeed the involvement of a capture entity and a transportation entity within the chain suggests a degree of vertical integration, although the storage targets in the early years of the project will likely be a variety of EOR projects owned by different and independent parties.

D. A hybrid model

We conclude by canvassing a hybrid model drawing on the complaint-based nature of both the common order model, the AGTL model, and the proposed UK regime. The model has three elements. The first element would require any party proposing to construct a CO_2 pipeline or to apply for a scheme approval for a storage project to advertise the application describing the main elements of the application and effectively conducting an open season, during which a third party might suggest changes to project configuration. If the applicant and the third party cannot reach an agreement, the proposed scheme would allow the third party to apply to the regulator (in Alberta it would be the ERCB) for a relevant order.

The second element would apply where the infrastructure is already in place and a third party seeks access. As with the common orders and the UK approach the assumption would be that the parties would endeavour to negotiate access on commercial terms. However, failing that, the third party could apply to the regulator (likely the ERCB) for an access order. The third party would also be able to apply to have the operator of the infrastructure modify or expand the facility on terms to be established by the regulator. In making a decision, the regulator would take into account the sorts of factors listed in the UK regulations. There would be no automatic or even presumptive prorating of available capacity.

The third element would draw directly from the UK proposals and require operators of CO_2 transportation and injection infrastructure to publish annual data on capacity utilization. The model would need to be implemented through new legislation, likely an amendment to the OGCA.[93]

[92] Press statement, Government of Alberta, 16 February 2011, available at <http://alberta.ca/acn/201102/299142F005558-0884-CE61-EBD4608FE8C36216.html> (accessed 2 October 2011). The agreement is not yet available to the public but is expected to be once agreements have been executed with all four project proponents.

[93] It might be possible to achieve some of these results through the terms of sequestration permits and leases. See Roggenkamp, 'Concept' (above n 43) at 295 describing this as an option in the Netherlands under the terms of the Mining Act. See also Victoria's Greenhouse Gas Geological Sequestration Act, 2008 ss 113–18, which contemplates that: (1) minister may direct the holder of an injection licence to inject and store a GHG substance for a third party; and (2) compensation to be determined by an administrative tribunal in the event that the parties cannot agree.

VI. Conclusions

This chapter has dealt with a new form of infrastructure—the carbon infrastructure, the transportation and injection and storage facilities—required to implement CCS technology in the interests of mitigating greenhouse gas emissions. The chapter addressed two questions: (1) the case for economic regulations of these two links in the CCS chain; and (2) the form that such regulation might take. The chapter concludes that there is a case for economic regulation principally in order to realize the economies of scale that might be achieved through more coordinated planning but also in order to avoid problems of market power that might emerge as the industry grows. However, the chapter also concludes that there is no case for heavy-handed prospective rate-making rules. Indeed, heavy-handed regulation would be a significant disincentive to companies considering investments in carbon infrastructure. Instead, this chapter argues in favour of requiring some form of open season requirement to facilitate the development of appropriately sized facilities plus a complaint-based jurisdiction to resolve access disputes where parties cannot agree on reasonable commercial terms. In reaching these conclusions, this chapter has drawn on the experience in both Europe (the Directive and the UK implementing legislation) and Alberta.

PART III

MARKET LIBERALIZATION AND CHALLENGES FOR NETWORK INVESTMENTS AND PLANNING

14

Transportation Regulation as an Instrument for Developing Natural Gas Networks in Brazil

Yanko Marcius de Alencar Xavier and Anderson Souza da Silva Lanzillo

I. Introduction

Over the past decade, global capitalism has promoted several changes concerning state participation in the economy. Among the set of reforms, regulation of competition in the infrastructure sector stands out.[1] Under this new concept, the gas industry was subject to several institutional transformations envisaging the progressive liberalization of the state monopoly and allowing for an increase in the number of agents in the market.

In recent decades, Brazil has made significant changes in its regulation of the natural gas industry. By approving amendments to the 1988 Brazilian Federal Constitution, the state liberalized the monopoly in the gas industry. This allowed public- and private-sector agents to act in the market and compete with Petrobras, the sole service provider until that time.[2]

[1] With regard to regulation of an infrastructure sector, one may say that a 'complete regulatory system of public infrastructure should function based on three sets of instruments that act simultaneously: (1) laws, codes, decrees, regulations, orders, authorizations, licenses, orders, etc, which constitute specific legislation for the regulated sector; (2) concession contracts to exploit a public service or equipment, or even invest in public infrastructure; (3) agencies responsible for implementing the regulation, or operational branches of the system'. (F. Froes, 'Infra-estrutura e Serviços Públicos: Princípios da Regulação Geral e Econômica' in J.E.M. Cardozo (ed), *Curso de Direito Administrativo Econômico* (São Paulo: Malheiros, 2006) 513.

[2] 'As per the wording of s 1 in article 177 of Constitutional Amendment No. 9/95, Petrobras ceased to be the sole service provider of the Federal oil and natural gas monopoly. Norms regarding the participation of other companies in such activities were subsequently regulated by Act No. 9478 of 6 August 1997. The approval process of Constitutional Amendment No. 9/95 began with a proposal to alter article 177, jointly elaborated by the Ministers of Justice, Treasury, Planning and Budget, Social Security, and Mines and Energy. Together with the proposal, the Executive Branch sent 'Justification No. 39'. It expressed Government concern about liberalizing the oil monopoly, allowing the Union to hire third parties to render related services, which had been performed exclusively by Petrobras until that time' (P. Valois, *A Evolução do Monopólio Estatal do Petróleo* (Rio de Janeiro: Lumen Juris, 2000) 118–19). Similarly: 'As can be observed, the promulgation of Constitutional Amendment No. 9, of 9 November 1995, maintained the Federal Union as owner of the monopoly in *petroleum reserves and exploration*. However, it permitted other state-owned, private, national or foreign companies to take part in exploration by means of a contract with the Union. In other words, the monopoly remains with the state and Petrobras maintains an extremely privileged position, yet the Constitutional Amendment

In this context, a need for sustainable growth in the natural gas industry in terms of reserves, supply, and trade was identified and competition was determined as a potential instrument for development. As such, in the set of regulations formulated and applicable to the natural gas industry, transportation regulation was devised in order to enhance competition and achieve the objective of sustainable growth.

Gas transportation involves two distinct contracts: one associated with the trade of gas and the other related specifically to its transportation. In the traditional model, no contractual distinction is made between these services. One measure to introduce competition is to separate transportation and trade contracts and agents (unbundling). It not only divides and distinguishes contractual bonds, but also specifies the role of the shipper handling the transportation contract, and the company that 'owns' the gas and is responsible for its purchase or sale. Unbundling prevents the shipper from participating in the gas trade other than for his own consumption. This leaves its purchase and sale to the owner, who may opt for the corporate separation of such activities.

Nevertheless, transportation regulation in order to aid development in the natural gas industry has prompted frequent debate among market agents and sector specialists. Despite acknowledgement of its importance, the mere promotion of competition began to be viewed not as a global development factor for the industry, but rather as an impediment that lacked regulation that would promote effective and sustainable growth. These discussions resulted in the Gas Act 2009 (Act 11909/2009).

This chapter aims to evaluate the transportation regulation in the Brazilian natural gas industry. It will give a brief institutional overview of the sector in Brazil, outline changes incurred in transportation regulation to the current stage of the Gas Act, and reflect on whether competition is still considered a development instrument in this market and the extent to which it is a proper instrument to plan and invest in the construction and maintenance of gas pipelines.

II. The Brazilian Gas Industry: Regulation History and Current Situation

A. The natural gas industry in the state monopoly era

Contrary to the oil industry, development in the natural gas industry is recent in Brazil. Historically, the Brazilian search for oil was concerned with ensuring self-sufficiency in supplying fuel and its derivatives to the internal market.

enables the Union to delegate *oil exploration* to private initiatives or other state-owned companies. For example, refinement of national or foreign oil may be carried out by private or state-owned companies from other countries. The Constitutional Amendment also opens the sector to foreign capital since there are no restrictions on participation by multinational companies. Nevertheless, the government prefers to refer to this measure as flexibilization of the monopoly. This is because the Federal Union maintains control in the sector despite allowing the participation of private companies' (A.L.M. Silva, *Introdução ao Direito Econômico*, Rio de Janeiro: Forense, 2002, 194–5).

The Brazilian gas industry dates back to the distribution of piped gas to large cities, namely Rio de Janeiro (1854) and São Paulo (1873). This gas, however, was manufactured from coal and its main use was in providing electricity. Primary development in the sector was aborted due to growth in the Brazilian hydroelectric power industry. At the same time, there was a movement to the state monopoly of the oil and gas industry.

The creation of Petrobras in 1953 led to a government monopoly in the Brazilian natural gas industry and has determined various elements of the sector's structure and exploration method to date. In addition to instituting the Brazilian Petroleum Corporation and defining new attributions of the National Petroleum Council, Act No. 2004/53 regulated industry-related activities in the federal monopoly. These included research, exploration, processing, and maritime transportation of oil, derivatives, and 'rare gases', among them natural gas.[3]

The 1967 Constitution elevated only the monopoly in oil mining and production to constitutional level,[4] while Federal Union monopoly in the gas industry remained governed by statute. This remained unchanged in Constitutional Amendment No. 01/1969.[5] As to such gas, Petrobras also held a monopoly in the exploration, production, import, export, and transportation of gas. It also distributed to all Brazilian states, except Rio de Janeiro, supplied by CEG, and São Paulo, which received supplies from COMGÁS.[6]

Exploration of natural gas in Brazil began in the 1960s. It grew stronger during the 1970s in the north-eastern states of Sergipe, Alagoas, and Bahia, as well as in the 1980s with the discovery of reserves in Rio Grande do Norte. The north-east was the first centre of natural gas production in Brazil, led by Bahia at Pólo de Camaçari, although its participation in the national energy matrix remained insubstantial.

From the 1980s onwards, with the discovery of Bacia de Campos in Rio de Janeiro, another industrial centre was created for natural gas production and exploration. Its proximity to large city centres made industrial usage feasible and it became Brazil's main gas production facility. Nevertheless, projection of gas in the Brazilian energy matrix remained low.[7]

Natural gas can be found in associated and non-associated forms. Associated gas is typically found in oil wells and can either be dissolved in the oil or take the form of a gas cap above it. Non-associated gas is the natural gas found in reservoirs, containing no oil or water.[8]

[3] P. Valois, *A Evolução do Monopólio Estatal do Petróleo* (Rio de Janeiro: Lumen Juris, 2000) 70–1.

[4] Art 162. In accordance with the law, the study and mining of oil in national territory is a monopoly of the Union.

[5] Art 169. According to law, the study and mining of oil in national territory are a monopoly of the Union.

[6] A. Wald, 'A Evolução do Setor de Gás Natural no Brasil e sua Regulação (do Monopólio à Abertura)' (2005), 50 *Cadernos de Direito Constitucional e Ciência Política* 13.

[7] J.C. Cechi (ed), *Indústria Brasileira de Gás Natural:Regulação Atual e Desafios Futuros* (Rio de Janeiro: ANP, 2001) 60–3.

[8] E.M. Santos, *Gas Natural—Estratégias para uma Energia Nova no Brasil* (São Paulo: Annablume, 2002) 73.

Most natural gas reserves found in Brazil are of the associated type.[9] Given that gas is often associated with oil and oil companies have long regarded natural gas as a sub-product, it was almost completely exhausted.

Since gas was mainly associated with oil, which was under state monopoly, the natural gas industry in Brazil was propelled by the oil industry and Petrobras investments.

During the 1980s, within the monopoly period, the Brazilian natural gas industry was supported by an initial Union policy to increase its participation rate in the Brazilian energy matrix by 10 per cent. Lack of better conditions made this a feasible goal at that time.[10] The role of gas accelerated when liberalization of the monopoly in the natural gas industry took place.

B Liberalization of the state monopoly in the national gas industry

Under 1988 Federal Constitution, the Brazilian natural gas industry remained a state monopoly. It was a dual monopoly because the chains of gas exploration, production, transportation, importation, and exportation are a government monopoly, while the distribution of natural gas to end-users is a monopoly of the member states. Private agents were not able to participate in these activities, not even as contracted service providers.[11]

Redefinition of the state in the economic domain and its administrative configuration throughout the 1990s promoted initiatives to liberalize the state monopoly over the natural gas chain. The term 'liberalization' is applied, since ownership was not transferred to the private sector. The sector was merely opened, allowing private industry to be hired by the Brazilian state to perform activities related to the gas industry. Ownership of the monopoly remained with government agencies.[12]

As such, Constitutional Amendment No. 05 of 1995 made it possible for private industry to exploit local piped gas services through means of concession contracts, previously a monopoly of member states. The amendment allowed member states to privatize state-owned gas exploration companies and draft a regulatory regime for the distribution of natural gas so as to retain ownership of the public service.[13]

[9] J.C. Cechi (ed), *Indústria Brasileira de Gás Natural: Regulação Atual e Desafios Futuros* (Rio de Janeiro: ANP, 2001) 40–5.

[10] H.Q. Pinto Jr, *Elementos para a Formação de uma Política de Preço para o Gás Natural no Brasil* (Rio de Janeiro: COPPE, 1987) 102–3; 109. This was established by the Ministry of Mines and Energy (MME) through Order No. 1061/1986.

[11] P. Valois, *A Evolução do Monopólio Estatal do Petróleo* (Rio de Janeiro: Lumen Juris, 2000) 113. Art 177, s 1 in its original wording: 'The monopoly defined in this article includes all risks and results derived from the activities mentioned herein, with the Union forbidden to award or allow any form of participation, in kind or value, in the exploration of oilfields or natural gas beds, except for the dispositions of art 20, s 1.'

[12] P. Valois, *A Evolução do Monopólio Estatal do Petróleo* (Rio de Janeiro: Lumen Juris, 2000) 118–21.

[13] C.A. Sundfeld, 'Regime Jurídico do Setor Petrolífero' in C. A. Sundfeld (ed), *Direito Administrativo Econômico* (São Paulo: Malheiros, 2002) 391–4.

Constitutional Amendment No. 09 of 1995 opened up the Union's monopoly. As with the Brazilian member states, the Union maintained monopoly ownership, allowing public and private companies to carry out activities related to the monopoly in the natural gas industry. This did not terminate the monopoly, since ownership was not transferred to the private initiative. It merely introduced the possibility of hiring their services, with no obligation to do so.[14]

To this end, Constitution authorized laws to define industry hiring terms and attribution conditions for the state regulatory agency.[15] This left implementation of the natural gas policy to the Ministry of Mines and Energy (MME).

During liberalization, the natural gas industry experienced significant market penetration. Validating the goal of raising gas participation in the Brazilian energy matrix to 10 per cent,[16] the industrial demand for gas grew significantly, primarily due to incentives and competitive gas prices. Another factor favouring development was the construction of the Brazil–Bolivia gas pipeline.[17] This increased gas availability, especially for the south-southeastern areas, in a scenario where the existing Brazilian supply could not promote sustainable growth for the sector.[18]

Needs in the electric sector also impelled the government's policy. Gas was made available as additional fuel for electric power generation, in light of the limits reached by hydroelectric generation in recent years and problems caused by energy crises, such as the blackout in 2001.[19]

In order to complete this institutional overview of the Brazilian natural gas industry, it is important to consider figures related to this sector.

[14] Art 177, s 1. The Union may commission private or state-owned companies to carry out activities foreseen in items I and IV of this article, respecting the conditions established by law.

[15] Art 177, s 2. The law to which s 1 refers concerns: I—guaranteeing oil derivatives supply throughout the national territory; II—contract terms; III—the structure and attributions of the Union's monopoly regulatory agency.

[16] J.C. Cechi (ed), *Indústria Brasileira de Gás Natural:Regulação Atual e Desafios Futuros* (Rio de Janeiro: ANP, 2001) 60.

[17] In recent years, the Brazil–Bolivia gas pipeline (GASBOL) has demonstrated the paradox of growth in the Brazilian gas industry. The massive gas supply promoted in the Brazilian market in 2005, mainly in the Southeast, indicated its fragility with the Bolivian nationalization and gas crisis, signaling the need to break dependency on a single source.

[18] J.C. Cechi (ed), *Indústria Brasileira de Gás Natural:Regulação Atual e Desafios Futuros* (Rio de Janeiro: ANP, 2001) 122.

[19] '[...] the government launched the Thermoelectric Priority Program (TPP), coordinated by the Ministry of Mines and Energy under Decree 3371 of February 24, 2000 envisaging increased participation of thermal power' in Santos, *Gas Natural—Estratégias para uma Energia Nova no Brasil* (São Paulo: Annablume, 2002) 276–301.

Energy Research Company, *Balanço Energético Nacional* (2010), available at <https://ben.epe.gov.br/downloads/Relatorio_Final_BEN_2010.pdf> (accessed 2 October 2011).

Oil, Natural Gas and Biofuels Agency, *Boletim Mensal do Gás Natural* (2010), available at <http://www.anp.gov.br> (accessed 2 October 2011).

M.C.P.P. Mathias, *El Sector del Gas Natural en Brasil y la Necesidad de un Nuevo Marco Regulatorio* (Rio de Janeiro: ANP, 2005), available at <http://www.anp.gov.br> (accessed 2 October 2011). A. Wald, 'A Evolução do Setor de Gás Natural no Brasil e sua Regulação (do Monopólio à Abertura)' (2005), 50 *Cadernos de Direito Constitucional e Ciência Política* 26. Edmilson Moutinho severely criticized the Brazilian thermoelectric programme based on natural gas due to its competitiveness and maturity in regard to hydroelectric power generation in Brazil. (E.M. Santos, *Gas Natural—Estratégias para uma Energia Nova no Brasil*. São Paulo: Annablume, 2002, 276–301).

The major final users of natural gas in Brazil are: (a) residential (261.10^6 m^3); (b) transportation (2106.10^6 m^3); (c) commercial/public (204.10^6 m^3); (d) highways (2106.10^6 m^3); (e) industrial (8137.10^6 m^3); (f) cement (17.10^6 m^3); (g) pig-iron and steel (985.10^6 m^3); (h) iron alloys (2.10^6 m^3); (i) mining/pelletization (272.10^6 m^3); (j) non-ferrous/other metallurgical (748.10^6 m^3); (k) chemical (2002.10^6 m^3); (l) foods and beverages (635.10^6 m^3); (m) textiles (327.10^6 m^3); (n) paper and pulp (692.10^6 m^3); and (o) ceramics (1137.10^6 m^3).[20]

Brazil has proven gas reserves of 358.1 billion m^3, of which 18.3 per cent are onshore and 81.7 per cent offshore. Brazil's net domestic gas production in 2010 was 33,975 thousand cubic metres per day (m^3/d), while imports (through pipelines and regasification) reached 34,682 thousand m^3/d, resulting in a total gas supply of 68,657 thousand m^3/d.

The current transportation structure in the country extends 9,131.3km, with a capacity of 218.9 million m3/d. Petrobras/Transpetro is responsible for the operation of 68.2 per cent of existing gas transportation pipelines. The largest transportation network is the Bolivia–Brazil pipeline with a length of 3,150km, of which Gaspetro S.A., a subsidiary of Petrobras, is the main shareholder.[21] Brazilian gas pipelines are mainly located along the coastline, following the development pattern of its large cities.[22]

Contrary to other countries such as the United States and the European Union, Brazil has no need for natural gas that demands the intensive use of this energy source (eg heating and electricity generation by combined cycle gas turbines). Moreover, natural gas in Brazil faces intense competition with fuel oil and liquefied petroleum gas (LPG) in the industrial sector; with hydroelectricity in the electric power sector; and with diesel and alcohol in the road transportation sector.

In addition to this competition among energy resources, there is a lack of gas culture which could lead to a more efficient end-use of gas as opposed to other energy sources. Natural gas is often thought of as a 'safety net' in energy crises. Thermoelectric plants were designed to avoid potential blackouts and ration electric power, disregarding issues such as energetic efficiency and the penetration of hydroelectricity in the Brazilian energy matrix.[23]

These data shed light on the relationship between competition and development in gas transportation regulation and in understanding the direction of regulatory system development in relation to the national needs.

[20] Energy Research Company, *Balanço Energético Nacional* (2010), available at <https://ben.epe.gov.br/downloads/Relatorio_Final_BEN_2010.pdf> (accessed 2 October 2011).

[21] Oil, Natural Gas, and Biofuels Agency, *Boletim Mensal do Gás Natural* (2010), available at <http://www.anp.gov.br> (accessed 2 October 2011).

[22] M.C.P.P. Mathias, *El Sector del Gas Natural en Brasil y la Necesidad de un Nuevo Marco Regulatorio* (Rio de Janeiro: ANP, 2005), available at <http://www.anp.gov.br> (accessed 2 October 2011).

[23] F. Anuatti et al, *Desafios da regulação do mercado de gás natural no estado de São Paulo* (Rio de Janeiro: Rio Oil and Gas, 2006) 03.

III. Regulation of Transportation Activity in Law 9,478/97 (the Oil Law) and by the ANP

In order to understand the current regulation of natural gas, one must first understand the initial regulatory view of the problem and adjustments made until the Gas Act came into force in 2009. This initial view is set forth in the Oil Law of 1997 (or 'Oil Act') and in the regulations. This regulatory framework regulated both the oil and gas industry, primarily according to oil industry needs.

Thus, the first major issue of transportation at that moment was the clear absence of difference between oil and gas industry needs. That was the primary problem in the planning and developing of new gas pipelines.

A. Transportation activity ruling

Transportation regulation was established in article 56 of the Oil Law. The article refers to the type of authorization given to companies or consortiums organized under Brazilian laws. The conditions to be attached to authorizations are granted through Order No. 170/98 of the Brazilian Oil, Natural Gas, and Biofuels Agency (Agência Nacional do Petróleo, Gás Natural e Biocombustíveis, or ANP). This Order governs the authorization for building, expanding, and operating transportation and transfer facilities such as natural gas pipelines.

In Brazilian administrative law, an authorization is considered a unilateral and discretionary administrative act, addressing an activity reserved for public use or generally prohibited. It is granted provisionally[24] and is an expression of state policing power. ANP does not have the power to revoke authorization for any reason other than those specified in regulation.

With regard to the transportation activity ruling, it is important to note that at the time of the Oil Law, the transportation company is not required to own transportation assets; it may be only an individual renting the premises. In the gas industry, this means the company merely conducts activities. Rights are defined by a lease agreement, without transferring ownership. Moreover, according to the Oil Law, a transportation company can transport the natural gas only, or can also exploit the networks.

The most conclusive point about the transportation ruling is that the transportation company is almost exclusively responsible for the task of involving the building, expansion, and operation of gas pipelines. This was under the liberal premise of free initiative or reliance on free market forces.

[24] 'Thus, administrative authorization can be defined, in a broad sense, as the unilateral, discretionary and provisional act by which the Administration gives the private entity the choice of using a public property, performing a material activity or carrying out a practice that, without such consent, would be prohibited by law' (M.S.Z. Pietro, *Direito Administrativo* (São Paulo: Atlas, 2002) 218).

Then, in this system, free market forces determine the expansion of natural gas networks instead of the state. The state has no regulatory power to direct the building and the expansion of gas networks.

B. Free access characteristics in the transportation activity

Originally, article 58 of the Oil Law defined the rules adopted in Brazil on free access to the transportation networks. It also set out the limits for ANP actions regulating transportation. According to the article, free access is an option given to any party interested in using either existing transportation facilities or those to be built, upon adequate compensation to the owner.

In order to evaluate the free access model adopted, two parameters are considered: (a) form of access; and (b) separation of the transportation activity from that of commercialization of gas.

Access established through this law was negotiated. Allocation of the gas capacity is a process of direct negotiation between the owner of the transportation networks and the supplier. This defines whether there is capacity to allocate terms of the transportation agreement to the future carrier (daily volume dispatched by the company, point of receipt, and point of delivery, among others), as well as the duration of the agreement. It differs from regulated free access, which, in summary, is a public procedure of capacity allocation through bidding.[25]

The Oil Law did not determine standards to be complied with in contractual negotiation of access to the transportation infrastructure. This is pertinent given that the characteristics of the natural monopoly in operating gas pipelines grant the owner of the networks a dominant position which could be abused. One form of abuse is the unjustified denial of access, which could also be considered illegal under Brazilian competition laws (Law No. 8,884/94).

Although the provisions of Oil Law concerning free access were of the negotiated type, the regulation enforced by ANP was driven in a contrary direction. ANP Resolution No. 27/2005's approach is contrary to that of the Oil Law regarding how access will occur. The Oil Law refers to a negotiated access model, while the Resolution establishes a model of regulated access, since it will take place through Open Competition for Capacity Allocation (CPAC). CPAC works as a form of bidding where the transporter advises the market that it has some transmission capacity available, as well as its allocation terms.

With regard to ANP powers, the first paragraph of article 58 of the Oil Law states that the ANP may intervene in contractual relationships as an arbiter in order to establish an amount when there is no agreement between the parties or when it determines abuse by charging amounts incompatible with the market. ANP Order No. 254/2001 regulated the resolution of conflict regarding compensation

25 At the time of Oil Law, the European Union already had regulated access. Regulated access is defined as the process in which fees and other conditions of network use are public and established by government or government agency (Commission Européenne, *ouvrir la voie au Choix—lancement du marché unique du gaz Européene*, Luxembourg: Office des publications officielles des Communautés europénnes, 2000).

of the transportation company set forth in article 58 of the Oil Law. The procedure is similar to a legal proceeding, ensuring the adversary full defense rights. Subordinated appeal to the Ministry of Mines and Energy is not established, only appeal to the ANP Board.

The ANP was not given any other general jurisdiction to regulate conditions of access, although ANP Resolution No. 27/2005 sets out definitions on contractual relationships and establishes the possibilities of using free access. Thus, the ANP would not be responsible for regulating matters of competition related to free access as a form of activity structure or for repressing conduct. This is the responsibility of the Administrative Council for Economic Defense (CADE).[26]

The second point in the regulation of free access is the separation between transportation and commercialization. It is important to note that the Oil Law does not treat commercialization as a specific authorization title.[27] This separation was not because of the separate establishment of commercialization of gas, but rather due to the use of different titles for each stage of the chain. Since transportation activity exists and, given the classification of pipelines supporting it, it is of 'general interest', the separation model (unbundling) would not truly conceive commercialization activity. Instead, it defines the contractual position of the transportation company as a facilitating intermediate of the gas purchase agreements entered into by the carriers. These would be participants in the natural gas chain defined by law (eg gas producer and distributor, producer, and exporter).[28]

Article 58, paragraph 2 of the Oil Law discusses regulation by the ANP of the preference for handling the transportation company's own products in order to optimize transportation capacity. It does not distinguish oil from natural gas and by-products. By referring to the transportation company as handling its own products, one admits it owns the natural gas.

C. Tariff regulation in natural gas transportation

The third point of transportation regulation is tariff regulation. Use of transportation services cannot be free of charge, since this would disregard corporate free initiative and ownership rights. For this reason, article 58 of the Oil Law and its items originally ensured carriers paid compensation to the owner of the premises.

Inconsistencies exist in fee regulation with regard to powers of the agency. Article 58, item VI, provides jurisdiction for the ANP to establish criteria for the calculating of pipeline transportation fees and to determine amounts in cases under the Oil Law. On the other hand, article 58, paragraph 1, only establishes

[26] Ana Maria de Oliveira Nusdeo ('Agências Reguladoras e Concorrência' in C.A. Sundfeld (ed), *Direito Administrativo Econômico* (São Paulo: Malheiros, 2002) 183) highlights that the provisions of article 10 of the Oil Law only requires the ANP to inform the Administrative Council for Economic Defense—CADE—on competition violations, with no derogation of the Antitrust Law (Law No. 8,884/94). Thus the agency would have minimal power in the antitrust area.

[27] Although ANP Order No. 170/98 and ANP Resolution No. 27/2005 enforced a regulation in a contrary direction determining the separation of transportation and commercialization.

[28] M.A.C. Menezello, *Comentários à lei do petróleo* (São Paulo: Atlas, 2000) 152.

ANP normative powers when there is no agreement or to determine whether the amount agreed is compatible with the market.

Then, in the Oil Act, tariff regulation is delegated to the private agents, the transporter, and the carrier. The financial aspects of natural gas networks (building, expansion, and operation) are determined by the forces of the free market.

However, this can give a fake impression that private agents were responsible for the expansion of natural gas networks in Brazil in recent years. That is not true. In fact, it must be remembered that Petrobras is the major agent in the Brazilian natural gas industry, as pointed out before in this study. Since Petrobras is a company that mixes government and private capital, in practice the policy and tariff regulation was determined by a semi-public regulation and not a private regulation, as the neoliberal tone of the Oil Law states. Thus, both in the monopoly era and after the monopoly era under Oil Law, government was, and still is, the major investor in natural gas network building and expansion.

IV. Transportation Regulation in the Gas Law (Federal Law 11.909/2009)

A. Origins of the Gas Law

As noted earlier, the Oil Law and ANP normative acts contain legal grounds that do not allow more appropriate regulatory status for the natural gas industry. This is because regulation of the gas industry under the Oil Law is flawed due to its scope, covering both oil and gas, treating gas as any energy product together with oil. The natural gas regulation under the Oil Law was not adapted to natural gas industry needs.

As a result, calls for a Natural Gas Act to correct the regulatory distortions existing in the Brazilian scenario began to emerge in the industry. Consequently, Law 11,909 of 4 March 2009 (the Gas Law) was enacted, significantly changing the regulation of natural gas industry. The Gas Law brought a series of changes to the Brazilian legal and regulatory environment.

Legislation in the Oil Law provided a less interventionist profile for state participation in gas industry regulation. As such, investments in industry and prices of services depended on transportation companies in the sector. With the Gas Law, however, mechanisms were created to increase planning in the national supply of natural gas, competitiveness, and a guarantee of return to investors, although considerably reducing the freedom existing in the previous system. Moreover, new regulation deals exhaustively with links in the natural gas industry chain, although exploration and production of natural gas is still regulated by Law 9478/97.

This section therefore examines the major differences brought by the new law on transport regulation and access to gas pipelines. We do not include an extensive review of the Gas Law, merely of those sections that relate to development of natural gas networks.

B. Transportation activity ruling[29]

Among changes promoted by the current regulatory framework, arrangements for transportation of natural gas received more attention, occupying most of the new law.

Substantial modification appears in legal provision for the activity: initially, natural gas transportation was exercised upon authorization; nowadays this activity is conducted primarily through concession.[30] In Brazil, concession is generally understood as a kind of administrative contract between a private or public agent and a government or government agency to explore a public utility,[31] but in Gas Law it is applied to explore an economic activity (the Gas Law does not consider gas transportation as public utility). The concession system applies to gas pipelines defined as of public interest. Duration of the concession contract will be 30 years and may be extended for an equal period.

Completion of the concession contract will be preceded by two instruments: a call for public bidding, and contracting capacity. The public call for contraction is a tool that aims to identify potential carriers and actual gas demand. It is an instrument of activity planning and of promoting development of the general natural gas sector, since the carrier is determined based on interest expressed, not according to the will of the carrier, as in the previous system.

Bidding is a instrument to be applied prior to the concession contract. The demand for bidding is a basic requirement imposed by the Brazilian Constitution and laws in general to carry out an administrative contract. This same is true for the Gas Law when applying this instrument. Bidding occurs by decree, where the lowest annual revenue is used as a criterion to select the winning bid.

The decree must contain a series of standard clauses (as a draft contract, but addressed to a general public) that will be present in the final form of the concession contract. The Gas Law stipulates a series of mandatory clauses. These must be included in the bidding notice and final contract to be signed by the carrier. The possibility of including further clauses is not excluded should these be consistent with the law and the principles covered by the contract (principles of legality). Mandatory clauses are:

(i) route of the transportation pipeline under lease, delivery and receiving points, as well as projected transport capacity and criteria used to calculate it;

(ii) maximum annual transportation income and criteria employed for its calculation;

(iii) competitor requirements and criteria for pre-qualification, when this procedure is adopted;

[29] Arts 3–31.

[30] The Gas Law maintained its forecast of applying for authorization. In this case, residual application is limited to pipelines involving international agreements and transportation on existing transport pipelines. Authorization will last 30 years and be extended for an equal period. Note that the stated period of authorization applies only if they are conceived by the new law.

[31] M.S.Z. Pietro, *Direito Administrativo* (São Paulo: Atlas, 2001) 273.

(iv) a list of necessary documents and norms to be followed for the evaluation of technical ability and financial suitability, as well as the legal and tax status of stakeholders for technical and economic-financial assessment of the proposal;

(v) explicit indication that the concession holder will be responsible for compensation payment due to expropriation or services necessary for fulfillment of the contract, in addition to obtaining licences through the appropriate organizations, including those of an environmental nature;

(vi) deadline, time, and place for information provided to interested parties on data, studies, and other elements required to elaborate the proposal, as well as the cost of its acquisition;

(vii) period of exclusivity for initial carriers to exploit contracted capacity for new gas pipelines; and

(viii) duration of concession and possibility of extension, if applicable.

Similar to the Oil Law, the Gas Law allows transport activity to be exploited in partnership with the following requirements:

(i) evidence of commitment to the partnership, either public or private, signed by partners;

(ii) indication of the leading company in the partnership, which is responsible for operations without affecting the joint liability of other association members;

(iii) presentation by each partner enterprise of documents required to evaluate technical and economic-financial qualifications in the partnership;

(iv) a ban on participation by the same company in another association, or sole bidding for transportation on the same pipeline;

(v) a conditional grant for the successful association, based on registration of component instrument of the association, as per the Sole Paragraph of article 279 for Law 6,404 of 15 December 1976. Allowances are also made in the Gas Law for foreign company participation in bidding which, if successful, should be in compliance with Brazilian laws, requiring headquarters and administration in Brazil.

On completion of bidding, a winner is declared and the concession contract is drafted. It should reflect the tender document and contain several clauses, namely:

(i) description of the gas pipeline;

(ii) the list of necessary goods and facilities to execute transportation, as well as rules specifying the withdrawal and return of areas and removal of equipment, in addition to detailing situations where these will be ceded to the federal government, in case of grant extinction;

(iii) duration of concession and, when applicable, conditions for its extension;

(iv) implementation schedule, minimum investment provided, and pipeline expansion possibilities;

(v) annual income and criteria for readjustment;
(vi) warranties provided by the concessionaire, including implementation of proposed investment;
(vii) specification of rules on eviction and the return of areas, including removal of equipment and incorporation of goods to the Union;
(viii) procedures for monitoring and supervising concession activities and contract inspection;
(ix) concessionaire obligation with regard to ANP reports, data, and information relating to developed activities;
(x) rules for access by any interested party shipper to the pipeline under concession, as per the present Act;
(xi) rules on dispute settlement related to the contract and its implementation, including conciliation and judgment;
(xii) of cancellation and dissolution of contracts;
(xiii) penalties applicable in case of concessionaire default in contractual obligations;
(xiv) period of exclusivity for initial shippers to exploit contracted capacity of new pipelines.

Furthermore, a series of contractual obligations for the concessionaire are defined, relating to incidents that may occur in operation, civil liability, adoption of the best Natural Gas Industry practices, providing information on facilities, rates, available capacity, and contracts.

Reflecting a still greater government role in regulation of transport activity, the Gas Law also requires reports of technical, accounting, operational, and economic-financial information, maintaining separate accounting for transport and storage of natural gas and submission for prior ANP approval of the draft of the standard contract to be signed with carriers.

The Gas Law also addresses the system of existing pipelines at the time of enactment, consisting of gas pipelines under current authorization provided by the Oil Law. The Act ratified existing rights granted in permits, while giving permit holders specific obligations and duties, as discussed above.

From a regulatory competence standpoint, the Gas Law has also innovated by providing the MME with wider and more general powers to set policy for transport activity. The Gas Law also delegated to the ANP the role of executor of this policy. The MME responsibilities are to: (a) propose transportation gas pipelines for construction or expansion; (b) establish guidelines for hiring of transport capacity; (c) determine concessions or authorization; (d) define exclusivity period of initial shippers. The ANP must: (a) promote the process of public tenders for capacity contraction; (b) set the maximum tax to be applied to shippers; (c) grant permission to be a shipper; (d) prepare bidding documents, promote bidding process, and conclude the concession contract (in this case, upon delegation of the Ministry of Mines and Energy).

With these new rules, natural gas transportation is no longer considered within the patterns initially set by the Oil Law. The development of transport activity is no longer based on free enterprise and competition (market forces). Planning now rests in the hands of the state in the new model, which has the power even to propose the construction of pipeline sections. Consequently, natural gas industry planning became a state task, no longer an action proposed by transporters.

C. Free access characteristics in the transportation activity[32]

Following the same logic as in regulation of transport activity, regulation of free access to the pipeline network by third parties has also been remodeled. In Oil Law both for oil and gas pipelines, the model of freely negotiated access was established, where parties reach an agreement on the use of pipelines. Failing agreement, the ANP was responsible for defining the situation. In the Gas Law, the regime of free access is regulated, where, by law, the regulator and its regulatory acts will define the entire process.

Moreover, unlike the Oil Law, the Gas Law expressly stipulates trading activity for natural gas, which should be formalized by contract and registered with the ANP.[33]

The first step for a third party seeking access to pipeline network is hiring a transport service. In this sense the prior tendency to separate transport, purchase, and sale remained in the new Act, the only requirement compatible with free access.

Furthermore, the Gas Law established a new classification for gas pipelines:[34]

(a) Transfer: relative to a pipeline for specific and exclusive interest of its owner;
(b) Transport: carries natural gas between other gas pipelines, processing facilities, storage and delivery of gas to gas distribution;
(c) Outflow: pipeline destined to move gas from producing fields to treatment or liquefaction facilities.

Definitions presented are far more enlightening than those contained in the Oil Law. The Gas Law clarifies that there is no application of free access to cases of outflow and transfer pipelines.

The transport service is presented in three forms: (a) constant, consisting of a transport service where the carrier commits to schedule and deliver the daily volume of natural gas required by the shipper until transporting an established capacity under contract with the shipper; (b) interruptible, which consists of a transport service that may be interrupted by the carrier due to priority of the constant transport service; (c) extraordinary, which consists of a modality of hiring capacity available, whose transportation service can be interrupted at any time.

Third party access by hiring transport services should first occur according to available capacity. When hiring a constant transport service, a public tender promoted by the ANP under MME direction is demanded. In case of interruptible

[32] Arts 32 to 35. [33] Art 47. [34] Art 2, XVII, XVIII, and XIX.

and extraordinary transportation, access will occur as regulated by law. Moreover, assigning capacity for pipeline transport is guaranteed, and operations will be regulated by ANP.

The most significant change presented by the Gas Law is the establishment of a period of exclusivity. This period modifies the initial idea of competition promoted by free access as a precept of the gas industry. The Brazilian case requires an exclusive period for anyone hiring the service with the first carrier (referred to in Brazilian law as 'initial carrier'). As such, the Gas Law establishes[35] an exclusivity period of 10 years (from the onset of commercial operation on a pipeline) for existing pipelines, that is, those already granted operational authorization, as well as for new gas pipelines under bidding. In the form of a concession contract, the law stipulates that the Ministry of Mines and Energy, in conjunction with the ANP, will determine the initial period of exclusivity.

D. Tariff regulation in natural gas transportation[36]

Gas Law has changed substantially the regulation system. Now the state has more powers to apply standards to the fees due in reason of the use of the services provided by the transporters to gas carriers.

According to the Gas Law, in the process of public bidding for contracting capacity, ANP must apply the maximum rates for the services to be offered by the transporters to the gas carriers. The maximum rates are applied by ANP in combination with two criteria: (1) maximum annual income, defined in bidding decree; and (2) annual income, defined in bidding process. Moreover, a concession contract must have provisions about the annual income and criteria for readjustment, as well as the obligation of publishing the existent tariffs for the gas market.

The analysis of these provisions brings the conclusions that Gas Law has adopted price-cap regulation. In this regulation the tariff is not focused on the costs or the profits, as it can be found in rate of return or cost of the service methodologies, but in the prices applied in a market. The justification is that price-cap regulation encourages cost efficiency and avoids regulatory capture.[37] Besides the gas sector, the telecommunications and electricity sectors in Brazil use price-cap regulation, but this can be an obstacle in the natural gas industry, since 'market' prices are defined by Petrobras today.

V. Prospects and Challenges for Transportation Regulation in Brazil in the Development of Natural Gas Networks

The analysis of transportation regulation in the Brazilian gas industry and the changes from the Oil Law to the current Gas Law in this study indicate that the

[35] Art 32; art 30, s 3; art 3, s 2. [36] Art 5; 13–14; 21–2.

[37] S. Cowan, 'Price-cap Regulation' (2002), 9 *Swedish Economic Policy Review* 171, available at <http://www.regeringen.se/content/1/c6/09/52/69/bc4473b0.pdf> (accessed 2 October 2011).

Brazilian system has modified the understanding about the role of competition and its relationship with pipeline network development, although the regulation had still not resolved problems with the real monopoly power of Petrobras.

In fact, the major complaint by agents (read 'Petrobras') of the different sectors is that the transportation regulation through authorization and the unrestricted implementation of free access would pose a risk to the expansion of gas pipelines. This is because a possible carrier may benefit from the rules of free access, which would put investment at risk, especially in the absence of a guarantee of constant revenue.

First, it is important to remember that the natural gas industry is organized as a network-bound industry. With natural gas in its natural state, energy flow must be continuous. Should the consumer require gas for end use, the producer must have gas immediately available, since the storage options are few and costly in Brazil.[38] This creates a need for clear scaling of supply and demand for available gas, in addition to synergy along the supply chain to maintain continuous gas flow and avoid the risk of supply disruption. In addition, instability exists in the projection of future demand when there are no captive uses for natural gas, with strong competition from other energy sources.[39]

Second, the natural gas chain consists of tangible and intangible assets specific to the natural gas industry which cannot be reused for application in other economic sectors. This leads to a situation of very specific assets, significantly increasing dependence among industry agents and, simultaneously, raising the information imbalance between agents. Specificity enables an asset-holder to enjoy considerable market power.

Third, the scale of demand and cost of bringing gas to end-consumers makes transportation, in its broadest sense, not very susceptible to competition, increasing information asymmetry and the power of transportation asset owners.

According to the technical and economic characteristics listed, there is a tendency towards natural gas activity in the form of a natural monopoly. This is due either to efficiency problems in exploiting the structure (structure duplication through exploitation of the same activity by several agents would increase irrecoverable costs for individual firms) or to the specificity and inflexibility of assets involved (its network character creates a need for coordination between the various links in chains). Thus, transaction costs in the natural gas industry are traditionally very high. In this context, the main strategy adopted by companies in the sector is the formation of a governing structure for all contracts in the chain in order to achieve optimum efficiency throughout the chain. For this purpose they coordinate and direct each link, or apply the vertical integration of assets under the control of a single company.

Thus, direction of the chain does not depend on market structures (supply, demand, price), but on the contracts or internal structure of the company. This

[38] Gas can be stored in the form of underground salt and coal storage, as well as in compressed metallic structures. The Gas Law regulates gas storage (arts 37–42).

[39] As before mentioned in this study, natural gas competes with gasoline, alcohol, and diesel in the transport branch, fuel oil in the industrial branch, and hydroelectricity in the electric branch.

promotes company loyalty throughout the supply chain. Alternatively, a single company may operate in all links, preventing sudden disruptions in gas supply that unbalance the chain as a whole.[40]

Given this situation, contractual mechanisms linking parties paying for gas in order to cope with the uncertainties of exploration and gas production, as well as with investments in the expansion of transmission and distribution[41] networks, are prevalent in this industry. Thus, traditional features of contracts in the industry follow these necessities: (a) long-term contracts for investment amortization in transport and distribution; (b) payment by an independently contracted volume of gas actually consumed during a period to ensure development and return on investment (take-or-pay clauses);[42] and (c) price-indexing clauses and contract reviews to allow, even paradoxically, rigidity against uncertainties and adaptation to new gas market circumstances.[43]

Considering these characteristics and complaints from industry agents, the Brazilian state modified the regulation to recognize the predominance of traditional structures in the natural gas industry, understanding that a period of exclusivity should be guaranteed. These changes were incorporated into the Gas Law, which ultimately incorporated several provisions of previous ANP resolutions on the subject.

On the other hand, in the current regulation the state has more powers to define the tariffs to be applied to the services to be offered by transporters to gas carriers. However, in fact there is no true regulation about this matter yet, because only in December 2010 was Ordinance No. 7382/2010 published. This ordinance was much awaited in the Brazilian gas market because the tariff system would be applied only with this legal instrument. Then, there is no regulatory practice by the government on tariff issues, only several private contracts that apply several criteria without uniform standards.[44] Thus, the absence of transparency and the lack of a regulated tariff system in transport regulation can be added as factors that contribute to an underdevelopment of natural gas networks.

As a summary of the changes, competition in transport regulation is not viewed as a major factor in the development of natural gas networks. Nevertheless, this understanding does not mean that the importance of competitive market is not recognized, rather that its initial formulation did not contribute to the development of the gas pipeline networks.

[40] F.H.G.C. Laureano, *A Indústria do Gás Natural e as Relações Contratuais—uma Análise do Caso Brasileiro* (Rio de Janeiro: COPPE, 2005) 01.

[41] F.H.G.C. Laureano, *A Indústria do Gás Natural e as Relações Contratuais—uma Análise do Caso Brasileiro* (Rio de Janeiro: COPPE, 2005) 02.

[42] Take-or-pay clauses offer important protection to the gas seller, primarily in long-term contracts. Clauses ensure sellers will receive a minimum income during the contract term (or at least during investment consignation), even if the buyer does not use the minimum amount of delivered gas, as established by the contract.

[43] J.M. Martin, *Économie et Politique d'Énergie.* (Paris: Amand Colin, 1992) 87–8.

[44] The tariffs on transportation services are now published by ANP, but they are related to one established by the transporters in Oil Law period. Available at < http://www.anp.gov.br/?pg=44589&m=&t1=&t2=&t3=&t4=&ar=&ps=&cachebust=1308609854023> (accessed 2 October 2011).

This change in understanding demonstrates how the scope of specific goals such as development can be surrounded by several suggestions and questions. Introduction of a regulatory structure often leads to inconsistent application of principles, since two states of affairs generally cannot exist simultaneously without leading to the non-performance of one or the other.

Indeed, the central theme for reform in previously highly regulated infrastructure sectors is free competition. Competition is a means and instrument to encourage development. In support of this idea, institutionalization occurs along two branches: (1) removal of institutional barriers, clearing the way for a large-scale presence of several economic agents (operation of free and natural market forces); and (2) creating institutional mechanisms that promote competition. Through regulation, means are established for the functioning of market forms or redirecting market power for balanced coordination with all of the natural gas chain (market value as established by the regulatory framework, market as an institution).[45] In transportation regulation, free access to gas pipelines represents the fulfillment of this second type of policy.

Contrary to the more limited neoliberal belief, the regulatory practice in the Brazilian gas industry indicates that it is not enough to institutionally ensure competition in order to develop natural gas networks. Despite debate as to the weaknesses of the regulatory system, this chapter also shows that there are other institutional and technological constraints that have an effect on an efficient way of attracting investments to natural gas network through transportation regulation.

It is apparent that discussions about the role of free competition in developing natural gas networks give the impression that competition is an insurmountable obstacle to attract new investments. However, it is not necessarily an unbridgeable conflict, but rather a design problem in this relationship for the gas industry.[46]

[45] The idea of the market created as a result of regulation, with the primary goal of economic development, is highly emphasized in neoclassical theory (C.F. Salomão, 'Regulação e Desenvolvimento', in C.F. Salomão (ed), *Regulação e Desenvolvimento*, São Paulo: Malheiros Editores, 2002, 29–30). It is important to emphasize the notion of market mechanisms as a development factor in the neoclassical moulds that form the foundation for the dominant regulatory concept in the country.

[46] 'Without doubt, granting market power to a specific party constitutes, at first glance, a strong incentive for said party to invest in specific assets, such as those characterizing the infrastructure of gas transportation infrastructure. This is because market power and, ultimately, the monopoly—in cases where there is an absence of free access, whether temporary or not—guarantee the totality of market demand to the agent under their own conditions, as well as acquisition of extra-economic profit. This occurs due to a lack of order that is provided by competition. This, however, is a socially undesirable mechanism since it transfers the income of society as a whole to the monopoly, thereby demotivating the quest for economic efficiency and restricting market expansion. In light of unwieldy sizes, there is no reason to abandon the effort of seeking alternatives to removing competition as a means of neutralizing incentives towards opportunistic conduct, which would paralyse investment efforts. Close consideration of the North-American experience demonstrates that the key issue lies in financial compensations obtained by the investor at odds with adherence to free access regulations. These compensations are also necessary in order to minimize regulatory risk. Indeed, in the absence of compensation, apprehension in the face of regulatory risk—represented by changes in game rules as a result of learning from past experience—may slow or even paralyse development investment in this market' (D.P. Pedra and L.H. Salgado, *Indústria de gás natural no Brasil: quadro regulatório e perspectivas*, Rio de Janeiro: International Seminar for the Restructure and Regulation of the Electrical Energy and Natural Gas Sectors, (2006), 07). Even the ideas of these authors indicate that the concept of development is the notion of competition as a means of promoting progress and of transaction costs linked to the uncertain stability of regulatory rules.

Discussion on transportation regulation presents an opportunity to consider not only the role of competition in the Brazilian natural gas industry and the establishment of development principles, but also the role of gas itself in national energy policies. Debate on development in this area can be expanded, primarily by realizing the need to control public policies in the energy sector. This awareness is pertinent in the light of current energy efficiency needs and the use of clean sources in a country where gas policy appears to be that of making money and solving energy shortages, without considering such essential guidelines for modern times.

Developing gas networks is a challenge in a country with large areas, such as Brazil, and when faced with current considerable domestic and industrial energy demands. To meet this challenge, providing rules on open access is not enough. These rules must not only take into account aspects of the gas supply market or an annual income, but they must also take into account the costs and the investments needed to build the networks.

15

Law and Regulation for Energy Networks in New Zealand

*Barry Barton**

I. Introduction

New Zealand's physical character presents particular policy challenges for the energy transmission and distribution networks that are the subject of this book. There are no international connections, obviously, but the country is long and narrow, its terrain is difficult, and its population density is low. Natural gas, produced in Taranaki, mostly offshore, is carried by pipeline to centres in the North Island but not the South. About 55 per cent of electricity is produced from hydroelectric generation (although it varies with rainfall), 13 per cent from geothermal, and 4 per cent from wind; the rest is mostly natural gas with some coal.[1] Renewable resources have required long electricity transmission lines—the national grid is described as 'stringy'. Much of the electricity generation is in the south of the South Island, while most of the population and industry is in the north of the North Island. The two islands were connected across the difficult waters of Cook Strait in 1965 by a 40km submarine power cable that is part of a 500km-long long high-voltage direct current (HVDC) system. By 2014, the total capacity of the link will be 1200 megawatts (MW).

The small population (4.4 million) of New Zealand and its small economy mean that energy and energy transport markets are less likely than in a larger nation to see new entrants exerting competitive pressure on incumbents. These circumstances would suggest that in New Zealand the law and regulation for energy networks would emphasize the control and coordination of monopolies, rather than the possibility of competition. That, however, has not been the case until 2008—much later than in most countries. The reason lies in the strength of the grip of neoliberal thinking on law and policy in New Zealand since 1984. The country was early in its liberalization of markets for electricity and natural gas, as for other parts of the economy,

* The author thanks Mr Ian Wilson of the Gas Industry Company and Mr Bart van Campen of the University of Auckland Energy Centre for their thoughtful comments on a draft. Any errors are those of the author.

[1] Ministry of Economic Development, *New Zealand Energy Quarterly*, March Quarter 2011.

and the reforms went very deep. In no small measure, the making of law and policy in recent years has been dominated by the need to establish the regulatory legislation and institutions that liberalization had avoided, but which prove to be necessary for coordination and for the protection of the public interest in energy networks.

II. The Background to 2003: From Public Service to Liberalization and Light-handed Regulation

It follows that the historical context is essential.[2] Before 1992, the provision of electricity was a matter for the public sector. The generation and transmission of electricity was the responsibility of a government department. Local distribution and sales were carried out by elected electric power boards or by municipal electricity departments. Decisions about construction projects and about prices were made by these public bodies, accountable to the electorate. There was no separate regulator to scrutinize these decisions. These arrangements were all affected by the wave of institutional and economic reform of the late 1980s and the early 1990s. This reform, inspired by neoliberal political thinking, sought to reduce the role of the state in many sectors and to allow market forces to operate.[3] The reform did reduce the role of the state in petroleum development and in natural gas production, but it did not lead to immediate privatization in the electricity sector. Its initial emphasis was to give enterprises a corporate form and purpose, rather than those of a government department or local body. Corporate form was promoted because it would produce businesslike operations, customer focus, market competition, and capital market discipline. In addition, legal monopolies were removed in order to allow new entrants, at least in theory, to challenge the established incumbent operators in the market. The Electricity Corporation of New Zealand Ltd (ECNZ) was established under the State-Owned Enterprises Act 1986 with a clear single objective of operating as a successful business. In 1994 Transpower NZ Ltd was separated out of ECNZ to take the national grid and the transmission and system operator functions. That left ECNZ in the field of electricity generation, where competitive behaviour was more possible; and to that end the company was split between 1995 and 1998 into four separate companies, one of which was sold.

Corporate form was also required of the electric power boards, municipal electricity departments, and municipal gas departments, by the Energy Companies Act 1992. Each publicly owned entity was required to produce an establishment plan for its transfer to a company under ordinary corporations legislation and a share allocation plan for the ownership of the company. Consultation and then ministerial approval was necessary, but the establishing authorities had a reasonably free

[2] Generally see B. Barton, 'From Public Service to Market Commodity: Electricity and Gas Law in New Zealand' (1998), 16 *JERL* 351; L.T. Evans and R.B. Meade, *Alternating Currents or Counter-Revolution? Contemporary Electricity Reform in New Zealand* (Wellington: 2005).

[3] A Sharp (ed), *Leap into the Dark: The Changing Role of the State in New Zealand since 1984* (Auckland: 1994).

hand, and a wide diversity of ownership patterns emerged. In electricity, these new companies initially ran their local distribution networks and ran the retail operation which sold electricity to customers. However, competition in the electricity retail market was slow to emerge. Just as with the split of ECNZ, structural reform was preferred to regulatory reform. The Electricity Industry Reform Act 1998 obliged the companies to divest themselves of either the distribution function or the retail function; a strict ownership split was imposed. Most companies held on to their distribution functions, and sold their retail operations to the generator companies. A period of restructuring and consolidation followed. The larger of the 28 present companies are owned by a mix of investors, local bodies, consumer trusts, and community trusts. The smaller ones are mainly owned by trusts.

In 1996 the New Zealand Electricity Market began to operate as a self-regulating wholesale market for electricity. Its framework was one of multilateral contracting, not statutory regulation. There was still no regulatory agency for electricity or gas. There was no price control of networks or of energy supply. 'Light-handed regulation' was the policy preference. It comprised the general competition law of the Commerce Act 1986, information disclosure as to transmission and distribution tariffs (to facilitate third party access), and at the same time, the threat of further regulation.[4] The first of these three elements, the general competition law, was primarily section 36 of the Commerce Act 1986, the prohibition of taking advantage of a substantial degree of market power for the purpose of restricting market competition. But it was ineffective.[5] The interpretation given to it by the courts has been much of the reason, but there is an inherent tension in monopolization law: we do not want the large firm to stifle its rivals or to entrench itself as the incumbent; but we do want it to engage in healthy competition and investment that may produce greater innovation, efficiency, and choice. There was little evidence that any element of light-handed regulation affected corporate behaviour, except, perhaps, in securing basic access to essential facilities for third parties. There was evidence of substantial excess profits in electricity distribution businesses and in electricity wholesale markets.[6]

[4] As noted in *Power NZ Ltd v Mercury Energy Ltd* [1996] 1 NZLR 686 and *Air New Zealand Ltd v Wellington International Airport Ltd* [2009] 3 NZLR 713 (CA). A.E. Bollard and M. Pickford, 'New Zealand's "Light-Handed" Approach to Utility Regulation' (1995), 2 *Agenda* 411 described the policy, and noted that its success would have to be gauged against the probability that natural monopoly is more widespread in New Zealand, where markets are typically small, than in larger countries.

[5] R. Adhar, 'The Unfulfilled Promise of New Zealand's Monopolisation Law: Sources, Symptoms and Solutions' (2008), 16 *Competition and Consumer L J* 291, particularly discussing *Telecom Corp NZ Ltd v Clear Communications Ltd* [1995] 1 NZLR 385 (PC).

[6] G. Bertram and D. Twaddle, 'Price–Cost Margins and Profit Rates in New Zealand Electricity Distribution Networks Since 1994: the Cost of Light Handed Regulation' (2005), 27 *Journal of Regulatory Economics* 281, estimated the difference between profits realized between 1994 and 2003 and the profits that would have been allowed under a rate of return regulatory framework to be about $200 million a year. As to wholesale market, F. Wolak, *An Assessment of the Performance of the New Zealand Wholesale Electricity Market* (Report for the New Zealand Commerce Commission, 2009) suggested that over a period of some 6.5 years the generators exercised their market power to earn market rents estimated conservatively to be $4.3 billion, which averaged 18 per cent of the total wholesale market revenues received by all generators over the period.

III. First Steps in Formal Regulation of Energy Networks 2003 to 2008

Self-regulation seemed successful in managing the wholesale market, and in conforming with the concept of light-handed regulation without a specialist regulator. From 2000 the government had an agenda of further matters for it to pursue, including the long-vexed question of transmission pricing methodology and a number of other items such as a consumer complaints system and demand-side management measures. The transmission pricing methodology (TPM) states how Transpower is to set prices for the use of different parts of the network (see also section V of this chapter). However, it became clear that this agenda could not be fitted into self-regulation—Transpower among others opposed it—and, on top of two scares about security of supply because of low hydro inflows, a statutory Electricity Commission was established in 2003 under an amendment of the Electricity Act in 2001 made with this possibility in mind.[7]

The other significant change in the law at that time was the Commerce Amendment Act 2001, which introduced Part 4A of the Commerce Act 1986. The Commerce Commission is New Zealand's agency for the regulation and enforcement of market competition. It had long had general powers to impose price control, but had hardly ever used them. The Act of 2001 brought the Commerce Commission explicitly into the field of energy for the first time, and authorized it to impose 'targeted control' of the prices of the electricity distribution companies. The intention was to avoid a situation where price control was the norm. Mistrust of regulatory interference ran very deep.[8] The Commission would set thresholds for companies as to their price and quality performance. Only if a company crossed a threshold would the Commission begin an investigation to decide whether it should be subject to *ex post* price control. In the Court of Appeal it was noted in 2008 that there was little overseas case law that was relevant to such matters, because in most comparable jurisdictions the issue for the regulator was not whether there should be regulation of monopolistic utilities companies, but only how regulation should be implemented; whereas in New Zealand the inquiry started at the 'should' stage.[9]

One therefore sees that one of the chief characteristics of the formal regulation of transmission and distribution in New Zealand is its recency. The Electricity Commission's Electricity Governance Rules only came into effect on 1 March

[7] B. Barton, 'Reaching the Limits of What the Market will Provide: Energy Security in New Zealand' in B. Barton, C. Redgwell, A. Rønne, and D. Zillman (eds), *Energy Security: Managing Risk in a Dynamic Legal and Regulatory Environment* (Oxford: OUP, 2004) p 373; D. Caygill, 'Why Did Electricity Self-Regulation Fail?' (2004), 7 *NZ Yearbook of Jurisprudence* 20; B. Barton, 'Self-Regulation, State Regulation, and Co-Regulation in Energy Law in New Zealand' in B. Barton, L. Barrera-Hernández, A. Lucas, A. Rønne (eds) *Regulating Energy and Natural Resources* (Oxford: Oxford University Press, 2006) p 137.

[8] *Inquiry into the Electricity Industry June 2000 Report to the Minister of Energy* (D Caygill chairperson, Ministry of Economic Development) paras 191–5.

[9] *Powerco Ltd v Commerce Commission* [2008] NZCA 289 para 47.

2004, and its transmission investment and pricing methodology on 28 May 2004. The Commerce Commission's first electricity lines control notice was issued in 2003. This recency may explain some of the turmoil that occurred in the first five years; the regulatory commissions, their counterparts in the regulated companies, ministers of the Crown, and the financial community were all coming to grips with regulation and regulatory pressure that was in truth new to the country.[10] Two controversies broke out. The first concerned a proposal for a new 400 kV transmission to run between Whakamaru in the central North Island and Otahuhu in Auckland. This was Transpower's first large project for many years. Transpower submitted it to the Electricity Commission. In its draft decision in April 2006, the Commission compared the project with feasible alternative projects, under the Grid Investment Test, and concluded that they would provide the required improvements in grid reliability at a much lower price. Transpower did not accept this, and the Deputy Prime Minister and Minister of Energy intervened to 'bang heads together'. The government issued two directive policy statements and removed the Chairperson of the Commission. Transpower submitted an amended proposal for the transmission line, and the Commission approved it in July 2007 without demur.[11]

The second controversy concerned the imposition of price control by the Commerce Commission. A price–path threshold was introduced for electricity lines companies under the 2001 Amendment, using the basic CPI-X formula applied to the average prices of a distribution company, where CPI is the consumer price index and X is a factor that represents the required annual reduction of prices in real terms.[12] If a company went over its threshold, it would be subject to a post-breach inquiry that might lead to control. The Commerce Commission fought off one judicial challenge to its thresholds that had argued that the character of the thresholds had to be connected to control. Instead, the Supreme Court of New Zealand emphasized the extent of the Commission's power to set thresholds.[13] However, the Commission received a great deal of criticism for its abrupt announcement of its intention to impose control on Vector Ltd, the country's largest electricity and gas distribution company. The breach of the price–path threshold was trivial (NZ$77,000 on a turnover of $477,000,000) and unrelated to the issues about overcharging that the Commission later decided were important; indeed, the Commission variously offered three quite different issues for its inquiries. The announcement of the intention to impose control came as a bombshell and caused a sharp drop in the company's share price. The company did not go to

[10] For the events recounted in this paragraph and the next, see B Barton, 'Electricity Regulation in New Zealand: the Early Stages of a New Regime' (2008), 26 *JERL* 207.

[11] A challenge by disaffected landowners on the grounds of predetermination, bias, illegality, and unreasonableness was unsuccessful: *New Era Energy Inc v Electricity Commission* [2010] NZRMA 63 (HC).

[12] Natural gas was a lesser concern, but company operations had caused the Commission to carry out a Part 4 inquiry, with the result that in 2005 price control was imposed on Powerco and Vector, the two main natural gas distribution companies. The companies were operating under authorizations granted by the Commission. For the background of CPI-X regulation in Britain, see Aileen McHarg's chapter (chapter 17) in this volume, 'Evolution and Revolution in British Energy Network Regulation: From RPI-X to RIIO'.

[13] *Unison Networks Ltd v Commerce Commission* [2008] 1 NZLR 42.

court, but the experience contributed to a sense among policymakers that targeted control was unpredictable and uncertain. Thresholds were the subject of extensive consultation but were not subject to challenge except by judicial review or in later enforcement proceedings. A threshold could be set low and it was easy to go into breach of it. The threshold did not need to be linked to control by providing a screen or filter for behaviour raising the concerns identified in the objectives of Part 4A. Once an investigation got under way it could go wider than the threshold matter.[14] The hopes held for the policy of targeted price control were therefore not fulfilled; perhaps the mechanisms for keeping it light-handed made it too complicated and unpredictable.

IV. Network Price Control

These perceptions informed a review of the price control that the government initiated in 2007.[15] The review took place in the context of a broader policy process of development of the New Zealand Energy Strategy 2007, and the institution of the country's first Emissions Trading Scheme by amendment in 2008 of the Climate Change Response Act 2002.[16] The review led to law reform in the Commerce Amendment Act 2008, which replaced the old Parts 4 and 4A of the Commerce Act 1986 with a new Part 4. The main change was to dispense with the targeting of control. The old legislative purpose was to promote the efficient operation of markets directly related to electricity distribution and transmission services through targeted control for the long-term benefit of consumers.[17] The new legislation declares that it provides for the regulation of the price and quality of goods or services in markets where there is little or no competition and little or no likelihood of a substantial increase in competition, and that its purpose is to promote the long-term benefit of consumers in markets by promoting outcomes that are consistent with outcomes produced in competitive markets. This is explicit recognition that in energy networks and their like there is little competition and little likelihood of it; a self-evident truth, perhaps, but in New Zealand law something of a breakthrough.[18]

The Commerce Act now says that all electricity lines services, including Transpower, and all gas pipeline services are regulated.[19] There is no suspension

[14] Barton (2008, above n 10) at 226.

[15] Ministry of Economic Development, 'Review of Regulatory Control Provisions under the Commerce Act 1986: Discussion Document' (April 2007); Minister of Commerce and Minister of Energy, 'Review of Parts 4 and 4A of the Commerce Act' (Cabinet paper, published 22 November 2007).

[16] Ministry of Economic Development, *New Zealand Energy Strategy to 2050: Powering Our Future*, October 2007; Ministry for the Environment, *New Zealand's Climate Change Solutions: An Overview*, September 2007.

[17] Commerce Act 1986 s 57E (repealed).

[18] For a useful discussion of the evolution of New Zealand competition law, and of the new Part 4, see M Sumpter, *New Zealand Competition Law and Policy* (Auckland: 2010).

[19] Commerce Act 1986 ss 54E, 55B, also as to certain airports s 56B. For other goods and services to be regulated, an inquiry must first be held: s 52G.

of regulation pending the effect of targeting. The particular type of regulation (except for Transpower) is 'default/customised price–quality path regulation'.[20] The starting point is that the Commission sets default price–quality paths that apply for a period (normally five years), stating maximum prices and/or revenues, and quality standards. The alternative that a company may apply for is a customized price–quality path, which may suit where a company wants approval in advance of substantial new investment. The Commission makes determinations to specify how the default or customised regulation applies, including price–path starting prices, rates of change, quality requirements, time frames, and input methodologies.[21]

An electricity lines company is exempt from price–quality control if it is wholly owned by a consumer trust, community trust, or co-operative where the trustees or directors are elected by consumers, and if the company has fewer than 150,000 consumer connections.[22] Twelve companies have declared themselves exempt. The theory is that consumers as owners can control, or at least influence, the company's rate of return and the trade-offs it makes between price and quality of service. However, all electricity lines services and gas pipeline services are subject to information disclosure,[23] and consumers can petition for price/quality control to be imposed notwithstanding the exemption.[24]

The 2008 Amendment introduced procedures for the setting of 'input methodologies' separately from particular price-control decisions. The Commerce Commission must develop and publish the input methodologies that say how it will evaluate cost of capital, valuation of assets, allocation of common costs, tax treatment, pricing methodologies, regulatory processes, and rules for the price-quality regulation that controls the non-exempt electricity lines (ie distribution) companies and the gas companies.[25] The methodologies are therefore the basic general rules for price control, especially for the difficult question of the weighted average cost of capital. A draft methodology is to be made available for comment and the holding of a conference before finalization. These procedures were a significant change, because under the previous system the only way to challenge the Commission on the subject was in dealing with a company's breach of a threshold, which was backward-looking. The new procedures should allow disagreements

[20] Commerce Act 1986 ss 54G, 55D. The details are provided in ss 53K–53ZB. The specific information about prices, rates of change in prices, and quality standards are stated in section 52P determinations: s 53O–53P.

[21] Commerce Act 1986 ss 52P, 54I–54K, 55E–55F. See Commerce Commission, *Commerce Act (Electricity Distribution Default Price–Quality Path) Determination 2010*, Decision 685, amended by Decisions 704 and 722, and in force on 1 April 2010.

[22] Commerce Act 1986 ss 54G(2), 54D (definition).

[23] Commerce Act 1986 ss 54F, 55C. The Commission has work under way for a determination on information disclosure requirements. Information disclosure was formerly required by the Electricity Act 1992.

[24] Commerce Act 1986 s 54H.

[25] Commerce Act ss 52T–52V. See Commerce Commission, *Commerce Act (Electricity Distribution Services Input Methodologies) Determination 2010,* Decision 710, *Commerce Act (Gas Distribution Services Input Methodologies) Determination 2010*, Decision 711, and *Commerce Act (Transpower Input Methodologies) Determination 2010*, Decision 713.

about methods to be thrashed out by all parties involved, looking to the future in which they are to be applied in particular cases.

The Commerce Commission also makes what are called section 52P determinations in order to decide how the relevant forms of regulation apply to companies. In particular, for electricity and gas lines services suppliers, a section 52P determination is where the Commission sets the price–quality path for each supplier, drawing on the previously settled input methodologies.[26] Section 52P determinations also prescribe the information disclosure requirements. Once the requirements are set, the Commission's functions are monitoring and enforcement. It can address price–quality paths and information disclosure requirements with pecuniary penalties, compensation orders, injunctions, and prosecutions.

The Commerce Commission recommendation that from 1 April 2011, Transpower be subject to individual price–quality regulation was accepted and given effect by an order in June 2010, and the price path itself was determined soon after.[27] The input methodologies are distinct from those applying to distribution company networks, and involve the Electricity Authority, as will be discussed below.

The question of appeals was much discussed during the law reform process.[28] The general position under the Commerce Act has long been that parties have a right of appeal from a Commerce Commission determination to the High Court, as a general appeal by way of rehearing.[29] The High Court generally sits with one or more lay members qualified in a relevant discipline, usually economics. Parliament's Commerce Committee supported appeals on input methodologies, otherwise the only challenges would be in the context of final price path decisions, which would cause delay and complication.[30] The appeal provided in the Act is by way of rehearing, and so broadly the same as other appeals under the Act, but with three special features.[31] First, the appeal is conducted solely on the basis of the documentary information and views that were before the Commission, and no new material may be introduced. Second, if the High Court chooses to allow an appeal, it must be satisfied that the result it proposes will be materially better. Third, the Court must sit with two lay members, unless it considers that only one is required, as is common in other appeals. Input methodologies will certainly present some novel challenges to the Court. However, the High Court has a more confined role in relation to appeals against section 52P decisions about the regulatory

[26] Commerce Act 1986 ss 54I–54K.

[27] Commerce Commission, *Commerce Act (Transpower Individual Price–Quality Path) Determination 2010*, Decision 714. The decision on the nature of the control was made pursuant to Commerce Act 1986 ss 52N and 54M(3).

[28] The IEA had weighed in, recommending appeal on the merits: International Energy Agency, *New Zealand: 2006 Review* (2006) p 11. See Barton (2008 above n 10) at 229.

[29] Commerce Act 1986 ss 77, 91(1), 93–4.

[30] Commerce Amendment Bill (201–2) (select committee report) 28 July 2008 at 5.

[31] Commerce Act 1986 ss 52Z, 52ZA, 91(1)(b). A certain amount of discussion of the matter has described this as a 'merits review', eg see Sumpter (above n 18) 365; but that may overstate the difference between this appeal and other appeals under the Commerce Act or under New Zealand law generally. J Every-Palmer, 'The State and Monopolies: New Zealand's Experience' (2010), 12 *Otago L Rev* 227 at 242 also notes this.

requirements on individual companies; for them the appeal is only on a question of law.[32]

V. Regulating Electricity Networks

A. Electricity Authority

After the election of 8 November 2008, a new government took office and began a ministerial review of electricity market performance. Less far-reaching reviews had occurred in 2006 and (after another dry winter threatened hydro generation) in 2008. The new government's review was motivated by concern that prices had risen substantially in the past decade, concern about the reliability of supply, and dissatisfaction with the institutional arrangements for the regulation of the industry. There was also dissatisfaction with the lack of competition in the wholesale market. The Commerce Commission had begun an investigation in late 2005 after complaints about wholesale electricity prices and allegations of abuse of market power and collusion. The Commission concluded that the four main generator companies had a substantial degree of market power in the wholesale market, and that the current market structure had serious systemic problems. However, it did not reach a conclusion that the market power had been used for any unlawful anti-competitive purpose.[33]

The review was carried out by officials and an Electricity Technical Advisory Group appointed by the Minister of Energy in 2009, and the resultant decisions were approved by Cabinet at the end of that year.[34] The Electricity Industry Bill was introduced in December 2009 to give effect to the resultant decisions, and enacted as the Electricity Industry Act 2010. A number of the reforms concern electricity generation or retailing, rather than the transmission and distribution that is the focus of this chapter; the main changes are asset swaps and long-term hedges by the three state-owned enterprises to improve competition, compensation for consumers for energy savings during any conservation campaign when hydro output is low, encouragement of demand-side participation in the wholesale market, and encouragement of retail customers to compare and switch suppliers.

The Electricity Industry Act 2010 disestablished the Electricity Commission and replaced it with the Electricity Authority ('the Authority'). It states the purpose of the Authority to be 'to promote competition in, reliable supply by, and the efficient operation of, the electricity industry for the long-term benefit of consumers'.[35] This reflects the conclusion that the Commission had been pursuing too many functions and objectives. Several of these functions were shifted elsewhere. Section 8

[32] Commerce Act 1986 ss 91(1)(a), 91(1B).

[33] Commerce Commission, *Commerce Act 1986 s 27, s 30 and s 36 Electricity Investigation*, (21 May 2009), relying on Wolak (above n 6).

[34] Cabinet Paper, *Ministerial Review of the Electricity Market*, adopted by Cabinet 7 December 2009.

[35] Electricity Industry Act 2010 s 15. The Act repeals parts of the Electricity Act 1992, returning it to a focus almost entirely on safety and access to land for company works.

shifted information provision and short- and medium-term of security of supply to Transpower, and also the duty to manage supply emergencies. The promotion of energy efficiency in electricity was transferred to the Energy Efficiency and Conservation Authority. The approval of transmission grid upgrades and the application of the grid investment test were transferred to the Commerce Commission, which would have overall regulatory control of Transpower's revenues and expenditure. The making of statements of opportunities as part of grid investment planning was transferred to the Ministry of Economic Development.

On the other hand, the Electricity Authority will focus more specifically on making the wholesale market work. The Authority and not the Minister now has power to amend the Electricity Industry Participation Code, which governs the wholesale market. Several parts of the Code (based on the previous Electricity Governance Rules) particularly concern transmission, including transmission agreements, grid reliability, the transmission pricing methodology, financial transmission rights, and interconnection asset services.[36] It states the obligations of Transpower as the system operator. It provides for metering standards, and rules for the connection of distributed generation. The Code can require Transpower and others to enter into transmission agreements for connection, including payment in accordance with the transmission pricing methodology. It can require electricity distributors to standardize their tariff structures and use-of-system agreements, to reduce an impediment to retail competition. The Code is binding on industry participants.

The Rulings Panel deals with complaints about breaches of the Electricity Industry Participation Code that are referred to it by the Authority or an industry participant, it determines appeals against certain decisions of the Authority under the Code, and it resolves certain disputes between industry participants relating to the Code. The Rulings Panel is constituted by the Act, which is better than the previous arrangement, where it was under regulations. The Panel can make orders for financial penalties, compensation, compliance directions, suspension, and termination from the industry.[37]

The independence of the electricity regulator from the government had been much disputed in the previous round of reform that established the Electricity Commission.[38] Arguments in 2003 and 2004 against ministerial control and intervention were met by the answer from the Minister that the government would be held responsible for electricity outcomes, for keeping the lights on, and should therefore set the Commission's direction. The Commission was therefore constituted as a Crown agent within the meaning of the Crown Entities Act 2004, so that it was obliged to give effect to government policy when so directed by the responsible minister.[39] In addition, members of the Commission were appointed for three-year terms only. The opposite point of view has now prevailed. Under the Electricity Industry Act, the Electricity Authority is an 'independent Crown entity'

[36] Commerce Act 1986 ss 38, 42(2), 44.

[37] Electricity Industry Act 2010 ss 23–6, 50–62. Also relevant are the Electricity Industry (Enforcement) Regulations 2010.

[38] Barton (2008 above n 10) at 218.

[39] Crown Entities Act 2004 s 103.

subject to ministerial direction only as provided in that Act; and the Minister must first consult the Authority, then can make a statement of government policy, which the Authority must have regard to, rather than give effect to.[40] Authority members are normally appointed for five years, and in addition to their general duties as members of such public bodies, they are prohibited from representing or promoting the interests or views of any organization or industry participant.[41] These arrangements will be tested the next time that a minister feels political criticism for a possible electricity shortage or a price increase, and tries to explain that he or she cannot tell the governing Authority what to do.

B. Electricity distribution network regulation

The other major reform carried out by the Electricity Industry Act 2010 was to repeal the Electricity Industry Reform Act 1998, which had imposed the ownership split that prevented an electricity distribution company from engaging in electricity generation and retailing. The split had been moderated in subsequent years to allow local distribution companies back in to certain generation and retail operations; first for non-traditional renewable electricity, then for generation and retail outside their network areas, and then for any kind of renewable generation. The split performs an important function in preventing a distribution company from discriminating between its own retail operations and those of its competitors, so suppressing retail competition. Nonetheless, the 2009 review decided that retail competition would be stimulated, especially in smaller remote areas, if the restrictions were removed or reduced. There are three main layers to that effect in the 2010 Act. First, the ownership separation remains in force between distribution and the large generator companies (with more than 250 MW and connected to the national grid). Second, there must be a corporate separation (different companies) and compliance with arm's length rules between distributors and generators or retailers, if the generator has more than 50 MW capacity connected to the distributor's network, or if the retailer sells more than 75 GWh of electricity in a year to customers on the distributor's network. Third, a distributor must disclose its use-of-system agreement if it has generation of more than 10 MW capacity or retailing of more than 5 GWh a year.[42] This regime presents new opportunities for local distribution companies, although it is by no means sure that small companies will take the risk of entering the competitive retail market to such an extent as to make a difference to electricity prices.[43] Nor is it clear that the regime will be enough to prevent anti-competitive use of monopoly distribution networks.

Another restriction on electricity distribution companies that was relaxed in 2010 was their duty to maintain the connections of customers on their networks. Business logic would suggest that a company would not wish to continue expensive

[40] Electricity Industry Act 2010 ss 12 and 17, Crown Entities Act 2004 ss 7, 14, 105, and Pt 3 of Schedule 1.

[41] Electricity Industry Act 2010 s 14, Crown Entities Act 2004 s 32.

[42] Electricity Industry Act 2010, ss 75, 76, and 77, respectively.

[43] C. van Leuven and A. Booth, 'Shake Up of New Zealand's Electricity Industry Likely' (2009), 28 *Aust. Resources and Energy L J* 373.

maintenance of a long line to an isolated customer over difficult terrain. In the Electricity Act 1992 these customers were only protected from disconnection until 2013, but in 2008 the government decided to continue to protect those customers indefinitely, and the Bill introduced for that purpose became part of the Electricity Industry Act 2010.

C. Electricity transmission network regulation

How Transpower should recover the cost of new investments in the high-transmission network had long been a difficulty. During the period of industry self-regulation between 1994 and 2003 there was no statutory obligation for users of the network to pay for it; it was thought that contracts between Transpower and its users would be enough. However, if users did not agree to those contracts there was very little that could be done about it, and the deadlock caused Transpower difficulty in committing to new projects. When the Electricity Commission was established, one of the reasons for its statutory mandate was to break the deadlock and impose rules, which it did in Part F of the Electricity Governance Rules 2003. Part F included the transmission pricing methodology, the grid investment test and the procedure for approving grid upgrade plans.[44] A new Part F was negotiated and became effective on 1 April 2008.

The ministerial review of 2009 criticized the dual regulation of Transpower by the Electricity Commission and Commerce Commission. It noted that there was a backlog of investment in transmission that would take some years to clear, so that improved procedures were required for transmission upgrades.[45] One of its main reforms, therefore, was that the Electricity Industry Act 2010 transferred from the Electricity Commission to the Commerce Commission all the functions of approving grid upgrade plan proposals by Transpower.[46] Part F is to be replaced by an input methodology for Transpower capital expenditure proposals.

The Electricity Authority, however, still retains a substantial regulatory presence in transmission. Its extent is not immediately apparent, because the Authority's powers to enact Code provisions are broadly expressed.[47] Grid reliability standards were thought in the review of 2009 best to stay with the industry-specific regulator because they strongly interrelate with the overall operation and efficiency of the market. They are therefore not affected by the transfer of other powers to the Commerce Commission, and are part of the Code.[48] Similarly, quality standards for Transpower may be in the Authority's Code, and the Commerce Commission is directed to base its quality requirements for Transpower on the

[44] An unsuccessful challenge of the application of the grid investment test and related rules was made in *Major Electricity Users' Group Inc v Electricity Commission* [2008] NZCA 536.

[45] The International Energy Agency, *Energy Policies of IEA Countries: New Zealand 2010 Review* (2011) p 102 notes that investment 1995–2005 appears to have been inadequate (below NZD 100 million a year).

[46] Electricity Industry Act 2010 ss 155, Commerce Act 1986 ss 54R–54S.

[47] Electricity Industry Act 2010 s 32.

[48] Cabinet paper 2009 (above n 34) 30; Commerce Act 1986 s 54R; Code Part 12, Transport.

Authority's standards.[49] As to pricing, the Commerce Commission makes the ultimate decisions about capital expenditure and return on investment. However, the Electricity Authority was again thought to be the proper place for the transmission pricing methodology (TPM) to be determined. The TPM is important because it determines how total transmission charges—which may not exceed the maximum allowable return set by the Commerce Commission—are divided up among the contracting parties using the network, and the basis for those charges such as their share of peak usage. The Authority's work in the TPM is not affected by the transfer of other powers to the Commerce Commission. The TPM is initially developed by Transpower, and sent to the Authority for approval or return to Transpower. With certain exceptions, Transpower is required to charge for its transmission services only in accordance with the TPM.[50] In relation to both the transmission network and local distribution networks, the Electricity Authority continues to set pricing methodologies.[51] Transmission was an aspect of an 'unacceptable trading situation' on 26 March 2011, which has presented the Authority with a difficult question about wholesale market behaviour.[52] Finally, one notes that the 2010 Act established a Security and Reliability Council to provide independent advice to the Authority on the performance of the electricity system and the system operator, and on reliability of supply issues. Overall, Transpower is still regulated by both the Commerce Commission and the Electricity Authority.

Section 54V requires the Commerce Commission and the Electricity Authority to work together in various ways. The Authority must consult the Commission about amending the Code, and advise it about certain actions in its administration. The Commission must take into account the Code and certain Authority actions, and reconsider determinations concerning such matters if the Authority requests. Quality standards for Transpower that the Commission sets in a section 52P determination must be based on, and consistent with, the quality standards that the Authority sets. The two agencies have signed a memorandum of understanding.

A different aspect of transmission that is under the jurisdiction of the Electricity Authority is transmission hedges. Transmission constraints apparently discourage retailers from business in areas where they do not have generation capacity, thereby reducing retail competition. The remedy that has long been discussed (assuming that transmission network upgrades will not always be possible) is a system for retailers to buy hedges against transmission risk. Work on transmission hedges had been under way, but the review of 2009 felt that the amount of risk and money involved could lead to paralysis, so the Act requires the Authority to deal with it in the Code within twelve months.[53]

[49] Electricity Industry Act 2010 s 32(2); Commerce Act 1986 s 54V(6); Code Part 8, Common Quality.

[50] Cabinet paper 2009 (above n 34) 30; Commerce Act 1986 s 54R; Code Part 12, Transport, 12.84, 12.91, 12.95.

[51] Commerce Act 1986 s 52T, Electricity Industry Act 2010 s 32(2).

[52] Electricity Commission, *Draft decision pursuant to Part 5 of the Electricity Industry Participation Code 2010 regarding an Alleged Undesirable Trading Situation on 26 March 2011*, 6 May 2011.

[53] Cabinet paper 2009 (above n 34) 6; Electricity Industry Act 2010 ss 42–3.

VI. Natural Gas Network Regulation

The regulation of natural gas networks in New Zealand has lagged behind electricity, and has had a lower profile. What evolved, after the electricity experience in the period 2000 to 2003, was 'co-regulation', where the industry was given considerable latitude to develop self-regulatory systems for governance, but within a legislative framework that expressed aspects of the public interest and that, as a quid pro quo, gave statutory backing to the system. Thus the Gas Act 1992, as amended in 2004, provides for co-regulation by the government and an industry body that satisfies the Minister that it is broadly inclusive of industry participants. The Gas Industry Company Ltd was formed for this purpose and obtained the necessary approval, and therefore has the legal power to recommend statutory rules for the industry to the Minister, and indeed the effectively sole legal right to do so.[54] The Company has objectives and outcomes stated for it in a government policy statement.[55] The present statement echoes statutory objectives such as access to essential infrastructure, particularly transmission and distribution pipelines. The Company has completed work on systems for switching customers, reconciling quantities in the downstream market, managing critical contingencies, and handling consumer complaints.[56]

The main policy work of the Gas Industry Company now is on third party access to pipelines and other infrastructure, along with balancing and interconnection. Vector introduced an open access regime for its network in 1997 and revised it in 2007 in the form of the Vector Transmission Code. Open access to the large Maui pipeline from Taranaki to Huntly only began under the Maui Pipeline Operating Code of 2005. These access regimes are working as intended, but they are requiring attention from stakeholders for concerns that have arisen. The Maui balancing arrangements are not working as intended; there is a lack of transparency and other characteristics necessary for genuinely open third party access.[57] In addition, a 2009 study identified a number of aspects of Vector's transmission arrangements that militated against competition and efficient pricing.[58] Further, in 2009 Vector announced that it could not provide new firm capacity on the key pipeline that runs from Huntly to Auckland, because the pipeline is at the limit of its physical capacity, but the commercial arrangements for the pipeline meant that Vector cannot recall capacity from incumbent retailers who are entitled to renew the same level of capacity from one year to the next. This presents a barrier to new entrants wishing to enter the market to sell gas. Thus, for the third party access regimes in

[54] Barton et al (2006 above n 7) 137. Apart from co-regulation with the Gas Industry Company, one must note that in 2005 the Minister of Energy imposed price control on the gas distribution services of Powerco and Vector, which required the companies to obtain authorizations under the old Part 4 of the Commerce Act 1986.

[55] Gas Act 1992 s 43ZO; Ministry of Economic Development, *Government Policy Statement on Gas Governance April 2008*.

[56] Gas Industry Company, *Annual Report 2009–10*.

[57] Gas Industry Company, *Transmission Pipeline Balancing* (research paper) April 2008.

[58] Creative Energy, 'Review of Vector Capacity Arrangements' (research paper for Gas Industry Co, 2009).

the Codes to be efficient, work is needed on pipeline balancing and the allocation of scarce capacity rights.

Also a concern is the regime for future investment in gas transmission pipelines, whether by upgrading or by building new pipelines.[59] There are a number of obstacles to such investment, including regulatory ones. The present system of gas co-regulation has no equivalent of electricity's statements of opportunities, the grid investment test, grid upgrade proposals and the like, which may seem cumbersome, but provide a path for decisions on proposals to invest capital in a shared infrastructure project and on payment for its use. Without them there is a real risk of paralysis, whether for the upgrading of the Auckland pipeline or for the construction of a new pipeline if natural gas is discovered elsewhere than in Taranaki.

The IEA Country Review of 2010 sees New Zealand's governance arrangements for natural gas as complex and ad hoc in their development.[60] They suffer from a lack of coordination, timeliness, and general effectiveness. It doubts whether co-regulation, dominated by the industry, will effectively undertake the rulemaking, compliance, and management of market power that would be more suitably undertaken by a government regulator. It recommended a review to that effect, and recommended increased information and access arrangements for competition and transparency.

VII. Reorientation of Energy Networks for Renewables and Energy Efficiency

It is possible to identify elements of New Zealand's legal regime for networks that seek a greater use of renewable sources of energy and a greater degree of energy efficiency and demand-side management, with attendant decarbonization of the energy system; but they are not many, and not a major part of the policy mixture. The Electricity Authority is not required to pursue the objectives of its predecessor, the Electricity Commission for environmental sustainability, efficient use of electricity, and energy efficiency; energy efficiency went to the Energy Efficiency and Conservation Authority; and there has been less policy emphasis on increasing renewables.[61]

It was noted at the beginning of this chapter that most of New Zealand's electricity generation is already from renewable sources. The proportion dropped in 2003 to 65 per cent, but more recently it has varied between 70 and 79 per cent,

[59] Gas Industry Company, *Proposed Gas Transmission Investment Project*, April 2011.

[60] International Energy Agency, *Energy Policies of IEA Countries: New Zealand 2010 Review* (Paris: IEA/OECD, 2011) p 88.

[61] Electricity Act 1992 s 172N (repealed in 2010). The New Zealand Energy Strategy 2007, confirmed in the Government Policy Statement on Electricity Governance 2008, declared a target of 90 per cent renewables by 2025, but a new Strategy without that target is under consideration and the Policy Statement has been revoked. A National Policy Statement on Renewable Energy under the general environmental law, the Resource Management Act 1991, does not state the 90 per cent target.

subject to rainfall. There has been a shift in the past ten years from adding natural gas-generation capacity to renewables, for three reasons; natural gas became more expensive and less plentiful as the Maui gas field wound down; there was a new realization of the high quality of wind energy resources; and the legal regime for geothermal generation became clearer. The Electricity Commission made a series of recommendations on wind integration, mostly on forecasting for the purposes of having accurate pre-dispatch information in the wholesale market, and on the ability of the power system to handle large-scale unpredicted changes in wind-generation output. It was thought that existing procedures for grid upgrade plans would be satisfactory. Indeed, much of the new wind and new geothermal capacity is close to the backbone of the national grid, and some is close to load centres such as the city of Wellington. Wind and other renewables have therefore not turned energy network planning on its head.

Efforts have been made to provide a favourable climate for distributed generation; small generation facilities connected not to the national grid network but to the local distribution network. It is usually small-scale, even micro-scale, and often in the form of co-generation or renewable energy generation. At the simplest level, a farmer with electricity generation from biogas, for example, may wish to sell electricity into the local network when the farm has a surplus, and to buy it when it has a deficit. However, distribution companies are often unenthusiastic. The Electricity Industry Participation Code provides a framework for the connection of distributed generation to local distribution networks, on regulated terms and subject to prescribed pricing principles.[62]

Smart meters, or advanced metering infrastructure, offer a number of benefits in load control and demand-side management. Suitable meters allow users to understand and modify their use of electricity. In New Zealand meters are installed and owned by retailers, while the preferred approach in many developed countries is for meters to be owned by distributor companies.[63] The retailers are carrying out programmes to renew their meter installations. The meters that they are installing will allow for remote reading of power consumption by the company, but they will not have the capacity to connect to a home area network and communicate with energy-using devices so as to control load, nor to enable a real-time in-home display. The result is that the retailers will no longer need to pay meter-readers to visit each customer's meter, but the possible benefits to the consumer and to the environment will not be reaped. The Parliamentary Commissioner for the Environment recommended regulations be made to require the companies to instal genuinely smart meters, but the Electricity Commission and the government declined.[64] The IEA sees smart meters as an important part of energy policy and environmental policy,

[62] Electricity Governance (Connection of Distributed Generation) Regulations 2007, included in the Electricity Industry Participation Code 2010: see Electricity Industry Act 2010 s 34.

[63] International Energy Agency (2011, above n 60) 113.

[64] Parliamentary Commissioner for the Environment, *Smart Electricity Meters: How Households and the Environment Can Benefit* (2009); Electricity Commission, *Advanced Metering Infrastructure in New Zealand: Roll-out and Requirements* (2009); Minister of Energy and Resources, 'Minister Agrees to Smart Meter Recommendations' (press release, 9 March 2010).

and has urged a review of this decision so as not to lock in a sub-optimal technology.[65] It is unfortunate that the national electricity system is not being made more adaptable to demand-side management and energy efficiency opportunities.

These measures do not amount to a vigorous re-orientation of energy networks to a new energy future, but a new provision in an unexpected place, the Commerce Act, is full of far-reaching implications. Section 54Q says that the Commerce Commission must, in carrying out price and quality control regulation, promote incentives (and avoid disincentives) for lines companies to invest in energy efficiency and demand-side management, and to reduce energy losses. This is a positive obligation, to promote, and imposed on an influential established agency. It will be interesting to see how the Commission interprets its duty to promote incentives for distribution companies to invest in energy efficiency. The incentives could appear in input methodologies by requiring companies to engage with energy consumers in order to improve their energy efficiency, to shift peak load, and to reduce demand generally as part of the price–quality path.

VIII. Conclusion

Apart from this mixed record of reorientation to a new energy future, what does one make of the New Zealand experience in shaping the legal framework for energy networks? One can identify different periods in the nation's recent legal history in liberalization and regulation. The dates are approximate, and the changes in price control in 2001 and 2008 came two years before the other relevant changes.

To 1988	Government ownership of the whole electricity sector and much of natural gas sector.
1988 to 2003	Liberalization of energy networks, deregulation, structural reform. Light-handed regulation, electricity market self-regulation.
2003 to 2010	First steps in formal regulation of monopoly energy networks; targeted price and quality control. State regulation of electricity industry; co-regulation in natural gas.
2010–	Second stage of reform of regulation of energy networks. Revision of electricity industry regulation.

In being influenced by neoliberalism in economic law and policy, and then by a movement away from it, New Zealand is in the mainstream of thinking in developed countries. However, it embraced neoliberalism more completely than many countries, and for many years it left monopoly energy networks unregulated as to prices, third party access, and the like; the presumption that price control was unnecessary lasted, as we have seen, until 2008. The second phase in establishing formal regulatory control of networks corrected the shortcomings of the first phase, and recognized the inefficacy of light-handed regulation. There has therefore been

[65] International Energy Agency (2011, above n 60) 113.

a shift of emphasis from the possibility of competition to the control of monopolies, and a shift towards the coordination of infrastructure investment and development. There is a logic and coherence in the evolution that has occurred, even if it has not been planned.[66] Indeed, a number of countries have witnessed similar changes, impelled first by deregulation and self-regulation, then shifting to a more balanced combination of regulation and market forces.[67]

These developments make New Zealand's regulation of energy networks look more like those of other countries. As to the regulation of the tariffs and quality offered by monopoly providers, the law will probably be stronger now for starting with an acceptance of the fact of monopoly in electricity and natural gas networks, although litigation about its application will no doubt echo the neoliberal cast of mind for some time. This new starting point, along with the possibility of customized price–quality paths, should facilitate investment in new network projects. As to the regulation of industry activity apart from price control, the Electricity Authority has a clearly focused jurisdiction on transmission, distribution, and market matters. Third party access and related matters such as interconnection and use of system agreements are all firmly within the Authority's control. Less clear is the division of responsibility between the Authority and the Commerce Commission for electricity transmission system projects; but at least a framework is in place for the coordination and management of system expansion.[68] As to natural gas, there is real doubt whether the industry's co-regulation is adequate for the coordination and management of new transmission pipeline investments, and it is unclear whether co-regulation can make third party access work as efficiently as it should. The project of reform is not yet over.

[66] It is therefore difficult to agree with J. Every-Palmer, 'The State and Monopolies: New Zealand's Experience' (2010), 12 *Otago L Rev* 227, that the changes have been ad hoc; but it is easier to agree with him about the importance of the changes in the second stage that are not yet fully appreciated: 'a much quieter revolution' p 234.

[67] Barton et al (2006 above n 7) collect and appraise studies that demonstrate this evolution.

[68] The International Energy Agency (2011, above n 60) 107, 112–14, saw the country's electricity regulation as somewhat complex, and encouraged measures to streamline and simplify the arrangements. It approved of the continuous efforts to maintain arms-length regulation, but warned that sometimes it can lead to sub-optimal results. It did not see the logic of moving parts of electricity industry regulation to the Commerce Commission: it recommended '[i]ntegrate all regulatory tasks into one Electricity Authority and let the Commerce Commission focus solely on competition issues in the energy sector'.

16

Electricity Network Development: New Challenges for Australia

Lee Godden and Anne Kallies[*]

I. Introduction

The law and regulation of energy networks in the electricity industry in Australia has undergone substantial change over the past two decades. After energy market reform, the regulatory model governing electricity transmission and distribution is now largely premised on assumptions of competition theory and economic efficiency. Similar reforms are evident in other jurisdictions, as discussed in this volume. A national energy regulator determines performance and service levels for the electricity industry, including setting standards for network infrastructure development and replacement. However, changing community needs for energy outcomes, including the necessity of responding to climate change impacts, have revealed the difficulties of successfully incorporating public interest requirements into this 'least-cost' efficiency-based regulatory model. In this light, the chapter examines two contemporary inter-related challenges for the governance of electricity network development in Australia. First, the chapter examines how the expected influx of new, renewable generation will place pressures upon the highly centralized, existing electricity networks.[1] This pressure reveals that the existing network configuration does not readily allow for electricity sourced from renewable energy. Second, the chapter explores the need to replace bushfire-prone network distribution infrastructure in light of heightened bushfire risk.

Within this general framing of contemporary challenges for law and regulation of transmission and distribution networks in Australia, the chapter outlines the electricity law reforms in Australia, before providing a detailed analysis of the current electricity market regulatory models. By reference to this model, the chapter

* The authors acknowledge the excellent research assistance of Ms Lisa Caripis and the support provided by Australian Research Council Grant DP0987850 Responding to Climate Change: Australia's Environmental Law and Regulatory Framework. The law is stated as at October 2011. There is ongoing litigation in respect of fire-related damage, as well as prospective regulatory reform in the transmission context.

[1] Similar issues are covered in other chapters of this book. See, for example, chapter 20, where the situation in Germany is analysed.

explores its implications for electricity network infrastructure provision, suggesting that the regulations create particular 'path dependencies' that are difficult to realign around public interest objectives. The chapter concludes that there are particular barriers to the achievement of future environmental and safety objectives for transmission and distribution networks under the existing regulatory model that will hamper structural change in the energy networks.

II. Background for Australia

A. The characteristics of electricity networks in Australia

Australia, in concert with comparable jurisdictions in Europe and North America, has introduced major changes to the legal and policy frameworks that regulate its energy sector. Since the 1980s, a twofold reform process has emerged in utility services like electricity and gas. The process basically comprises a physical and legal 'restructuring' to separate the components of the utility sector where competition rules operate from the 'natural monopoly' elements (transmission, distribution). There has also been a contemporaneous transfer of assets and service provision activities from public to private ownership.[2] In certain instances, there was rapid privatization. Overlapping with this sequence within Australian utility sectors have been institutional arrangements to regulate the electricity markets and associated networks on a national basis. These laws apply in the eastern states of New South Wales, Queensland, South Australia, Tasmania, and Victoria. Western Australia operates the same arrangements but with an internal state regulator. The Northern Territory, with several small grids, has its own institutional arrangements outside the national framework.

Turning to the specific electricity network arrangements, in Australia, as elsewhere, electricity networks are composed of transmission networks and distribution networks. Differentiated regimes can apply to the transmission and distribution components. While transmission networks are the high-voltage lines that carry electricity long distances and provide cross-border connection between the Australian states participating in the National Electricity Market (NEM),[3] it is the low-voltage distribution lines that carry electricity to the final consumers.[4] Australia's electricity networks are principally operated by corporate entities.[5] While generation and retail are competitive markets, networks generally remain regulated monopolies due to the natural monopoly nature of electricity networks. It is argued that given the corporatized nature of network providers in Australia, in the absence of regulation, company behaviour is likely to lead to over-pricing and

[2] J. Quiggin, 'Market-oriented Reform in the Australian Electricity Industry' (2001), 12(1) *Economic and Labour Relations Review* 126 at 136–8.

[3] Australian Energy Regulator (AER), *State of the Energy Market 2010* 47.

[4] Ibid.

[5] Currently all of Victoria's transmission and distribution networks are privatized; the South Australian transmission networks are privatized and the distribution network is privately leased. The networks in Queensland, New South Wales, and Tasmania are government-owned: ibid.

under-providing, because network providers do not have competitive market pressures to keep their prices at efficient levels.[6]

The restructuring of the Australian electricity network industries to derive a separation of activities has thus given rise to a complex regulatory arrangement that differs in important aspects from the older models of vertically integrated generation and network operations. Earlier public enterprise models, such as State Electricity Commissions, were modelled on UK institutions.[7] Network infrastructure during this earlier period was driven by public service provision objectives and 'nation building' paradigms. Typically, the network systems in this era were developed around large, centralized generation facilities with limited decentralization. One advantage of this period though was attention to public safety requirements.[8] In Australia, however, significant problems of bushfire risk were posed by electricity networks even in the earlier period,[9] and these have continued.[10] There have been further legal and policy reforms over recent decades to introduce major structural changes to energy utility sectors to drive 'efficient and effective' provision of electricity generation in Australia. Yet the challenges of ensuring competitive access for renewable energy to networks and the replacement of ageing or ineffective infrastructure for transmission and distribution networks have only recently received explicit attention.[11]

B. Drivers for transformation in the electricity sector

Australia now faces regulatory and technological transformations in its energy sector that will impact network scope, configuration, access, and the rapidity of infrastructure renewal. Clearly, climate change is a prominent driver of change:

> Climate change policy will have profound impacts on the composition of future electricity supply. Putting a price on carbon as well as other policies...will change the underlying dynamics of generation...This is likely to result in changes in dispatch, generation, location, exit and entry decisions, and affect the prevailing network flows.[12]

Overwhelmingly, Australia produces electricity from brown and black coal,[13] with stationary energy production accounting for 53.9 per cent of all of Australia's CO_2

[6] Australian Energy Market Commission (AEMC), 'Strategic Priorities for Energy Market Development' (Discussion paper, 2011) 48.

[7] S. King and R. Maddock, *Unlocking the Infrastructure: The Reform of Public Utilities in Australia* (St. Leonards: Allen & Unwin, 1996) chs 1 and 2.

[8] Ibid.

[9] See 'History of Electricity-caused Ignitions, Electricity Caused Fire' in Parliament of Victoria, *2009 Victorian Bushfires Royal Commission* (2010, vol 2) 148.

[10] S. Solomon et al (eds) *Climate Change 2007: Physical Science Basis. Contribution of Working Group I to the Fourth Assessment Report of the Intergovernmental Panel on Climate Change* (Cambridge: Cambridge University Press, 2007).

[11] R. Garnaut, 'Transforming the Electricity Sector' in *Garnaut Climate Review Update 2011* (Update Paper 8) 39–46.

[12] Ibid at 29.

[13] Australian Bureau of Agricultural and Resource Economics and Sciences (ABARES), *Energy in Australia 2011* (2011) 21.

emissions in 2009.[14] Stationary energy generation is the largest source of emissions, but it has proven a very contentious sector to regulate. Ninety per cent of Australia's total emissions increase since 1990 has resulted from energy combustion,[15] and this trend is forecast to continue.[16]

On the other hand, Australia has an unconditional pledge to reduce greenhouse gas emissions by 5 per cent on 2000 levels by 2020, with additional reductions of up to 15 per cent and 25 per cent possible, depending upon action taken by other nations.[17] The pledge has recently been raised to an 80 per cent reduction on 2000 emissions levels by 2050.[18] The Australian federal government has announced a 'Clean Energy Package' and from July 2012, Australia will introduce a price on carbon to reduce domestic emissions. The carbon pricing mechanism has two phases: the first is a fixed-price period of three years, with a second phase starting in 2015 which will 'mature' into a full emissions trading scheme.[19] The revenue earned from the scheme will cover compensation for households and impacted industries, including the power generation sector, where facilities that have scope 1 emissions of 25,000 tons of carbon dioxide equivalent (CO_2-e) a year or more, will fall under the scheme.[20] This policy has been introduced amid significant concessions for stationary power generation, underscoring the difficulties that governments face in regulating a largely privatized power sector.

Along with carbon pricing, the other major plank of climate change policy is a commitment to ensure that 20 per cent of energy is generated from renewable sources by 2020. Latest data shows that Australia's renewable share in energy consumption is currently only five per cent, providing about seven per cent of electricity generation.[21] Substantial new renewable power generation capacity will be needed to meet even the relatively modest 2020 target. The Clean Energy Package contains significant financial incentives to stimulate investment in renewable energy power generation in Australia.

Most increases in renewable energy generation will be achieved by intermittent generation, sourced from wind and solar, much of it sourced from locations remote to the grid.[22] To successfully integrate large amounts of intermittent, renewable generation into the electricity grid, highly interconnected, meshed, and

[14] Department of Climate Change and Energy Efficiency, *National Greenhouse Gas Inventory* (5 May 2011) 9.

[15] pitt&sherry, 'pitt & sherry Carbon Emissions Index CEDEX™' (February 2011), available at <http://www.pittsh.com.au/documents/201103_ps_cedex.pdf > (accessed 2 October 2011) 3.

[16] Department of Climate Change and Energy Efficiency, 'Stationary Energy—Australia's emissions projections 2010–12, available at <http://www.climatechange.gov.au/-/media/publications/projections/australias-emissions-projections-2010.pdf> (accessed 2 October 2011).

[17] Australia, Quantified Economy Wide Emissions Targets for 2020 (2010), available at <http://unfccc.int/files/meetings/application/pdf/australiacphaccord_app1.pdf> (accessed 2 October 2011).

[18] Australian Government, *Securing a Clean Energy Future: The Australian Government's Plan* (2011) 15.

[19] Ibid at 103.

[20] Ibid at 105.

[21] Ibid at 31.

[22] See Australian Energy Market Operator (AEMO) '2010 National Transmission Network Development Plan' (15 December 2010) (NTDP), 10, available at <http://www.aemo.com.au/planning/ntndp.html> (accessed 2 October 2011).

modernized grids are necessary.[23] Australia's existing grid is unique compared to most developed countries, in that it is long and linear, stretching 5000km along the eastern coast of Australia,[24] thereby providing power to six jurisdictions and most of Australia's population. This makes it the largest alternating current system in the world.[25]

Additionally, the Australian grid is characterized by limited cross-border connection capacity and little cross-connection within the Australian states.[26] The large distances between population centres makes Australia's grid expensive to run and network charges comprise a high percentage of the final consumer electricity price compared to many developed countries.[27]

Concurrent with climate change mitigation initiatives, much of Australia's electricity transmission and distribution network is reaching the end of its lifespan. For example, much of Victoria's network infrastructure will need replacing within the next 10 to 15 years.[28] The general safety risks associated with an ageing electricity transmission and distribution network are exacerbated by climate change. In southern Australia, bushfires have caused massive loss of life and property in recent years, and it is very likely that extreme fire weather will occur more often, with longer, more intense fire seasons under climate change conditions.[29]

Yet, increased investment in electricity networks to replace assets and to meet rising energy demands is a major contributor to rising retail prices.[30] With the introduction of a carbon price set to increase retail electricity costs further, this situation will limit the willingness of the industry sector to undertake any 'additional' costs for network provision and of government to compel that expenditure. Litigation has been used strategically to highlight inadequate law-making activity and to prompt wider policy change.[31] As yet, emerging awareness of public safety concerns have not substantially altered the underpinning assumptions upon which the electricity market is regulated and the trends in network development in Australia. The following sections trace the policy and legal developments in electricity regulation that produced the current model for electricity network provision in Australia.

[23] Melbourne Energy Institute, *Zero Carbon Australia 2020 Stationary Energy Plan* (Melbourne: University of Melbourne, 2010) 89–93.

[24] Ibid at 89.

[25] Ibid.

[26] Ibid. As two-thirds of the estimated 22 million inhabitants live in capital cities, the grid is mostly built from a generation site to major urban centres.

[27] Australian network charges amount to between 38 per cent and 52 per cent of the consumer retail bill (NEM), see Roam Consulting, 'Impact of Renewable Energy and Carbon Pricing Policies on Retail Electricity Pricing', Report to Clean Energy Council, 11 March 2011, 13–14. Network charges in Germany, for example, amount to only 24 per cent of the electricity bill: Federal Ministry of Economics and Technology, 'Electricity', available at <http://www.energie-verstehen.de/Energieportal/Navigation/Energiepreise/strom,did=309424.html?view=renderPrint> (accessed 2 October 2011).

[28] See above n 9, 151–3.

[29] Commonwealth Scientific and Industrial Research Organisation (CSIRO), *Climate Change in Australia: Executive Summary* (Canberra 2005) 11.

[30] Above n 14, at 11–12 and 38.

[31] See J. Peel, 'The Role of Climate Change Litigation in Australia's Response to Global Warming' (2007), 24 *Environmental and Planning Law J* 90–105.

III. How is Electricity Network Development Regulated in Australia?

A. Regulating electricity networks within the national electricity market

Australia's transmission and distribution network development and strategic planning issues are regulated under the National Electricity Rules (NER). These regulations are enabled by the National Electricity Law, governing the National Electricity Market (NEM). This regulatory structure was established through the National Electricity (South Australia) Act 1996, which is valid in all jurisdictions covered by the NEM through enabling legislation.[32] Electricity in all Australian states apart from Western Australia and the Northern Territory is traded through the NEM, which is a wholesale market for electricity supply to retailers and consumers and commenced operations in 1998.

Investment in electricity networks is required to develop new network infrastructure to accommodate new and remote generators; and to augment the network in order to replace ageing or insufficiently sized infrastructure to avoid network congestion. To secure network investment in a competitive market, regulators face the challenge of providing the right incentives to the regulated network service providers to ensure the providers both deliver reliable and future-oriented network services while at the same time minimizing costs. These aspects have a critical influence on how investment occurs in the sector, and whether, if at all, public interest goals such as increased renewable generation access and reduced bushfire risk are provided for.

Apart from the electricity market participants (generators, retailers, and network service providers), several institutions have a central role in the electricity market regulatory framework. The main policy maker is the Ministerial Council on Energy (MCE) within the Council of Australian Governments (COAG), consisting of all energy ministers in States and Territories and the Federal minister (since July 2011 the MCE is part of the new COAG Standing Committee on Energy and Resources). The Australian Energy Market Agreement (AEMA) sets out the main function of the MCE: 'to provide national oversight and coordination of policy development'.[33] This includes responsibility for the legislative and regulatory framework of the NEM.[34]

The Australian Energy Market Commission (AEMC) is the statutory authority, 'responsible for rule making and energy market development on a national level...including the National Electricity Rules'.[35] Market enforcement is the responsibility of the Australian Energy Regulator (AER), a body established as part of the Australian Competition and Consumer Commission under Part IIIAA

[32] Electricity (National Scheme) Act 1997 (ACT) s 5, Electricity—National Scheme (Tasmania) Act 1999 TAS s 6; Electricity—National Scheme (Queensland) Act 1997 Qld s 6; National Electricity (Victoria) Act 2005 Vic s 6; National Electricity (New South Wales) Act 1997 NSW s 6.

[33] Australian Energy Market Agreement (as amended) between the Commonwealth of Australia, the State of New South Wales, the State of Victoria, the State of Queensland, the State of Western Australia, the State of South Australia, the State of Tasmania, the Northern Territory of Australia, and the Australian Capital Territory (2 July 2009) ('AEMA') recitals A.

[34] Ibid 4.3.d.

[35] Ibid 5(1)(a).

of the Competition and Consumer Act 2010. The AER is responsible for the economic regulation of the wholesale electricity market and transmission networks and the enforcement of the National Electricity Law and the NER. This includes determination of revenues that the transmission and distribution network service providers receive for infrastructure augmentation and extension. These determinations play a crucial role in the allocation of funding and the income to be derived by network transmission and distribution businesses. Moreover, the regulator cannot compel investment in infrastructure not otherwise funded under the determination. Again, this situation of enforcement relying on price determination has specific ramifications for whether investment occurs in areas not seen as economically efficient.

Finally, the Australian Electricity Market Operator (AEMO), a corporate entity, is responsible for the operation of the market as well as recently resuming a national planning function, which includes preparing, maintaining, and publishing a National Transmission Network Development Plan.[36] The first plan was published in December 2010.[37]

B. Genesis of the current regime and the national electricity objective

The current market regime is the result of far-reaching microeconomic reforms since the 1990s. Reforms followed a worldwide paradigm change in the nature of electricity provision. Electricity was no longer considered to be best delivered by integrated state monopolies, but by a competitive market-based framework to deliver energy more efficiently.[38] These reforms, variously called energy market liberalization, deregulation, or restructuring, had economic efficiency as their primary objective.[39] Reforms to electricity sectors during this period shared several basic features: 'unbundling' (the factual separation of the functions of generation, transmission, distribution and retail); open access to networks and infrastructure; opening the non-monopoly components of generation and retail to full competition; and the introduction of independent regulators.[40]

The drive to deregulate the economy was guided by two underlying principles which continue to underpin the design of today's electricity market: that of 'least cost' and the 'unfettered market'. According to the latter, the market cannot accommodate non-economic objectives and its ability to function effectively and achieve the efficient allocation of resources is compromised by intervention attempting to require it to do so. Any objectives contrary to competitive market outcomes should therefore be excluded. These principles were implemented in structural change measures in the Australian economy designed to minimize state

[36] See s 49(2) of the National Electricity Law.

[37] Australian Energy Market Operator (AEMO), '2010 National Transmission Network Development Plan' (15 December 2010) ii.

[38] See, eg, P. Cameron, 'Reforming Energy Markets' (2000), 18 *Journal of Energy and Natural Resources Law* 353.

[39] International Energy Agency (IEA), *Power Generation Investment in Electricity Markets* (2003) 21.

[40] Cameron, above n 39 at 357; IEA above n 39 at 21.

intervention in the industrial and commercial sectors.[41] In Australia, the impetus for reform which led to the restructuring of the electricity market began with several influential reports in the early 1990s, culminating in the *Hilmer Review*.[42]

The electricity sector was initially run by state monopolies for several reasons: 'the significance of scale economies; the inability of the private sector to finance the large investments required; government concerns about the exploitation of market power; and government decisions to use electricity pricing to pursue social and development objectives'.[43] The monopoly control over all aspects of electricity provision allowed uniform customer charges independent of location and costs to provide electricity as an essential service, amounting to a cross-subsidization of remote, rural customers. Provision of a relatively low-cost wire distribution system to rural areas under this cross-subsidization is now of concern due to significant fire-safety implications.

In 1991, an Industry Commission Report recommended that the sole objective of the operation of the utility sector should be commercial performance.[44] This recommendation came when performance in the electricity sector was marked by inefficiency. The requirement that utilities undertake non-commercial functions was regarded as compromising accountability. Further, it gave rise to conflicts of interest where the utilities regulated activities in which they were involved, making it difficult to monitor their performance. A single, simply-stated commercial objective was regarded as preferable to multiple, ill-defined objectives. Such reforms would improve accountability and performance monitoring.[45] The report concluded that each utility should be driven by the imperative to supply electricity or gas (but not both) in the most economically efficient manner. The influence of the principles of 'least cost' and the 'unfettered market' driving the microeconomic reform is clear.

Consequently, in July 1991, Australian leaders agreed to establish a National Grid Management Council (NGMC). The NGMC was the first of several independent bodies to drive regulatory reform in the electricity sector, and began developing rules to create the NEM.[46] The publication in 1993 of the *National Competition Policy Review*, or *Hilmer Review* accelerated reforms and greatly influenced the formation of the NEM.[47] The *Hilmer Review*[48] was broad in scope, making recommendations regarding anti-competitive behaviour, unjustifiable regulatory restrictions on competition, restraints on monopoly pricing behaviour,

[41] Quiggin, above n 2 at 7; in regard to microeconomic reforms, see J. Quiggin, *Great Expectations: Microeconomic Reform and Australia* (Sydney: Allen and Unwin, 1996); Productivity Commission, 'Microeconomic Reforms and Australian Productivity: Exploring the Links; (Commission Research Paper, 1999, vol 1).

[42] Fred Hilmer (Chairman), *National Competition Policy Review: Report* (1993).

[43] M. Abbott, 'The Performance of an Electricity Utility: The Case of the State Electricity Commission of Victoria, 1925–93' (2006), 46(1) *Australian Economic History Review* 23 at 25.

[44] Industry Commission, *Inquiry Report: Energy Generation and Distribution* (Report No. 11, vol 2, 17 May 1991) 4.

[45] Above n 3 at 4.

[46] Special Premiers' Conference, Communiqué, Sydney, 30 July 1991.

[47] A Kallies, 'The Impact of Electricity Market Design on Access to the Grid and Transmission Planning for Renewable Energy in Australia: Can Overseas Examples Provide Guidance?' (2011), 2 *Journal of Renewable Energy Law and Policy* 147.

[48] Above, n 42 v.

ways to foster 'competitive neutrality' between government and private businesses and, significantly for this analysis, the scope for reforming the structure of public monopolies to facilitate competition.[49] Economic efficiency, the *Hilmer Review* made clear, was to be viewed 'as the primary element of public benefit'.[50] The Review led to far-reaching reforms to the general regulatory environment influencing the structure and objectives of the newly created NEM to create an electricity market driven foremost by efficiency concerns.[51] Nowhere is this more evident that in the current National Electricity Objective: 'The objective of this Law is to promote efficient investment in, and efficient operation and use of, electricity services for the long term interests of consumers of electricity. . . '[52]

C. How does electricity network investment work in practice?

The dominance of economic efficiency objectives influences network infrastructure development in Australia. There are two main aspects to the regulatory process for electricity network investment. First, network providers are required to submit a proposal forecasting their capital expenditure for a regulatory period of five years to the AER and they must obtain approval for the investment proposal. In effect, this gives network providers a 'pool of funds' for capital expenditure for the network. Second, each individual investment project proposal for network infrastructure must also pass a regulatory investment test based upon economic efficiency criteria.[53]

The NER seeks to ensure that network service providers do not charge monopoly rents by using a method which employs a 'building block' approach for the determination of network provider revenue.[54] This determination is based on a mixture of an assessment of the asset base of the service provider and a capital expenditure test.[55] The network provider also submits revenue proposals to the AER, which then determines the maximum allowed revenue for the regulatory period.[56] If the network providers manage to deliver the relevant service standards for less than this revenue cap, they can capture cost savings made in the regulatory period.

[49] Ibid at xvii.

[50] Parliamentary Research Service, *National Competition Policy: Overview and Assessment* (Research Paper No. 1, 21 January 1994) iv.

[51] See R, Cantley-Smith, 'A Changing Legal Environment for the National Electricity Market' in W. Gumley and T.D. Winterbottom (eds), *Climate Change Law: Comparative, Contractual and Regulatory Considerations* (Sydney: Thomson Reuters, 2009) 15; A. Bradbrook and R. Lyster, *Energy Law and the Environment* (Melbourne: Cambridge University Press, 2006) 114; C. Hamilton and R. Denniss, 'Generating Emissions? The Impact of Microeconomic Reform on the Electricity Industry' (2001), 20(3) *Economic Papers: A Journal of Applied Economics and Policy* 15.

[52] *National Electricity Law* s 7.

[53] See above, n 3, 52.

[54] National Energy Rules (NER) 6A.5.4 for transmission network providers, and NER 6.4.3 for distribution network providers.

[55] In detail, the test includes a calculation comprising: existing assets, forecasted operating expenditure to maintain the asset base, a market-based rate of return on capital and depreciation and taxation expenses.

[56] NER 6A.3 for transmission network providers, See also the rules for distribution network providers in 6.1, 6.2, especially part E rules around a building block approach.

All new investment decisions for network infrastructure that exceed a capital expenditure of AUD$1 million (including replacement/augmentation of ageing infrastructure), are subjected to a regulatory test set by the AER.[57] Its purpose is to identify new network investments or non-network alternative options that:

(1) maximize the net economic benefit to all those who produce, consume and transport electricity in the market; or

(2) in the event the option is necessitated to meet the service standards linked to the technical requirements... minimize the present value of the costs of meeting those requirements.[58]

These two elements of the test are called the reliability and the market benefits limb, respectively. In both cases a cost–benefit analysis is required. The rules expressly state that it is the 'net economic benefit'[59] that must be maximized. Importantly, investment in transmission networks is subject to the Regulatory Investment Test for Transmission (RIT-T),[60] and investment in distribution networks, to the Regulatory Test, version 3.[61] While the enhanced RIT-T now applies similarly to investments that are reliability-driven and to deliver market benefits, the test applying to distribution networks distinguishes between market benefit investments and those based on network reliability standards. The consequence of applying the test is that any proposed investment must achieve its objectives at least cost.[62] This has ramifications for the replacement of infrastructure on the basis of safety or environmental considerations. The national regulations demand a complex test to meet market benefits which must:

(1) be based on a cost-benefit analysis of the future (which includes assessment of reasonable scenarios of future supply and demand conditions):

 (i) were the *new network investment* to take place, compared to the likely alternative option or options,

 (ii) were the *new network investment* not to take place.[63]

While the test is highly technical, it imports several assumptions as to what constitutes 'reasonable scenarios' as alternatives. Where a particular investment 'option' is mandated to meet the service standards linked to technical requirements, or in *applicable regulatory instruments*, it must minimize the present value of the costs of meeting those requirements.[64] In essence, the tests adopt least-cost efficiency criteria for determining how present networks should be upgraded against a range of options to achieve a projected future outcome for the infrastructure. Given discount rates in such tests, typically future values, such as meeting renewable energy

[57] NER 5.6.5A. [58] NER 5.6.5A(b). [59] NER 5.6.5A(b)(1).

[60] AER, *Regulatory Investment Test for Transmission*, 2010 (in effect since 1 August 2010).

[61] AER, *Regulatory Test, Version 3*, 2007; the distribution network providers are required to apply the test according to clause 5.6.2(g) NER.

[62] Note a new RIT-D is in discussion—for the current test, see AEMC, *Review of National Framework for Electricity Distribution Network Planning and Expansion, Final Report*, 2009.

[63] NER 5.6. A. [64] 5.6.5A(b)(2) NER.

targets, will have reduced value alongside present-day needs. Further, the reliability test for new investment in distribution networks covers all investments 'necessitated to meet service standards'.[65] While safety standards are separate to reliability standards, and administered under a separate state legal regime, safety standards can, nevertheless, form part of a capital expenditure proposal.

The reliability standards remain regionally specific. Thus the applicable standards are mandated by state legislation, although currently a national approach for transmission standards is being considered, with a new national transmission planning role for AEMO.[66] Regulatory tests undertaken so far have been based on reliability standards. Tests based solely on the market benefit limb have not been undertaken. Thus while a least-cost paradigm is pervasive, the tests for investment in network infrastructure must still take into account how networks will meet given service provision standards.

D. Transmission planning in the regulatory framework

Closely connected to new infrastructure investment in electricity networks is strategic transmission planning, which remains a difficult policy dilemma. How can future investment be financed without the guarantee that the matching generation capacity actually will be built? Strategic planning in transmission investment is now the task of the AEMO. While nationalization of transmission planning is a welcome move,[67] the AEMO's plans are not binding upon network providers.[68] Undoubtedly, the existence of a national transmission plan will allow the Australian energy regulation authority to take a more national approach when approving individual investment proposals. However, the regulator will still rely on the initiative of a network to make proposals that reflect the public interest, rather than having specific powers to independently set such objectives with which providers must comply. As reconfirmed by the federal minister, '[t]he establishment of the regulatory regime for the network sector is deliberately at arms' length to government in order to maintain investor confidence'.[69]

The regulatory model, with its highly technical rules governing the electricity market, the presence of a third party regulator, and the function of price determinations in investment decisions, strongly influences the future development of electricity transmission and distribution networks in Australia. Such a model has broad similarities with regulatory models in liberalized energy markets in many

[65] Note 5.6.5 A (b) (20 as above, 'linked to the technical requirements of schedule 5.1 or in applicable regulatory instruments').

[66] AEMC, *Transmission Reliability Standards Review*, Updated Final Report (2010), 3; ACIL Tasman, *Energy Transmission Network Planning: The Emerging Role of the Australian Energy Market Operator* (2010).

[67] Interconnections between states reduce congestion and take advantage of the fluctuating nature of renewable generation.

[68] This is different for Victoria, where AEMO is also the transmission planner.

[69] Federal Minister for Resources and Energy, Strategic Priorities for Energy Market Development Speech delivered on 1 April 2011, available at <http://minister.ret.gov.au/MediaCentre/Speeches/Pages/StrategicPrioritiesforEnergyMarketDevelopment.aspx> (accessed 2 October 2011).

comparable jurisdictions.[70] However, to ground an understanding of this legal model and its challenges in the specific Australian context, the chapter considers two examples of how contemporary public interest objectives—increasing renewable energy generation and replacement of infrastructure to reduce bushfire risks—face difficulties under the current model.

IV. Challenges in Integrating Public Interest Objectives into Network Regulation

Public interest benefits may be compromised by the current electricity and network regulation, where frameworks centre on a pre-eminent market efficiency objective. Indeed, there are different standpoints on whether the regulation of electricity networks achieves policy goals. The relevant federal minister described the regulatory framework as 'leading edge' and affirmed the commitment to microeconomic reform.[71] Alternatively, an inability to integrate 'non-market' social and environmental objectives successfully into the legal framework demonstrates some inherent limits of deregulatory reform. The replacement of ageing infrastructure is an endemic problem of electricity utility provision in most jurisdictions across the world, but it is given particular pressure under the increased fire risk in southern Australia. Similar climate change imperatives drive the need for greater access to the grid by renewable energy power generation, particularly wind and solar power.

A. Renewable energy and electricity networks

The regulation of electricity transmission and distribution can present problems for renewable energy generators. Generators, depending on the size of the installation, can connect to either a distribution or transmission network.[72] Generators face several hurdles that derive from the introduction of a market efficiency regulatory framework, well before the rise of renewable technologies. This regulatory model instituted particular 'path dependencies'[73] for infrastructure development that now pose barriers where there are strong policy drivers for incorporating renewable energy sources. While the national regulations for transmission and distribution that were outlined above equally apply to all new generation facilities, they specifically disadvantage renewable energy generators in a practical sense. The viability of renewable generators, even with the projected carbon price, is very much dependent on grid access and the extension of the existing network.

[70] See generally, L. Philipson and W. H. Lee, 'Regulation and De-Regulation' in *Understanding Electric Utilities and De-Regulation* (2nd edn, Boca Raton: CRC Press, 2006) vol 10, 261.

[71] Above, n 69.

[72] Large-sized generators usually connect to the transmission network, whereas smaller-sized embedded generation is connected to the distribution network. See Parliament of Victoria, *Inquiry into the Approvals Process for Renewable Energy Projects in Victoria* (2009) ch 9.

[73] For discussion of path dependencies in the resource and energy fields, see T. Daintith, *Finders Keepers: How the Law of Capture Shaped the World Oil Industry* (New York: Earthscan 2010) ch 1.

Renewable energy facilities, for the most part are dependent on their source location for generation; typically, at some distance from existing network infrastructure. Barriers for renewable generators occur on several levels of network grid access. These include direct network access connection, as well as infrastructure augmentation and extension; and crucially which entities will be required to pay for these developments.

1. *Grid access*

Currently, network access is a costly process for a new generator in Australia. New connection rules are governed by chapter 5 of the NER. The NEM relies on an open-access regime, but the cost of grid access permits can be forbidding, especially for smaller-scale generators. A recent inquiry found that the network access process is characterized by a serious information imbalance between generator and the monopoly network provider. Further, the required grid connection study, financed by the generator, can be very costly.[74] These factors disadvantage small and medium-sized generators.

Additionally, Australia's unique grid layout, which reflects substantial public investments made in an earlier era of fossil fuel dependency, means many high-grade renewable resources are remote to the electricity grid. Thus grid connection costs can become the determining factor for new network investment, rather than an assessment of the best energy resource for power generation. While it is important to heed location factors in commercial 'least-cost' decision-making, the public interest in instituting a low-carbon economy should also be taken into account. Therefore, to realize the potential of remote electricity production sources, modifications to the regulatory model are needed to reduce practical disadvantages for renewable energy generation.

2. *Grid augmentation and extension*

A related issue is electricity grid augmentation and capacity extension to allow for more renewable energy. In Australian grid regulation, the need to balance reliability with cost considerations leaves little space for the 'greening' of the grid. Expanding and strengthening the network to facilitate renewable energy has proven difficult for a regulatory framework which emphasizes market competition, and where a transmission and distribution network was instituted before renewables featured in the power generation mix. Physical infrastructure framework was designed around centralized and constant generation facilities, with new conventional fossil fuel generation coming onto the grid in the vicinity of existing coal mines and accompanying grid infrastructure. This leads to a dilemma for decision makers as:

> Building transmission infrastructure is capital intensive, complex and time consuming and tends to be a stumbling block for renewable energy project developers. Conversely if transmission infrastructure building costs were to be recovered by the [network service

[74] See above, n 71 at 226–8.

provider] from all end-users, this could distort the operation of the market by encouraging generation connections in areas that are inefficient or more expensive than elsewhere.[75]

Typically now, renewable energy generators simply connect to the closest suitable grid, as there is no incentive to group together to capture economies of scale by building larger, closer-connection infrastructure. Moreover, the problem of stranding expensive transmission network assets when extending transmission lines into areas of excellent renewable resources has been recognized.[76] Regulatory rule change has been intitiated, including a regulatory framework for Scale Efficient Network Extension (SENE). Initially, one possibility discussed, was upfront partial funding of grid extensions by network providers, to then be borne by all network customers.[77] This model would have addressed the danger of stranding assets, and the 'first movers' disadvantage for generators. In addition, it would have facilitated the capture of economies of scale in transmission investment. After extensive stakeholder and community consultation, the resulting rule departed significantly from initial options by expressly allocating risks and costs of a network extension to the market participants, not consumers.[78] The new rule requires transmission businesses to undertake and publish on request specific locational studies to allow market participants to 'make informed, commercial decisions to fund a SENE, having weighed the potential gains from coordinated, efficient generator connection arrangements against the potential costs of assets not being fully used'.[79]

The final decision to fund, construct, operate and connect to a SENE is made within the existing framework of rules. The new rule addresses some information imbalances and it may reduce risks of stranded infrastructure assets. However, the risks for the first investor to invest in scale-efficient network extensions that may not be recouped from later generators using the same extension remains high. With such inherent disincentives still at play, the likelihood that the rule will overcome barriers for renewable generators seeking 'competitive' grid augmentation and extension is uncertain.

3. *Transmission planning for renewable energy*

Transmission planning is crucial for introducing more renewable power generation. Most commercially viable, mature renewable generation in Australia (largely wind and solar) is characterized by the intermittent nature of the generation capacity. Intermittent generation capacity requires robust transmission networks if extensive new conventional generation 'backup' capacity, sourced in gas or coal, is to be avoided. The Transmission Network Development Plan sets out models for

[75] Above, n 72 at 227. Indeed, in some European countries, a recent EWEA study found that connection of unsuitable generators 'clogs' the grid for more efficient generators; see European Wind Energy Association, *WindBarriers: Administrative and Grid Access Barriers to Wind Power* (July 2010).

[76] Australian Energy Market Commission Final Report: *Review of Energy Market Frameworks in Light of Climate Change Policies* (2009) ch 2, especially 2.4.

[77] Ibid at 17.

[78] Australian Energy Market Commission, Scale Efficient Network Extensions, Final Rule Determination, 30 June 2011. The new rule only considers augmentations of existing lines as SENEs, extensions into new resource areas are not covered by the framework.

[79] Ibid.

several potential future-generation scenarios. In the plan, the AEMO gives guidance regarding investment decisions for new transmission infrastructure developments by providing extensive modelling of future-generation potential. Yet the final planning and investment decisions remain with network providers. Lack of regulatory power to compel certain forms of investment militates against renewable generation gaining grid access.

Unless renewable energy generators have strong incentives, as assumed in AEMO modelling scenarios, little investment targeted to renewables may be forthcoming. The main value of the plan may lie in its potential to provide for 'credible scenarios' against which the regulatory investment test for infrastructure price determinations can be assessed. There is no independent 'triggering' to instigate network extension. Additionally, successful planning for increased transmission interconnection between Australian jurisdictions, to better absorb a higher percentage of renewable energy,[80] is not adequately provided for in regulation. Network providers operate within state borders, and have little incentive to fund interconnections across jurisdictions. Cross-jurisdictional interconnection could assist remote renewable energy generators by increasing potential market share.

Finally, the forecast undertaken by AEMO for transmission planning limits all the possible investment scenarios to a 20 per cent renewable energy target instituted by the current federal government.[81] It is anticipated that this legislatively mandated target will be achieved mostly by wind generation, accompanied by gas-fired back-up generation. All modelling for grid planning and energy mixes are undertaken on this assumption. Ultimately, this assumption may lead to strategic network planning that cannot react to more than 20 per cent renewable energy in the power generation post-2020 (which could occur when driven by a carbon price). As such, a further 'lock in' to a particular generation scenario where renewable technologies remain a minor power generation source may be the result.

In summary, strategic national planning currently does not necessarily translate into infrastructure development according to any mandated plan for network development. The network providers, driven by efficiency and competition imperatives, retain the final power to make investment decisions. To date, this industry sector has few incentives to invest into renewable-friendly infrastructure under current legal models.

Renewable generators in Australia face serious barriers inherent in the electricity market and its regulation. The National Electricity Market hampers the development of renewable energy compatible transmission infrastructure development by limiting the extent to which environmental objectives can be considered in investment decisions. Within the regulatory framework, there is little support for renewable generation, as a central part of Australia's climate change mitigation effort or to stimulate investment in a low-carbon future. The market structure ultimately

[80] For example, Melbourne Energy Institute, *Zero Carbon Australia Stationary Energy Plan* 2010, Melbourne.

[81] The Renewable Energy (Electricity) Act 2000 (Cth) establishes the renewable energy target. As part of the federal government's Clean Energy Future Plan, it is planned that AEMO will have to expand its future planning scenarios to include 100 per cent renewable energy.

leads to a 'lock-in' or path dependency in existing generation and transmission patterns by disadvantaging renewable generators in respect of network connection, extension, and augmentation.

B. Bushfire risk in a climate change era

While facilitating strategic investment and planning beyond efficiency parameters are critical to greening of the grid in Australia, network regulation also influences whether the electricity network can reduce the risks of bushfire ignition into the future. Bushfire hazard reduction requires attention to existing network coverage and is concerned more with localized distribution facilities than renewable energy access to the transmission network. Nonetheless, any future planning for electricity network infrastructure which does not conjoin a consideration of both public interest factors will continue systemic inadequacies. Extension of network infrastructure to promote greater renewable generation access should ideally embed prominent public safety imperatives. However, the prospects for both criteria being prioritized under the immediate regulatory framework seems unlikely.

1. The nature of the bushfire risk

In 2005, Australian research institutes produced a report assessing fire-weather risk associated with climate change.[82] The study focused upon south-east Australia, an area predicted to be hotter and drier under climate change, and already one of the three most fire-prone areas in the world. A key finding was a heightened fire risk in all major metropolitan areas. Whatever the success of mitigation efforts, some degree of climate change is already 'locked in'. As such, measures to respond to the resulting environmental and social effects are imperative.[83] Within Australia, as in other countries, the momentum to implement effective adaptation measures is growing.[84] Adaptation in Australia will require increasing attention to the risks of bushfire ignition caused by faulty electricity network infrastructure.

The extent of the risk is graphically illustrated below:

> The bushfires of Black Saturday, 7 February 2009, caused the death of 173 people. Black Saturday wrote itself into Victoria's history with record-breaking weather conditions and bushfires of a scale and ferocity that tested human endurance…The recommendations we make give priority to protecting human life, and they are designed to reflect the shared responsibility…for minimising the prospect of a tragedy of this scale ever happening again.[85]

[82] K Hennessey et al, *Climate Change Impacts on Fire-weather in South-east Australia* (CSIRO, 2006) 5.

[83] N. Leary et al (eds), *Climate Change and Adaptation* (London: Earthscan, 2008), at 1–27.

[84] T. Bonyhady, A. Macintosh, and J. McDonald (eds) *Adaptation to Climate Change—Law and Policy* (Sydney: The Federation Press, 2010).

[85] Parliament of Victoria, *2009 Victorian Bushfires Royal Commission, Final Report Summary* (Parliament of Victoria, 2010), available at <http://www.royalcommission.vic.gov.au/finaldocuments/summary/PF/VBRC_Summary_PF.pdf> (accessed 2 October 2011).

The extract comes from the Victorian Bushfire Royal Commission 2010 on the horrific fires that engulfed towns on the perimeter of the metropolitan capital of over four million people, as well as raging in regional areas. The fires caused the highest number of civilian deaths in recent Australian history,[86] and seem far removed from the prosaic issues of regulating electricity networks. However, the link between these extreme events and the ageing electricity networks is only too apparent.

Key recommendations emanating from the Victorian Bushfire Commission were to amend Victoria's Electricity Safety Act 1998 and to give effect to:

> the progressive replacement of all SWER (single-wire earth return) power lines in Victoria with aerial bundled cable, underground cabling or other technology that delivers greatly reduced bushfire risk. The replacement program should be completed in the areas of highest bushfire risk within 10 years and should continue in areas of lower bushfire risk as the lines reach the end of their engineering lives... Priority should be given to distribution feeders in the areas of highest bushfire risk.[87]

2. *Ageing electricity networks and enhanced bushfire risk in Victoria*

Victoria is a highly bushfire-prone area with a long record of electricity assets causing major bushfires through infrastructure failures.[88] As a 1977 Victorian Inquiry found that, although 'it was claimed that 1.5 per cent of the total annual fire ignitions in "normal circumstances" were caused by electricity assets, this was in sharp contrast to the high incidence of fires ignited by electricity assets on days of extreme conditions'.[89] A similar pattern was repeated on 7 February 2009, when five of the 11 major fires that began that day—a day of extreme gusting winds and temperatures of 46 degrees Celsius in the Melbourne metropolitan area—were caused by failed electricity assets. The causes of those five fires comprised 'systemic factors associated with the reliability and safety of Victoria's electricity distribution networks'.[90] Such systemic factors led the Commission to recommend that major changes be made to Victoria's electricity distribution infrastructure, and its operation and management, to substantially reduce the risk to human life posed by bushfires on catastrophic fire days.[91] The Commission was emphatic about dealing with the risks, notwithstanding the substantial costs involved in replacement of the distribution network. Businesses operating distribution networks acknowledge that electricity assets have the capacity to start fires, especially in rural areas.[92] While steps to mitigate that risk are mandated by the Electricity Safety Act 1998, Energy Safety Victoria, the responsible regulatory agency, was found to have a 'confused mandate' in respect of bushfire risk reduction,[93] and to have carried out its functions in a perfunctory way as a form of 'compliance ritualism'.[94] Accordingly, the Commission recommended 'extending Energy Safe Victoria's mandate and

[86] For statistics on bushfire-related deaths, see Appendix Band C ibid, (vol 1) 10 at 347 and 350.

[87] Above, n 9 at 159.

[88] Above, n 9 at 148.

[89] Report as cited in Parliament of Victoria *2009 Victorian Bushfires Royal Commission*, (2010, vol 2) 179.

[90] Ibid.

[91] Parliament of Victoria *2009 Victorian Bushfires Royal Commission*, (2010, vol 1) 29–30.

[92] Ibid at 150.

[93] Ibid at 175.

[94] Ibid at 177.

resources to require and enable it to play a more active role in reducing the risk of electricity distribution infrastructure causing bushfires through strengthening its regulatory capacity'.[95] While a stronger regulatory presence can assist, the core problem of replacement of the electricity distribution network remains.

The 22 kV distribution feeders and SWER 'line' system electricity distribution network, the source of several fires on Black Saturday were introduced in the 1950s by the state-owned electricity commission of the time. Evidence before the Commission revealed that the age of the infrastructure contributed to three fires, including the East Kilmore fire that caused the most damage and loss of life. A witness to the Commission, noted that the capacity of electricity distribution businesses, 'to respond to an ageing network is constrained by the existing regime for the industry's economic regulation'.[96] Further, that regime tended to entrench the status quo and impeded 'step change reform'.[97] That regime for economic regulation has been outlined above, but particular aspects are worth reinforcing here:

> Victorian electricity distribution businesses are subject to an incentive-based regulatory regime whereby an economic regulator sets the total amount of revenue each distribution business may receive in a specified period. Under the regime the economic regulator makes a revenue determination on the basis of submissions the distribution businesses make in relation to their forecast capital and operating expenditure. If the distribution businesses deliver their services at a cost that is lower than the revenue cap set by the regulator, they are rewarded with an increased return. This gives them an incentive to conduct their business efficiently.[98]

Since 2011, the economic regulator is the Australian Energy Regulator.[99] The underlying idea is that the standards provide an incentive scheme. The Bushfire Commission noted that this 'deters the distribution businesses from reducing expenditure at the expense of service and reliability'.[100] The strength of the incentive-based economic regulation must be questioned in the light of the serious failure to replace and adequately maintain critical network distribution infrastructure over many years.

The Australian Energy Regulator witness who appeared before the Commission observed that a network provider's 'case for safety related investment, as with any other investment, needs to be considered by reference to its efficiency and prudency'.[101] Significantly, the witness acknowledged the lack of integration of public benefit considerations in investment decisions, stating that the Australian Energy Regulator 'is not aware of (nor does it use) a specific model which could immediately be applied to measure community benefit'. Ultimately, 'community benefits are policy questions to be considered by individual jurisdictions, in the context of their regulatory regimes'.[102] Presumably under such a model, the

[95] Ibid at 148. [96] Ibid at 151. [97] Ibid. [98] Ibid at 156.

[99] *Essential Services Commission Act 2001*, s 8; powers conferred by s 23 *National Electricity (Victoria) Act 2005* by *National Electricity (Victoria) Amendment Act 2007*, s 5.

[100] Above, n 9 at 156.

[101] Witness Statement of Chris Pastas, 16 February 2009, 13/14, above n 9.

[102] Ibid, at 14. See also above, n 9 at 157.

relevant regimes do not include the actual regulation of the electricity industry and networks themselves.

3. *Who pays for replacing ageing network infrastructure?*

This omission is even more telling given other evidence to the effect that there were a significant number of assets of over 61 years in age (where a replacement plan ideally should be 'in the pipeline' if not already implemented) and of assets in the 55–60 year bracket (where replacement plans were warranted in the very near future).[103] The Bushfire Commission heard that one distribution business, Powercorp, had made a submission in the course of price determinations in 2004/5 requesting funds for undergrounding overhead distribution infrastructure in bushfire-prone areas. This request was rejected by the Essential Services Commission (the regulator at that time). The Bushfire Commission noted that the company was not precluded from undertaking the investment in underground distribution facilities itself.[104] However, the Essential Services Commission acknowledged that if a regulator does not approve a particular investment proposal, the distribution business is unlikely to implement it. As distribution businesses do not control their own price setting, this constrains the extent to which they invest in activities aimed at reducing bushfire risk.[105]

Clearly, this regulatory model has constraints when incorporating public benefit safety outcomes that deviate from a least-cost conception of regulation. This point was emphasized by the Bushfire Commission: '[t]he Australian Energy Regulator's failure to factor in the costs to human life and property arising from bushfire as part of its cost–benefit equation means that real and substantial costs to the community imposed by bushfire are left out of the price determination process'.[106] Against this background, the Commission adopted the above recommendation that the replacement of ageing infrastructure was an urgent matter. Costs were to be borne in part by the Victorian community at large through government contributions, and by network distribution businesses, under a 'price pass through' mechanism.[107] Currently, there are ongoing negotiations around exactly which costs of the replacement of network infrastructure will be borne by the community and which by private network providers.

Liability for negligence may also prove a spur to amending the regulatory model. Several class actions have been initiated in the Victorian Supreme Court, on behalf of groups of bushfire victims affected by the devastating Black Saturday bush fires, including specifically the Kilmore East, Horsham, and Coleraine fires.[108] An alleged cause of these fires were faulty electricity distribution lines. While the Horsham proceedings have recently been settled out of court, with AUD$40 million expected to be awarded to the victims, the other actions are ongoing. In the

[103] Above, n 9 at 155. [104] Ibid at 156–7. [105] Ibid at 156.

[106] Ibid at 158. [107] Ibid at 157–8.

[108] See, eg, *Matthews v SPI Electricity and SPI Electricity Pty Ltd v Utility Services Corporation Ltd* (Supreme Court of Victoria, 4788/2009) for the Kilmore fires, *Thomas v Powercor Australia Limited* (Supreme Court of Victoria, 9166/2009) for the Horsham fires, *Perry & Anor v Powercor Australia Limited* (Supreme Court of Victoria, 330/2009) for the Coleraine fires.

action regarding the especially devastating Kilmore East fire, several power companies, together with a company responsible for checking the safety of the electricity lines, were joined as tortfeasors in the class action with the State of Victoria. The actions highlight the potential liability of companies even in a privatized electricity market where infrastructure is ageing and potentially unsafe. The substantive matters are yet to be heard by the court, but given the rise in litigation around climate change issues, there is potential for more claims of this nature to be initiated. Similarly, the class action raises issues of the extent of the statutory duty of regulators to ensure community safety in situations where the risks of ageing network infrastructure are well recognized.

4. Summary

Given the escalation in extreme fire-weather events under climate change conditions, the entrenched problems of a continuing lack of investment in new distribution infrastructure heightens the systemic problems of ensuring public safety outcomes in a deregulated and largely privatized electricity sector. When combined with the endemic problems of stimulating new network investment, it highlights inherent difficulties in combining economic efficiency with public interest benefits in one regulatory regime. The Bushfire Commission made it clear that '[p]rotection of human life must become the priority when evaluating distribution businesses' expenditure proposals. The economic regulatory regime must include mechanisms for ensuring that safety-related matters are properly reviewed so as to minimize the risk of bushfire being caused by the failure of electrical assets.'[109] To date, there is no widespread legal reform of the basic electricity network regulation models to give effect to these proposals.

V. Conclusion

The divergence of aims between public interest environmental and safety-related goals and the prevailing market-oriented efficiency parameters for electricity transmission and distribution networks is evident from the case studies. Despite this divergence, and potentially conflicting policy outcomes, many in government still see a virtue in maintaining the current regulatory model whereby 'this Law is to promote efficient investment in, and efficient operation and use of, electricity services for the long term interests of consumers of electricity'.[110]

However, electricity consumers are also affected by climate change impacts that have accelerated the risks of bushfire hazards in Victoria with immense costs in economic and human terms when fires do occur. These 'externalities' remain unincorporated in any least-cost formula for price determinations, even given the reliability tests and the service provision 'standards'. Similarly, it might be argued that structural reforms that allow a higher percentage of renewable energy generation

[109] Above, n 9 at 158. [110] *National Electricity Law*, s 7.

to access the electricity grid are also in the long-term interests of consumers by reducing long-term dependency on fossil fuel generation. Yet the efficiency and least-cost regulatory model is entrenched, and given the technicalities of the various 'tests', it is unlikely to be significantly modified in the near future.

Significantly, consumers are facing a series of electricity price rises in Australia. In these circumstances, it is unlikely that the costs for augmenting and extending the transmission lines and replacing and/or undergrounding the distribution lines in bushfire-prone areas will be borne primarily by consumers (as opposed to taxpayers!), especially as electricity industry investors under existing regulatory models cannot be compelled to make 'public benefit' investments. Governments may see virtue in 'governing at a distance' under institutional arrangements which mean that the Australian Energy Regulator remains an interface between government responsibility and electricity price rises.

Clearly, there remain significant barriers to the pursuit of non-market public interest outcomes under the regulatory model for electricity networks in Australia. Moreover, it is questionable whether other statutory regimes, such as those mandating specific levels of renewable energy generation increases or requiring electricity industry safety standards, will be able to penetrate the highly technical operation and implementation of governing rules around the electricity markets to interpose standards for network bushfire risk reduction and grid access 'beyond compliance'. Litigation may have some capacity to highlight individual deficiencies and gaps in the laws. However, the capacity to facilitate widespread, systemic transformation in the manner in which electricity network development occurs, remains constrained by an understanding of the legal aspects of energy networks that gives pre-eminence to 'efficiency', 'least-cost', and an 'unfettered market'.

17

Evolution and Revolution in British Energy Network Regulation: From RPI-X to RIIO

Aileen McHarg

I. Introduction

If the United States' major contribution to the theory and practice of economic regulation is the independent regulatory agency, then that of the United Kingdom (UK) is incentive regulation. Although incentive regulation has a long pedigree and can take many forms,[1] the primary method currently employed in the UK is price cap regulation, also known by the regulatory formula RPI-X. First proposed by Professor Stephen Littlechild for the regulation of British Telecommunications when it was privatized in 1984,[2] it was subsequently adopted for other privatized utilities in the UK, and has since spread to many network industries worldwide. In the energy industries, price caps were applied to British Gas when it was sold in 1986, and to the National Grid Company, the 12 English and Welsh Regional Electricity Companies, and the two vertically integrated Scottish electricity undertakings when the British electricity industry was privatized in 1990–91.[3] Today, following wholesale and retail market liberalization, and the disaggregation of monopoly and competitive businesses, price regulation is limited to the natural monopoly transmission and distribution functions in both gas and electricity.

According to Joskow, 'the regulatory institutions and mechanisms that have evolved [in the UK] represent the gold standard of effective incentive or performance-based network regulation'.[4] Nevertheless, in October 2010, following a two-year

[1] See T. Jamasb and M. Pollitt, 'Incentive Regulation of Electricity Distribution Networks: Lessons of Experience from Britain' (2007), 35 *Energy Policy* 6163.

[2] S. Littlechild, *The Regulation of British Telecommunications' Profitability* (London: Department of Industry, 1983). See also S. Littlechild, *Economic Regulation of Privatized Water Authorities* (London: HMSO, 1986).

[3] NB the discussion is limited to regulation of the British energy networks. Energy is a devolved function in Northern Ireland and the Northern Irish energy utilities are regulated separately.

[4] P.L. Joskow, 'Lessons Learned from Electricity Market Regulation', in R.J. Green and M.G. Pollitt (eds), *The Energy Journal: Special Issue: The Future of Electricity: Papers in Honours of David Newbery* (Cleveland, Ohio: IAEE, 2008) 9 at 23.

review (the RPI-X@20 review), the energy regulator, Ofgem (the Office of Gas and Electricity Markets), announced that RPI-X is to be replaced by a new approach to network regulation: Sustainable Network Regulation, or RIIO.[5] While RPI-X has been successful in the past, it argued, it is no longer 'fit for purpose' given the radically changed economic, technical, and policy context in which the industries are now situated. Not only is substantial investment required to replace network assets reaching the end of their working lives, and thus to ensure continued security of supply, but the energy networks are central to the delivery of environmental policy objectives. In order to meet the UK's stringent, legally binding greenhouse gas emissions reduction targets,[6] the Committee on Climate Change has advised that the electricity system should be almost completely decarbonized by 2030.[7] This requires not only further capital investment, but also changed operational practices to facilitate the expected influx of intermittent renewable generation (also the subject of binding targets),[8] increased micro-generation[9] and distributed generation, and the development of 'smart grids'.[10] An estimated £32 billion of investment is required by 2020,[11] and this needs to be accommodated by the regulatory system, which hitherto has focused primarily on incentivizing cost reduction rather than long-term investment. At the same time, it is extremely uncertain how exactly either the gas or electricity network will develop in future.[12] Thus, according to Ofgem, the regulatory system also needs to be flexible in order to avoid prematurely locking the industries into a particular development path, as well as to encourage innovation to deliver the necessary changes at least cost to consumers.

[5] Ofgem, RIIO, *A New Way to Regulate Energy Networks: Final Decision* (2010c).

[6] The Climate Change Act 2008 commits the UK to an 80 per cent reduction in greenhouse gas emissions by 2050, compared with 1990 levels, as well as to legally binding interim targets set via a carbon budgeting process. The recently agreed fourth carbon budget, for the period 2023–2027, commits the UK to a 50 per cent reduction on 1990 levels by 2027—the Carbon Budgets Order 2011, SI 1603.

[7] Committee on Climate Change, *Building a Low Carbon Economy—the UK's Contribution to Tackling Climate Change* (2008), ch 5.

[8] Under European Union law, 15 per cent of the UK's overall energy consumption must come from renewable sources by 2020 (Directive 2009/28/EC of the European Parliament and of the Council on the Promotion of the Use of Energy from Renewable Sources, OJ L140/16, 23 April 2009). In practice, given constraints on the development of renewable heat and transport, this is likely to require a renewable electricity share of c. 30 per cent by 2020—Department of Energy and Climate Change, The UK Renewable Energy Strategy, Cm 7686 (2009).

[9] The British Microgeneration Strategy (adopted under the Energy Act 2004, s 82) sets a (non-binding) target of meeting 30–40 per cent of electricity consumption from microgeneration by 2050—Department of Trade and Industry, *Our Energy Challenge: Microgeneration Strategy: Power From the People* (2006).

[10] See, eg, Department of Energy and Climate Change, *Smarter Grids: The Opportunity* (2009).

[11] Ofgem, *Regulating Energy Networks for the Future: RPI-X@20 Recommendations: Consultation* (2010b) 2.

[12] Ofgem has identified five distinct, but equally plausible development paths for the electricity industry by 2050, ranging from large-scale, centralized transmission and distribution grids, as at present, to micro-grids, in which the consumer is at the centre of network activity—*Long Term Electricity Networks Scenarios (LENS)—Final Report* (2008). The future shape of the gas network is also unclear, given uncertainty about levels of demand for gas for heating and electricity generation, the nature and location of sources of wholesale gas, and the potential use of gas networks for carbon capture and storage—Ofgem, *Regulating Energy Networks for the Future: RPI-X@20: Principles, Process and Issues* (2009a) 30; Ofgem 2010b (above, n 11) 59.

Regulation of network prices—ie how much revenue network operators are permitted to raise from network users, and for what purposes—is by no means the only determinant of whether Britain's energy networks will deliver long-term security, sustainability, and value for money. Issues such as network access policy, charging methodologies, and land-use planning are also important, and are similarly the subject of reform.[13] Nevertheless, price regulation is crucial. Since energy networks are highly capital-intensive, with very long asset lives, and large sunk costs, potential investors require reassurance that they will be able to recover their reasonable costs over the long term. At the same time, because the networks are natural monopolies, users require protection against potential exploitation. How to balance investment risk and the risk of monopoly abuse has therefore long been the central problematic for the design of network policy, although the balance has been struck in different ways and in different places, depending on the perceived relative importance of the two objectives. The current level of uncertainty surrounding future network development is merely a complicating factor.

In fact, in some ways, Ofgem's new approach to network regulation is recognizably an evolution from existing practice, which has become considerably more complex in the quarter-century since RPI-X regulation was first introduced. Sustainability and security of supply objectives have been increasing in importance in energy law and policy,[14] including Ofgem's statutory duties,[15] since the turn of the century, and recent price control reviews have allowed for price rises to fund investment and included mechanisms to stimulate innovation and the transition to lower carbon networks. Moreover, RIIO retains a number of features of the RPI-X model.

In other ways, though, the change is revolutionary. RPI-X was more than merely a technical formula; it embodied a philosophy of regulation—that it should be light-handed, pro-competition, and focused on economic rather than social regulatory goals—with which its abandonment signals a decisive break. Whilst UK energy policy remains market-based, the regulatory system appears to have reached a tipping point where measures to promote goals such as security of supply and decarbonization are no longer just add-ons, but central to the design of regulatory and market systems.[16] Reform of network regulation should thus be seen in the context of other recent changes, such as the creation of a combined Energy and Climate Change department in October 2008, the downgrading of Ofgem's duty

[13] See, eg, Ofgem, *Tranmission Access Review—Final Report* (2008); Ofgem's Project TransmiT, available at <http://www.ofgem.gov.uk/Networks/Trans/PT/Pages/ProjectTransmiT.aspx> (accessed 2 October 2011); Department of Energy and Climate Change, *Overarching National Policy Statement for Energy* (2011).

[14] On UK energy policy, see Department of Trade and Industry, *Our Energy Future: Creating a Low Carbon Economy*, Cm 5761(2003); *Meeting the Energy Challenge: A White Paper on Energy*, Cm 7124 (2007). EU energy policy similarly includes supply security and environmental objectives alongside market objectives—Treaty on the Functioning of the European Union, Art 194.

[15] See Gas Act 1986, s 4AA and Electricity Act 1989, s 3A, as amended by Sustainable Energy Act 2003, s 6, Energy Act 2004, s 172, Energy Act 2008, s 83, Energy Act 2010, s 17.

[16] On the relationship between social and economic regulation, see generally T. Prosser, *The Regulatory Enterprise: Government, Regulation and Legitimacy* (Oxford: Oxford University Press, 2010) ch 9.

to promote competition by the Energy Act 2010,[17] and proposed reform of the electricity wholesale market to improve security of supply and secure a level playing field for low-carbon generators.[18]

The aim of this chapter is to explain and evaluate Britain's new system of network regulation. Given that many other countries are facing similar investment and innovation challenges, together with the 'market-leading' nature of British energy regulation, this is an issue which should be of broad interest to the readers of this volume. In order to provide the context necessary to understand and assess the new system, the chapter first discusses the theory and practice of RPI-X regulation, and Ofgem's reasons for making the change.

II. RPI-X Regulation: Theory and Practice

A. The current network price caps

There are at present distinct price cap arrangements in place for gas and electricity transmission network operators (TNOs), gas distribution network operators (GDNOs), and electricity distribution network operators (EDNOs).[19] Because British Gas was privatized as a vertically integrated monopoly, there was initially a single price control for the entire gas business. However, separate controls were introduced for gas transportation and storage, on the one hand, and gas supply on the other, in 1994, with further unbundling of gas storage in 1997, system operation in 1999, and gas distribution networks in 2002. Electricity TNOs and DNOs (distribution network operators) have been separately regulated since privatization, although separate metering caps were introduced for DNOs in 1994 and separate system operator incentives for TNOs in 2002. Slightly different arrangements apply to independent gas transporters and independent electricity DNOs. These operators predominantly serve network extensions, rather than particular geographic areas, and are regulated through relative price controls, by reference to the relevant gas or electricity distribution network control. Offshore electricity transmission networks are also regulated separately, through individual competitive tendering arrangements. Interconnectors are not currently subject to any price regulation, although separate proposals are being developed for a 'cap and floor' regime for new interconnectors.[20] The RPI-X@20 review proposals apply only to the regulation of the core, onshore transmission, and distribution networks.

Price controls are contained in operating licences, rather than primary legislation, and can therefore periodically be revised. Until recently, licence amendment was by

[17] Gas Act 1986, s 4AA(1C) and Electricity Act 1989, s 3A(1C) (as amended).

[18] Department of Energy and Climate Change, *Planning Our Electric Future: A White Paper for Secure, Affordable and Low-Carbon Electricity*, Cm 8099 (2011).

[19] For detailed discussion of the evolution of network price caps, see Ofgem, *Regulating Energy Networks for the Future: RPI-X@20: History of Energy Network Regulation* (2009b).

[20] Ofgem/CREG, *Cap and Floor Regime for Regulation of Project NEMO and Future Subsea Interconnectors* (2011).

agreement between Ofgem and the relevant licence holder, subject to a reference to the Competition Commission if agreement could not be reached.[21] There were limited formal participation rights for third parties, but in practice the regulator followed increasingly elaborate consultation arrangements before new price controls were adopted. However, in response to the third European Union (EU) Internal Energy Market package,[22] the government has amended the licence modification process, so that changes are made by the regulator independently, but subject to a right of appeal to the Competition Commission for licence holders and directly affected third parties, including the statutory consumer body, Consumer Focus.[23]

The current TNO and GDNO price controls run from April 2007 until April 2013, while the EDNO controls run from April 2010 until April 2015.

B. RPI-X regulation in theory

The operation of price cap regulation is very simple in theory: it limits the rate at which prices may rise (usually by reference to average prices or a basket of prices) to inflation (RPI = retail price index) minus an efficiency factor (X), sometimes with provision for pass through of costs (Y) which are beyond the company's control. If X is given a negative value, prices may rise above the level of inflation, but are still subject to limits. Because price levels are fixed, typically for a period of five years, the company has an incentive to 'beat the cap' by making greater efficiency savings than assumed by the regulator. The company is allowed to retain any additional profits until the price cap is revised, at which time the benefits of its efficiency efforts can be shared with consumers in the form of a one-off price reduction and/or a tighter cap for the next price control period. In effect, regulation mimics the effects of competition, first, by making the company a price-taker and, second, by harnessing the profit motive to act as a discovery mechanism, thereby overcoming the regulator's informational disadvantage relative to the company regarding where there is scope for cost cuts.

Price cap regulation was recommended by Littlechild for the regulation of privatized monopolies in preference to rate of return, or cost of service, regulation as it was practised in the United States at the time. Whereas RPI-X regulation sets prices ex ante, rate of return regulation adjusts prices ex post according to the costs actually incurred by the regulated firm, with the aim of preventing it from earning excess profits. For Littlechild, rate of return regulation was not only ineffective in protecting consumers against monopoly power, but also imposed excessive burdens on regulators.[24] Because firms could earn no more than the allowed rate of profit,

[21] Gas Act 1986, ss 23, 24; Electricity Act 1989, ss 11, 12.

[22] Directive 2009/72/EC of the European Parliament and of the Council of 13 July 2009 Concerning Common Rules for the Internal Market in Electricity, OJ L211/55, 14 August 2009; Directive 2009/73/EC of the European Parliament and of the Council of 13 July 2009 Concerning Common Rules for the Internal Market in Natural Gas, OJ L 211/94, 14 August 2009.

[23] Electricity and Gas (Internal Markets) Regulations 2011, SI 2704, Part 9.

[24] Littlechild 1983 (above, n 2); Littlechild 1986 (above, n 2). For criticism of rate of return regulation, see also C.D. Foster, *Privatization, Public Ownership and the Regulation of Natural Monopoly* (Oxford: Blackwell, 1992) ch 6.

and had to pass on any cost reductions in full to consumers, they had no incentive to seek efficiency savings. They could, however, increase profits by inflating the capital base on which returns were calculated, potentially leading to inefficient investment decisions.[25] For both reasons, prices might remain artificially high. To counteract these tendencies, regulators had to become heavily involved in 'second-guessing' managerial decisions—determining allowable expenses and appropriate depreciation policies, approving capital expenditure, and allocating joint costs, deciding what rate of return is 'fair'—and this process had to be repeated each time the firm wished to raise its prices. As a consequence, Littlechild argued, price setting under rate of return regulation was time-consuming and expensive, prone to information failure, and, because of the extent of regulatory discretion, vulnerable to lobbying and capture.

By contrast, Littlechild considered RPI-X regulation to be superior in a number of respects. By focusing on prices rather than profits, it would give better protection to consumers. By harnessing rather than dampening the profit motive, it would provide greater incentives to efficiency and would transfer some of the burden of monitoring firm performance from the regulator to the capital market. Because the regulator would not have to scrutinize managerial decision-making, the system would be transparent, straightforward, and cheap to operate. For the same reason, it would be non-discretionary, hence decisions would be less likely to be distorted by bureaucratic incompetence, political lobbying, or industry capture, and the system would also be more stable, thereby reducing regulatory risk and the cost of capital. Finally, a light-touch approach to price control would free regulators to concentrate on the promotion and policing of competition, with the aim of removing the need for regulation wherever possible. In Littlechild's famous phrase, regulation was intended merely to be 'a means of "holding the fort" until competition arrives'.[26]

C. RPI-X regulation in practice

In practice, price cap regulation has not worked quite as Littlechild expected, and it has increasingly diverged from the theoretical ideal as time has gone on.

1. RPI-X and rate of return regulation compared[27]

In reality, price cap regulation is not as radically different from rate of return regulation as it appears. First, the difference between the efficiency incentive properties of the two schemes is one of degree, rather than kind. This is because, under

[25] The so-called 'Averch-Johnson effect': H. Averch and L.L. Johnson 'The Behaviour of the Firm Under Regulatory Constraint' (1962), 52 *American Economic Review* 1052.

[26] Littlechild 1983 (above, n 2) para 4.11.

[27] There is a large literature on the relative merits of incentive and rate of return regulation. See, eg, Littlechild 1986 (above, n 2); J. Vickers & G. Yarrow, *Privatization: An Economic Analysis* (Cambridge, MA: The MIT Press, 1988), 85–7, 207–8, 240–1; Foster (above, n 24) ch 6; D. Helm, 'British Utility Regulation: Theory, Practice and Reform' (1994), 10 *Oxford Review of Economic Policy* 17; J.-J. Laffont and J. Tirole, *A Theory of Incentives in Procurement and Regulation* (Cambridge,

rate of return regulation, there is typically a 'regulatory lag' between changes in costs and tariff revisions, which, as under RPI-X, allows firms to retain additional profits temporarily. RPI-X still provides superior efficiency incentives because the intervals between price changes are longer and fixed in advance, which creates greater regulatory commitment and stability. However, as far as the British system is concerned, this is only guaranteed by regulatory self-restraint. Contrary to Littlechild's assumption, the regulatory scheme is in fact highly discretionary in legal terms. Ofgem does therefore have the power to revise price controls early, and it may come under pressure to do so should costs and prices move too far out of line. So far, it has not exercised this power.[28] Nevertheless, recent price controls have incorporated specific 're-opener' provisions to deal with unpredictable costs, such as pension costs, or anticipated changes in statutory requirements. In any case, even with greater regulatory commitment, the incentive properties of RPI-X regulation appear to weaken as the price control period nears its end. To counteract this, Ofgem has further refined the regulatory formula by introducing so-called 'rolling' efficiency incentives. These permit licensees to retain the benefit of additional savings for five years, irrespective of when during the price control period they were achieved.

The second reason why the two schemes are similar is the need for periodic revision of price caps. Because Littlechild expected regulation to be temporary, this was an issue to which he paid little attention in his report on telecommunications, although his later report on water regulation did recognize that regulation was likely to be permanent.[29] Similarly, in the core energy networks, while competition has made some inroads at the margins—for instance, in the provision of connections and metering and, via competitive tendering, of offshore transmission lines—there is no realistic prospect of regulation becoming redundant. Under price cap regulation, with the passage of time, prices are likely to move increasingly out of line with costs and profits, producing allocative inefficiency between consumers and shareholders. If prices are too high, the situation is likely to become politically unacceptable; if they are too low, the financial viability of the networks might be threatened. The process of resetting price controls is, though, much the same as tariff revision under rate of return regulation. Ofgem employs a 'building blocks' approach, which requires it to take a view on the accuracy of the network operators' forecasts of operating and capital expenditure (opex and capex), as well as to determine an appropriate rate of return on capital and the relevant 'regulatory asset value' on which returns are calculated. Indeed, because of the ex ante nature of price cap regulation, the information demands it places on the regulator are even greater than under rate of return regulation, arguably producing more serious problems of information asymmetry and regulatory gaming.

MA: The MIT Press, 1993); M. Armstrong et al, *Regulatory Reform: Economic Analysis and British Experience* (Cambridge, MA: The MIT Press, 1994).

[28] In 1995, the first EDNO price control review was reopened because new information came to light about the extent of the companies' profitability. However, this took place within the statutory period of notice and comment before new control came into effect.

[29] Littlechild 1986 (above, n 2) ch 10.

Because of the need for revision, the supposed benefits of RPI-X regulation in terms of transparency, simplicity, lack of discretion, and consistency have largely turned out to be illusory. Over the years, price control reviews have become increasingly lengthy and expensive affairs, as the regulator has sought to respond to criticisms of lack of transparency and accountability, and to deal with information asymmetry. Taking two years on average, they now involve multiple consultation rounds, the commissioning of consultants' reports, and academic and stakeholder workshops. In pursuit of greater accuracy in assessing required costs, the detailed methodologies underlying the controls have changed from review to review, and Ofgem has also employed additional regulatory tools to improve information quality. For instance, yardstick regulation—which involves comparing the performance of similar networks, and setting price controls by reference to the median or average network, such that poorly performing firms are penalized and good performers rewarded—has been used from the outset for the English and Welsh EDNOs,[30] and Ofgem has subsequently sought to align the timing of price reviews for other network operators in order to increase the use of comparative data. More recently, it has introduced the so-called Information Quality Incentive (IQI) in order to improve the accuracy of DNOs' capex forecasts. This employs a 'menu of contracts' approach, whereby DNOs are allowed to choose from a range of capex allowances, each with varying rates of return and efficiency incentive rates. Firms choosing high capex allowances (relative to the amount Ofgem's consultants consider reasonable) are penalized with relatively low rates of return and incentive rates; those choosing lower capex allowances are correspondingly rewarded. This scheme has been described by Ofgem's own Chief Executive as 'virtually unfathomable to those outside the cognoscenti'.[31] Regulatory reporting requirements have also increased over time, and the regulator has powers (which it has invoked) to impose substantial fines on companies found to have breached these requirements.[32]

2. *Quality of service*

A second problem with RPI-X regulation, which Littlechild also recognized in his report on the water industry,[33] is that companies can earn increased profits not only by improving efficiency, but also by reducing service quality. Accordingly, statutory powers to regulate quality standards were included in the Electricity Act 1989, and were extended to the gas industry by the Competition and Service (Utilities) Act 1992.[34] The initial approach to quality regulation for the distribution networks took the form of guaranteed and overall standards of service, the

[30] This was a solution which Littlechild had recommended for the water industry—ibid paras 10.16–10.20.

[31] A. Buchanan, 'Ofgem's "RPI at 20" Project', Speech at SBGI, 6 March 2008, 6, available at: <http://www.ofgem.gov.uk/Networks/rpix20/Factsheets/Documents1/AB%20March%2008.pdf (accessed 2 October 2011).

[32] Gas Act 1986, ss 30A–30F; Electricity Act 1989, ss 27A–27F.

[33] Littlechild 1986 (above, n 2) chs 7 and 8.

[34] In fact, the regulator had already acquired powers to regulate quality through amendment of British Gas's licence.

former triggering payments to individual consumers if breached, and the latter enforced via formal regulatory enforcement orders. The overall quality standards have, however, been superseded by a further incentive scheme—the Information and Incentives Project, introduced in 1999. This is a system of pre-determined rewards and penalties for performance against agreed targets, covering customer satisfaction and the number and duration of supply interruptions.

Quality regulation has also expanded as social and environmental objectives have become more prominent. Whereas Littlechild and other free market economists envisaged that sectoral regulation would be restricted to economic objectives (ie preventing monopoly abuse and promotion of competition), with social and environmental goals being pursued, if at all, outside the regulatory system,[35] over time they have increasingly been incorporated within the price control framework. Thus, a range of additional fixed and discretionary reward schemes have been established, for example, to reduce transmission and distribution losses and gas shrinkage, reduce greenhouse gas emissions from the networks, extend gas networks to fuel poor areas, improve gas safety, expand services to vulnerable consumers, and increase corporate social responsibility initiatives.

3. *Investment and innovation*

Again, because it was expected to be temporary, there was little consideration of the impact of RPI-X regulation on long-term investment; Littlechild's concern was to prevent over-investment, not to avoid under-investment. Nevertheless, there are reasons to think that pure price cap regulation is inappropriate in circumstances where substantial investment is required. First, the cost of capital under RPI-X is higher than under rate of return regulation, because shareholders carry all the risk of cost shocks under the former, whereas risks lie with consumers under the latter.[36] Second, although the empirical evidence is mixed,[37] price cap regulation may act as a disincentive to investment because of the mismatch between the long-term nature of network assets, but uncertainty as to the treatment of capital costs in future price control periods. In other words, there is a risk that assets may become 'stranded'. Third, although firms have strong incentives to innovate to reduce short-term operating costs, there is no incentive to engage in innovation where the returns are more speculative. Indeed, research, development, and demonstration (RD&D) is an area in which short-term cost savings can be made, and RD&D expenditure by the energy network operators has fallen dramatically since privatization.

Investment and innovation funding was not a major issue in the early years post-privatization: no significant capital expenditure was required, there was

[35] See, eg, Littlechild 1983 (above, n 2) para 14.7; Foster (above, n 24) 7, 318, 323; C. Robinson, *Energy Policy: Errors, Illusions and Market Realities* (London: Institute of Economic Affairs, 1993) 57–60.

[36] D. Helm, 'Utility Regulation, the RAB and the Cost of Capital', Competition Commission Spring Lecture 2009, 10, available at <http://www.dieterhelm.co.uk/sites/default/files/Helm_CC_060509.pdf> (accessed 2 October 2011).

[37] K. Petrov et al, 'Regulatory Incentives for Investments in Electricity Networks' (2010), available at <http://www.jcrni.org/extranet/public/cp137.pdf> (accessed 2 October 2011).

no 'disruptive' technological change, and the investment climate was benign.[38] However, as already noted, circumstances have changed. Thus, in recent price controls not only have prices been allowed to rise in real terms, but again additional mechanisms have been employed to stimulate investment and innovation. Investment incentives have included: so-called 'revenue drivers', which permit allowed revenues to vary in accordance with, for instance, the number of customers connected to a distribution network, or the amount of new renewable generation connected to the grid; allowances for undergrounding transmission lines; mechanisms to prevent capex underspend; and the use of long-term auctions for new gas entry capacity. Innovation stimuli have included so-called Registered Power Zones (RPZs), which provided an incentive under the 2005–2010 EDNO price control to develop new and more cost-effective ways of connecting distributed generation, by increasing the revenue driver for connection projects which demonstrated genuine innovation. The Innovation Funding Incentive, introduced at the same time, but subsequently extended to all transmission and distribution network operators, similarly allows partial pass-through of the costs of RD&D projects aimed at producing long-term improvements in supply quality, safety, or environmental performance. Most recently, a Low Carbon Networks (LCN) Fund has been introduced, superseding RPZs, which provides EDNOs with ring-fenced funding, allocated on a competitive basis, for specific innovation projects which advance sustainability goals. As investment requirements have grown, it has become increasingly difficult for the regulator to avoid becoming involved in the approval of specific expenditures.[39]

In sum, the current British system of energy network regulation is far from the 'light-handed' scheme initially envisaged by Littlechild. To describe it as RPI-X regulation is also a misnomer. In truth, RPI-X is but one of several different types of incentive schemes and other regulatory instruments employed to stimulate a range of different behaviours. The net effect is substantial complexity, leading to potential distortions in decision-making[40] and blunting of the incentive properties of the regulatory system as a whole.[41]

III. The RPI-X@20 Review

The RPI-X@20 review was announced in March 2008.[42] There seems to have been no single trigger for its launch. Rather, it appears to have been prompted by a general feeling that the challenges facing network regulation had changed significantly since the regime was introduced. The focus had shifted away from

[38] Ofgem 2009a (above, n 12) 18.

[39] See, eg, S. Cowan, 'Network Regulation' (2006), 22 *Oxford Review of Economic Policy* 248 at 254.

[40] Ofgem 2009a (above, n 12) 6.

[41] M. Pollitt, 'Network Regulation in the Light of the Climate Change Act' (no date), 7, available at: <http://www.eprg.group.cam.ac.uk/wp-content/uploads/2009/05/michael-pollitt.pdf> (accessed 2 October 2011).

[42] Buchanan (above, n 31).

efficiency towards climate change and security of supply objectives, and there were also greater concerns about fuel poverty and financeability, due to the economic downturn, as well as uncertainty about the future size and shape of the networks.[43] These changes were reflected in increasing academic criticism of the continued suitability of the regulatory regime, and in concerns expressed by the network companies.[44] Accordingly, in the light of its obligations to promote sustainability and the interests of future consumers, as well as to follow 'better regulation' principles,[45] Ofgem considered it appropriate to undertake a 'regulatory MOT'.[46]

The review emphatically did not signify that Ofgem considered RPI-X regulation to have been a failure. On the contrary, the regulator argued, it had been extremely successful in achieving its main objective of reducing costs and prices. At the same time, quality of service had improved, network capacity and investment levels had increased, and the cost of capital had fallen.[47]

Nevertheless, what began in a spirit of 'good housekeeping', focused on a fairly narrow set of technical issues,[48] soon became more ambitious, as both the scale of the investment challenge and the degree of uncertainty facing the networks became clearer as the review proceeded. The enactment of the Climate Change Act in November 2008, which created a legally binding obligation on the UK Government to reduce greenhouse gas emissions by 80 per cent compared with 1990 levels by 2050, may well have been a significant factor. By the time the first consultation paper was issued in February 2009,[49] Ofgem was signalling a much more fundamental review of the entire regulatory system. Ultimately, the review process itself was a major undertaking, involving three rounds of consultation, informed by numerous supporting papers, consultants' reports, academic and stakeholder workshops, an interactive web forum, and specialist advisors.[50]

The review identified three main problems with the existing regulatory regime. First, RPI-X was unlikely to deliver the same results in the future as it had in the past. There was limited scope for further significant cost savings and investment requirements were set to increase dramatically, hence it was almost inevitable that network charges would rise.[51]

Second, the regime had produced network operators who were:[52]

- low risk and potentially risk-averse, willing to undertake investment only when commitment was provided by users or the regulator;
- focused on allowed revenue and less concerned about the structure of network charges, although the latter could significantly affect the need for new investment and the costs faced by consumers;

[43] Ofgem 2009a (above, n 12) 1, 18. [44] Ibid 6–7.
[45] Gas Act 1986, s 4AA(1), (2)(c), (5A); Electricity Act 1989, s 3A(1), (2)(c), (5A).
[46] Buchanan (above, n 31) 8.
[47] See Ofgem, *Regulating Energy Networks for the Future: RPI-X@20: Performance of the Networks Under RPI-X* (2009c). [48] Ofgem 2009a (above, n 12) 7.
[49] Ibid.
[50] All the review materials are available at <http://www.ofgem.gov.uk/Networks/rpix20/Pages/RPIX20.aspx> (accessed 2 October 2011).
[51] Ofgem 2009a (above, n 12) 20–1. [52] Ibid 21.

- more focused on Ofgem and how to 'beat the regulatory contract' than understanding their own consumers' needs and being rewarded for improving consumer service;
- reactive to developments in government policy, rather than proactive;
- reluctant to innovate; and
- focused on their own businesses, rather than interactions with markets or other networks.

As a result, key investment decisions were not being made, or at least not in a timely fashion,[53] and such companies were unlikely to be able to meet the challenges of delivering sustainable energy networks and long-term value for money in conditions of substantial uncertainty.[54]

Third, as noted above, the regulatory system had become extremely complex. This was problematic because it reduced its transparency and accessibility to anyone other than the regulator and network operators.[55] Yet, as the task of network regulation became more difficult, Ofgem needed to maintain the legitimacy and accountability of its decisions by enabling stakeholders to become more fully involved in the price control process.[56]

In the light of these problems, Ofgem's objectives were 'to design a new regulatory process for price control reviews that is more streamlined, accessible and transparent' and to develop a new regulatory framework that would:

encourage network companies to focus on the longer term and:

- Play a much greater role in facilitating the delivery of a sustainable energy sector whilst continuing to facilitate competition;
- Deliver continuous, long-term improvements in outputs and efficiency;
- Take more responsibility for developing solutions that are best value for present and future consumers;
- Manage uncertainty, taking on risk where appropriate and keeping options open where cost-effective;
- Engage more effectively with all stakeholders, responding to existing and anticipated needs of consumers of network services; and
- Be more innovative, looking for new and better ways of delivering and adapting over time as they learn what works best.[57]

The decision to move from RPI-X to RIIO, or Sustainable Network Regulation, was formally endorsed by Ofgem's governing body in October 2010. However, the new system will not be implemented until the next round of price controls commencing in April 2013 for TNOs and GDNOs, and April 2015 for

[53] Ofgem, *RPI-X@20 Working Paper 3: Delivering Desired Outcomes: Who Decides What Energy Networks of the Future Look Like?* (2009d) para 2.8.

[54] Ofgem, *Regulating Energy Networks for the Future: RPI-X@20: Emerging Thinking* (2010a) 8–9.

[55] Ofgem 2009a (above, n 12) 7, 24.

[56] Ofgem 2010b (above, n 11) 3.

[57] Ofgem 2010a (above, n 54) 1.

EDNOs.[58] As noted above, there is the possibility of appeal to the Competition Commission against price control revisions, so there is no guarantee that the RPI-X@20 review proposals will be adopted. In any case, some of the details have been left to be worked out in the course of the individual price reviews. Nevertheless, the elaborate process followed by Ofgem during the review was designed to build consensus around the principles and key elements of the new system, and it therefore seems unlikely that there will be any major objection to or departure from them in practice.

IV. Towards Sustainable Network Regulation: RIIO[59]

RIIO stands for setting Revenue using Incentives to deliver Innovation and Outputs. Although presented as a new regulatory formula, it is in reality a collection of regulatory instruments and techniques (a 'regulatory toolkit') intended to meet a range of regulatory objectives in (as far as possible) a complementary and mutually reinforcing manner.

The new system will have a number of features in common with RPI-X regulation. It will continue to be an ex ante, incentive-based, and inflation-linked price control, based on the same 'building blocks'—ie the expected efficient costs of delivery plus an allowed return on capital. The same principles will also be used to calculate the return on capital. Ofgem considered this to be necessary, in the light of its statutory duty to ensure that licence holders can continue to finance their activities,[60] in order to reassure the capital markets that the changes to the regime would not increase regulatory risk.

In other respects, however, the RIIO model will differ substantially.

A. An output-led approach

One obvious novelty is that the default price control period will increase from five to eight years. More importantly, though, the whole basis on which required revenues are calculated will change. The starting point for the price control process under the new system will be the setting of the outputs that each network operator is expected to deliver. Outputs will be in six categories, relating to customer satisfaction, reliability and availability, safety, connection terms, environmental impact, and social obligations (although these will be of varying relevance to different categories of network operator). A limited number of primary outputs will

[58] Although some elements of the new approach have already been incorporated in the 2010–2015 EDNO price control. Work on the new TNO and GDNO price controls has already commenced—see <http://www.ofgem.gov.uk/Networks/Trans/PriceControls/RIIO-T1/Pages/RIIO-T1.aspx> and <http://www.ofgem.gov.uk/NETWORKS/GASDISTR/RIIO-GD1/Pages/RIIO-GD1.aspx>, both accessed 7 October 2011.

[59] Unless otherwise indicated, this account draws upon: Ofgem 2010b (above, n 11); Ofgem 2010c (above, n 5); and Ofgem, *Handbook for Implementing the RIIO Model* (2010d).

[60] Gas Act 1986, s 4AA(2)(b); Electricity Act 1989, s 3A(2)(b).

be set in each category, including both legal and voluntary obligations. Where possible these will remain in place over a number of price control periods, with the expected level of performance against each output determined for each period. In addition, so-called 'secondary deliverables' will be set, representing the means by which the primary outputs are to be delivered in both the current and future price control periods. In other words, network operators will be permitted to raise revenue for long-term investments or innovation projects where these are expected to deliver improved outputs in future, even if there is no immediate benefit for consumers. Secondary deliverables will, however, be less fixed than primary outputs, with operators being permitted to change their plans in the course of a price control period if this is likely to produce better long-term value for money.

Having agreed primary outputs and secondary deliverables, these will then be used to calculate the company's 'base revenue' for the price control period (the expected efficient costs of delivery plus the cost of capital). Total allowed revenue from consumers will consist of base revenue, adjusted in the light of performance indicators, and also adjusted for uncertainty mechanisms. The former will include rewards and penalties for performance against outputs (although a range of performance incentives may be used, including purely reputational incentives, depending upon the output in question) as well as upfront efficiency incentives. The latter will include inflation indexation plus a variety of revenue drivers and reopeners. Outputs will be reviewed mid-way through the price control period and adjustments made to base revenue if necessary. Other aspects of revenue will remain fixed, unless the subject of specific reopeners, which will only be invoked in accordance with transparent principles (and subject to the regulator's duty to ensure that network operators can continue to finance their activities).[61]

The output-led nature of the RIIO model marks a significant departure from RPI-X regulation. Whereas quality controls are currently an add-on to the price control framework, RIIO makes them central, and takes a much more comprehensive approach to capturing desired outcomes. Ofgem argues that this will make it clearer what consumers are getting for their money, and better promote both long- and short-term value for money, by making it easier to distinguish between cost reductions that reflect genuine efficiency savings and those achieved at the expense of reduced delivery. In addition, it should enable the network companies to be held to account for delivery without bias towards any particular delivery methods, thereby providing strong incentives for innovation to drive efficient outcomes. For the same reason, penalties for non-delivery of particular secondary deliverables are unlikely to be applied automatically; instead, assessments will be made on the reasonableness of decisions at the time they were taken. Nor will any retrospective adjustments to allowed revenues be made if investments fail to deliver their expected benefits.

[61] NB although this has never been tested in court, Ofgem interprets this duty as being to ensure that an efficiently run licence holder could finance its activities, not as an absolute guarantee against bankruptcy.

B. Risk-sharing

Also central to the design of the new price control is the desire to strike a balance between facilitating long-term decision-making (in particular, investment in network enhancement in anticipation of demand) and ensuring value for money for consumers. For instance, the increased length of the price control period is balanced by new efficiency incentive arrangements. In place of RPI-X regulation, a profit-sharing scheme has been adopted. This means that the benefits of any efficiency efforts which reduce costs below anticipated levels will be shared with consumers immediately in a fixed proportion. However, the scheme will work symmetrically, which means that cost over-runs will also be shared with consumers (unless they are clearly unreasonable). In other words, consumers will have to share the risk that some assets will become stranded. Thus, the RIIO model aims to reduce investment risks, whilst maintaining efficiency incentives and offering consumers better protection against excessive profits. To further reduce investment risks, Ofgem has undertaken to commit to the principles underpinning the price control process on a long-term basis, especially those relating to financeability, thus avoiding the methodological inconsistencies experienced under RPI-X regulation.

C. 'Smart planning' and stakeholder engagement

The output-led nature of the new scheme represents a decisive shift away from the notion of regulation as a competition-substitute towards a more planned approach. This is in line with developments in EU law,[62] although the new British system appears to have emerged independently. Planning under RIIO is not, however, of the traditional, centralized, top-down, and directive kind. Rather, it might best be described as 'smart planning'. To begin with, the planning process will be substantially decentralized and bottom-up, since Ofgem's assessment of the outputs that the networks are obliged to deliver and the revenue they need in order to do so will be informed to a large degree by long-term business plans that the companies will be required to submit at the start of the price review process. Second, there is considerable emphasis on participation. The network operators will be required to demonstrate 'constructive engagement' with a wide range of stakeholders in drawing up their business plans, and Ofgem, similarly, will undertake enhanced stakeholder engagement when assessing corporate plans and developing price control proposals. The government is nevertheless expected to play a role in the stakeholder engagement process, by providing guidance on policy goals to inform the choice of primary and secondary outputs.

Arguments were put forward during the RPI-X@20 review in favour of stronger stakeholder engagement models, such as negotiated price settlements,[63] but Ofgem ultimately rejected them as being incompatible with its duty to protect consumers.

62 See I. del Guayo and C Pielow, 'Electricity and Gas Infrastructure Planning in the European Union', chapter 19 of this volume.

63 See, eg, S. Littlechild and N. Cornwall (2009), *Potential Scope for User Participation in the GB Energy Regulatory Framework, with Particular Reference to the Next Transmission Price Control Review,*

Nevertheless, it hopes that enhanced engagement will increase the transparency and legitimacy of the price control process, as well as promoting innovation. It may also be hoped that greater participation at an early stage will help to reduce later opposition to new network infrastructure when it enters the land-use planning process, planning delays having been a major problem in securing the connection of new renewable generation. In order to facilitate stakeholder engagement, Ofgem has committed to providing more accessible documentation. It also made proposals to enhance third party rights to trigger Competition Commission references in respect of price control revisions, but these have been overtaken by the more fundamental changes to the licence modification process outlined above.[64]

D. Proportionate regulation

Constructive engagement is also one of the keys to a further distinctive feature of the RIIO model: the adoption of a proportionate approach to regulation. This means that both the intensity of the price control review process, and regulatory rewards and penalties, will vary according to each company's performance. As regards the former, the degree to which Ofgem scrutinizes network operators' business plans will differ according both to the quality of the plan—assessed, inter alia, by reference to the effectiveness of engagement with stakeholders—and previous performance in delivering outputs and value for money. The plans and revenue forecasts of the best-performing companies will be subject to relatively light scrutiny and they may be able to complete the price control review process early; by contrast, the poorest performers will be subjected to much more searching scrutiny. The aim is to harness reputational incentives to encourage the companies to meet performance targets and to produce well-justified business plans. As regards regulatory rewards and penalties, the IQI will continue to be used to encourage companies to submit accurate expenditure forecasts, which means that incentive rates will continue to vary from firm to firm. In addition, Ofgem will employ more rigorous sanctions for poorly performing companies. This could include enforced market testing of elements of their delivery proposals, enforcement action for failure to deliver agreed outputs or, in extreme cases, even licence revocation.

E. Encouraging innovation

Another important element of the RIIO model is its much greater emphasis on encouraging innovation, not merely to reduce short-term costs, but also to meet sustainability objectives and deliver lower costs over the long term. A number of aspects of the new framework have been designed to promote innovation, including, as already noted, its output-led nature, the emphasis on constructive engagement and well-justified business plans, longer price control periods, and risk-sharing and

available at <http://www.ofgem.gov.uk/Pages/MoreInformation.aspx?docid=2&refer=Networks/rpix20/ConsultReports> (accessed 2 October 2011).

[64] See nn 22–3 above and accompanying text.

non-retroactivity, so that companies are not penalized for unsuccessful innovation. In addition, a common efficiency rate will be used for all network costs, to remove potential distortions from differential treatment of capex and opex. Ofgem is particularly keen to ensure that network operators explore a variety of solutions to deliver required outputs, such as new operational practices, new business structures, new commercial relationships, or new charging or financial arrangements, and not merely investment in new technologies or infrastructure.

The expectation is that these measures will in time provide sufficient incentives towards innovation. However, companies may not respond to the new incentives immediately and, given that innovation has some public good characteristics, may not always recognize the commercial benefits to be derived from innovation.[65] Hence, given the short timescale in which substantial innovation is required, a time-limited innovation stimulus has been added to the regulatory package, building on the LCN Fund included in the 2010–2015 EDNO price control. This will consist of two separate funds—one each for gas and electricity—which will award support for particular innovation projects through regular open competitions, judged by an independent panel. The innovation stimulus funding will be raised from consumers via network use of system charges, subject to a cap set by Ofgem for each price control period, and the sums transferred to successful bidders. As under the LCN Fund, successful projects will be obliged to share resulting intellectual property and lessons learned within the industry, to the ultimate benefit of consumers. However, in addition to applying to all networks, there will be two key differences from the LCN Fund. First, funding will be available for a much wider range of innovation activities, ranging from research and development, through trialling and demonstration projects, and will be available for commercial as well as technological innovation. Second, the fund will be open to bids not only from network companies, but also from third parties. A new category of licensable activity will have to be created for this purpose, and non-network bidders will be required to hold an innovation licence in order to be eligible to bid for funding.

F. Competitive tendering

The rationale for bringing third parties within the scope of the innovation stimulus package is, presumably, that they will be a source of fresh ideas and a means of overcoming obstacles to change from within the network companies. Similar thinking seems to be behind the final notable aspect of the new regulatory framework, namely the decision to include the option of using competitive tendering within the regulatory toolkit. As well as being used as a sanction for poor performance, companies will be expected to consider in their business plans the option of outsourcing aspects of delivery. In addition, Ofgem may decide to employ competitive tendering for the delivery of key projects, where these are large-scale and separable from other network assets. This is similar to the approach already used for offshore transmission lines.

[65] Ofgem 2010a (above, n 54) 34.

V. Conclusion

There are undoubtedly very serious challenges facing energy networks in Britain, as elsewhere. Ofgem is therefore to be congratulated for having grasped the sustainable network regulation nettle; in other words, for having attempted to redesign the regulatory system to deliver long-term security and environmental objectives in conditions of considerable uncertainty, whilst still ensuring value for money. In so doing, it appears, once again, to be leading the field internationally in the development of energy regulation.

It has been argued in this chapter that, in some ways, the RIIO model is an evolutionary development from the existing system of network regulation, especially since there have already been considerable changes to the regime since the gas and electricity industries were first privatized—partly in response to the shortcomings of RPI-X regulation and partly in response to new regulatory objectives. In addition, although some aspects of the new system are highly innovative—in particular, its output-led nature—it shows evidence of learning from the considerable developments in, and much greater sophistication of, both regulatory theory and practice since RPI-X regulation was first devised.[66] For example, the use of multiple regulatory instruments is consistent with the insights of 'smart regulation', which suggests that combinations of regulatory instruments, if carefully designed, are likely to be more effective than single instruments used in isolation.[67] Ofgem's proportionate approach to regulation echoes the idea of 'responsive regulation',[68] which recommends tailoring enforcement strategies to the level of compliance demonstrated by particular firms, saving the regulator's 'big guns' for the times when they are really needed. Similarly, the role of business planning, constructive engagement, and reputational incentives suggest the influence of theories of 'reflexive regulation',[69] which attempt to address informational, cognitive, and other resource constraints by decentralizing standard-setting and compliance. Constructive engagement models have also been used by the UK airports regulator, the Civil Aviation Authority, and by regulators in other countries.[70]

At the same time, though, given the centrality of RPI-X regulation to the British regulatory model, the reform of energy network regulation is a remarkable example of regulatory reinvention, especially for a mature regulatory agency, and one which has hitherto been closely associated with what might be termed a 'market fundamentalist' approach. The final abandonment of the RPI-X formula represents a decisive break with the notion that the regulator is no more than a 'competition-substitute',

[66] For a survey, see B Barton, 'The Theoretical Context of Regulation', in B Barton et al (eds) *Regulating Energy and Natural Resources* (Oxford: Oxford University Press, 2006).

[67] N. Gunningham, P. Grabovsky, and D. Sinclair, *Smart Regulation: Designing Environmental Policy* (Oxford: Oxford University Press, 1998).

[68] I. Ayres and J Braithwaite, *Responsive Regulation: Transcending the Deregulation Debate* (Oxford: Oxford University Press, 1992).

[69] G. Teubner, 'Substantive and Reflexive Elements in Modern Law' (1982), 117 *Law and Society Review* 239; 'After Legal Instrumentalism? Strategic Models of Post-Regulatory Law' (1984), 12 *International Journal of the Sociology of Law* 375.

[70] See n 63 above.

engaged in a limited and technical task. Rather, the new system clearly acknowledges that the regulator's role is to drive forward a positive policy agenda, and that this inevitably involves balancing a range of sometimes conflicting goals and values. This has been accompanied by Ofgem's somewhat belated conversion to a stakeholder model of regulatory accountability,[71] seeking to legitimize regulatory decisions through a transparent and participatory process. One can only speculate here as to the reasons for this change in attitude, but it probably results from a mixture of factors, including the cumulative impact of repeated changes to its statutory duties over the past decade, cultural changes within the organization brought about by increased responsibility for the delivery of environmental programmes, and the galvanizing effect of strong government policy commitments in the form of legally binding climate change and renewables targets. The fact that energy prices are now rising rather than falling probably also makes a process-based model of legitimacy more appealing than an outcome-based one.

It is obviously too early to tell whether the RIIO model will be successful. There are, however, some grounds for concern. The first relates to timing. Since the new price controls will not be implemented until 2013 or 2015, there will be less than one price control period in which to achieve the turnaround necessary to meet the UK's 2020 renewables targets, and only two periods in which to achieve full decarbonization of the electricity industry, required by 2030. Second, whilst the regulatory system may be seeking to induce the networks to adopt a longer-term focus, as privately owned companies, their core incentive is still to make profits, and the UK capital markets are notorious for their short-termism. Third, it is questionable whether unbundled network operators and decentralized planning are capable of producing a sufficiently holistic approach to the future development of energy network infrastructure. Finally, there are likely to be limits to the effectiveness of stakeholder engagement. For one thing, effective engagement places heavy cognitive and resource demands on potential participants. Although Ofgem has sought to make the new regulatory system simpler and more accessible than the old one, it is questionable how far this aim has been achieved—or, indeed, how far it could be achieved, given the complexity of the issues involved in contemporary network regulation. In addition, there is particular cause for concern about effective consumer engagement. The statutory consumer body for the energy sector, Consumer Focus, is to be abolished as part of an austerity-driven 'bonfire of the quangos'. Although it is currently intended to transfer its statutory functions in the energy sector to the voluntary sector organizations, Citizens Advice and Citizens Advice Scotland,[72] these are primarily local consumer advice bodies, with little experience of engagement with the kind of issues involved in infrastructure planning and network price regulation.

[71] These ideas first gained popularity during the 1990s—see, eg, D. Souter, 'A Stakeholder Approach to Regulation', in D. Corry, D. Souter, and M. Waterson (eds), *Regulating Our Utilities* (London: IPPR, 1994).

[72] Department for Business, Innovation and Skills, *Empowering and Protecting Consumers: Consultation on Institutional Changes for Provision of Consumer Information, Advice, Education, Advocacy and Enforcement* (2011).

For all these reasons, it may be that what is needed is less smart planning and more old-fashioned planning, with greater central direction as to the future development of the energy networks, in order to reduce uncertainty and provide greater impetus for change. There have in fact been moves from government to provide greater guidance as to the direction of future energy policy, for instance, through national planning statements,[73] the analysis of options for achieving the 2050 climate change target,[74] and, most recently, via a commitment to publish a new Strategy and Policy Statement for the energy industries, which is intended both to improve regulatory accountability and to increase regulatory certainty.[75] So far, however, the government has been unwilling to commit to any particular vision of what future energy networks should look like. There are undoubtedly risks involved in premature commitment to a particular development path, especially given the very long timescales involved.[76] At the same time, though, there is a clear danger of the best being the enemy of the good. In other words, unwillingness to commit to particular options for fear that sustainability objectives might not be met in the most effective manner may mean that these goals are missed altogether. As renewable energy and climate change targets have demonstrated, clear and credible policy commitments can provide the kind of impetus for change that regulatory constraints and incentives—no matter how cleverly designed—cannot achieve by themselves.

[73] Above n 13.

[74] Department for Energy and Climate Change, *2050 Pathways Analysis* (2010).

[75] DECC, *Ofgem Review: Summary of Conclusions* (2011).

[76] For discussion, see Ofgem 2009d (above, n 53).

18

Third Party Access Exemption Policy in the EU Gas and Electricity Sectors: Finding the Right Balance between Competition and Investments

Tjarda van der Vijver

I. Introduction

The liberalization of the gas and electricity markets has been a top priority of the European Union (EU) for many years now. Achieving this goal is not an easy task. On the one hand undistorted competition and market contestability is essential. On the other hand, as delivery of gas and electricity is bound to a network, these markets can only function properly if the relevant networks are adequately maintained and expanded where necessary. Stimulating investments in the networks is therefore of crucial importance as well.

Unfortunately, the interests between free competition and upholding investments in energy networks may not always be aligned. New infrastructures can be particularly susceptible to this dilemma, as investors often prefer limited or no contestability on their networks. The European regulatory framework for gas and electricity tries to create a balance in this situation. It demands access for third parties (which should support competition) while at the same time enabling an exemption for new infrastructures from this requirement if certain conditions have been met (which should encourage investment).

This chapter examines how the relevant legislation and decisional practice deals with this conundrum. I shall focus in particular on the balance the European Commission ('the Commission') attempts to strike between luring investments and ensuring unimpaired competition.[1] This analysis attempts to discern common themes and developments in the Commission's decisional practice.

[1] This chapter makes use of earlier articles; see 'Exemptions to Third Party Access for New Infrastructures in the European Community Gas Sector—The Exception that Defies the Rule?' (2008), 29 *European Competition Law Review* 229; 'Commission Policy on Third Party Access Exemption Requests for New Gas Infrastructure' in M.M. Roggenkamp and U. Hammer (eds), *European Energy Law Report VI* (Antwerp: Intersentia, 2009) 115.

II. Third Party Access

Gas and electricity are so-called 'network-bound' sectors. Competition and market contestability largely depend on effective possibilities for new entrants to gain access to storage (for gas), transmission and/or distribution infrastructures. This issue has brought to life the concept of third party access (TPA). In essence TPA is a legal norm that prescribes network owners or operators to allow (potential) competitors access to their networks. The norm may apply to existing and new infrastructures alike.

Access can be required by competition law or by regulation.[2] Experience has shown that the *ex post* rules of competition law are particularly cumbersome to apply, however.[3] In the EU context it requires a finding of an abuse of dominance; a difficult threshold to meet. The Commission seems to have been more influential by its adoption of several commitment decisions in the wake of the 2007 energy sector inquiry. Several European energy behemoths, ranging from GDF Suez to ENI and E.ON, have agreed to divest (parts of) their networks in order to facilitate access for competitors.[4] Although the Commission did not reach a conclusion as to an infringement of the competition rules, these commitment decisions have been instrumental in taking the European energy liberalization process a step further. A downside of the widespread use of commitment decisions, however, is its fragmented approach.

A less case-by-case approach to TPA is provided by *ex ante* regulation. Regulatory rules can be made much more specific than the often vague principles of competition law. As a consequence, regulation often yields a greater impact. In terms of market access, regulation is often more potent than competition law. Competition law regards TPA as the exception to the rule that companies—even dominant ones—are *not* required to proactively aid their competitors. Regulation, however, establishes TPA as the principle with a limited number of possible exemptions.

It is clear that TPA benefits market access. However, a strict application of TPA may have downsides. The construction of new gas and electricity infrastructures entails enormous costs. Far-going access provisions may decrease incentives to finance costly new infrastructure projects, as it reduces investor certainty as to its

[2] In this chapter I don't delve into the possibilities that *ex ante* merger control might allow. Such a review might lead the competition authority to adopt undertakings that determine that the merging parties must provide access under certain conditions. However, as behavioural remedies are cumbersome to enforce, they are rarely imposed.

[3] In essence, competition law can only mandate access where the conditions of the so-called 'Essential Facilities Doctrine' have been met. It has been put forward by the European Court of Justice in judgments such as ECJ 26 November 1998, Case C-7/97, *Oscar Bronner GmbH & Co KG v Mediaprint Zeitungs- und Zeitschriftenverlag GmbH & Co KG*, [1998] ECR I-7791; ECJ 29 April 2004, Case C-418/01, *IMS Health GmbH & Co OHG v NDC Health GmbH & Co KG*, [2004] ECR I-5039.

[4] See, eg, the *ENI* case. On 29 September 2010, the Commission announced that ENI would divest its shares in three transport pipelines to Italy. This will ensure that TPA requests will be dealt with by an entity independent of ENI (press release IP/10/1197).

use.[5] There is thus a tension between TPA (risking underinvestment) on the one hand and exclusivity (risking foreclosure) on the other. A balance must therefore be struck between TPA and the need for investments.[6] The sector-specific gas and electricity legislation enables National Regulatory Authorities (NRAs) to exempt new gas infrastructures from TPA, provided that certain conditions have been met and that the Commission gives its consent. The following section will explore the legal framework.

III. The Exemption Regime of the Gas and Electricity Directives

A. Introduction

EU member states have already implemented two major strands of energy regulation, also referred to as 'packages'. Alas, the first two packages have been widely considered to be insufficiently effective to establish a 'true' EU-wide internal market for gas and electricity. This explains the adoption of the so-called third energy package in 2009.[7] Both the European electricity and gas legislation determine that regulated TPA must be provided.[8] This means that member states should publish and regulate access tariffs. Both legislative tools contain the possibility to exempt new infrastructures from TPA. These shall be explained below.

The gas and electricity exemption regimes are basically governed by the same procedure. The exemption is initially reviewed and decided upon by the competent NRA in the relevant member state. Subsequently, the Commission scrutinizes the NRA exemption decision. The Commission has three options: it may consent to the decision, demand its amendment, or demand its withdrawal.

B. Gas

Article 32 of the third gas directive establishes the general rule that TPA must be provided. Article 33 is a *lex specialis* on the access regime vis-à-vis gas storage. Article 36 determines that 'major new gas infrastructures' may, upon request, be exempted for a defined period of time from article 32. A 'major new gas infrastructure' is understood as any of the following three infrastructures: interconnectors, liquefied natural gas (LNG) facilities, and storage facilities. According to article

[5] See E. Cross, B. Delvaux, L. Hancher, P.J. Slot, G. Van Calster, and W. Vandenberghe, 'EU Energy Law' in M.M. Roggenkamp, A. Rønne, C. Redgwell, and I. Del Guayo (eds), *Energy Law in Europe* (2nd edn, Oxford: Oxford University Press, 2007) 260.

[6] It should be noted that TPA is by no means the only factor which can provide a disincentive for investors; other issues such as uncertainty on future regulatory policies or the level of tariffs are obviously relevant as well.

[7] Available at <http://ec.europa.eu/energy/strategies/2007/2007_09_package_electricity_gas_en.htm> (accessed 17 October 2011).

[8] Directive 2009/72/EC and Regulation 714/2009 apply to the electricity market. Directive 2009/73/EC applies to the gas market.

36, an exemption of TPA shall be granted if the new gas infrastructure meets the following conditions:

- the investment must enhance competition in gas supply and enhance security of supply;
- the level of risk attached to the investment must be such that the investment would not take place unless an exemption was granted;
- the infrastructure must be owned by a natural or legal person which is separate at least in terms of its legal form from the system operators in whose systems that infrastructure will be built;
- charges must be levied on users of that infrastructure; and
- the exemption must not be detrimental to competition or the effective functioning of the internal market in natural gas, or the efficient functioning of the regulated system to which the infrastructure is connected.

C. Electricity

For the electricity sector the relevant piece of legislation is Regulation 714/2009.[9] This regulation deals with cross-border access issues in the electricity sector. Article 17 of Regulation 714/2009 determines that 'new direct current interconnectors' may be eligible for an exemption from TPA. The exemption criteria are similar to those for new gas infrastructures, but there are also a number of specific conditions:

- the investment must enhance competition in electricity supply;
- the level of risk attached to the investment is such that the investment would not take place unless an exemption is granted;
- the interconnector must be owned by a natural or legal person which is separate at least in terms of its legal form from the system operators in whose systems that interconnector will be built;
- charges are levied on users of that interconnector;
- since the partial market opening referred to in article 19 of Directive 96/92/EC of the European Parliament and of the Council of 19 December 1996 concerning common rules for the internal market in electricity, no part of the capital or operating costs of the interconnector has been recovered from any component of charges made for the use of transmission or distribution systems linked by the interconnector; and
- the exemption must not be to the detriment of competition or the effective functioning of the internal market in electricity, or the efficient functioning of the regulated system to which the interconnector is linked.

[9] OJ [2009] L 211/15.

D. Explanation of the criteria

The Commission's 2004 Interpretation Note offers a first set of guidelines that should be taken into account while examining an exemption request.[10] This document stresses that exemptions shall only be granted exceptionally. Since TPA is a legal principle, exemptions should only be awarded *if* and *insofar* as they are necessary. This requires a necessity and proportionality test. Moreover, an exemption should not be granted to a dominant undertaking if it is likely to have a significant amount of its capacity allocated in one of the markets affected.[11]

As will be apparent from the conditions laid down above, the criteria are broadly similar for both gas and electricity TPA exemptions. The new infrastructure must enhance competition, cannot be detrimental to competition, and the level of risk associated with the infrastructure is sufficiently high as to necessitate an exemption. These conditions will usually be the key hurdles in obtaining an exemption.

There are a number of differences between the conditions for gas and electricity infrastructures. The most important difference is that an exemption request for an electricity interconnector does not require an improvement of the security of supply, whereas a gas exemption does. This difference might originate from historic experience that the supply of gas can be volatile, as it is more sensitive to geopolitical pressure than electricity supply. In addition, security of supply may be more relevant for gas, as it is a primary source of energy. In the electricity sector one source of power generation can be more easily replaced by another.

E. An overview of the NRA and Commission's decisional practice

1. *Introduction*

As regards cross-border electricity interconnectors, exemptions have been requested, *inter alia*, for the BritNed, Estlink, East–West Cables and Tarvisio–Arnoldstein projects.[12] These cases shall be examined in section V of this chapter.

As regards gas infrastructures, exemptions have been requested, *inter alia*, for the construction of various LNG terminals,[13] one gas storage facility (Dambořice), as well as gas interconnectors between the UK and the Netherlands (the Bacton–Balgzand Line, or BBL), between Italy and Greece (Interconnection Greece–Italy, or IGI; also named Poseidon), between Germany and the Czech Republic (Opal), between Bulgaria, Romania, Hungary, and Austria (Nabucco), and within the Czech Republic (Gazelle). Section IV discusses the developments that can be discerned from the *Nabucco* and *Opal* decisions, compared to the earlier exemption decisions. Before entering into the examination of these relatively recent cases,

[10] See the Interpretation Note of DG Energy & Transport on Directives 2003/54–55 and Regulation 1228/03 in the electricity and gas internal market, 'Exemptions from certain provision of the third party access regime' (30 January 2004), available at <www.ec.europa.eu/energy/electricity/legislation/doc/notes_for_implementation_2004/exemptions_tpa_en.pdf> (accessed 17 October 2011).

[11] Ibid, at 4–5.

[12] Available at <http://ec.europa.eu/energy/infrastructure/exemptions/doc/exemption_decisions.pdf> (accessed 17 October 2011).

[13] Ibid.

however, it is useful to go through the main observations and conclusions of the earlier exemption decisions in the gas sector.

2. *The early gas exemption decisions*

In previous articles I have argued that the early gas exemption requests did not lead to particularly strict reviews by the NRAs.[14] Most of these NRA decisions suggested that competition in the gas market will benefit directly from an increase in total network capacity, which consequently justified a TPA exemption. This approach can be observed in early cases in the UK,[15] the Netherlands,[16] and Italy.[17] This seems a somewhat circular way of reasoning: since *any* new infrastructure will enlarge capacity, this key criterion is often considered to be met with relative ease. In a similar vein, the security of supply is believed to benefit from an increase in capacity brought about by the new infrastructure.

The lenient stance by NRAs can be explained in several ways. Most new infrastructures are believed to improve the integration of an EU gas sector that is still largely partitioned along national lines. New infrastructures are also likely to improve the diversification of energy supply.[18] Besides, most of the early exemptions were requested by a new entrant. It is understandable that NRAs and the Commission attempt to encourage their entry into the market, as this is likely to decrease market concentration and increase competition.

Still, the reviews of the early exemption decisions could have been more thorough. Since TPA is a legal principle, exemptions should only be awarded *if* and *insofar* as they are necessary. A necessity and proportionality test should therefore be applied with regard to the duration of the exemption and the percentage of capacity to which it applies. The early exemption decisions do not show a clear adherence to such a review.

3. *Moving to a stricter regime*

Notwithstanding the early approach described above, over the years the exemption decisions started to show that the assessments in the gas sector were gradually getting stricter. The Italian NRA has granted exemptions relating to 80 per cent of the capacity instead of the full capacity of the Rovigo and Brindisi LNG terminals.[19] In addition, the Commission seems more likely to impose various conditions in order to ensure that capacity hoarding will not take place. For instance, in the

[14] See the articles mentioned earlier, above n 1.

[15] See, eg, the *Grain* decision, UK NRA Ofgem final views of December 2004, para 2.34 (available at <http://www.ofgem.gov.uk>, (accessed 17 October 2011).

[16] See, eg, the *BBL* decision, Netherlands NRA DTe advice (public version) based on the initial exemption request of 25 November 2003, paras 22–4 (available at <http://www.nma.nl/images/12_1463922-149992.pdf>, (accessed 17 October 2011)).

[17] See, eg, the *Rovigo* decision, Italian NRA AEEG advice of 23 November 2004, available at <http://www.autorita.energia.it/docs/04/206-04.htm> (accessed 17 October 2011)).

[18] See, for instance, the *Brindisi* LNG project in Italy, which would attract gas from Libya.

[19] The NRA assessment of *Brindisi*'s is available at <http://www.autorita.energia.it/docs/05/046-05.htm> (accessed 17 October 2011).

case of the Anglo–Dutch gas pipeline BBL,[20] the Commission stressed that the exemption ought only to apply in respect of the duration and capacity covered by the initial contracts. In addition, the Commission demanded that the exemption would not apply to the reverse flow nominations.

In May 2009, the Commission published a staff working document explaining its policy on gas and electricity exemption decisions.[21] The document was meant to alleviate the problem that the exemption conditions were insufficiently clear.[22] The Commission states the need that exceptions have to clearly meet the necessity requirement.[23] Exemptions should thus be strictly limited to what is necessary, both as regards scope and duration. Partial exemptions are generally to be preferred over full ones. The Commission suggests that NRAs should consider to what extent a planned project would constitute a natural monopoly. In such a case the access regime must be particularly strict; indicating that an exemption from TPA will not be given easily.[24]

To ensure a proper analysis, the Commission stresses that project promoters should perform a market test to measure demand before they can obtain an exemption.[25] The Commission's preference seems to go out to an Open Season procedure, but other means are not excluded *a priori*. Project promoters should not only test the market demand, but should also act accordingly. This may even entail a requirement on project developers to make changes in their plans if the market demand proves greater than they had first anticipated.

Dominant parties are likely to receive a particularly high level of scrutiny.[26] If a dominant undertaking is the beneficiary of an exemption, effective counterbalancing measures should ensure that competition is not impaired.[27] The Commission argues that such exempted infrastructure should, at a minimum, provide significantly improved possibilities for non-dominant undertakings to enter the concerned markets or to expand their market position. In addition, the Commission goes as far as to state that 'the new infrastructure should have the effect of *diluting the market power of the dominant undertaking*' [emphasis added by author].[28]

The Commission suggests that the share of capacity in the new infrastructure held by a dominant firm should be substantially lower than its market share. It is not clear how far the Commission is willing to take this statement—at first sight it seems that this would disqualify any exemption request from a dominant firm with a shareholding in the project that is higher than its current market share. It is submitted that one must be careful with focusing too much on the relation between a dominant firm's present market share and its share of capacity in the new infrastructure. It could lead to a paradoxical situation in which a (quasi-)monopolist

[20] Commission amendment request in the *BBL* case of 12 July 2005, available at <http://ec.europa.eu/energy/gas/infrastructure/doc/bbl_decision_ec.pdf> (accessed 17 October 2011). One should take into account, however, that the *BBL* decision was somewhat specific, as the authorization procedure had already started before the adoption of Directive 2003/55.

[21] Commission staff working document on Article 22 of Directive (EC) No. 2003/55 and article 7 Regulation (EC) No. 1228/2003, available at <http://ec.europa.eu/energy/gas_electricity/interpretative_notes/doc/sec_2009-642.pdf> (accessed 17 October 2011).

[22] Ibid at para 5.

[23] Ibid at para 17.

[24] Ibid at para 21.

[25] Ibid at para 28.3.

[26] Ibid at para 33.

[27] Ibid at para 34.

[28] Ibid.

would be allowed to book a much higher share of capacity than a firm that barely meets the dominance threshold—even though the rules should be toughest on those with the highest degree of market power.

Dominance is evidently a major theme. Apart from the Commission's observation that competition law principles should be applied,[29] it is not clear how this concept should be interpreted in the context of new energy infrastructure. The Commission suggests that the competition assessment should have a strong forward-looking element.[30] This may lead to analytical difficulties if there are strong indications that a firm's current dominance may fade in the (near) future. What has the priority in such a case, the use of *ex post* dominance analysis or a forward-looking approach? Perhaps this issue can be resolved by using a merger control-style approach: future events, including the creation or strengthening of a dominant position, could then be sufficiently taken into account.

Another difficulty could arise from the Commission's statement that the new infrastructure should dilute market power. Although exemptions should obviously not be given lightheartedly to dominant undertakings, it must also be acknowledged that such firms are often important players when it comes to setting up new infrastructure. An overly strict approach to dominant firms entails the risk of greatly decreasing incentives to invest in new infrastructures; which may be to the detriment of the market as a whole.

IV. Case Study—Gas

A. Introduction

This paragraph will examine the *Nabucco*, *Opal*, *Gazelle,* and *Dambořice* exemption decisions in the gas sector, as these have not been (fully) discussed in my previous articles.[31]

B. Nabucco

The Commission's stance in the *Nabucco* case offers a strong indication that its review is relatively strict.[32] The Nabucco project entails a gas pipeline stretching from Turkey to Austria. It has a planned maximum capacity of 31 billion cubic

29 Ibid at para 32. 30 Ibid at para 35.

31 See for a more recent analysis also L. Hancher, 'Cross Border Infrastructure Projects: The EU Exemption Regime', *TILEC Discussion Paper No. 2011-006*, available at <http://papers.ssrn.com/sol3/papers.cfm?abstract_id=1749117> (accessed 17 October 2011). I shall not examine the various exemption decisions with regard to LNG terminals, as these do provide little added value vis-à-vis the cases discussed in this chapter and in previous articles (above n 1).

32 Commission exemption decision in the *Nabucco* case (Austrian section) of 8 February 2008; Commission exemption decisions in the *Nabucco* case (Bulgarian and Hungarian sections) of 20 April 2009, available via <http://ec.europa.eu/energy/infrastructure/exemptions/doc/exemption_decisions.pdf> (accessed 17 October 2011). Further references shall be made to the last exemption decision, namely vis-à-vis the Romanian section of *Nabucco* (below n 34).

metres per year. Nabucco is projected to deliver gas from alternative sources such as Azerbaijan.[33] The company requesting the exemption, Nabucco International, is a joint venture of various energy companies on whose territory the pipeline will be built: BOTAŞ (Turkey); Bulgargaz (Bulgaria); TRANSGAZ (Romania); MOL (Hungary); and OMV (Austria). Apart from Turkey, all countries through which Nabucco runs are EU member states where exemptions must be requested. I shall examine the common themes that run through the Commission decisions as regards the Austrian, Romanian, Bulgarian, and Hungarian sections of Nabucco. The first request was lodged in Austria, followed by the other three countries. The Commission's approach is essentially the same on all the sections of Nabucco.

The relevant NRAs consented to the TPA exemption requests. The Commission agrees with a partial TPA exemption for Nabucco for 25 years. In practice, it comes down to the following. The pipeline owners have the first option to use 50 per cent of Nabucco's capacity. If they do not use this capacity, the principle of 'use it or lose it' (UIOLI) applies, meaning that the capacity will have to be offered to the market. The other 50 per cent of Nabucco's capacity will be subject to TPA.

The Commission took due account of the major technical, economic, and political risks that the Nabucco pipeline bears.[34] This indicates that an exemption request must be seen in its full context; the risks involved cannot be reduced to a mere equation of expected costs and revenues. In this case, a delay in the conclusion of an international agreement applicable to Nabucco was thought to be to the detriment of the project. The Commission also took into account the worsening of Nabucco's business case as a result of severe delays of the start of the exploration of a gas field that is meant to supply Nabucco.[35]

In all the markets concerned, incumbents dominate the upstream trade and supply of gas.[36] In addition, these countries are particularly dependent on gas imports from Russia, which creates a dependence on the Russian export monopolist Gazprom. The Commission perceives a risk that the project may strengthen one or even more dominant positions in the markets concerned. It is unclear, however, from the Commission's analysis why the position of Gazprom is particularly relevant. As Nabucco is meant to supply gas from other sources than Russia, it would seem that the project is more likely to weaken, rather than strengthen Gazprom's position on the Austrian upstream gas supply market.

In a worst-case scenario, it is conceivable that the (current) dominant undertakings could become the capacity holder of a large part of the exit points in their respective countries. A single dominant market player may be able to book between 50 per cent and 100 per cent of the annual domestic exit capacity, *even* if there is an interest by non-dominant firms, partly because there are no clear rules on how the capacity in the general Open Season will be allocated in the case of overbooking. The fact that the Commission takes into account a worst-case scenario from a

[33] As opposed to the EU's traditional gas suppliers, such as Russia.

[34] Commission exemption decision in the *Nabucco* case (Romanian section) of 23 June 2009, available via <http://ec.europa.eu/energy/infrastructure/exemptions/doc/exemption_decisions.pdf>, (accessed 17 October 2011), at para 34.

[35] Ibid.

[36] Ibid at para 40 *et seq.*

competition point of view seems to entail a requirement for the NRAs to include such a scenario when reviewing an exemption application. In the Commission's view, the anti-hoarding and liquidity increasing measures proposed by the NRAs were insufficient to alleviate these worst-case scenario risks.[37]

When dealing with the security of supply criterion, the Commission takes a rather broad view. It seems to take geo-political considerations into account by considering that the EU will have to compete for gas with other parts of the world, such as China.[38] It also states that the Nabucco pipeline will run through countries that are politically relatively stable; enhancing the security of supply. Nabucco also provides a better interconnection and higher import capacity for European countries that are currently not well connected.[39] The Commission considers the security of supply criterion to be met.

The Commission referred to various factors that meant that the level of risk is sufficient to merit an exemption. It mentioned regulatory risks, the sheer financial scale of the project, as well as the long-term commitment necessary to obtain upstream supply.[40] The Commission does not provide a lengthy analysis on the specific risks involved, or on the causal relationship between the lowering of the risks and the provision of the exemption. The ownership and charges conditions, which I do not consider to be particularly difficult to fulfill, are also considered to be met.

As to the condition that the exemption given may not be detrimental to competition, the Commission argues that Nabucco is important for the development of the internal gas market by increasing liquidity resulting from additional import capacity and short-term capacity. The Commission considers the condition to be met, since the exemption does not foreclose the realization of possible alternative projects to be built without the need of an exemption.[41] If this is the underlying criterion, the threshold of article 36's fifth condition is relatively low; as an exemption of a new infrastructure is unlikely to create a virtual impossibility to build another infrastructure.

The Commission puts forward a number of amendments in order to alleviate the competitive risks. The Commission limits the capacity that a single dominant firm may hold to a maximum of 50 per cent.[42] This applies to the total of all exit points of the four EU member states through which the Nabucco pipeline runs. The Commission acknowledges that the capacity caps do risk preventing new investments into capacity expansion. Therefore, the Commission allows derogation from the caps if there is insufficient interest from third parties to book capacity. Even though this would allow dominant firms to book more than what would normally be allowed based on the capacity cap, the additional gas volume must be offered to the market in an open, transparent, and non-discriminatory procedure subject to the approval of the NRA.

[37] Ibid at para 50. [38] Ibid at para 53. [39] Ibid at para 55. [40] Ibid at para 62.

[41] Ibid at para 67.

[42] Ibid at para 70. See similarly para 71 of the Commission decision in the *Shannon* case of 27 July 2010, available at <http://ec.europa.eu/energy/infrastructure/exemptions/doc/doc/gas/2010_shannon_decision_en.pdf> (accessed 17 October 2011).

Another obligation that the Commission imposes as regards the Romanian, Bulgarian, and Hungarian sections is to increase the capacity according to demand. Such an expansion can be achieved by building additional pipeline capacity or by raising the quantity of gas that can be transported by increasing the gas pressure in the pipeline. In order to avoid an overly burdensome system, this requirement will apply only if binding capacity requests are made which amount to an additional aggregated transport capacity of 1.0 bcm/year.[43] Of course, the Commission's specific obligations still leave intact the usual requirements to allocate all capacity in a transparent and non-discriminatory procedure and to inform the competent regulatory authority accordingly.[44]

C. Opal

The Opal project is a joint venture by WINGAS and E.ON. Opal is a gas pipeline creating a link between the Nord Stream pipeline (which runs from Russia straight to Germany) from Northern Germany to the Czech border. In March 2009 the German NRA BNetzA granted a TPA exemption for 22 years.[45] The exemption applies to all gas flows from Germany to the Czech Republic. In its decision, BNetzA made TPA conditional upon various requirements with the aim to prevent capacity hoarding and ensure proper congestion management.

In June 2009 the Commission approved the national decision while imposing various amendments.[46] The Commission considered that Opal is beneficial to the security of supply, partly because it will be supplied by Nord Stream.[47] However, the Commission was not yet convinced that the project would be beneficial to competition.[48] Similar to the *Nabucco* case, the Commission was mainly concerned about the risk that dominant firms would be able to book a high percentage of capacity. The Commission laid down that a 50 per cent cap be imposed on the capacity that any dominant firm may acquire. The cap can be exceeded if at least three billion cubic metres per year are released to the market.[49]

The Commission's risk assessment seems to relate largely to the position of Gazprom. It would have been instructive if the Commission would have offered a more elaborate reasoning as to the dominance of Gazprom. By virtue of its export monopoly, Gazprom has a dominant position as to the supply of Russian gas. However, it is less clear from the Commission's decision how this position impacts the Opal pipeline. Surely, Nord Stream's gas will come from Russia (and therefore from Gazprom), but why does that necessarily mean Gazprom will remain a dominant player throughout the various segments of the Opal pipeline? How far

[43] Ibid at para 84. [44] Ibid at para 85.

[45] The national exemption decision in the *Opal* case is available at <http://www.bundesnetzagentur.de/DE/DieBundesnetzagentur/Beschlusskammern/1BK-Geschaeftszeichen-Datenbank/BK7-GZ/2008/2008_001bis100/BK7-08-009_BKV/BK7-08-009_Beschluss_vom_25022009.pdf?__blob=publicationFile> (accessed 17 October 2011).

[46] Commission exemption decision in the *Opal* case of 12 June 2009, available at <http://ec.europa.eu/energy/infrastructure/exemptions/doc/doc/gas/2009_opal_decision_de.pdf> (accessed 17 October 2011). [47] Ibid at para 37.

[48] Ibid at para 78. [49] Ibid at para 89.

downstream must one go before Gazprom loses its dominance? It would be worthwhile if future analysis by the Commission delves into these matters, in order to provide more legal certainty.

D. Gazelle

In May 2011 the Commission issued its decision with regard to the Gazelle interconnector. Gazelle is a project by the Czech TSO and has a capacity of approximately 30 billion cubic metres per year. It forms part of the wider Nord Stream project, as it connects to the OPAL pipeline.[50] The Czech NRA ERU agreed with an exemption of Gazelle's full direct forward-flow capacity for a period of 23 years.

The Commission agrees with ERU that the construction of Gazelle will have an overall positive effect on the Czech gas markets. In terms of the upstream production markets, Gazelle will not strengthen Gazprom's market power. The Commission observes that there will be a change of gas transits to OPAL, which means that the construction of Gazelle ensures that competitors of Gazprom still have sufficient import capacities.[51] The Commission also holds that the project does not lead to a strengthening of the market position of RWE Transgas; the most important player on the Czech wholesale and retail markets.[52]

The Commission does impose a number of amendments, however. Gazelle's operator will have to offer adequate capacity for bi-directional gas flows, ensuring that Gazelle enhances the security of supply.[53] Another amendment is that ERU must ensure that infrastructure users are entitled to trade their contracted capacities on the secondary market.[54] Finally, the Commission stresses that the risks surrounding Gazelle are closely linked to OPAL, and therefore aligns the exemption's duration of these two projects.[55]

E. Dambořice

In June 2011 the Commission issued its decision with regard to the Dambořice Underground Gas Storage (UGS) facility. Dambořice is a 580 million cubic metre infrastructure in the Czech Republic.[56] It will be indirectly owned by MND, formerly a state-owned company. Including Dambořice MND will control approximately 25 per cent of the total storage capacity in the Czech Republic. The Czech

[50] Commission exemption decision in the *Gazelle* case of 22 May 2011, available at <http://ec.europa.eu/energy/infrastructure/exemptions/doc/doc/gas/2011_gazelle_decision_en.pdf> (accessed 17 October 2011), para 8.

[51] Ibid at para 33.

[52] Ibid at para 34.

[53] Ibid at para 44 and 51, referring to article 7 of Regulation No. 994/2010.

[54] See article 5 of the *Gazelle* decision (ibid).

[55] Ibid at paras 55–56. This means the exemption applies for 22 years instead of 23 years.

[56] Commission exemption decision in the *Dambořice* case of 27 June 2011, available at <http://ec.europa.eu/energy/infrastructure/exemptions/doc/doc/gas/2011_damborice_decision_en.pdf> (accessed 17 October 2011).

Ministry of Industry and Trade allowed a 15-year-long exemption relating to 90 per cent of Dambořice's capacity, and notified its decision to the Commission.

The Commission does not consent with the notified decision. It is skeptical of the project's competitive effects vis-à-vis the Czech wholesale gas markets and its benefits as regards the security of supply.[57] The Commission's key criticism, however, relates to the level of risk involved in the project. There is insufficient evidence that investments in the Dambořice project would not take place unless an exemption is granted. The domestic decision suggests that long-term contracts are necessary for investor approval, and that TPA provisions preclude such contracts. The Commission has a different reading of the relevant provisions. It argues that domestic rules allow storage operators to sell up to 90 per cent of new capacity to one single bidder and up to 15 storage years under certain minimum procedural requirements.[58] Consequently, even without an exemption Dambořice's operator should be able to find a reliable long-term business partner and satisfy the demands of its investors. The Commission concludes that the domestic exemption decision must be withdrawn.

F. Conclusion

Early exemption decisions in the gas sector often show a relatively lenient approach by NRAs and the Commission vis-à-vis granting TPA exemptions. It is submitted that recent policy documents (such as the Commission staff working document) and recent exemption decisions show an increasingly strict approach. This means that partial exemptions, both with regard to scope and duration, may become increasingly widespread in the future. The Commission does not shy away from requesting far-reaching amendments of national exemption decisions and has even demanded the withdrawal of the domestic *Dambořice* decision. Finally, the Commission has developed a particularly strict approach as to dominant firms requesting an exemption. Such firms can expect considerable hurdles in the exemption process.

Generally speaking this development should be welcomed. Exemptions to TPA should only be applied if and insofar as they are needed. The more stringent approach and the Commission's willingness to impose amendments, however, do ask for a more elaborate reasoning in the exemption decisions of the NRAs and the Commission. This is most important in order to guarantee legal certainty. Last, although the Commission is right to enforce particularly high scrutiny on dominant firms, it should not be taken too far. It should not significantly decrease incentives to build much-needed gas infrastructure.

[57] Ibid at paras 58 and 60–67.

[58] Ibid at paras 74–78. The Commission suggests that negotiated TPA is possible for UGS facilities, referring to the recent Haidach storage project in Austria (ibid at para 80).

V. Case Study—Electricity

A. Introduction

There have been fewer TPA exemption cases in the electricity sector than in the gas sector. In this chapter, I shall discuss the recent *BritNed*, *Estlink*, *East–West Cables*, and *Tarvisio–Arnoldstein* exemption decisions. I shall not deal with exemption decisions where no (elaborate) substantive examination has been published, such as the *NorGer* interconnection.

These decisions reveal that the Commission has been relatively strict in its scrutiny and forceful in imposing various types of remedies to alleviate competition issues. Perhaps it indicates that policy on exemption decisions in the electricity sector has generally been stricter on a more consistent basis when compared to the gas sector.

B. BritNed

In October 2007 the Commission issued its decision in *BritNed*. BritNed is a joint project by the UK and Dutch transmission system operators (TSOs) to construct a cross-border interconnector with a capacity of 1.000 MW and a peak capacity of 1.320 MW. BritNed started operations in April 2011.[59]

Although unusual for an exemption request, BritNed's business plan is actually based on providing TPA.[60] An exemption would, however, provide BritNed more leeway in its business operations. Contracting the capacity to one undertaking could be considered if, for instance, the capacity uptake by third parties is too low. BritNed believed that such a fallback option is necessary to provide a better balance between risks and rewards for its investors.[61] BritNed argued that, under the applicable regulatory regime, returns to investors would be capped if the project proves to be successful, whereas no compensation mechanism exists if it proves to be unsuccessful. The exemption thus facilitates that investors will face less risk if demand proves to be low, but will be able to benefit from the potential upside if demand proves to be higher.

According to the Commission, the BritNed project will enhance competition by setting up the first electricity interconnection between the UK and the Netherlands.[62] It must be seen in the context of a wider integration of the electricity markets in north-western Europe. Notwithstanding the relatively limited capacity of BritNed, it may contribute to the convergence of marginal electricity prices, increase liquidity, and lead to less volatile and lower average prices.

Since the UK and Dutch TSOs are not vertically integrated with any of the national energy incumbents, the Commission finds they have no incentive to

[59] Commission exemption decision in the *BritNed* case of 18 October 2007, available at <http://ec.europa.eu/energy/infrastructure/exemptions/doc/doc/electricity/2007_britned_decision_en.pdf> (accessed 17 October 2011); see also <http://www.britned.com/> (accessed 17 October 2011).

[60] Ibid at para 9.

[61] Ibid.

[62] Ibid at para 11(a).

favour one supplier over the other or to foreclose market access.[63] Indeed, as revenues are determined by their sales of interconnector capacity, the TSOs have an incentive to maximize the capacity sold. The financial risks involved relate mainly to the difficulty in predicting the differences in electricity prices between the UK and the Netherlands. According to the Commission, the calculations on the impact of various price differences made by the requesting parties were incomplete. The calculations were based on a relatively slow rate of amortization. However, in more favourable conditions (ie higher price differences that BritNed could use in its favour), the investment could be amortized more quickly. The exemption request did not take into account such a scenario.

According to the Commission, there is a risk that the proposed interconnector has insufficient capacity to cope with a potential future rise in demand.[64] In order to deal with this risk, the Commission demands that the NRAs conduct a careful review, whether the capacity chosen by the project promoters avoids monopolistic behaviour and is the most desirable from the perspective of market integration and consumer benefit. In the *BritNed* case, the NRAs ought to have examined whether it would have been optimal to build more capacity. The approach above shows the Commission's willingness to carry out a strict and in-depth examination in order to maximize the infrastructure's value for the market as a whole.

The Commission considers that the suboptimal capacity may lead to higher than expected profits. In order to mitigate the risk of an overly lengthy exemption the Commission constructs a regulatory safety valve in order to assess the true costs and revenues of the project after a specified duration. BritNed must provide a report ten years after the start of operations.[65] If the estimated internal rate of return for the entire project is more than one percentage point higher than what was claimed when filing the exemption request, BritNed has two options:

1. Increase the interconnector capacity to such an extent that the initially estimated rate of return is met. Such an addition will not automatically be covered by an exemption.
2. BritNed accepts that the profits exceeding the initially estimated rate of return by more than one percentage point are capped and used, at equal parts, to finance the regulated asset base in the UK and the Netherlands.

These options show how proactive the Commission is willing to be in order to make sure the exemption does not have long-term detrimental effects on competition. It is submitted that the Commission's strict review is better in line with the applicable legislation than a more lenient one. Nevertheless, this approach could raise issues as regards companies that have purchased capacity on the basis of price levels that are deemed to be supra-competitive. Such firms may not be able to benefit from these changes if (i) they no longer procure capacity from BritNed; or (ii) they do not make use of the regulated asset base benefiting from the extra capital injection. On balance, however, the Commission's approach does seem to be the most appropriate mechanism to ensure that supra-competitive profits are invested back into the network.

[63] Ibid at para 11(b). [64] Ibid at para 11(f). [65] Ibid at para 13.

C. Estlink

The Estlink project concerns a 350 MW electricity cable between Estonia and Finland. The cable has been operational since January 2007. It is a project of the Estonian electricity incumbent, as well as energy companies from Latvia, Lithuania, and Finland.

Initially the investors were allowed to book all of Estlink's capacity for a certain period of time. After that initial period, but no later than 31 December 2013, cable ownership and operations will be transferred to the TSOs of Estonia, Latvia, Lithuania, and Finland. The exemption was requested vis-à-vis the initial period. The relevant NRAs—from Finland and Estonia—believed the project met all the exemption requirements. Although the Commission did not request the NRAs to amend or revoke their exemption decision, it did put forward its views.[66] In the Commission's opinion, the project has pro-competitive effects, offering the possibility for trade between the Baltic countries (where electricity is relatively cheap) and the Nordic countries (where electricity is more expensive). In addition, the Commission argues that the possibility of a reverse flow would be able to keep downward pressure on prices in the Baltics as well.[67]

Nevertheless, the Commission did have concerns about the capacity bookings already in place.[68] As all capacity towards the Baltic countries was booked by the Baltic energy incumbents, the Commission feared that this would prevent entry by potential competitors in the Baltic market. It could also provide incentives for the Baltic energy incumbents to hoard their capacity. It is not clear from the decision why the Commission emphasized this particular part; as it seems that the cable's main purpose was to serve Nordic countries from the Baltic countries, not the other way around. Perhaps this shows that the exemption decision may also take into account long-term developments that are relatively difficult to gauge.

In the Commission's view, as long as there is a price difference between the two regions, it is unlikely that competitors from Finland would be barred from selling into the Baltic market.[69] In addition, as the investors have to pay a fee regardless of whether the cable is in use, the Commission believes there is a strong incentive for them to use all available capacity instead of hoarding it. In order to reduce the risk of capacity hoarding even further, the Commission recommended imposing transparency requirements on the investors. This would have to be done by way of forecasts and evaluations of the actual cross-border trade via Estlink.[70]

The Commission's decision could have elaborated on its argument as to the importance of the level of the fee. If the expected gain of capacity hoarding is higher than the expected fee for providing access, the Commission's argument does not hold. Also, it is unclear how the transparency requirement as such may be able to lower the risk of capacity hoarding. Perhaps the Commission hopes that, in

[66] Commission exemption decision in the *Estlink* case of 27 April 2005, available at <http://ec.europa.eu/energy/infrastructure/exemptions/doc/doc/electricity/2005_estlink_decision_en.pdf> (accessed 17 October 2011), at 1.

[67] Ibid at 7.

[68] Ibid at 2.

[69] Ibid at 7.

[70] Ibid at 3.

the event capacity hoarding would take place, it will be noticed as a result of the transparency requirements. Possibly 'naming and shaming' will do the rest.

D. East–West Cables

The East–West Cables project seeks to construct an electricity interconnection between Ireland and Great Britain.[71] The project is an initiative by Imera, a new entrant in the Irish and British energy markets. In April 2008, Imera applied for an exemption for two 350 MW cables.

The Irish and UK NRAs consented with Imera's request for an exemption relating to the cable's full capacity for 25 years (the first cable) and 20 years (the second cable), respectively.[72] In its request, Imera attempts to reassure the regulators that no foreclosure will take place. It offers to facilitate a secondary market on which capacities would be auctioned on a short-term basis. Imera also proposed to cap the capacity offered to any one firm at 70 per cent, and in the case of the incumbent Irish electricity supplier ESB, at 40 per cent. If a firm reaches the cap, the NRA would subsequently be in the position to undertake a competition analysis before the firm is allowed to purchase any further capacity.

The Commission seems to have a positive stance towards the project, as it would be the first electricity interconnection between Ireland and the UK.[73] The East–West Cables project would enable market participants to enter each other's markets, thereby facilitating the creation of a regional market.[74] The project would also introduce a new entrant on the market. These features are especially valuable considering the small size of Ireland's electricity market.[75] In its decision the Commission also took into account the Irish Government's target to get 40 per cent of its electricity demand from renewable sources. As this entails an increased use of wind-generated power, there is great future need for import and export capacity in case of lows and spikes in power generation.[76] In its decision, the Commission refers to the fact that another interconnector was being planned, built by the Irish TSO Eirgrid. TPA fully applies to this project, as no exemption has been requested. In the Commission's view, the fact that the East–West Cables project has to compete with Eirgrid raises the uncertainty on investors that the East–West Cables project could be run profitably. In the Commission's view, there is scope for an exemption for East–West Cables as long as the Eirgrid interconnector is continued as planned.[77]

The condition that the Eirgrid interconnection must continue as planned is somewhat obscure. It seems to be inspired by the staff working document that suggests that alternative infrastructure makes an exemption more likely.[78] In the context of a decision by an administrative body such as the Commission, it seems

[71] Commission exemption decision in the *East–West Cables* case of 19 December 2008, available at <http://ec.europa.eu/energy/infrastructure/exemptions/doc/doc/electricity/2008_east_west_cable_decision_uk_en.pdf> (accessed 17 October 2011).
[72] Ibid at para 1.
[73] Ibid at para 6.
[74] Ibid at para 13.
[75] Ibid at paras 9–10.
[76] Ibid at para 18.
[77] Ibid at para 23.
[78] See, eg, para 43 of the Commission staff working document; above n 21.

unwise to lay down a condition that is not under the control of the requesting party. If the Eirgrid interconnector cannot go forward as planned, why should that under all circumstances be used to the detriment of Imera by revoking the exemption? Finally, as a general policy matter it may send out a wrong message if this is understood in such a way that projects are particularly likely to receive an exemption if an alternative is already being built. Indeed, new projects will often add the most value exactly on the sites where no alternative exists or is in the making.

Although the Commission was willing to grant the requested exemption, it did request the NRAs to amend their exemption decisions in a number of ways.[79] First, as suggested by Imera, a capacity cap of 40 per cent shall apply to dominant undertakings. Second, before the interconnector goes into operation, effective congestion management rules shall be implemented. Third, the NRAs shall assess the effectiveness of the secondary trading and UIOLI provisions on providing access.

E. Tarvisio–Arnoldstein interconnector

The most recent Commission exemption decision in the electricity field stems from early 2011. Unfortunately, the decision leaves quite a few confidential parts blank, which hampers a comprehensive reading of the text. The project concerns a 132 kV electricity interconnector between Tarvisio in Italy and Arnoldstein in Austria.[80]

The Italian NRA accepted the exemption request in September 2010. The exemption is granted for a period not exceeding 16 years. From the eighth year onwards, the Italian and Austrian NRAs will conduct an annual review of the exemption. This review will examine whether the actual revenues and costs incurred allow early recovery of investments and therefore a reduction in the exemption period. Fifty per cent of the exempted capacity must be returned to the market if it remains unused. After the exemption period, the network will be sold to the Italian TSO.

The Commission appears sympathetic towards the project, regarding the interconnection as a necessary addition to the European electricity grid. A report by the European network of electricity TSOs suggests that an enhancement of cross-border capacity through the Alps was needed to improve safety and reliability and reduce trade barriers.[81] The Commission also considers the risk criterion to be met.[82] The risk is higher than usual, in particular as the success of the projects partly depends on future investments already committed to by third parties, such as the Italian TSO. Still, the Commission does not agree with the NRA's risk assessment that supports exempting 50 per cent of capacity.[83] According to the Commission, the expected revenues from auctioning the capacity should be sufficient to adequately address the risk associated with the investment.

The Commission suggests that the necessity criterion has not been fulfilled, as the exemption went further than necessary.[84] The Commission is also critical of

[79] Commission exemption decision in the *East–West Cables* case, above n 71, at para 56.

[80] Commission exemption decision in the *Tarvisio–Arnoldstein* case, available at <http://ec.europa.eu/energy/infrastructure/exemptions/doc/doc/electricity/2011_tarvisio_decision_it.pdf> (accessed 17 October 2011), at para 3.

[81] Ibid at para 12.

[82] Ibid at para 16 *et seq.*

[83] Ibid at para 22.

[84] Ibid at para 27.

the capacity allocation scheme outside of the regular legal framework. Based on these observations the Commission decides that the exemption from TPA must be withdrawn.[85] This means that the full capacity must be allocated through an auction procedure. Capacity must be allocated in line with the general European and domestic rules of procedure. Any exemption for a significant capacity increase must be notified to the Commission.[86] The NRA exemption decision will lose its effectiveness if the cable has not become operational five years after the adoption of the exemption decision. This could be different if the NRA and the Commission agree that the delay is due to circumstances beyond the control of the applicant.

The Commission's decision in the *Tarvisio–Arnoldstein* interconnector case is yet another indication of its willingness to conduct an in-depth examination of the exemption requests. It also shows that the Commission attaches great value to the necessity criterion. As the Commission did not consider the necessity criterion to be met, it imposes far-reaching amendments to the NRA decision.

F. Conclusion

Overall, the examination above shows that the Commission is willing to engage in a strict review of TPA exemption decisions in the electricity sector, and has seemed willing to do so from the earliest of cases. In *BritNed*, the Commission articulated its worries as regard sub-optimal infrastructure capacity. The Commission proactively gave two options to remedy any potential lack of adequate capacity. In *Tarvisio–Arnoldstein*, the Commission did not agree with the exemption being granted. Overall the Commission's strict review is better in line with the applicable legislation than a more lenient one. It does mean, however, that the Commission will have to explain its way of reasoning thoroughly in order to strengthen legal certainty. In my view, the Commission should also tread carefully in imposing overly restrictive conditions. As said, the unclear condition in the *East–West Cables* case referring to the completion of another interconnector by another firm has a great risk of muddying the waters.

VI. Comparison and Conclusion

This examination of gas and electricity exemption decisions gives an opportunity to provide a short comparison by way of conclusion. The different nature of the gas and electricity sectors caution against drawing overly strong parallels. However, some common themes are discernable and merit attention.

Although the early exemption decisions in the gas sector were relatively lenient, more recent decisions show an elevated level of scrutiny. The NRAs, and especially the Commission, seem to take a stronger stance upholding the necessity and proportionality conditions. They have reduced the exemption's scope and/or duration to a level where competition is believed not to be impaired.

[85] Ibid at para 37.

[86] Ibid at para 40.

In both the gas and electricity exemption decisions the Commission is relatively lenient when the exemption is requested by a new entrant. The same applies if the infrastructure links together two markets which previously lacked such a link (see, for examples *East–West Cables*). As the Commission assesses the full context of the case before it, an exemption request appears to be more likely to be favourably received if the exemption request already includes provisions that are meant to lower the risk of foreclosure. This could include an Open Season procedure, effective congestion management rules or a capacity cap that any individual firm may use. The Commission appears more willing to demand amendments; especially compared to the early gas decisions. Some of these amendments can be quite far-reaching. For example, in *BritNed* the Commission laid down rules that aim to reduce the risk of insufficient capacity.

Finally, there is an increased focus on the possibility that dominant firms must not be able to book a high percentage of capacity. Unfortunately, it is still unclear what the Commission actually means when it refers to dominant firms. It is also unclear how the gradual development of the internal energy market will impact upon the analysis given by the Commission—especially as the analysis necessarily entails a view on future developments. The wider the geographic market becomes, the less likely it is that a firm can be qualified as being dominant.

Notwithstanding the similarities, there are many differences as well. Overall exemption decisions in the field of electricity seem to have been stricter than those in gas on a more consistent basis. For instance, in the field of gas, the Commission has often made a strong link between the augmentation of capacity and the subsequent rise in competition. In its electricity exemption decisions, the Commission places much less emphasis on that link. In addition, the amendments that the Commission has put forward in electricity cases appear more far-reaching than those in the gas sector. In the *Dambořice* and *Tarvisio–Arnoldstein* cases the Commission even refused to give an exemption. Finally, a difference exists as regards the possibility that dominant firms should be barred from strengthening their position. This is understandable if one has regard to the often higher levels of concentration in the gas sector compared to the electricity sector.

Aligning different interests at the same time, such as stimulating investments and strengthening competition, is never easy. It is clear, however, that a proper balance can only be struck if the NRAs and the Commission are willing to conduct an in-depth case-by-case analysis of exemption requests. The Commission should be commended for its recent approach in the field of TPA exemptions, notwithstanding the inevitable room for improvement. Hopefully it means that TPA is now firmly secured as the leading principle regarding energy infrastructure.

19

Electricity and Gas Infrastructure Planning in the European Union

Iñigo del Guayo and Johann-Christian Pielow

I. Introduction—Market Liberalization, Investment, and Infrastructure Planning

This chapter analyses some recent legal provisions at European Union (EU) level related to electricity and gas infrastructure planning in the so-called 2009 Third Energy Package, as well as in subsequent legislation. The inclusion of detailed rules on infrastructure or network planning within a number of EU Directives and Regulations is a remarkable recent development of EU energy law.

In general terms, compulsory energy infrastructure planning drafted and approved by central governmental institutions is something of the past. Compulsory planning, as opposed to indicative planning, implied a governmental decision on by whom, when, and where investments in new infrastructure or in expanding the existing infrastructure had to be made. Companies had to comply with those decisions—not only companies owned by governments (which were the majority, in post-Second World War Europe), but also private companies. It was one of the core characteristics of the industrial energy paradigm set up in Europe after the Second World War.[1] A new economic and legal paradigm that emerged in the early 1980s was characterized by market liberalization, privatization, and competition. This new scenario went against old compulsory planning and in favour of allowing market forces to satisfy energy needs. In that paradigm, it is for the market to determine, where, when, and how much investment must be done in terms of infrastructure. In that context, EU efforts were not directed at guaranteeing investment in infrastructure via governmental compulsory planning, but rather to open markets, break monopolies, unbundle integrated energy companies, and introduce third party access (TPA) obligations. Somehow, planning was seen as an inefficient endeavour imposing unnecessary extra costs on the energy supply

[1] T. Daintith and L. Hancher, *Energy Strategy in Europe: The Legal Framework* (1st edn, Berlin: De Gruyter, 1986) 36–8.

chain, since governmental forecasts were usually wrong (unreasonably based only on security of supply), too rigid, and leading to infrastructure over capacity.[2]

In 2009, 25 years after the start of efforts to create an internal market for both electricity and gas, planning is again central in the process of building a European energy market. It is an interesting development, showing how planning must play a role within the regulatory and legal framework. This development should not be seen as a mere result of a pendulum-style motion, swinging from a free market for energy towards a new planned energy model. This development opens a space for planning within liberalization and competition, two legal principles which led to the inception and development of an internal European energy market. It is a new way of planning, far from the old planning of the Keynesian paradigm that was in effect from 1945 to 1980, where the energy sector was either owned or dominated by governments. New provisions on planning recognize that both electricity and gas supply take place by means of an infrastructure which has the characteristic of a natural monopoly, and that planning is one instrument, among others, to fight monopolies and ensure security of supplies. As will be explained below, failure to recognize this explains a number of shortages that some European electricity systems experienced after liberalization, as well as the decrease in proper maintenance and development of infrastructure.[3] Investment should not be the result of unrestricted free decisions of the owner of the infrastructure itself, but rather the result of a wide consultation procedure with a final fundamental decision by the regulator.

This new planning policy is not only linked to security of supply (though that is its main objective), but will undoubtedly help to achieve other equally fundamental aims of the internal market. The creation of an effective European market (electricity-to-electricity and gas-to-gas competition) depends on the existence of proper network interconnections, and the new planning policy is directed towards such interconnections.

II. The Third Energy Package and the Relevance of Infrastructure Planning

As a result of the set-up of a European Internal Energy Market, EU institutions have issued several packages of legislation enabling the application of general concepts of EU law to the energy sector. In the case of 'traditional' energy utilities—the electricity and gas supply sectors—the First and Second Energy Packages[4] (covering the sets of norms adopted in the periods 1996–98 and 2003–05, respectively) were followed in 2009 by the Third Energy Package. That package was adopted with a view to intensifying liberalization, integration, transparency, competition,

[2] For example, Spanish National Energy Plan 1979–1983 and 1983–1991.

[3] The Italian electricity system collapsed for several hours on 28 September, 2003, due to a lack of proper investment in the electricity lines connecting Italy with foreign countries.

[4] A set of legal norms on energy passed in 1998, and 2003, respectively.

and security of supply, and to adopting further measures on unbundling and on access to the networks. The relevant legislation related to planning includes:

(i) Directive 2009/72/EC of the European Parliament and of the Council of 13 July 2009 concerning common rules for the internal market in electricity and repealing Directive 2003/54/EC (the Electricity Directive, hereinafter referred to as the ED);

(ii) Directive 2009/73/EC of the European Parliament and of the Council of 13 July 2009 concerning common rules for the internal market in natural gas and repealing Directive 2003/55/EC (the Gas Directive, hereinafter referred to as the GD);

(iii) Regulation 713/2009 of the European Parliament and of the Council of 13 July 2009 establishing an Agency for the Cooperation of Energy Regulators (ACER) (hereinafter referred to as the ACER Regulation);

(iv) Regulation 714/2009 of the European Parliament and of the Council of 13 July 2009 on conditions for access to the network for cross-border exchanges in electricity *and repealing Regulation 1228/2003* (the Electricity Regulation, hereinafter referred to as ER)*;*

(v) Regulation 715/2009 of the European Parliament and of the Council of 13 July 2009 on conditions for access to the natural gas transmission networks and repealing Regulation 1775/2003 (the Gas Regulation, hereinafter referred to as the GR);[5] and

(vi) Regulation 994/2010 of the European Parliament and of the Council of 20 October 2010 concerning measures to safeguard security of gas supply and repealing Council Directive 2004/67/EC.[6] Directive 2005/89/EC of the European Parliament and of the Council of 18 January 2006 concerning measures to safeguard security of electricity supply and infrastructure investment, belonging to the second package, is still in force.[7]

Both the ED and the GD provide for an unbundling regime under the following three models: (i) the ownership unbundling model, or 'TSO' model, where producers and suppliers cannot own network companies and vice versa; (ii) the independent system operator (ISO), where the network owner does not run the network activity, but rather an ISO is designated to do so; and (iii) the independent transmission operator (ITO), where the network company remains within a vertically integrated company, but there are several and tough rules to assure that the production and supply branches do not interfere in the running of the networks. According to the *Interpretative Note*, the three models should create incentives for the necessary investments and guarantee the access of new market entrants under a transparent and efficient regulatory regime.[8] However, the European

[5] All these five norms were published in *OJ L* 211, 14 August 2009.

[6] *OJ L* 295, 12 November 2010.

[7] *OJ L* 33, 4 February 2006.

[8] Commission Staff Working Paper. *Interpretative Note on Directive 2009/72/EC Concerning Common Rules for the Internal Market in Electricity* and Directive 2009/73/EC concerning common rules form the internal market for natural gas. The unbundling regime, Brussels, 22 January

Commission makes it clear that the preferred option remains ownership unbundling.[9] Recital 10 of the ED states that only removal of the incentive for vertically integrated undertakings to discriminate against competitors as regards investment can ensure effective unbundling. Ownership unbundling is clearly an effective and stable way to solve the inherent conflict of interests and to ensure security of supply. For that reason, the European Parliament in its resolution of 10 July 2007 on prospects for the internal gas and electricity market referred to ownership unbundling as the most effective tool by which to promote investments in infrastructure in a non-discriminatory way. However, the fact that the existing laws contain planning provisions demonstrates that ownership unbundling is not enough, per se, to guarantee actual unbundling, but that planning is needed.

This chapter highlights that planning in infrastructure investment has a relevant role in the existing energy legislation. Planning is actually one of the novelties of recent EU legislation, and was introduced into EU energy law, quite unexpectedly, in the 2009 Third Energy Package. There were very limited or no previous references to planning in the First and Second Energy Packages. For example, whereas in the 2004 and 2006 Directives on security of electricity and natural gas supply, respectively, there is not a single reference to planning, the ED, the GD, the ER, and the GR have many such provisions, and the 2010 Regulation on security of natural gas supply (substituting the 2004 Directive) is based on planning.

III. Long-term Infrastructure Planning at National Level[10]

A. Article 22 of the 2009 Gas and Electricity Directives

Article 22 of both the ED and the GD addresses the security of supply aspects of long-term planning of investments in sufficient cross-border interconnection capacities and other infrastructures to ensure the long-term ability of the system to guarantee security of supply and meet reasonable demand. It deals, therefore, with network development and power to make investment decisions. Article 22 imposes upon transmission system operators (hereinafter referred to as TSOs) the obligation to approve a national ten-year network development plan (hereinafter referred to as NTYNDP) and to lay down rules on how to proceed when investments foreseen in the plan are not actually being made, empowering the National Regulatory Authorities (hereinafter referred to as NRAs) to impose the obligation to proceed with the investment upon the company responsible for transport in the area. According to article 22(5) of both the ED and the GD, NRAs shall also

2010. See, further, I. del Guayo, G. Kühne, and M.M. Roggenkamp, 'Ownership Unbundling and Property Rights in the EU Energy Sector', in A. McHarg, B. Barton, A. Bradbrook, and L. Godden (eds), *Property and the Law in Energy and Natural Resources* (Oxford: Oxford University Press, 2010) 326–59.

[9] Ibid.

[10] The authors are thankful to Christian and Philip Engels for having translated this section from German into English.

examine the coherence of NTYNDPs with the non-binding TYNDP), referred to in article 10 of both the ER and the GR.

B. Infrastructure planning versus land planning

The long-term planning mentioned in article 22 of the ED and the GD focuses primarily on the need for investment in generation and in networks. When defining long-term planning, article 2(25) of the ED refers to the need for investment in generation and transmission and distribution capacity, whereas article 2(30) of the GD refers to planning of supply and the transport capacity of natural gas undertakings. They address the problem of the planning of future investments in the electricity and gas networks, or in infrastructures which are linked to them, like interconnections at the internal borders of the EU. However, neither the ED nor the GD mention the physical plan, ie the exact and detailed location for the construction and extension of certain electricity or gas supply lines. This is the responsibility of member states, for example, for development and building planning and/or for technical planning of the asset, as opposed to planning of investment in infrastructure, subject to harmonization under EU laws. In Germany, for example, construction and extension of large electricity and gas pipelines requires a special planning approval in accordance to article 43 et seq. of the 2005 Energy Industry Act.[11] The EU lacks competence in this respect, which means that neither the ED nor the GD include guidelines for binding location planning. New investments have, undoubtedly, an impact on the (sovereign) location planning of member states, at least indirectly. It should also be up to the member states to coordinate the two planning levels through adjustments to their national laws. This could lead to problems if other authorities than the NRA (which, according to article 22 of the ED and the GD, has control over the four-year investment plan) have decision-making power over the location. In a federal system like the German one, regional authorities of the federal states (Länder) have decision-making power over planning approval. In cases such as these, new regulations are required to ensure coordination between the different authorities of the member states.[12]

C. One or several national network development plans?

Article 22 refers to a ten-year network development plan drafted by each TSO. Since there can be more than one TSO in each member state, there can be more than one plan. That explains that nowhere in the ED or the GD there is a reference to a single national ten-year network development plan. On the contrary, article 8 of both the ER and the GR talk in plural about the 'national investment plans' or the 'national ten-year network development plans'. It is common, however, to have only one TSO in each member state. Within the European Network

[11] *Energiewirtschaftsgesetz* (Energy Economy Act), of 7 July 2005 (BGBl. I S. 1970, 3621), as modified by article 4 of the Act of 7 March, 2011.

[12] See also and in more detail chapter 20 of this book.

of Transmission System Operators for Electricity (ENTSO-E), for example, there are 41 TSOs from 34 countries (note that there are TSOs in countries which are not, as yet, members of the EU): 31 members have only one TSO (such as France or Spain) and three members have more than one TSO (Germany has four, the UK has four, and Austria has two). Thus, in most circumstances (with the notable important exceptions of the UK, Germany, and Austria), there is only one TSO, one investment plan and, consequently, one 'single national plan'.

D. Which transmission system operators are subject to the obligation to draft a ten-year network development plan?

Both the ED and the GD contain an inconsistency in so far as the NTYNDP is regulated solely in the context of the ITO model, in chapters V (ED) and VI (GD). This could lead to the conclusion that those TSOs which are organized through one of the two other options of network unbundling (property unbundled TSO and ISO) have no planning obligations,[13] regulated, respectively, in articles 9 and 13 of the ED, and 9 and 14 of the GD. In the authors' opinion, the inconsistency is the result of a fault in the structure of both the ED and the GD. In both Directives, the expression TSO as used in article 22 would cover every kind of TSO, ie in any of the three possible forms. Both articles 13(2)(c) and 14(2)(c) of the ED and the GD, respectively, state that member states will appoint an ISO only when the candidate operator has undertaken to comply with a ten-year network development plan monitored by the regulatory authority. In addition, article 18(4) of the ED and article 20(1) of the GD, when addressing the ITO system in both electricity and gas, state that the vertically integrated undertaking shall not determine, directly or indirectly, the behaviour of the TSO in relation to activities necessary for the preparation of the NTYNDP developed pursuant to article 22. It is only within the part of the ED and the GD dealing with ownership unbundling that no reference is made to infrastructure planning. However, there would be no rational explanation to exclude TSOs (which are unbundled in ownership terms) from the obligation to draft a NTYNDP. The European Commission is of the opinion that ownership unbundling guarantees investment, and one might think that if that statement were true, there would be no need to impose upon the ownership-unbundled TSO any planning obligation. However, it is clear that the scope of article 22 of both the ED and the GD covers every TSO.

E. The national ten-year network development plan

Since planning of national network investments is carried out by TSOs, which normally are of a private nature,[14] the NTYNDP should not be regarded as a governmental productivity plan, based on past socialistic views of (public) economic

[13] See in this sense J. Baur, 'Chapter 12' in J.F. Baur, P. Salje, and M. Schmidt-Preuß (eds), *Regulation in the Energy Industry* (Cologne: Carl Heymanns Verlag, 2011), margin no. 48.

[14] Most TSOs in Europe are private companies, subject to governmental regulation.

order. Both the ED and the GD have the protection of property and the entrepreneurial freedom of network providers at their core.[15] Article 22(1) of both the ED and the GD specify, as a basic obligation of the TSOs, the development of a NTYNDP plan and its submission to the NRA, after consulting all the relevant stakeholders. The plan should contain efficient measures in order to guarantee the adequacy of the system and the security of supply.

The purpose of the NTYNDP is, in particular, and according to article 22(2) of both the ED and the GD, (i) to indicate to market participants the main transmission infrastructures that need to be built or upgraded over the next ten years; (ii) to record all the investments already decided on and identify new investments which have to be executed in the next three years; and (iii) to provide for a timeframe for all investment projects.

F. Nature of the plan

The NTYNDP has to be elaborated *every* year, but the investment forecast has to be made for ten years. This means that the TSO has to continuously adjust the plan to the prevailing technical and economic conditions of the grid. The requirement should not be misunderstood as a *legally* binding obligation to invest, which might derive from, for example, corresponding impositions by member states. Instead, one should again emphasize the general purpose of the NTYNDP, which consists in guaranteeing the adequacy of the system and the security of supply, which in contemporary regulatory models does not necessarily imply foreseeing a consequence to the lack of compliance.[16] The 'adequacy' of, and consequent necessary investments into, the system logically depend on the technical and economic situation of the existing network infrastructure and on demand in underserved areas, which has to be assessed first and foremost by the TSO. The network operator should therefore enjoy a broad scope of discretionary decision-making.

The plan relies on 'reasonable assumptions' about the development of generation, supply, consumption, and exchange of electricity or gas with other countries, and shall also take into account investment plans for regional and EU-wide networks, as well as investment plans for storage facilities and liquefied natural gas (LNG) regasification facilities for the gas sector.[17] The latter requirement, ie the need to consider existing regional plans (among bordering member states) or the EU-wide TYNDP, can be understood in the sense of the so-called 'counterflow principle' ('Gegenstromprinzip') of the German land-use planning law, according to which planning on a lower (regional, local) level has to take into account the planning at a higher (eg national) level, and vice versa.[18] This suggests that the EU-wide network development plan will indeed—despite its explicit denomination as 'non-binding'—have at least indirect influence on the investment planning at the

[15] See article 6(1) of the EU Treaty and articles 16 and 17 of the Charter of Fundamental Rights of the European Union.

[16] See article 22(1), 2nd sentence, of both the ED and the GD.

[17] Article 22(3) of both the ED and the GD.

[18] See article 1(3) of the German *Raumordnungsgesetz* (Act on land-use planning) from 2008.

national level. On the other hand, neither the ED nor the GD clearly states what is deemed as 'reasonable' according to article 22(3) of both Directives. It is for the NRA to assess reasonableness in accordance with the result of consultations with other stakeholders, and to which extent the data basis and the forecasts of the plan may be accepted. However, taking into account that the assumptions made by the TSO cover, besides the supply and transport capacity of natural gas undertakings, such sensitive aspects as demand for natural gas from the system and diversification of sources, one has to assume that the appreciation of those assumptions by the NRA is anything but easy. NRAs are obliged to solve an equation with many unknowns in the shape of extremely vague legal terms which may give rise to legal debates and even to lawsuits.[19]

G. Consultation procedure

The TSO doesn't assess on its own which assumptions should be considered as 'adequate' and, consequently, which investments are 'efficient'. Instead, the plan has to be consulted with all relevant stakeholders (see article 22(1), first sentence, of both the ED and the GD). The consultation must be organized by the NRA in an open and transparent manner and with all actual or potential system users (article 22(4) of both the ED and the GD). Article 22(1) of both the ED and the GD does not clearly state whether the TSO has a separate obligation to consult all the relevant stakeholders before implementing the plan. According to the wording of article 22(4), the NRA could also be obliged to consult the stakeholders after the plan is implemented. In article 2(18) of the ED and 2(23) of the GD, system users are defined as natural or legal persons supplying to, or being supplied by, the system. In order to avoid third party interference, potential system users of gas or electricity networks are obliged to prove their status as a potential network system user (article 22(4), second sentence, of both the ED and the GD). However, the Directives do not clearly state under which circumstances a party is a potential system user. It will be interesting to see how member states interpret these uncertainties in a practical way through implementation. This is particularly true for the case of Germany, taking into account that the German gas economy already has parallel transmission networks, which makes it difficult to determine if there is a potential network user or not.

H. Measures to guarantee the implementation of investments planned

NRAs have to examine the need for detailed investment measures, which have been financially calculated by the TSO, taking into account the information provided by the NTYNDP and the results of the consultation process. According

[19] In Germany it is controversially debated whether the NRA disposes of a margin of discretion and to what extent it is subject to judicial control. See C. Pielow, 'Chapter 56' in J.F. Baur, P. Salje, and M. Schmidt-Preuß (eds), *Regulation in the Energy Industry* (Cologne: Karl Heymans Verlag, 2011).

to article 22(6) of both the ED and the GD, NRAs shall monitor and evaluate the implementation of the NTYNDP. This can lead to a decision concerning the extent of implementation of the investment measures.

Article 22(7) of both the ED and the GD states that when the TSO, other than for overriding reasons beyond its control, does not execute an investment, which, under the NTYNDP, was to be executed in the following three years, and the investment has not become irrelevant on the basis of the most recent NTYNDP, member states are required to obligate their NRAs to take at least one of three possible measures to ensure that the investment in question is made. These measures, which have to be selected according to the principle of proportionality, in particular to their necessity, are included in article 22(7) of both the ED and the GD, as follows: (a) a requirement on the TSO to execute the investments in question; (b) the introduction of a tender procedure open to all investors for the purpose of implementing or financing the investment in question through a third party or through the TSO itself, provided it participates and its tender is successful,[20] and then only with the approval of the regulatory authority; and (c) the obligation of the TSO to accept a capital increase to finance the necessary investments and allow independent investors to participate in the capital. Neither the ED nor the GD clearly states whether the NRAs are obliged to carry out one of these three measures or if they can choose between them. Taking into account the effectiveness and proportionality principles of EU law, the latter should be the most appropriate interpretation. The wording 'at least one of the following measures' indicates that the NRA can also adopt several measures. However, the NRA does not seem to have any discretion to decide whether measures are to be taken or not.

Neither the ED nor the GD point out explicitly who is to bear the subsequent costs. At least for the second option, when another investor is charged with the execution or even with financing of the infrastructure measure after a call for tenders, article 22(4) of the Directives requires detailed financial arrangements on who is going to pay for the infrastructure, which shall be approved by the NRA. It can be assumed that the costs of all mentioned measures have to be financed by all electricity or gas system users. Generally, ie without differentiation of individual measures, article 22(8) of both the ED and the GD requires the relevant tariff regulations to cover the costs of the investments in question in accordance with the decision of the NRA.[21] In this scheme, the costs for network investments are usually passed on to the system users who, in turn, impose the costs on the final consumers.

[20] According to both the ED and the GD, the TSO itself can participate in the tender process. This makes sense because it seems possible that there is no cheaper provider and the TSO should then undertake the construction project, possibly with the help of a third party at a price it can afford. On the other hand, the tender process can force the TSO to consider its costs and provide the investment at a cheaper price. In the end, it is decisive that the investment in the project is actually made.

[21] See Recitals nos 32 of the ED and 23 of the GD.

IV. Long-term Infrastructure Planning at the European Union Level

A. The European Network of Transmission System Operators for Electricity and for Gas

The ER and the GR impose upon all TSOs the obligation to cooperate at the Community level through the European Network of Transmission System Operators for Electricity (ENTSO-E) or through the European Network of Transmission System Operators for Gas (ENTSO-G). According to both the ER and the GR, one of the transmission activities is the representation of TSOs within the ENTSO-E or the ENTSO-G. In cooperating, members of ENTSOs should seek the following aims: (i) promotion of the completion and functioning of the internal market in electricity and gas; (ii) promotion of cross-border trade and supply; and (iii) ensuring optimal management, coordinated operation, and sound technical evolution of the European electricity and gas transmission network.[22]

The 2009 ER and GR mandate ENTSO-E and ENTSO-G to fulfil seven tasks: (i) draft network codes in twelve areas; (ii) adopt a non-binding Community-wide TYNDP every two years; (iii) adopt an annual work programme; (iv) adopt common network operation tools to ensure coordination of network operation in normal and emergency conditions, including a common incidents classification scale and research plans; (v) provide annual summer and winter generation adequacy outlooks; (vi) adopt recommendations related to the coordination of technical cooperation between Community and third party country TSOs; and (vii) adopt an annual report.[23]

The tasks of the ENTSOs should be carried out in compliance with Community competition rules (which are applicable to the decisions of both ENTSOs) and should be well-defined in the instruments mentioned (network codes and work programme). Their working method should ensure efficiency, transparency, and the representative nature of the network, and should not replace the TSOs they represent in fulfilling their obligations.[24]

The 2009 ER and GR asked TSOs to submit to the Commission and to the Agency for the Cooperation of Energy Regulators, by 3 March 2011, the draft statutes of the ENTSOs, a list of members and draft rules of procedure, including the rules of procedure on stakeholder consultation. The ER and the GR contain further procedural rules for setting up both bodies.[25]

The work to set up ENTSO-E was remarkably ahead of the provisions of the ER, since it was already established as an association (AISBL) under Belgian law on 19 December 2008 and became fully operational on 1 July 2009, ie even before

[22] Recital no. 7 of the ER; Recital no. 16 and article 4 of the GR; article 17(2)(b) of the ED, and article 17(2) (b) of the GD.

[23] Article 8 of the ER and article 8 of the GR.

[24] Recital no. 7 of the ER and Recital no. 16 GR.

[25] Article 5 of the ER and article 5 of the GR.

the publication of the ER in the Official Journal.[26] This can be explained by the fact that the current 41 members of ENTSO-E from 34 countries continue a long tradition of successful TSO coordination work by its six predecessor associations.[27] ENTSO-G was created on 1 Dec 2009 and has 33 members from 22 countries.[28]

In the report referred to by article 51 of the 2009 Gas Directive, the Commission, in consultation with the ENTSO-G, may also consider the feasibility of the creation, by TSOs, of a single European TSO, a provision which does not exist in the ED.

B. Regulation by cooperation and coordination and regulation of self-regulation

Under normal circumstances cooperation is based on the parties' free will, but the ER and the GR impose upon members of both ENTSOs the obligation to cooperate and to increase coordination. If market integration is based on cooperation, there is the need to promote it or, going further, to impose cooperation. The obligation imposed on the ER and the GR is not a mere suggestion, nor a moral commandment, but rather a legal obligation as such, since a number of decisions can only be adopted through a cooperation procedure. Therefore, it is important to analyse which aims must be reached by means of cooperation, since those aims define the limits of the obligation. We must recognize, however, that the areas in which European TSOs must cooperate are as wide as the entire electricity and gas sectors. Take, for example, the aim of promoting the completion and functioning of the internal market in electricity and gas. If one considers that the internal market in electricity and gas covers every single aspect addressed by the ED, the ER, the GD, and the GR, it is clear that cooperation of TSOs is expected to work in every aspect of TSO activities.[29]

Regulation through cooperation is a characteristic of our times. There is not one single (and central) power, but there are rather several institutions, in both horizontal and vertical levels, sharing power. There is no longer only one instance which commands and controls, but rather several instances which must cooperate among themselves and whose role is not confined to the traditional limits of command-and-control-type regulation, as is the case with modern regulatory bodies. The cooperative regulators are expected nowadays to deploy a great variety of regulatory techniques, which, in turn, are not exclusively, nor mainly, of a coercive nature, but rather of a facilitative and incentivizing character.

Another new characteristic of EU planning is that it is made by private entities (ENTSOs), but subject to strict control by the NRAs and the Agency for the Cooperation of Energy Regulators. This is an example of *regulation of*

[26] See S. de Moel, and F. Melchior, 'Cooperation between TSOs: Background, Organisation and Netcodes' in M.M. Roggenkamp and U Hammer (eds), *European Energy Law Report VIII* (Cambridge: Intersentia, 2011) 19–40.

[27] Such as the Union for the Coordination of Production and Transmission of Electricity (UCPTE), and the Union for the Coordination of Transmission of Electricity (UCTE).

[28] See <http://www.entsoe.eu> and <http://www.entsog.eu> (accessed 2 October 2011).

[29] Recital no. 6 and article 4 of the ER and Recital no. 15 and article 4 of the GR.

self-regulation: assuming planning is a regulatory technique, its drafting takes place through private entities (with no public regulatory powers), but under the control of regulatory powers.[30]

C. Community-wide ten-year network development plan

Increased cooperation and coordination among TSOs is needed to ensure coordinated and forward-looking planning. Consequently, both ENTSOs must draw up, adopt, publish, and regularly update a non-binding Community-wide TYNDP every two years (though it is good for ten years). Both the ER and the GR stress that in elaborating the TYNDP an effective consultation process is essential, and that NRAs and the ACER, as well as the Union for the Coordination of Transmission of Electricity (one of the predecessors of ENTSO-E, hereinafter UCTE) and the European Association for the Streamlining of Energy Exchange (one of the predecessors of ENTSO-G, hereinafter EASEE) must have an important role.[31]

The main objective of the TYNDP is security of supply. Security of supply is achieved, in turn, by investment in new infrastructure. The link between security and investment is so tight that some legal norms of the Third Energy Package refer to the NTYNDP as the investment plans of the TSOs. Actually, the Gas Security Regulation (GSR)—mainly devoted to guaranteeing the existence of proper infrastructure—states that the TYNDP for gas 'is a fundamental tool to identify the required investments needed at EU level, *inter alia*, in order to implement the infrastructure requirements laid down' in the GSR itself (Recital 23 of the GSR).[32]

However, one of the novelties of the new planning policy is that plans are not only directed at ensuring security of supply, with little or no focus on other aspects of the energy sector, but, rather, plans are considered to be useful instruments to reach several goals of the EU legislation. In particular, infrastructure planning is a key instrument to extend interconnection capacity and to allow cross-border trade to grow and consolidate. This explains that planning must take into account efficiency criteria. An efficient investment is that which actually responds to the demand of the market. When security of supply is the only concern of planning, and no consideration is paid to efficiency, planning leads to investment decisions which increase the costs of supply, since customers will have to finance the new infrastructure, which satisfies security concerns but adds unnecessary costs. One has to bear in mind that access tariffs to infrastructure must be calculated in a way that investment is recovered. If the investment only took into account security, it

30 In accordance to the ACER Regulation, the Agency was formally set up on 3 March 2011. Its purpose is to assist NRAs (referred to in the ED and the GD) in exercising, at Community level, the regulatory tasks performed in the member states and, where necessary, to coordinate their action. It has a relevant role in the design and monitor and of the TYNDP in reviewing its consistency with NTYNDPs: article 1 of the ACER Regulation.

31 Recitals nos 6, 8, and 9, and article 8(3)(b) and (10) (first subparagraph) of the RE, and Recitals nos 15, 17, and 18, and article 8(3)(b) and (10) (first subparagraph) of the RG.

32 See, for example, article 37(1)(g) of the ED and article 41(1)(g) of the GD; and Recital no. 23 of the GSR.

might be higher than if other aspects (such as efficiency), were also considered, and therefore would discourage access and competition. As a result of maximizing security, competition and liberalization are at risk. We read in both the ER and the GR, for example, that 'viable electricity and gas transmission networks and necessary regional interconnections, relevant from a commercial or security of supply point of view, should be included in the TYNDP'.[33]

Both ENTSOs are required to submit a draft TYNDP, including the information regarding the consultation process, to ACER. Within two months from the day of receipt of the draft TYNDPs, ACER has to provide a duly reasoned opinion, based on matters of fact, as well as recommendations to the ENTSOs, the Council, the European Parliament, and the Commission when it considers that the draft TYNDP submitted by the ENTSOs does not contribute to non-discrimination, effective competition, the efficient and secure functioning of the market, or a sufficient level of cross-border interconnection open to TPA, or that it does not comply with the ED or the GD. This could be taken to mean that plans must not only refrain from disturbing liberalization, but must contribute positively to it.[34]

In designing these plans (in terms of subjects involved, procedural matters and objectives which the plan must satisfy) and in bringing into force their provisions, regard must be given to related provisions of the network codes drafted by the ENTSOs and of the guidelines adopted by the European Commission. For example, planning seeks, among other objectives, to overcome problems arising from congestion and from discouraging access tariffs to infrastructure that do not include an appropriate return on investment. As a result, plans must take into account guidelines on congestion management and on TPA tariffs. On the other hand, networks codes and guidelines must have taken the content of plans into consideration.[35]

In fact, the *2010 Communication, Energy Infrastructure Priorities for 2020 and Beyond: A Blueprint for an Integrated European Energy Network*[36] states that the market will deliver most of the investment but estimates that obstacles remain, creating a gap of about €60 billion. The Communication recognizes that:

> tariff setting remains nationally focussed and key decisions on infrastructure interconnection projects are taken at national level. NRAs traditionally have aimed mainly at minimising tariffs, and thus tend not to approve the necessary rate of return for projects with higher regional benefit or difficult cost-allocation across borders, projects applying innovative technologies or projects fulfilling only security of supply purposes.[37]

To overcome those difficulties, the Commission proposes two things: (i) leveraging private sources through improved cost allocation;[38] and (ii) optimizing the

[33] Recital no. 9 of the ER and Recital no. 18 of the GR.

[34] Article 9(2) of the ER and article 9(2) of the GR, and article 6(3)(b) and article 6(4) of the ACER Regulation.

[35] Articles 6, 7, and 18 of the ER and articles 6, 7, and 23 of the RG.

[36] Brussels, 17.11.2010, COM(2010) 677 final.

[37] Ibid p 8.

[38] It envisages putting forward in 2011 guidelines for a legislative proposal, to address cost allocation of major technological complexities or cross-border projects through tariff and investment rules.

leverage of public and private sources by mitigating risk to investors.[39] Indeed, Mr Van Rompuy, President of the European Council, in his recent speech to the Strategic Energy Forum (*Challenges and Priorities for EU Energy Policy*), referred to the level of investment needed as follows:

> Estimations of the Commission show that one trillion euro[s] will be needed in the coming decade for investment in production and transport infrastructure. We expect the private sector to take the lead. But public authorities [member states and the EU] have to stimulate investments through intelligent budgetary choices. First, through the member state budgets [...] The leverage effect is spectacular. For the 2.3 billion euro[s] invested in interconnections, for example, the European Commission has calculated that this will leverage investments—national and private—worth some 22 billion euro[s] for the coming 3–5 years. A multiplication by 10![40]

The content of the TYNDP includes: (i) modelling of the integrated network; (ii) scenario development; (iii) an assessment of the resilience of the system; and (iv) a European generation (in the case of electricity) or supply (in the case of gas) adequacy outlook. This outlook must cover the overall adequacy of the electricity and gas systems to supply current and projected demands for electricity and gas for the following five-year period, as well as for the period between five and fifteen years, or ten years in the case of gas, from the date of that outlook; it must build on national generation adequacy outlooks prepared by each individual TSO. The TYNDP shall, in particular:

(i) build on national investment plans, taking into account regional investment plans and, if appropriate, Community aspects of network planning, including the guidelines for trans-European energy networks in accordance with Decision 1364/2006/EC of the European Parliament and of the Council, of 6 September;[41]

(ii) regarding cross-border interconnections, also build on the reasonable needs of different system users and integrate long-term commitments from investors referred to in the ED and the GD;[42] and

(iii) identify investment gaps, notably with respect to cross-border capacities. A review of barriers to the increase of cross-border capacity of the network arising from different approval procedures or practices may be annexed.[43]

39 *Communication from the Commission to the European Parliament, the Council, the European Economic and Social Committee and the Committee of the Regions. Energy Infrastructure Priorities for 2020 and Beyond: A Blueprint for an Integrated European Energy Network*, Brussels, 17.11.2010, COM (2010) 677 final.

40 11 May 2011, available at <http://www.consilium.europa.eu> (accessed 2 October 2011).

41 *OJ L* 262, 22 September 2006.

42 Article 8 and articles 13 and 22 of the ED, and articles 14 and 22 of the GD.

43 Article 8(3)(b), (4), and (10) (first and second subparagraphs), of the RE and article 8(3)(b), (4) and (10) (first and second subparagraphs) of the RG.

D. The plans in process

Due to the importance and urgency of the work ahead, both ENTSOs started working on the TYNDP before the formal adoption of the ER and GR (December 2008), acting as if ACER were already in place. ENTSO-E released its pilot TYNDP on 1 June 2010 and the next update of the plan is already in preparation and is expected to be finalized by June 2012. ENTSO-G published its first TYNDP 2010–2019 in December 2009, and adopted and published the second TYNDP 2011–2020, on 17 February 2011.[44] This move seems to indicate the relevance of private impetus in planning and a readiness to take the lead to avoid too much disruptive interference of NRAs.

E. Consistency between the Community-wide network development plans and their implementation

ACER is tasked to monitor the implementation of the TYNDPs. If it identifies inconsistencies between plan and implementation, it shall investigate the reasons for those inconsistencies and make recommendations to the TSOs, NRAs, or other competent bodies concerned, with a view to implementing the investments in accordance with the TYNDPs.[45]

F. Consistency between the Community-wide network development plans and the national ten-year network development plans, as reviewed by the Agency

Whereas the RE states that the ACER shall provide an opinion on the NTYNDPs to assess their consistency with the TYNDP, the RG states that the ACER shall review NTYNDPs to assess their consistency with the TYNDP.[46] However, no legal relevance should be given to the different wording, since both the opinion given (electricity) and the review conducted (gas) lead to a similar result: if the ACER identifies inconsistencies between a NTYNDP and TYNDP, it shall recommend amending the NTYNDP or TYNDP as appropriate. The TYNDP is of a non-binding nature, which explains why if there is an inconsistency between a NTYNDP and the TYNDP, the ACER is not asked either by the ER or by the GR to recommend the amendment of the NTYNDP, but rather is asked to recommend the amendment of either the NTYNDP or of the TYNDP. It rests on the discretion of the ACER to do one thing or the other, subject to the obligation to give a reasoned decision. If the ACER identifies an inconsistency, it addresses a non-binding recommendation to the author of the plan.[47] If the ACER recommends the amendment of the TYNDP, such recommendation should be addressed to ENTSO-E or to ENTSO-G. If, on the contrary, the ACER recommends the

[44] See, further <http://www.entsoe.eu> and <http://www.entsog.eu> (both accessed 2 October 2011).

[45] Article 6(8) of the ACER Regulation.

[46] Articles 8 and 11 of the ER, and articles 8 and 11 of the GR.

[47] Article 8(11) of the ER and Article 8 (11) of the GR.

amendment of a NTYND, it is not clear from the wording of the ER and the GR who is to be the addressee of the recommendation, either the TSO or the NRA.

Indeed, existing provisions of the Third Energy Package are unclear about a number of issues, particularly when comparing the ED and the GD with the ER and the RG:

(i) There is no place in either the ED or the GD where it is expressly said who is to adopt or approve the NTYNDP. Both the ER and the GR seem to assume that NTYNDPs are adopted by the NRA. That may not be the case, since there are sufficient elements in both the ED and the GD, and concordant regulations that would allow one to conclude that NTYNDPs are adopted by each TSO. However, the ED and the GD confer upon NRAs relevant powers within the procedure which lead to the adoption of the NTYNDP, as well as powers to monitor and evaluate.[48]

(ii) The ED and the GD impose upon the NRAs an obligation to provide in their annual reports an assessment of the investment plans of the TSOs as regards their consistency with the TYNDP referred to in the ER and the GR. When the NRA finds an inconsistency, three levels of consequences can be identified in the existing legal norms:

   a) include recommendations to amend the NTYNDP;
   b) require the TSO to amend its NTYNDP; and
   c) amend the NTYNDP, subject to a prior recommendation of the ACER.[49]

   The consequence listed under letter b) above is more coherent than that contained in the ER and the GR (listed under letter c)) since, in case of an inadequacy, it is not for the NRA to amend the NTYNDP (as the ER and the GR state), but rather to require the TSO to amend it. This is also consistent with the consequence provided for in both the ED and the GD (listed under letter a), above) by which the NRA, in its annual assessment, 'may include recommendations to amend the investment plans' (ie the NTYNDP), taking into account that it is for the TSO to approve the NTYNDP. The option listed under letter c) is contestable, since the ER and the GR empower the NRA to directly amend it.

(iii) The ER and the GR may be mistakenly referring to paragraph 7 of article 22 of the ED and the GD instead of paragraph 5, as paragraph 7 does not deal with the amendment of the NTYNDP, whereas paragraph 5 does.

G. Regional cooperation and planning

Given that more effective progress may be achieved through an approach at the regional level, since investment decisions are taken by instances which are closer to where the needs are, TSOs must set up regional structures within the overall

[48] Article 8(11) of both the ER and the GR and Article 22 of both the ED and the GD.

[49] Article 8(11) of the ER; article 8(11) of the GR; article 22(5) and (6) of the ED; article 22(5) and (6) of the GD; article 37(1)(g) of the ED and article 41(1)(g) of the GD.

cooperation structure, whilst ensuring that results at regional level are compatible with the TYNDP. It should be mentioned that in the context of the EU, 'regional' refers to two or more member states. Member states should promote cooperation and monitor the effectiveness of the network at a regional level. In particular, regional structures must, as a requirement, publish a regional investment plan every two years, and may take investment decisions based on that regional investment plan.[50] For this purpose, the existing regional structures can be used, such as the existing regional structures listed in Annex I of both the RE and the RG (for example, Northern Europe, ie Denmark, Sweden, Finland, Germany, and Poland).

On regional cooperation of gas TSOs, the GR specifically states that cooperation within such regional structures presupposes effective unbundling of network activities from production and supply. In the absence of such unbundling, regional cooperation between transmission system operators may result in anti-competitive conduct insofar as only unbundled activities guarantee that production and supply activities do not interfere with investment decisions. It sounds as if the EU was ready to tolerate an infringement of the GD. Cooperation at the regional level should be compatible with progress towards a competitive and efficient internal market in electricity and gas.[51] It must be noted that the Preventive Action and Emergency Plans of Regulation 994/2010, on measures to safeguard security of supply, may have a regional coverage.[52]

V. Conclusions

The Third Energy Package contains a new concept in planning quite removed from old conceptions. First, there is a new complex institutional structure involved in the drafting of the plans made up by governments (NRAs, TSOs, ACER, ENTSO-E, ENTSO-G, UCTE, and EASEE), as well as the need to give a voice to all stakeholders affected by planning, including energy companies and consumers. As a result of the application of the ER and the GR, some private associations which used to have marginal presence in the official EU documents, such as the UCTE, start to have an official recognition through the ENTSOs. Second, there are requirements for several planning instruments at the EU and national and regional levels, and a call for adequacy or compatibility among all of them. Third, is the introduction of new legal concepts which were not used in the past when planning was only a way of exercising the command-and-control type of regulatory power. Among the new concepts we find 'adequacy', 'monitor', 'cooperation', 'coordination', 'review', 'recommendation', 'consultation', etc. In previous times other types of expressions were used, such as 'order', 'decision', 'requirement', 'infringement', 'hierarchy'. All these features are in contrast with former governmental plans based

[50] Article 12(1) of the ER, and article 12(1) of the GR.
[51] Recital no. 7 and article 12(1) of the ER, and Recital no. 16 and article 12(1) of the GR.
[52] See section II of this chapter.

on the hierarchical position of the central executive power and on the assumption that planning was a public task. The new style of planning is a clear example of *regulated self-regulation*: the TYNDP is a non-binding plan, adopted by TSOs, which are not a branch of government and whose members are, mainly, private companies grouped under the ENTSOs, but under the regulatory supervision of NRAs and the ACER.

In addition to the adherence to a new type of planning policy, there also subsist clear elements of prior compulsory or binding energy planning, like those referring to the way an NRA must solve an infringement of an investment commitment within the NTYNDP, including requiring the TSO to execute the investments in question.

The EU has fostered unbundling of energy activities as a crucial tool for competition and liberalization, since it helps to set up a more transparent and efficient TPA system. By unbundling transport and distribution activities from production, generation and supply activities, TPA to transport and distribution facilities will be more transparent, since in the absence of integration between distribution and supply, for example, the distribution company will not be tempted to discriminate in favour of its parent supply company against other competing suppliers. When the Commission launched the proposal for a Third Energy Package, it stressed that property unbundling guaranteed security of supply, since it helps investments decisions to be adopted with no consideration of supply activities. Irrespective of the merits of the argument of the Commission, the content of the Third Package provides evidence (considering the amount of provisions devoted to planning) that, among a variety of means to achieve sufficient level of investment there is no better way than planning to guarantee investment in infrastructure in a liberalized market. The ED and the GD introduced effective unbundling (according to the European Commission), but if further measures introducing planning to guarantee investment were needed, such as those included in the ER and the GR, it becomes clear that property unbundling was not enough to guarantee investment.

In the past, within isolated European energy markets (as many as member states), legal certainty entailed clear national arrangements on who does what, on which crucial decisions of energy agents depend. In the era of integration and cooperation (among member states, but also among private and public institutions) there is the need to constantly refine institutional and procedural arrangements, so that energy agents can take investment decisions with confidence based on predictability of their consequences. With this new planning policy, there is the need not to soften legal certainty, as a key exigency of the State of Law, or, at least, the need to bear in mind that legal certainty must operate in a new context. Legal certainty means that laws are clear in imposing obligations upon somebody, and that are clear in establishing consequences to a lack of compliance, but the laws passed under the Third Energy Package lack said clarity. Take, for the example, the way to solve inconsistencies between the TYNDP and the NTYNDP, which are to be primarily solved by means of voluntary acceptance of recommendations, though in both the ER and the GR we seem to find a power of the NRA to require the TSO to change its NTYNDP.

20

Regulating the Extension of Electricity Networks: A German Perspective

Gunther Kühne

I. The Pivotal Role of Energy (Electricity) Infrastructure in Germany

This chapter analyses the pivotal role that energy (electricity) infrastructure, with particular emphasis on electricity networks, has assumed within the German energy system over the past decade. For several years this development was of a 'creeping' character, but foreseeable since the beginning of this century as a result of the changing pattern of electricity generation (reduction of bulk electricity production from fossil and nuclear installations and expansion of electricity generation from smaller and volatile renewable sources). This change was dramatically speeded up in the wake of the Fukushima nuclear catastrophe in March 2011.

This chapter discusses the stages of this development taking into account the most recent legislative measures in response to the Fukushima events and the legal concepts underlying these measures—concepts that even reach into constitutional and European spheres. As a matter of statutory law, the norms on construction and maintenance of energy facilities (generation and transportation) at first appear to be of a rather technical character. But the German example shows that they are deeply interwoven with basic tenets of energy law, such as security of supply and affordability of energy. Beyond this, the reshaping of energy infrastructure, especially grid infrastructure, has turned out to be one of the testing grounds for the viability of established decision-making procedures within the traditional democratic-representative political system.

II. Political and Factual Background

A. The present German landscape in energy policy with emphasis on electricity

The political and legal regime regarding energy policy in Germany during recent months has gone through a state of turmoil. The latest turbulence in this drama was undoubtedly caused by the nuclear reactor catastrophe that occurred in Fukushima, Japan, on 11 March 2011.

Under the impact of the fundamental EU legislative initiatives, in particular the first and second generation of Electricity and Gas Directives 1996/98 and 2003 respectively, the political and legal framework in Germany, over the past 15 years, has undergone significant changes. Originally, the shift from the old integrated monopolistic regime to the new competitive system was almost exclusively motivated by the objective to reduce energy prices through the instrument of competition. Third party access to the energy grids and its conditions were the dominating issues during the first phase of liberalization.[1] Even though the first generation of EU Directives already contained regulatory elements, their implementation in Germany was perceived mainly as a matter of competition law and its enforcement.

It was in 2000 when massive blackouts struck California's electricity system, and in 2003 with severe disruptions of electricity supply in parts of Europe, that the perspective of energy supply was substantially widened. Security of supply became an issue. With it, awareness grew that reorganizing the electricity system means more than the injection of competition into energy supply markets. It became apparent that increased emphasis on one of the three cornerstones within the 'magic triangle' (affordability, security, and environmental friendliness) of the energy system has an impact on the others. And, at the same time, changing the rules on one of the rungs of the value-adding chain (generation, transport, distribution of electricity) also has an impact on the other rungs. The second generation of EU Energy Directives (2003) with its tight regulatory approach is a reflection of this more systemic approach to the electricity supply regime. In 2005 Germany followed suit, replacing its competition-law model with a comprehensive regulatory framework.[2] This development was accompanied by a specific German aspect—ie the dynamic expansion of electricity generation through renewable energy sources (in particular wind energy), as advocated by the Red–Green coalition Government in power from 1998 to 2005.[3] As part of the legislative framework the pertinent statute guaranteed preferential grid access to all electricity generated by renewable energies and subsidized feed-in prices. More generally, decentralized electricity generation gained growing acceptance in the political arena and began to delegitimize large-scale centralized forms of electricity generation. In 2002 the Atomic Energy Act was amended to the effect of abandoning nuclear energy on the basis of a phase-out scheme. Under the decarbonization philosophy, coal and lignite power stations are increasingly coming under attack for running contrary to climate protection policies. Taking these developments into account, a deficit in

[1] See G. Kühne, 'Conventional Regulation Renascent but Changing—Alternatives to Regulation: The German Experience' in B. Barton, L.K. Barrera-Hernández, A. Lucas, and A. Rønne (eds), *Regulating Energy and Natural Resources* (Oxford: Oxford University Press, 2006) 203–21.

[2] Gesetz über die Elektrizitäts- und Gasversorgung (Energiewirtschaftsgesetz—EnWG) of 7 July 2005, *Bundesgesetzblatt* (*BGBl.*) (Federal Gazette) I, p. 1970.

[3] See G. Kühne, 'Legal Incentives to Alternative Energies in German Energy Law' in P.D. Cameron and D. Zillman (eds), *Kyoto: From Principles to Practice* (The Hague/London/New York: Kluwer Law International, 2001) 61–72; G. Kühne, 'Regulating Offshore Wind Farms in Germany' in M. Roggenkamp and U. Hammer (eds), *European Energy Law Report I* (Antwerp-Oxford: intersentia, 2004) 147–59.

electricity-generating capacity is projected for the coming decades. Massive expansion of offshore wind generation (wind farms) is generally seen as the main potential contributor to meeting future electricity demand in Germany.

This, however, would mean that over the coming decades a geographical gap between the centre of electricity generation on the one hand and the centres of electricity consumption on the other is bound to open up. Under these circumstances such a pattern of electricity generation would entail the need for large-scale extension of the electricity grid infrastructure. High-voltage transmission lines will have to be installed in order to transport North Sea electricity to Central and Southern Germany. This shows that electricity network infrastructure has come to be the neuralgic point of Germany's electricity system. This is even more true in the light of the Fukushima reactor catastrophe of 11 March 2011 and its political, as well as legal, fall-out in Germany.

B. Electricity network infrastructure as the key factor in German energy policy for the next decades

There is widespread political consensus in today's Germany that electricity grid infrastructure and its extension during the coming decades will play a key role in energy policy. This complex is closely linked to each of the three elements of the energy policy triangle: affordability (expansion costs!), security of supply, and environment-friendliness.

It is therefore not surprising that since autumn 2010, significant initiatives have been taken to describe the political and factual contours of a future energy supply system. Network infrastructure and its future configurations play a key role in the scenarios. Those initiatives are (a) the 'Energy Concept' of the Federal German Government of September 2010;[4] (b) the EU Communications of November 2010;[5] and (c) the DENA Grid Study II of November 2010.[6]

1. The Energy Concept of the Federal German Government

On 28 September 2010 the Federal German Government adopted an 'Energy Concept for an Environmentally Sound, Reliable and Affordable Energy Supply' (hereafter, the Energy Concept). Its most striking feature is the reorientation of the electricity supply system towards renewable energies.

[4] 'Energy Concept for an Environmentally Sound, Reliable and Affordable Energy Supply', 28 September 2010 (edited in English translation by the Federal Ministries of Economics and Technology and for the Environment, Nature Conservation and Nuclear Safety), available at <http://www.bmwi.de> and <http://www.bmu.de> (both accessed 26 October 2011).

[5] Communication 'Energy 2020—A Strategy for Competitive, Sustainable and Secure Energy', COM (2010) 639 final, of 10 November 2010. Communication 'Energy Infrastructure Priorities for 2020 and Beyond—A Blueprint for an Integrated European Energy Network', COM (2010) 677 final, of 17 November 2010.

[6] DENA Grid Study II—Integration of Renewable Energy Sources in the German Power Supply System from 2015–2020 with an Outlook to 2025. Summary of the main results by the project steering group (in English) available at http://www.dena.de (accessed 26 October 2011).

As already agreed between the coalition partners in autumn 2009, greenhouse gas emissions are to be cut by 40 per cent by 2020 in line with the industrial nations' target of at least 80 per cent by 2050 (base year: 1990). To achieve that, the development path up to 2050 will be the following: a 55 per cent reduction by 2030, a 70 per cent reduction by 2040, and an 80–95 per cent reduction by 2050. By 2020 renewable energies are to account for 18 per cent of gross final energy consumption. After that the German Government aims to achieve the following share of renewable energies in gross final energy consumption: 30 per cent by 2030, 45 per cent by 2040, and 60 per cent by 2050. By 2020, electricity generated from renewable energies is to account for 35 per cent of gross electricity consumption. For the following decades the Federal Government's targets are: 50 per cent by 2030, 65 per cent by 2040, and 80 per cent by 2050.

These very ambitious goals require a massive expansion of the existing electricity network infrastructure. Accordingly, the 2010 Energy Concept also encompasses a catalogue of infrastructure goals:

- Development of 'Target Grid 2050' covering all essential aspects, in particular:
  - further development of the existing grid;
  - planning of an overlay grid ('electricity highways') and possible pilot routes;
  - North Sea grid and clustered offshore connection;
  - integration of the German grid into the European network;
- ten-year grid expansion plan to be agreed among all transmission grid operators and to be submitted annually;
- federal grid plan to be developed on the basis of the ten-year grid expansion plans and to be presented by the Federal Government;
- streamlining of planning and licensing procedures for grid expansion;
- recoverability of grid expansion expenditures under the incentive regulatory framework for grid access tariffs;
- development of 'smart grids';
- development of an offshore grid.

2. EU Communications of 10 and 17 November 2010

As part of its enhanced efforts in the field of energy, the EU Commission, in exercise of its new energy competence in article 194 TFEU, has adopted two initiatives in the energy sector with the objective of upgrading energy infrastructure within the EU: Communication 'Energy 2020—A Strategy for Competitive, Sustainable and Secure Energy' of 10 November 2010 and Communication 'Energy Infrastructure Priorities for 2020 and Beyond—A Blueprint for an Integrated European Energy Network' of 17 November 2010.[7]

[7] See above, n 5.

In the Communication of 10 November, the Commission outlines five priorities. Priority No. 2 is 'Building a truly pan-European integrated energy market', subdivided into four actions. Action 2 relates to infrastructure: *Establishing a blueprint of the European infrastructure for 2020–2030*. The Agency for the Cooperation of Energy Regulators (ACER) and the European Network of Transmission System Operators for Electricity (ENTSO-E) will be given a mandate to develop the blueprint of a European electricity grid on the horizon of 2020–2030.[8] In the Commission's perspective this should be followed by a longer-term vision on the basis of the energy 2050 roadmap to be presented in 2011.

In the Communication of 17 November 2010 the Commission proposed priority corridors for power grids. From a German perspective, the most relevant corridor among the proposals is the one for an offshore grid in the Northern Seas and connection to Northern and Central Europe to transport power produced by offshore wind farms to consumers in big cities and to store power in the hydro-electric power plants in the Alps and the Nordic countries.

3. *The DENA Grid Study II*

In November 2010 the German Energy Agency DENA (Deutsche Energie-Agentur) presented its Grid Study II. Its objective was to investigate suitable system solutions for the German power supply system (up to 2020, with an outlook to 2025), to fully integrate 39 per cent renewable energy in the power supply into the German power grid while guaranteeing the security of supply and taking into account the effects of the liberalized European energy market. The DENA Grid Study II was preceded by Study I in 2005. In Study I, a total of 850km in grid expansion length had been determined as necessary. By the time Grid Study II was completed, only 90km had been implemented.

Study II, on the basis of 39 per cent renewable energy electricity having to be integrated into the grid system, concludes that the grid expansion need up to 2020 will amount to 3,600km in length. The gap between the projection of 850km by Study I and the actual 90km expansion realized is sufficient to give an idea of how enormous the infrastructural challenge will be.

C. The post-Fukushima scenario

The scenario of 2010 as just outlined was considerably shaken on 11 March 2011 when the Fukushima reactor catastrophe occurred as the result of an earthquake and an ensuing tsunami. This event will undoubtedly have substantial effects on the grid via the changing composition of the electricity-generating portfolio in Germany. At the beginning of 2011, nuclear energy accounted for roughly 20 per cent of Germany's electricity production. In 2002 the Red–Green coalition Government had amended the Atomic Energy Act by limiting the residual life

[8] See also chapter 19 of this book.

span of the then existing nuclear reactors.[9] This limitation was not expressed in terms of residual operation time but in terms of residual production allowances. Under this scheme, nuclear energy production was projected to end in or around 2025. The Conservative–Liberal coalition that came to power in October 2009 again amended the Atomic Energy Act extending the life-span of reactors: those installed up to the end of 1980 by roughly eight years, the more recent ones by 14 years, both on average.[10] On 19 March 2011 the Federal Government suspended the extension under the amended Atomic Energy Act by declaring a three-month 'Moratorium'. It ordered the temporary shutdown of the 'old' reactors for a three-month period. The Federal Government charged two commissions—the Reactor Safety Commission and the specially established Ethics Commission 'Secure Energy Supply'—to investigate the technical and ethical sustainability of the further use of nuclear energy. The results were made public at the end of May 2011. From a technical point of view German nuclear reactors were evaluated as fulfilling a high standard, while, however, none of the reactors is secured against the crash of a larger civil aircraft.[11] The Ethics Commission held that the events in Fukushima reduce the ethically sustainable residual risk of nuclear energy.[12] It advocated the total abandonment of nuclear energy by 2022 at the latest.

On the basis of this assessment the Federal Government, in early June 2011, presented a Draft according to which the 'old' reactors which were temporarily shut down in March 2011, were to be closed indefinitely and the remaining plants to be phased out until 2022; the German Parliament adopted this proposal in early July 2011.[13] The operating utilities have announced that they will file constitutional complaints against the amendment invoking, in particular, violation of the property clause (article 14 of the Constitution). Before signing the Bill into law, the Federal President has checked and affirmed the constitutionality of the new law. Proceedings before the Federal Constitutional Court (*Bundesverfassungsgericht*) are certain to follow.

With nuclear energy being phased out more rapidly and fossil fuels being problematical from a climate protection point of view, Germany will have to embark on a massive expansion of electricity generation from renewable energy sources (like wind, sun, biomass). This means that the grid system will have to be extended and upgraded on a large scale. It therefore comes as no surprise that, together with the amendment of the Atomic Energy Act with its phase-out scheme, the German

[9] See 'Act on the Structured Phase-out of Nuclear Power for the Commercial Production of Electricity', *BGBl.* 2002I 1351. For a review see: A. Vorwerk, 'The 2002 Amendment to the German Atomic Energy Act Concerning the Phase-out of Nuclear Power', *Nuclear Law Bulletin* 69 (vol. 2002/1) 7. See the English translation of the Act in *Nuclear Law Bulletin* No. 70 (supplement) (vol. 2002/2).

[10] See *BGBl.* 2010 I 1814. See also the report in *Nuclear Law Bulletin* No. 86 (vol. 2010/2) 76.

[11] See Report by the Reactor Safety Commission 'Anlagenspezifische Sicherheitsüberprüfung—RSK-SÜ—deutscher Kernkraftwerke unter Berücksichtigung der Ereignisse in Fukushima—I—Japan—', of 16 May 2011, 15, available at <http://www.bundesregierung.de> (accessed 26 October 2011).

[12] See Report by the Ethics Commission 'Sichere Energieversorgung' of 30 May 2011, 11, available at <http://www.bmu.de> (accessed 26 October 2011).

[13] *BGBl.* 2011 I 1704.

Parliament, in early July 2011, adopted a whole set of statutory provisions which represent the legislative basis of the 'energy turn'. Apart from the amendment of the Renewable Energies Act designed to promote electricity production from renewable energy sources, the legislative package, in particular, comprises measures for extending the grid infrastructure. Pertinent provisions are to be found both in the Amendment of the Energy Act 2005[14] implementing the Third Energy Package and in the new Grid Expansion Acceleration Act (*Netzausbaubeschleunigungsgesetz*, or NABEG).[15] The main objective of both legislative steps is to speed up grid extension projects by introducing new planning instruments on the federal level and by creating federal administrative jurisdiction in cases of inter-state grid extension projects. It is, however, doubtful, to what extent both legislative initiatives will achieve their goal.

The now settled horizontal conflict between two party political camps will possibly be replaced by a vertical conflict between the political leaderships and the grassroots level. Throughout Germany already local politicians and civil action committees are vigorously opposing many infrastructure projects necessary for enhancing the use of renewable energies (construction of high-voltage transmission lines, windmills, biomass installations, pumped storage plants). This movement was and still is significantly boosted by recent popular 'uprisings' against other infrastructure projects. The most noticeable of them is project 'Stuttgart 21'—the conversion of the existing Stuttgart central terminal railroad station into a large underground through-station. The way this project will finally be handled after a Green–Red coalition came into power in Baden-Württemberg will probably have an effect on the 'infrastructure opposition' movement in general.

III. Basic Legal Concepts for the Attribution of Network Infrastructure Responsibility

A. State responsibility for ensuring a viable energy network infrastructure

1. European Law

The importance of a secure and viable electricity infrastructure for European energy policy finds a clear expression in the fact that this objective, over the past ten years, has found its way into European energy legislation: the Directive 2005/89/EC of 18 January 2006 'concerning measures to safeguard security of electricity supply and infrastructure investment'.[16] Article 3 para 2 lit. c), on member states' obligation to ensure a high level of security of electricity supply, stresses 'the need for regular maintenance and, where necessary, renewal of the transmission and distribution networks to maintain the performance of the network'. Article 6 para 1

[14] *BGBl.* 2011 I, 1554.
[15] *BGBl.* 2011 I, 1690.
[16] *OJ L* 33, 4.2.2006.

commits member states to establish a regulatory framework that (a) provides investment signals for both the transmission and distribution system network operators to develop their networks in order to meet foreseeable demand from the market; and (b) facilitates maintenance and, where necessary, renewal of their networks.

The Third EU Energy Package provides for an additional infrastructure investment tool. As part of the tightened three-optional unbundling regime (ownership unbundling, independent system operator (ISO), independent transmission operator (ITO)), it commits the ITO to submit to the regulatory authority every year a ten-year network development plan based on existing and forecast supply and demand after having consulted all the relevant stakeholders (article 22 para 1 Electricity Directive).[17] The regulatory authority shall examine whether the ten-year network development plan covers all investment needs identified during the consultation process, and whether it is consistent with the non-binding Community-wide ten-year network development plan (Community-wide network development plan) referred to in article 8 para 3 lit. b) of Regulation (EC) No. 714/2009[18] (article 22 para 5 Electricity Directive).

At first sight, it is surprising to note that the Directive requires such a ten-year plan only from the ITO, but not from a totally unbundled operator, nor from an ISO. The differentiation may be explained by the fact that the authors of the Directive harboured a mistrust of the ITO, since it is still linked to the vertically integrated undertaking and thus may be hampered in its infrastructure investment strategy.

In sum, it appears well settled that under EU law, member states are firmly committed to ensuring a viable network infrastructure.

2. *State infrastructure responsibility in German constitutional law*

For more than two decades the European and German economies, especially in sectors relying on infrastructural investments, have been subjected to waves of demonopolization and privatization. This development has triggered a debate in Germany on the question of who is finally responsible for the availability of infrastructure when the state transfers its operational responsibility to private entities. Who is responsible for the existence and operation of a functioning infrastructure in, for example, the energy industry? Under the old monopolistic system the Federal Constitutional Court (*Bundesverfassungsgericht*, or BVerfG) considered the idea of 'security of energy supply' to be a 'common good of utmost importance'.[19] It has stuck to this idea ever since. In a more recent decision, the Court stated that secure energy supply today is as important as the 'daily bread'.[20] This means that, on the basis of the 'social state principle' of article 20 of the Constitution (*Grundgesetz*, or GG), the state is under an obligation to ensure a secure system of energy supply.

[17] Directive 2009/72/EC, OJ L 211, 55.

[18] Regulation No. 714/2009, OJ L 211, 15.

[19] Decision of the BVerfG of 20 March 1984 (*'Enteignung'*), Amtliche Sammlung der Entscheidungen des BVerfG (Official Collection of the Decisions of the BVerfG), BVerfGE 66, 248.

[20] Decision of 11 October 1994 (*'Kohlepfennig'*), BVerfGE 91, 186, 206.

Whenever private entities are operating this system, the state is not relieved of its obligation. In such a situation the state's obligation is converted into the function of a guarantor. The state is committed to ensure that the (private) operators of the infrastructure fulfill their function to maintain a system of secure energy supply. Administrative courts have confirmed this state guarantor responsibility after 1998 when the new Energy Act implementing the EU Electricity Directive of 1996 was passed.[21]

B. The grid operator's responsibility for maintaining and extending energy grids in adjustment to demand

Whenever the state does not act as operator of the energy system, the operational responsibility rests upon the (private) operator. The operator is under the obligation to operate the grid in a secure, reliable, and capable way (s 11, para 1 Energy Act, EnWG). The same provision obligates the operator to optimize, upgrade, and extend the grid to the extent this is economically reasonable. Thus the Act obligates operators not only to invest in existing grids but also to develop new routes. Viewed from a market economy standpoint this far-reaching statutory obligation to invest is an alien element. It derives its legitimacy, however, from the high-ranking public interest in a secure energy supply system. That is why the statutory investment obligations are compatible with the basic rights under article 12 (freedom of profession) and article 14 (protection of property) of the Basic Law (Constitution) to which, generally, grid-operators like any other entrepreneurial entity are entitled. The statutory investment obligation is also enforceable through administrative order (s 65 EnWG). But, in general, the authority will not be able to attach sufficiently concrete measures to its order. Hitherto, this enforcement tool has never been used.

Section 11 para 1 EnWG speaks of an extension 'adjusted to demand'. This does not require that a demand for extension is already existing. The 'adjustment to demand' has to include medium-range demand. This requires a prognosis.[22] The demand prognosis has to be established in accordance with recognized procedures, the foremost of which is the inclusion of all relevant data. The prognosis has to be plausible. A prognosis is implausible whenever the data base of the prognosis does not support the forecast (eg for lack of methodological correctness).

In principle, it is the grid operator's responsibility to ascertain the future demand and establish the prognosis accordingly. In this context the question has been raised whether to grant the grid operator a margin of assessment which cannot be reviewed either administratively or judicially. One might be inclined to grant such a margin, arguing that this is the expression of a residual procedural privilege

[21] Decision of the Bundesverwaltungsgericht (BVerwG) of 11 July 2002 ('*Enteignung*'), Amtliche Sammlung der Entscheidungen des BVerwG (Official Collection of the Decisions of the BVerwG), BVerwGE 116, 365, 371/72.

[22] See K. Bourwieg in G. Britz, J. Hellermann, and G. Hermes (eds), *Kommentar zum Energiewirtschaftsgesetz (EnWG)* (2nd edn, Munich: C.H. Beck, 2010), marg. note 35 to s 11 EnWG.

attached to the property right. On the other hand, the demand assessment involves numerous societal factors and data that do not belong to the operator's sphere. Moreover, it has to be taken into account that recent legislation has created legislative tools working towards cooperation between the operators and the state in establishing demand-oriented planning proceedings. This will be even more so under the Third Energy Package and eventually in the proposals for tighter state planning mechanisms being realized. Given these strong external influences on the demand prognosis, the operator's privilege of assessment should be very limited.[23] It may be relevant where at least two demand scenarios occur that are almost equally valid.

The obligation to grid extension is conditioned upon its 'economic reasonableness'. This term certainly lacks clear contours. Nonetheless, two conclusions seem to be relatively undisputed: extension measures are always economically unreasonable when they do not stand the test of proportionality; for example, they are unnecessary. The second clear-cut situation of economic unreasonableness involves the situation where the extension measure in question would ruin the operator's business.

C. The grid-owner's responsibility for grid extension under the Renewable Energies Act

In addition to the general grid extension obligation as just discussed, s 9 para 1 Renewable Energies Act (*Erneuerbare-Energien-Gesetz*, or EEG) provides for a specific grid extension obligation in favour of renewable energies.

Under s 5, para 1 sent 1 EEG grid system operators shall, as a matter of priority, immediately connect installations generating electricity from renewable energy sources and from mine gas to that point in their grid system (grid connection point) which is suitable in terms of the voltage and which is at the shortest linear distance from the location of the installation if no other grid system has a technically and economically more favourable grid connection point. According to s 8 para 1 EEG, grid operators shall immediately, and as a priority, purchase, transmit, and distribute the entire available quantity of electricity from renewable energy sources and from mine gas. Lack of grid capacity is not an excuse. Under s 9 para 1, EEG grid operators, upon request of those interested in feeding in electricity, shall immediately optimize, boost, and expand their grid systems in accordance with the best available technology in order to guarantee the purchase, transmission, and distribution of the electricity generated from renewable energy sources or from mine gas. This obligation does not apply if its fulfillment is economically unreasonable. In general, the criterion of economic unreasonableness is subject to the same standards as in s 11 para 1 EnWG. But because of the more narrow scope of the extension obligation, a more concrete standard has been proposed: the ratio between the sum total of the feed-in tariffs from all generating facilities to be connected and the overall

[23] Ibid, marg. note 37.

expansion costs. In case the sum of all feed-in tariffs to be recovered does not significantly exceed the expansion costs, the expansion investment has to be deemed economically unreasonable.[24]

IV. Planning Mechanisms

A. General observations

Under the old monopolistic system, responsibility for an adequate network infrastructure was almost exclusively vested in the vertically integrated utilities. As a result of the traditional monopolistic character of electricity supply, the database for future demand forecast was relatively transparent, stable, and accessible. This changed once the closed markets were opened for competition. Future network capacity is much more difficult to assess due to the uncertainties inherent in competitive systems. Furthermore, the unconditional priority for grid access of all 'green electricity' is a destabilizing element for network infrastructure planning. Infrastructure design therefore needs to be conceptualized on a higher and more general level. Licensing procedures have to be embedded in higher level decision-making. This is, indeed, what we can observe in recent years both on the European and the national (German) level.

B. Decision No. 1364/2006/E[25]

This EU Decision laid down targets, priorities, and principles for actions to be taken within the area of trans-European energy networks (TENs) and identifies 'projects of common interest' (article 6). These 'projects of common interest' are determined according to the criteria mentioned in article 6 lit. a)–c):

- the project falls within the scope of article 2 (*inter alia* high-voltage lines);
- the project meets the objectives and priorities for action set out in articles 3 (*inter alia* development and connection of renewable energy resources, lit. a)) and 4 (*inter alia* trans-border connections solving the bottleneck problem, 1 lit. a));
- the project displays potential economic viability.

Annex III to the guidelines enumerates projects of common interest and their specifications, currently identified according to the criteria set out in Annex II (additional—geographically defined—criteria for identifying projects of common interest, as referred to in article 6 para 2). Annex I lists certain axes for priority projects and the priority projects appertaining thereto to benefit preferentially from Community aid (article 7) and Projects of European interest (cross-border nature) (article 8). German projects figure in the Annexes both as intra-German and German cross-border projects.

[24] See U. Ehricke in W. Frenz and H.J. Müggenborg (eds), *Kommentar zum Erneuerbare-Energien-Gesetz (EEG)* (Berlin: Erich Schmidt Verlag, 2010), marg. note 46 to s 9 EEG with reference to legislative materials.

[25] OJ L 262/1 of 22 September 2006.

C. The German Infrastructure Planning Acceleration Act (2006)

In 2006 Germany enacted the Infrastructure Planning Acceleration Act (*Infrastrukturplanungsbeschleunigungsgesetz*).[26] Its objective was to accelerate planning and licensing proceedings for a variety of infrastructurally relevant projects, ie for electricity network infrastructure. The Act inserted s 17 para 2a EnWG in order to improve the integration of offshore wind farms into the onshore grid system. Under the provision, transmission operators are under an obligation to build and to operate the connecting lines from the offshore installation to the technically and economically most favourable connecting point of the closest transmission or distribution network.

D. The Electricity Grid Expansion Act (2009)

In line with the TEN guidelines of 2006, Germany, in 2009, enacted the Electricity Grid Expansion Act (*Energieleitungsausbaugesetz*, or EnLAG).[27] The Act is designed to accelerate the realization of 24 high-priority grid construction projects involving extra-high-voltage transmission networks (380 kV). These projects serve the adjustment, upgrading, and expansion of transmission networks for the purpose of integrating renewable energy production facilities, of furthering interoperability of electricity grids within the EU, and of connecting new power plants or of avoiding structural bottlenecks within the transmission grids.

In s 1 para 2 the statute deems the 24 projects listed in the Annex as fulfilling the objectives of s 1 EnWG. This statutory determination of both the conformity with energy policies and the high-priority character of the projects has the legal effect that their necessity cannot be challenged in subsequent licensing proceedings. This is the principal accelerating factor of the Act.

In s 2 the Act deals with another very controversial point in the public controversy over new high-voltage routes: the use of underground cables. The Act authorizes the use of 380 kV underground cables within the framework of four pilot projects and—under certain conditions—the use of 110 kV underground cables. The additional expenditures caused by using underground cables will be shared pro rata by all transmission operators.

E. The Third Energy Package: Regulation 714/2009 and Directive 2009/72/EU and their implementation in Germany

1. *European law*

As part of the Third EU Energy Package the European legislature has provided for two tools for a regulatory intensification of grid construction and expansion:

(i) According to article 8 para 3 lit. (b) of Regulation (EC) No. 714/2009,[28] the ENTSO-E is to adopt every two years a non-binding Community-wide ten-year network development plan (Community-wide network development plan), including a European generation adequacy outlook.

[26] *BGBl.* 2006 I, 2833. [27] *BGBl.* 2009 I, 2870. [28] See above, n 18.

The Community-wide network development plan, under article 8 para 10, shall include the modeling of the integrated network, scenario development, a European generation adequacy outlook and an assessment of the resilience of the system. It is to build on national investment plans, taking into account regional investment plans and, if appropriate, Community aspects of network planning, including the TEN guidelines and—regarding cross-border interconnections—also on the reasonable needs of different system users and long-term commitments from investors. The network development plan shall also identify investment gaps, notably with respect to cross-border capacities.

(ii) Article 22 para 1 of the Electricity Directive[29] obliges the ITO to submit every year to the regulatory authority a ten-year network development plan based on existing and forecast supply and demand after having consulted all the relevant stakeholders. The development plan shall contain efficient measures in order to guarantee the adequacy of the system and the security of supply.

The development plan, in particular, is to indicate to market participants the main transmission infrastructure that needs to be built or upgraded over the next ten years (article 22 para 2 lit. a)), to contain all the investments already decided, and to identify new investments which have to be executed in the next three years (lit. b)), and to provide for a time-frame for all investment projects (lit. c)). When elaborating the ten-year network development plan, the transmission system operator shall make reasonable assumptions about the evolution of generation, supply, consumption, and exchanges with other countries, taking into account investment plans for regional and Community-wide networks (article 22 para 3). The regulatory authority has to consult all actual or potential system users on the ten-year network development plan in an open and transparent manner. Persons or undertakings claiming to be potential system users may be required to substantiate such claims. The regulatory authority shall publish the result of the consultation process, in particular possible needs for investments (article 22, para 4).

The two-tier approach to network planning raises problems of coordination between the European and the national network development plans. Both plans are different in character and attached to different control organs. The Community-wide plan is non-binding and does not require formal enactment; on the other hand, the national ten-year plans are drawn up by the ITO and confirmed by the national regulatory authority. One should, therefore, assume that in case of inconsistencies between the European and the national plan the latter takes precedence or that, at least, the initiative to coordinate both plans is vested in the national regulatory authority. Instead, the Third Energy Package follows another approach, with ACER filling the lead role.[30] Article 8 para 11 of the Regulation provides that ACER shall provide an opinion on the national ten-year network development plans to assess their consistency with the

[29] See above, n 17.

[30] See also J. Kühling and R. Pisal, 'Investitionspflichten beim Ausbau der Energieinfrastrukturen zwischen staatlicher Regulierung und nachfrageorientierter Netzwirtschaft' 15 (2011), *Zeitschrift für Neues Energierecht*, 13, 17.

Community-wide network development plan. If ACER identifies inconsistencies between a national ten-year network development plan and the Community-wide plan, it shall recommend amending the national plan or the Community-wide plan, as appropriate. National ten-year plans elaborated on the basis of article 22 Electricity Directive will be amended by the national regulatory authority.

If a ten-year national network development plan is under examination by the respective national regulatory authority and the latter is in doubt as to the consistency with the Community-wide network development plan, the regulatory authority shall consult ACER. The regulatory authority may require the ITO to amend its ten-year network development plan (article 22 para 5 Electricity Directive). The interplay of article 22 para 5 Electricity Directive and article 8 para 11 Regulation No. 714/2009 works in a way that the national regulatory authority can enforce an amendment of a national plan because of inconsistency with a European plan only after ACER has recommended such a measure.

2. *German law*

During recent months the legal foundations for planning energy (electricity) grids have undergone substantial changes. This is due to two factors: one, the implementation of the Third Energy Package, including its new instruments for long-term planning of grid infrastructure, and two, the urgent need for expanding Germany's grid infrastructure following the political decision to abandon nuclear energy by 2022 as a result of the Fukushima catastrophe and to switch to renewable energies as the emerging main source of energy.

The implementation statute transplanting the Third Energy Package into German law provides for the ten-year network development plan required under article 22 Electricity Directive for the ITO but not for the ISO, nor for the totally unbundled network company. This exceptional position of the ITO may be rationalized by the argument that this is the unbundling model which preserves the closest connection between the network branch and the other branches of the energy business (production, distribution). This may support the suspicion that an ITO, in its network investment policies, might be influenced by considerations that are not network-related.[31] All models of unbundling, however, are subjected to the same risks of error in handling their network investment strategies, eg misestimation of future demand. The German implementation statute, therefore, treats all models of unbundling alike. The statute offers all three models to the German transmission operators and requires the ten-year network development plan indiscriminately. This is nothing less than adequate because of the strong interdependencies of investment measures. The statute commits all operators of transmission networks to annually submit a *joint* national network development plan[32] to the regulatory authorities.

[31] See E. Cabau, 'Unbundling of Transmission System Operators' in C. Jones (ed), *EU Energy Law Vol. I: The Internal Energy Market, The Third Liberalization Package* (3rd edn, Deventer: Claeys & Casteels, 2010) n 4.234.

[32] The provisions on the network development plan are contained in s 12a–12d of the Energy Act 2011, *BGBl.* 2011 I, 1554.

The joint national network development plan must contain all effective measures to optimize, to strengthen, and to extend the grid required during the following decade for a secure and reliable grid operation. The development plan has to be elaborated by the grid operators on the basis of a joint scenario framework (*Szenariorahmen*). The framework has to encompass at least three development paths (scenarios) which, for the following ten years, cover the range of probable developments within the medium- and long-term energy policy targets of the Federal Government. The development plan has to be confirmed by the Regulatory Authority. Public participation according to the provisions of the Environmental Impact Assessment Act is part of the confirmation procedure. Going beyond the requirements under European law (Third Energy Package) the German legislature, in order to still further streamline grid extension, has invented another tool. The joint development plan serves as the basis for the Federal Grid Requirement Plan (*Bundesbedarfsplan*) to be elaborated by the Federal Government within a period of three years. The Requirement Plan has to be enacted by Parliament.

These legislative steps clearly demonstrate that investment planning in the field of energy networks and, with it an essential part of entrepreneurial functions meanwhile, has been transferred to the state.

V. Project-related Licensing Procedures

A. General observations

It is self-evident that projects of a dimension of a high-voltage grid line need tight public control. Such projects are of regional relevance. Their public control cannot only relate to aspects of energy policy, but equally must encompass technical and environmental issues. Such projects have to adjust to their regional neighbourhood. That is why such projects, as a rule, have to pass a Regional Planning Procedure (*Raumplanung*) before the licensing procedure on an individual level starts.

Traditionally, there is no Regional Planning Mechanism on the federal level. Site selection for high-voltage grid routes is a matter of planning on the state level. The states are authorized to select sites for grid lines as 'Objectives of Regional Planning' ('*Ziele der Raumordnung*'). These 'Objectives' have to be respected by all authorities in their capacity as planning bodies. In addition, regular 'Regional Planning Procedures' (*Raumordnungsverfahren*) take place on the state level. Their aim is to prepare concrete proceedings for an individual project (high-voltage grid). According to s 1 no. 14 Regional Planning Ordinance (*Raumordnungsverordnung*),[33] based on the Federal Regional Planning Act (*Raumordnungsgesetz*, or ROG),[34] normally a regional planning procedure has to take place for construction of a high-voltage overhead line with a nominal voltage of 110 kV or more. The objective is to determine whether the project conforms to the requirements of regional planning, how it can be executed under the aspects

[33] *BGBl.* 1990 I, 2766. [34] *BGBl.* 2008 I, 2986.

of regional planning, how the project can be accommodated with other planning measures and, finally, to determine the measures necessary to ensure the environmental compatibility of the project. The results of the regional planning procedure have to be taken into account by all the authorities that engage in public planning and licensing procedures. This is particularly true of planning licences—a highly relevant procedure in matters of high-voltage grids.

When deliberating on measures to speed up grid extension it did not appear reasonable to Parliament to create federal mechanisms like the Federal Grid Requirement Plan on the one hand and to leave their implementation to the state or regional level on the other. NABEG takes care of this situation. Those extra-high-voltage transmission lines that are designated in the Federal Grid Requirement Plan are subject to a sector planning (*Fachplanung*) procedure on the *federal* level. Federal sector planning means that the BNetzA investigates the route corridors of extra-high-voltage lines registered in the Requirement Plan as to their compatibility with regional planning principles as listed in ROG.[35] The sector planning procedure involves public participation (s 9 NABEG). It does, however, not yet give the green light for the project.

B. Planning licences

The final stage of administratively clearing the way for high-voltage grid lines is the licensing procedure that relates to the individual project. Under s 43 EnWG a planning licence (*Planfeststellung*) is required for the construction, operation, and alteration of:

- high-voltage overhead lines with a nominal voltage of 110kV or more;
- high-voltage lines serving as connecting lines from offshore installations to the closest transmission or distribution grid;
- certain trans-border high-voltage direct-current transmission lines.

The EnWG also provides for optional planning licences in two instances:

- underground cables in the coastal areas of the North Sea and the Baltic Sea with a nominal voltage of 110 kV; and
- underground cables under s 2 EnLAG.

As compared to regular licensing procedures the planning licence procedure is a very sophisticated administrative procedure which has to conform to a number of specific substantive and procedural criteria.

1. Substantive criteria

Substantive criteria are: planning justification, statutory requirements, and the balancing requirement.

Planning justification means that the envisaged project has to be supported by a need, has to be 'necessary'. This requirement is indispensable because many projects

[35] See n 34.

interfere with private property in an expropriatory way. Taking of property, under the Constitution (art. 14 Basic Law), requires a necessity. Dealing with this criterion is particularly difficult in energy projects because necessity depends on demand forecasts that have to be established in consideration of the objectives of the EnWG (s 1), namely energy security, affordability, and environmental friendliness.

During the planning licence proceedings all pertinent *statutory requirements* contained in Acts and regulations of all kinds have to be met. Among them are most notably, as described under section VA. above, the results of all higher-level pre-established planning procedures, environmental requirements as for example emission ceilings, water protection, nature conservation, and safety restrictions.

Finally and most importantly, the *balancing requirement* has to be met. It is expressed in s 43 para 1EnWG, where it reads that 'during the planning licence proceedings all public and private interests affected by the project have to be taken into consideration within the balancing process'. The balancing requirement means that, in a first step, the competent authority has to collect and ascertain all relevant interests. In a second step, the authority has to insert these interests into the balancing process and weigh them according to their specific relevance. In projects like construction of high-voltage grids specific importance is to be given to environmental interests, eg nature conservation, but also protection against detrimental effects of electromagnetic waves. The balancing process also has to consider alternative routes which may have less damaging effects on important interests. The authority enjoys a considerable margin of discretion in its decision-making.[36]

2. *Procedural criteria*

The planning licence procedure is a type of administrative procedure that German administrative law has developed for a number of projects with infrastructural relevance, such as construction of highways, waterways, railroads, and airports. For the formal requirements, therefore, s 43 EnWG refers to the general provisions of ss 72–78 of the Administrative Procedure Act (*Verwaltungsverfahrensgesetz*, or VwVfG).[37] According to ss 73 and 74 VwVfG, the planning licence procedure encompasses the following stages:

- submission of the plan to the authority by project operator;
- request of comments from other authorities concerned;
- public presentation in the municipalities concerned;
- notice of presentation;
- deposition of objections to the plan;
- public hearing and discussion; and
- decision by the competent authority.

[36] The balancing requirement (*Abwägungsgebot*) is generally regarded as *the* core element of planning (licence) procedures that live up to the standard of the rule-of-law principle (art. 20 para 3 Basic Law), see, eg, Decision of the Bundesverwaltungsgericht (BVerwG) of 14 February 1973, BVerwG 48, 56, 63.

[37] *BGBl.* 2003 I,102.

The Environmental Impact Assessment (EIA) (known in Germany as *Umweltverträglichkeitsprüfung*, or UVP) is an integral part of the planning licence procedure. According to s 3 para 1 sent 1, Environmental Impact Assessment Act (*Gesetz über die Umweltverträglichkeitsprüfung*, or the UVPG)[38] in conjunction with Annex 1 No. 19.1, construction and operation of high-voltage overhead lines are subject to an EIA under certain conditions, depending on length and voltage. The essential elements of the EIA have been integrated into the statutory planning licence prerequisites.

Those grid extension projects that have passed through federal sector planning procedures are also subject to clearance through the planning licence procedure. The relevant specific provisions are to be found in s 18 NABEG. Its basic structures are fairly similar to the general planning licence provisions in ss 72–78 VwVfG. Here again, administrative competence is vested in the *federal* level, ie BNetzA, the Federal Network Agency. One has to bear in mind that the projects under NABEG have been intensively scrutinized already within the preceding stage of federal sector planning and that s 15 para 1 sent 1 NABEG declares the sector planning decision as binding for the planning licence procedure. That is why the EIA within the licensing stage can be limited to additional or severe environmental effects (s 23 NABEG). In view of the binding character of the sector planning decision one may question whether the licensing procedure still involves the crucial element of weighing conflicting interests. But the very essence of the introduction of NABEG together with the amendment to the Energy Act was to transfer the route selection process to a higher, quasi-political level, thereby taking 'steam' out of the grassroots administrative decision-making procedure which, additionally, was tainted with the different philosophical and political approaches on the state level.

C. The dilemma of public participation vs procedural expeditiousness

The legal framework applicable in Germany relating to grid extensions is confronted with a dilemma which most recently has grown by dramatic proportions: the dilemma between public participation and procedural expeditiousness.

For some time now the long duration of licensing proceedings concerning infrastructural projects including grid expansion projects has been sharply criticized in the public arena. As a rule, major grid construction projects consume no less than eight years from the initial planning phase to a final licence. To a large degree this explains why, according to the DENA Grid Study II, only 90km out of 850km of new grid required by DENA Grid Study I had been realized in 2010. The critical issues in the licensing procedure are:

- the *planning justification*: the determination of the necessity of an investment measure is laden with considerable uncertainty as to the database required for the demand forecast in a way to withstand judicial review.
- the *balancing procedure*: in particular, the scrutiny of alternative sites and routes with all the necessary data to be collected is very time-consuming.

[38] *BGBl.* 2010 I, 94.

- the *public hearing and discussion* need to work through thousands of objections from lobby groups and citizens.

Over recent years, German legislature made attempts to streamline licensing procedures. In 2006 it amended the EnWG by inserting the provision s 43a. Under s 43a no. 5 EnWG the competent authority is authorized to waive the requirement of a public hearing. The Electricity Grid Expansion Act, EnLAG, of 2009 identified 24 Extra-High-Voltage-Line projects and explicitly recognized that these projects are supported by an energy policy need and are of urgent priority (s 1 para 2 EnLAG). This legislative action takes the issue of planning justification out of the controversies surrounding the individual project on the local and regional levels.

Since the middle of 2010 the pre-existing tension between the postulates of public participation and procedural expeditiousness has reached new proportions:

- Over the past months popular resistance against infrastructural projects, including grid construction, has significantly increased. The violent demonstrations at Stuttgart Central railroad station in this respect worked as a beacon. Demonstrations only started at a time when the whole project was definitely licensed and demolition activities were about to begin. This situation triggered a widespread feeling that, in licensing procedures, public participation comes too late, ie when the end result is already on the table and the project cannot be dropped or significantly changed, and if so at all, only at the price of damage suits.
- Especially with regard to grid line construction in the inter-state segment, state administrative competence for planning and licensing, and multiple public participation arrangements have turned out to be a major delaying factor. Under previous rules separate proceedings had to be implemented in each of the states concerned as to the respective part of the project. It is just this kind of extra-high-voltage grid lines that is urgently needed in a scenario with shrinking nuclear power and expanding renewable energy capacities, notably offshore in the North Sea.

The legislative package adopted by Parliament in July 2011 has the objective to cure the deficits of the preceding legal situation. Its focus is on establishing tighter planning mechanisms with the need for interaction between the undertakings and the state, and on vesting administrative competence in the *federal* level whenever inter-state projects are concerned.

VI. The Relevance of Grid Extension Investment Expenditures on Transit Tariffs

A. The system of incentive regulation

Traditionally, monopolistic tariff regulation was dominated by the 'cost +' method: costs + profit margin = tariff. Even in the era of competition and liberalization in the energy sector the grid remains a natural monopoly.

The only way to inject competitive elements into such a system is hypothetical competition ('as if competition'). Under s 21 para 2 EnWG, tariffs for grid access have to take into account operative costs 'which have to conform to those costs accruing to an efficient and structurally comparable grid operator'. This hypothetically competitive standard together with the *ex ante* approval (authority) requirement in the past have resulted in noticeably lower tariffs. But this system still has the disadvantage that external authorities have to evaluate the efficiency of grid operation. Incentive regulation, in force since 1 January 2009, was designed to cure this drawback.

The very idea of incentive regulation consists in vesting the duty to detect operational inefficiencies and, with it, the potential for cost reduction in the operator. In principle, the system of incentive regulation provides for the regulatory authority to impose proceeds ceilings on the operator. The level of these ceilings is determined by the efficiency targets for each undertaking. The efficiency targets prescribe a development path for the proceeds ceilings over a certain period of time. The efficiency targets have to take into account:

- the general rate of productivity increase;
- the specific company-related rate of productivity increase;
- the inflation rate.

The specific company-related rate of productivity increase is a result of the efficiency factors for each undertaking. The proceeds ceilings are fixed for a period of five years. In case an operator is more efficient than the data being used for the ceilings, he may keep the additional profit. This is why the system is called 'incentive regulation': better than expected cost-cutting is rewarded by additional profits. One factor in determining the proceeds ceiling, of course, is investment expenditures. Under certain circumstances, an additional bonus may be added to the proceeds ceiling (s 10 *Anreizregulierungsverordnung*, or Incentive Regulation Ordinance (ARegV))[39] as an 'expansion factor', or a lump-sum investment bonus (s 25 ARegV). But this only applies to distribution network operators, not to transmission network operators. For transmission operators, so-called 'investment budgets' are the relevant tool.

B. Investment budgets

Under s 23 para 1 ARegV, the Federal Network Agency BNetzA has to approve investment budgets for capital and operational costs which are necessary for expansion and restructuring investments in transmission grids. This applies to the extent that these investments are necessary for the stability of the overall system or for the connection with the national or international high-voltage grid, as well as for the demand-oriented expansion of the electricity grid under s 11 EnWG. This encompasses, in particular, investments for grid connections, integration of renewable energy facilities, connection of offshore wind farms, expansion investments for high-voltage grids, and high-voltage direct-current systems.

[39] Of 29 October 2007, *BGBl.* I, 2529.

The problem is to find the basis for the 'necessity' of such an investment. It makes sense to coordinate the criteria for 'necessity' in this sense with 'necessity' for planning and licensing procedures. If one were to decide otherwise, a divergence is bound to arise between 'necessity' from a planning and licensing point of view and 'necessity' in a tariff regulatory sense. But what 'plan' should be determinative? Projects appear in a variety of plans, ie the TEN guidelines and the development plans under the Third Energy Package (ten-year non-binding Community-wide plan, ten-year plan from transmission operators), then the Federal Grid Requirement Plan and, finally, the Network Expansion Model the BNetzA intends to draw up for its own purposes. The relevance of these plans for regulatory rate-making has still to be solved. At any rate, the BNetzA should be entitled to deny recognition only to those parts of the investment—as uncompetitive—which are not pre-determined by the relevant plan, whichever that is.

C. The dilemma of security of supply vs cheap energy prices

The regulatory treatment of investments and their impact on network transit tariffs is situated right at the heart of the dilemma between security of supply versus cheap, or affordable, energy prices. The more generously investments are approved by regulators, the higher tariffs will be. The only controlling standard is hypothetical competition. This would mean that regulatory agencies not only have to have a keen eye on whether an investment should be made at all, but also on the level of costs to be incurred. German authorities have been hesitant in this latter respect. The reasons may have been lack of manpower or reluctance to prolong the administrative procedure by going too much into the details of the investment project. Another critical aspect of the present approval procedure is that investment budget approvals only extend over one regulatory period (five years). The investment period itself, on the other hand, takes far longer. Here again, security of investment and the cheap energy postulate are in direct conflict. A longer period of validity would certainly enhance the security (of investment) level, while it would withdraw a longer part of the investment process from competitive scrutiny. A key aspect for promoting *both* security of supply and affordability of electricity will be the clarification of the relationship between the planning/licensing and the regulatory financial framework.

VII. Conclusions

Energy policy and, consequently, energy law are undergoing a rapid development in Germany at present. Their focus, during the past decade, has shifted quite remarkably. Questions of third party access—the key issue at the turn of the century—have been relegated to second priority. Energy infrastructure, especially in the electricity sector, is in pole position. Energy infrastructure encompasses both the pattern of generating infrastructure and grid infrastructure, issues that are closely linked at second sight. The change began during the period of the Red–Green coalition Government (1998–2005): a trend away from centralized bulk nuclear and fossil fuel production towards renewable energies and decentralized generation. This trend has gained rapid momentum since autumn 2010: according to its 'Energy Concept 2050' of September 2010 the Black–Yellow federal Government projects an 80 per cent supply through renewable energy generation by 2050. In the wake of the Fukushima disaster this trend is even more accentuated: the phase-out scheme of nuclear energy and its total abandonment by 2022, as adopted by the German Parliament in July 2011, will certainly increase pressure on still expanded electricity production from renewable energies. At the same time such a shift in the generation pattern is bound to exercise an enormous pressure on network infrastructure (especially need to install new North–South transmission lines).[40]

There is widespread agreement in Germany that the traditional legal tools are insufficient for coping with this policy shift. There is also agreement that both under European and German law the ultimate responsibility for an adequate energy infrastructure rests with the state. It is equally undisputed that massive expansion of electricity network infrastructure (DENA Grid Study II 2010, before Fukushima: 3.600km required until 2020) imperatively calls for improved planning and licensing procedures.

After the Fukushima disaster Germany, within a period of roughly three months, undertook a massive legislative effort to accelerate the change in the electricity generation pattern (reduction of large-scale nuclear and fossil fuel electricity production and expansion of renewables) already formulated in 2010, by putting in place the normative basis for a total abandonment of nuclear energy by 2022. In addition, legislative steps were also taken to improve the legal tools for expanding the extra-high-voltage grid especially in the North–South direction. Parliament had to solve the dilemma between acceleration of procedures on one side and enhanced public participation on the other. The recent statutory amendments provide for a shift of administrative competence from the states to the *federal* level for inter-state transmission routes, for a more active involvement of the state in the process of route planning and for public participation already at the planning stages preceding the final licensing procedure. The coming years will prove whether the legislative goals will be achieved.

The expansion of the electricity transmission grid will, however, be brought about only if a sufficient number of investors will be found who deem an investment

[40] When considering the relationship between the electricity generation pattern and network infrastructure one has to bear in mind still another aspect which could not be dealt with more thoroughly in the present context: the new German post-Fukushima energy policy intends to promote decentralized electricity production from onshore renewable energy sources (onshore wind farms, solar energy (photovoltaic), combined heat and power facilities). Such a strategy will undoubtedly pose serious challenges for the stability of distribution grids. Very significant investments will also have to be made in this respect—investments which cannot be quantified at this stage and which have so far been eclipsed by the urgent problems on the transmission level. The financial pressures on the distribution level will be much more onerous, mainly for two reasons, first, the owners/operators of distribution grids to a large extent are municipalities and therefore short of financial resources; and second, the current system of incentive regulation, in general, does not allow for investment budgets at the distribution level.

in grid construction sufficiently attractive. This concerns the role of grid access tariffs within the complex of grid extension.

In this regard the Incentive Regulation Ordinance is pertinent. Section 23 provides for so-called 'investment budgets' to be approved by the Federal Network Agency. Its objective is to raise the grid operator's proceeds ceiling in order to enable him to integrate the investment expenditures into the grid access tariffs. One problem is to coordinate the approval proceedings both with regard to the planning aspects and to the tariff decision-making process within the Federal Network Agency. The issue of coordination is threefold: one of substantive criteria, one of competence, and one of chronological sequence. In determining what is to be required for an investment to be 'necessary', the Federal Network Agency has to accommodate the goals of security of supply and affordability of energy (electricity).

Generally speaking, recent developments in European and German energy legislation[41] reflect two trends.

First, the 'liberalization' of European energy markets has resulted in a dense net of regulatory norms and constraints. Quite in contrast to its underlying initial philosophy, 'liberalization' has brought a resurgence of state interventionism at the expense of entrepreneurial autonomy. This is particularly true of Germany's recent energy policy, which has led to Germany being qualified as a 'planned economy' (*Planwirtschaft*).[42] The new legislative framework for grid expansion in Germany is just one example.

Second, the electricity generation pattern, as shown above, does have implications for the grid configuration. The lack of an integrated European electricity generation pattern is bound to result in diverging electricity production structures (Germany: renewables; France: nuclear) and grid policies. That, in turn, most probably will work as a hindrance to the emergence of a truly integrated European energy market.

[41] On recent developments in European and German energy law see G. Kühne, 'Leitplanken und Entwicklungsstränge des Energierechts' (2011), 3 KSzW (Kölner Schrift zum Wirtschaftsrecht) 219.

[42] See *Financial Times Deutschland*, 22 June 2011, 1 ('*Deutschland wird zur Planwirtschaft*') concerning proposals to subsidize the construction of gas-fuelled power stations in order to counter the deficit of electricity generation capacities as a result of the abandonment of nuclear energy.

21

The Development of Electricity and Gas Networks in Russia

Sergey S. Seliverstov and Ivan V. Gudkov

I. Introduction

This chapter provides an overview of the dramatic reforms which have been taking place in the Russian energy sector and focuses on the further development of the electricity and gas networks. It provides an example of how the network industries evolve in a very specific historical, political, and economical environment.

II. Electricity Sector Reform in Russia—Privatization and Liberalization

A. Reform process

After the dissolution of the Soviet Union the electricity sector of the Russian Federation, which was the largest among the Soviet republics, underwent substantial changes. Since the 1990s it has been subject to privatization followed by liberalization.[1]

The Soviet electricity system represented a single technological and economic complex, functioning under administrative, rather than market-based, principles. The electricity-generating facilities and the transmission lines were located in accordance with the centralized plans which were common for all the elements of the system. There was a single grid for the major part of the territory and a single System Operator. All the elements of the system were technologically and economically compatible with each other. However, the compatibility of the system corresponded to the needs of a planned economy, which of course changed in the market economy.

[1] For more details see S.S. Seliverstov, 'The Electricity Sector Reforms in Russia: European Legal Concepts and Russian Reality' in B. Delvaux, M. Hunt, and K. Talus (eds), *EU Energy Law and Policy Issues* (2nd edn, vol. 2, Brussels: Euroconfidentiel, 2010) 37.

B. Privatization in the electricity sector

The privatization processes were formally initiated in Russia in 1991 with the adoption of the Law on Privatization of State and Municipal Undertakings.[2] As a result of this law and, more importantly, the Decrees of the President of Russian Federation, privatization of certain parts of the national electricity system has taken place. The Decree on Peculiarities of the Reorganization of the Public Undertakings, Unions, Organizations of the Fuel and Energy Complex into the Joint Stock Companies[3] (hereinafter the Decree 922) dated 14 August 1992 envisaged the reorganization of the public electricity undertakings at that time controlled by the Ministry of Fuel and Energy into the joint stock companies by means of the specific Presidential Decrees.

The Decree on Organization of the Electricity Sector[4] (hereinafter the Decree 923) adopted the day after created the Russian Joint Stock Company of Electricity and Electrification (RAO UES). The share capital of RAO UES came in the form of the transfer of 49 per cent of shares in the public-owned electricity undertakings, ownership in transmission lines, and large power stations (more than 300 MW for hydropower stations and more than 1000 MWT for other power stations). The list of undertakings whose shares should have been contributed to the share capital of RAO UES included 491 undertakings, 50 electro stations, all the major transmission lines, and the System Operator.

The Decree of the President implementing the Decree 922[5] imposed certain requirements and limitations on privatization of RAO UES. The Decree stated that for three years after the establishment of RAO UES the share of the state in the company's share capital should not be less than 50 per cent.

The strategic importance of RAO UES was repeatedly recognized in 1998 through the enactment of the federal law limiting the amount of shares in RAO UES that could be owned by companies and individuals[6] (hereinafter the Law No. 74). With effect from May 1998 51 per cent of shares in RAO UES was owned by the federal government. Sale or other alienation of the shares in RAO UES owned by the federal government was allowed only on the authority of federal law. The voting rights corresponding to 33 per cent of shares in RAO UES were transferred

[2] Law dated 3 July 1991 No. 1531-1 'On Privatization of State and Municipal Undertakings in Russian Federation' (Legal database 'Consultant').

[3] Decree of the President of Russian Federation dated 14 August 1992 No. 922 'On Peculiarities of the Reorganization of the Public Undertakings, Unions, Organizations of the Fuel and Energy Complex into the Joint Stock Companies' (Legal database 'Consultant').

[4] Decree of the President of Russian Federation dated 15 August 1992 No. 923 'On Organization of the Electricity Sector during the Privatization' (Legal database 'Consultant').

[5] Decree of the President of Russian Federation dated 5 November 1992 No. 1334 'On the Implementation in the Electricity Industry of the Decree of the President of Russian Federation' dated 14 August 1992 No. 922 'On Peculiarities of the Reorganization of the Public Undertakings, Unions, Organizations of the Fuel and Energy Complex into the Joint Stock Companies' (Legal database 'Consultant').

[6] Federal Law dated 7 May 1998 No. 74-FZ 'On Peculiarities of Disposal of the Shares in the Russian Public Company of Energy and Electrification "Unified Electric Power System", that are in the Federal Ownership' (Legal database 'Consultant').

to the regional governments in accordance with the proportion of the electricity consumed by the respective regions.

Foreign companies (as well as foreign states and international organizations) and their affiliates alltogether could not own more than 25 per cent of shares in RAO UES. At the same time, as a result of the privatization process by 2000 about 48 per cent of the shares in RAO UES were owned by private investors. Therefore, it may be said that taking into account the strategic character of the asset, the electric sector was privatized to a limited extent only and that the state retained majority ownership and control over it.

C. Liberalization—incentives and outcome

RAO UES was functioning as a state-controlled monopoly, a national champion embracing all the major elements of the electricity sector until the early 2000s. By that time the Russian economy and its energy sector had severe problems with non-payments and cross-subsidies. The purchasers of electricity had accumulated large amounts of debt, which were difficult to recover due to administrative constraints, ineffective management, and lack of market traditions. Inflation was experienced almost constantly, which certainly aggravated the problem. Even if the payments were made with significant delays, their real economic value was much lower than the value at the time the payment was due. Therefore, the electricity generators (the major part of which were the subsidiaries of RAO UES) did not receive what they had expected at the time of the actual delivery.

The incentives for further investments and energy efficiency were relatively small. Lock-outs were experienced in some regions of Russia. The system of tariff regulation applied at that time could not solve the existing controversies nor introduce real competition into the sector. Reform was proposed in order to cut that Gordian knot of cross-subsidies, monopolization, and administrative methods of management.

There were more incentives to introduce competition in the electricity sector than in the natural gas sector. The main difference is the enormous amount of export in the gas sector compared to the relatively small export activity in the electricity sector. The demand for Russian electricity on the international market has never been as high as the demand for Russian natural gas.

This is logical, since the costs of transportation are higher for electricity. Moreover, natural gas is a fossil fuel, which is non-renewable. Therefore, bearing in mind the structure of the pan-European natural gas market, it is more profitable for Russia to maintain the export monopoly for natural gas and to have one major player (or maximum two players controlled by the government) inside the country. At the same time, the local electricity producers predominantly supply the Russian internal market. The introduction of competition in the electricity-generating sector may bring more direct positive results to the economy of the country and at the same time provide some opportunities for export.

In 2003 the RAO UES published a paper called the Concept of the Strategy of the RAO UES for 2003–2008 (hereinafter the Strategy). The Strategy served as an explanatory note to two federal laws—the federal law 'On Electricity Sector'[7] (hereinafter the Law on Electricity Sector) and the federal law 'On the Peculiarities of the Functioning of the Electricity Sector during the Transition Period and on Introduction of the Amendments to the Certain Legal Acts of the Russian Federation in Connection with the Adoption of the Federal Law "On Electricity Sector" '[8] (hereinafter the Law on Transition Period).

The Strategy (unlike the Law on Electricity Sector) declared concrete aims for the reform:

- division of the sector into natural monopolies (transmission, distribution, and dispatch) and competitive activities (production and supply);
- creation of an effective market relationship in the competitive areas;
- non-discriminatory access to the services provided by the natural monopolies; and
- efficient and just public regulation of the natural monopolies, which stimulates minimization of costs and increased investment.

The first and the last aims are of specific interest for our topic. It should also be noted here that state ownership of the monopolies was not proclaimed among the aims of the reform. However, it was one of its most important results.

One of the initial steps in the overall liberalization process was the liberalization of the market for shares in RAO UES. It started on 1 January 2005 with the abolition of the Federal Law No. 74 limiting the disposal of shares in RAO UES.[9] The Law on Transition period further provided for the mandatory reorganization of the RAO UES. Initially the time frames for the reorganization were not defined. Later, in accordance with the amendments introduced in the Law on Transition in 2007, it was decided that the structural reorganization of the electricity sector should be completed on 1 July 2008, when RAO UES should cease to exist.[10]

The decisions of the executive bodies of RAO UES concerning its reorganization could be adopted by a simple majority of votes.[11] Bearing in mind that the Russian Federation owned about 52 per cent of the shares in RAO UES, the said rule provided for the monopoly right of the state to define how RAO UES should be reorganized and divided. It is worth noting, however, that guarantees for investors, creditors, and shareholders during the reform process were proclaimed as one of the principles of the reform.

As a result of the reform, the state has not decreased its share in the companies owning the hydroelectric power stations or in the holdings owning such companies.

[7] Federal Law dated 26 March 2003 No. 35-FZ 'On Electricity Sector' (Legal database 'Consultant').

[8] Federal Law dated 26 March 2003 No. 36-FZ Federal Law 'On the Peculiarities of the Functioning of the Electricity Sector during the Transition Period and on Introduction of the Amendments to the Certain Legal Acts of the Russian Federation in Connection with the Adoption of the Federal Law "On Electricity Sector"' (Legal database 'Consultant').

[9] Art 1 of the Law on Transition Period (both initial and current versions).

[10] Art 4 of the Law on Transition Period (as amended on 11 November 2007 by the Federal Law No. 250-FZ).

[11] Art 4 of the Law on Transition Period (both initial and current versions).

The relevant provision was aimed at securing state control over the manoeuvre capacity. At the same time the nuclear power plants were to be exclusively owned by the state.[12] Nuclear power, which has very poor manoeuvre characteristics, is usually used to secure the base load. Therefore, the state retained control over the base load and peak load capacities. Competition could only be introduced in the remaining part of the production sector, since the competition in the base load and the peak load was limited by state ownership.

Moreover, in 2007 the amendments introduced into the Law on Transition Period allowed the state to retain control over the territorial network organizations *inter alia* by means of creating state-controlled holding companies. Until 1 January 2011, the share of the state in the territorial network organizations (IDGC Holding) could not be decreased.[13] Discussion continues as to whether, when, and to what extent the public ownership in those organizations should be changed.

The organizations created as the result of the reorganization of the RAO UES could not own the transmission assets that are part of the Grid and other transmission or distribution assets.[14] Since 1 April 2006 (in accordance with the initial version of the Law on Transition Period—since 1 January, 2005) it was prohibited for companies and individual entrepreneurs to combine transmission or dispatch activities with the production of electricity or sale and purchase of electricity. After the end of the transition period this ban includes the affiliated (legal and natural) persons as well. In order to implement these requirements in practice it was prohibited to own or dispose of the assets used for electricity transmission and dispatch, on the one hand, and the assets used for the production and the sale and purchase of electricity, on the other.[15]

The legal entities, which do not fulfil the above obligations, are subject to obligatory reorganization in the form of division into different companies. The above provisions clearly introduced legal (corporate) unbundling in 2006 and full ownership unbundling after the end of the transition period. The control powers with respect to the unbundling regime are granted to the Federal Anti-Monopoly Service (FAS). In a number of cases FAS has already applied the division requirement, and judicial practice seems to support that position.[16]

There were certain exemptions to the unbundling provisions. They did not apply to technologically isolated systems, to companies combining production, transmission, and dispatch activities solely for their own needs and to territorial transmission companies acquiring the status of the guarantee supplier. In all those situations, however, the accountancy unbundling rules should be implemented.[17]

The existence of the Unified National Electricity Grid (hereinafter the Grid) with the centralized dispatch prior to the starting date of the reform influenced

[12] Art 9 of the Law on Transition Period (both initial and current versions).

[13] Art 4 of the Law on Transition Period (as amended on November 11, 2007 by the Federal Law No. 250-FZ).

[14] Art 4 of the Law on Transition Period.

[15] Art 6 of the Law on Transition Period.

[16] Decision of the Federal Arbitration Court of the Volgo-Vjatski District dated 29 October 2008, Case No. A28–3924/2008-117/12 (Legal database 'Consultant').

[17] Art 43 of the Law on Electricity Sector.

and facilitated the reorganization of the RAO UES and, more generally, the entire reform. The organization (Joint Stock Company FGC UES, hereinafter the Grid Operator) responsible for the management of the Grid, and the System Operator (Joint Stock Company CDU UES), were created at the initial stage of the reorganization as the subsidiaries of the RAO UES. The ownership in the existing transmission networks was transferred to the Grid Operator.

Afterwards the shares in the Grid Operator and in the System Operator were transferred to the federal government. By means of the asset swaps and additional emission of shares the government increased its stake in the Grid Operator to 75 per cent plus one voting share and in the System Operator to 100 per cent shares.

III. Gas Sector Reform in Russia

A. General characteristic of the gas sector

The Russian gas market is characterized by the dominant position of state-controlled (currently 50.002 per cent of the share capital) open joint stock company Gazprom in all segments of the gas value chain—upstream, midstream, and downstream.

Gazprom is the major vertically integrated gas company in the world. It holds on its balance around 18 per cent of global proven gas reserves. Gazprom was initially founded as a Russian joint stock company (RAO) in 1993 on the basis of state gas concern Gazprom, a successor of the Ministry of Gas Industry of the USSR. When founded, 100 per cent of Gazprom's shares belonged to the state; however, privatization started immediately and by 1995 only 41 per cent of shares remained in the direct ownership of the state, 10 per cent belonged to Gazprom itself, and the rest were in the possession of private investors. In 1998 RAO Gazprom was transformed into its current form, which is an open joint stock company.

When founded, Gazprom received ownership over the so-called Unified Gas Supply system (hereinafter the UGS). Federal law 'On Gas Supply in the Russian Federation' (hereinafter the Law on Gas Supply) lays down general regulatory framework of all segments of the gas value chain and establishes the special status of the UGS and its owner determined by the following rules:[18]

- Indivisibility of the UGS: although the legal content of this principle is not defined, its political aim appears to be conservation of the existing vertically integrated structure of the gas market and prevention of the full ownership unbundling such as that implemented in the electricity sphere.
- 50 per cent +1 ordinary share of Gazprom shall always belong to the state (other shares are traded freely, being available for both domestic and foreign investors).

[18] Federal Law dated 31 March 1999 No. 69-FZ 'On gas supply in the Russian Federation' (Legal database 'Consultant').

- Liquidation of Gazprom is permitted only on the basis of a special federal law.
- All objects connected to the UGS, irrespective of who owns them, are centrally managed by Gazprom.
- There is the possibility of transfer pricing between Gazprom and its subsidiaries.

The UGS is a system centrally managed by Gazprom which unites objects (facilities and assets) throughout the whole gas value chain 'from well to burner' (fields, gas transmission systems, processing plants, underground storages). Apart from the UGS there are several regional gas supply systems independent of Gazprom but they do not play significant role as compared with the UGS. Gazprom manages the USG through a wide network of specialized subsidiaries.

Key segments of the UGS are as follows.

In Upstream (exploration and production) the objects are gas fields being developed by specialized Exploration and Production (E&P) subsidiaries. The majority of gas in Russia is currently produced from a few giant and super-giant gas fields of Western Siberia which were put into operation in the Soviet times and are reaching declining stage now. In the future the main regions of gas production will be Yamal Peninsula, Eastern Siberia, the Far East, and the Arctic shelf.

In Midstream the objects are processing plants, transmission networks, and storage facilities.

There are currently six gas processing plants within the UGS managed by a specialized subsidiary 'Gazprom pererabotka' and 25 underground storages (three more are under construction) all being managed by a specialized storage operator—subsidiary 'Gazprom-PHG'. The heart of the UGS is a unique gas transmission system (GTS)—a network of gas trunk pipelines of high pressure, which is over 160,000km long. GTS is managed by 20 specialized subsidiaries—gas transmission operators holding relevant network assets on the basis of long-term lease agreements with Gazprom.

In Downstream Gazprom consolidates regional gas distribution networks of low pressure managed by more than two hundred gas distribution operators under the umbrella of a specialized subsidiary 'Gazprom Regiongaz'. Gas supply for the internal market supply is effected through regional gas supply entities consolidated into a specialized subsidiary 'Gazprom Mejregiongaz'. On the external market the major part of export supplies is channeled through a specialized subsidiary 'Gazprom export', and part of volumes are supplied directly by Gazprom.

Besides Gazprom there are a number of independent gas producers/suppliers present on the market of which the major one is 'Novatek'. Other independent producers/suppliers include oil companies: state-owned 'Rosneft', 'Zarubejneft' and privately owned, eg 'Lukoil', 'TNK-BP', 'Surgutneftegas'.

The independent producers depend on access to the GTS owned by Gazprom. At the same time they have a significant competitive advantage: gas produced by them (in contrast to Gazprom's gas) is not subject to state price regulation and may be sold at free price on the domestic market.

B. Liberalization and reformation of the gas sector

Legislative reforms are underway in all segments of the gas chain. However, their aims differ depending on the specifics of a segment in question.

In the *Upstream segment* the reforms are primarily aimed at increase of state control over strategic fields, ie the opposite to liberalization. Amendments introduced in the federal law 'On Subsoil' in 2008 treat all off-shore gas and oil fields as being of 'federal importance' and introduce restrictions on access of foreign investors to them.[19] Moreover, the offshore fields located on a continental shelf can now be developed only by state-owned companies (Gazprom and Rosneft), which means that even Russian privately owned major businesses are not allowed to be subsoil users on the continental shelf.

As regards the *Midstream segment*, which is of particular importance for this article, the core aim of already performed and planned reforms is twofold: (i) to facilitate the third party access (TPA) to existing infrastructure; and (ii) to develop clearly defined conditions for creation of a new infrastructure (and expanding the existing one).

Mandatory TPA is required by federal law 'On Gas Supply' (article 27) and federal law 'On Natural Monopolies' (article 8).[20] These provisions oblige owners of gas transmission systems and gas distribution networks to provide non-discriminatory access to any customers (not only producers or suppliers) if there is enough spare capacity (ie if a technical possibility of access is in place). The tariff for transportation of gas via transmission systems and distribution networks is regulated by a state authority, Federal Tariff Service, on the basis of federal law 'On Natural Monopolies' and Governmental Regulation No. 1021.[21]

The conditions and order of the TPA are established in detail in the Governmental Regulation No. 858,[22] which sets out a procedure for gaining access to the gas transmission system of Gazprom. In particular, this Regulation defines the term 'spare capacity of gas transmission system', and outlines prerequisites and stages of conclusion of gas transmission agreements as well as capacity allocation rules. In certain important aspects the scope of this Regulation is limited: it does not regulate access to gas transmission systems which are not included in the UGS (but form part of regional gas supply systems); it does not define conditions of access to 'non-transmission' gas infrastructure (eg storage, liquefied natural gas (LNG)), it does not govern relations between the owner of a gas transmission system and its subsidiaries (but just with third parties); it does not provide for

[19] Federal Law dated 21 February 1992 No. 2395-1 'On Subsoil' (Legal database 'Consultant').

[20] Federal Law dated 17 August 1995 No. 147-FZ 'On Natural Monopolies' (Legal database 'Consultant').

[21] Regulation of the Government of the Russian Federation dated 29 December 2000 No. 1021 'On State Regulation of Prices for Gas and Tariffs for Services on its Transportation on the Territory of the Russian Federation' (Legal database 'Consultant').

[22] Regulation of the Government of the Russian Federation dated 14 July 1997 No. 858 'On Ensuring Access of Independent Entities to the Gas Transmission System of Gazprom' (Legal database 'Consultant').

anti-hoarding mechanisms; and it does not regulate connection to the system. In 2006 the Federal Anti-monopoly Service proposed a draft of a new act replacing Governmental Regulation No.858 with a declared aim to increase the efficiency of the TPA. This proposal is still under discussion.

In the first half of the 2000s, ideas were broadly discussed and backed by several state authorities (primarily by the Ministry of Economic Development) to 'support' mandatory TPA with structural measures aimed at unbundling of vertically integrated gas undertakings in order to definitely eliminate possibilities and incentives for such undertakings to discriminate against third parties. Among others there was a radical proposal of full ownership unbundling which would prohibit combination of competitive (production/supply) with natural monopoly (pipeline transmission) types of activity within one entity (or group of entities).

These ideas were never supported by the Russian legislator who sticks to the principle of indivisibility of the UGS proclaimed by the federal law 'On Gas Supply in the Russian Federation'. At the celebration of the 10-year anniversary of Gazprom in 2003 then-President Putin made an announcement that there would be no forced division of Gazprom. At the same time Gazprom, on the basis of its own 'self-reformation' policy, has already separated different types of activity within its group of companies by charging different subsidiaries with different business tasks, ie by implementing a sort of legal unbundling.

As far as rules for creation of new (and expanding existing) gas transmission systems are concerned, the major legislative initiative in this direction is a draft federal law 'On Trunk Pipeline Transport', which was presented to the State Duma back in 1999 but is still not adopted. Among the basic questions connected with this topic are: who finances and on what conditions the development of the network and who becomes the owner of the new systems and expansions of the existing ones. Part 5 of the article deals in more detail with this issue.

Finally, the *Downstream side* of the internal Russian gas market is currently under active reformation aimed at a gradual shift from regulated to market gas prices. Freeing up prices of gas (which remains the only primary energy source still under state price regulation) would lead to increase of investment attractiveness of gas business and eliminate disparity of Russian energy balance currently dominated by relatively cheap gas to the detriment of other energy sources (in particular, coal). The transitional period for price liberalization is 2011–2013 so that in 2014 the domestic gas prices become equal to export market gas prices (except for the price of gas supplied to domestic consumers, which will remain under state regulation). Once gas prices are liberalized, a uniform gas transmission tariff will be introduced applicable to both the independent producers and the owners of the gas transmission systems (and their affiliated companies).

As regards the external Russian gas market the major reform was introduced in 2006 when federal law 'On the Export of Gas' was adopted.[23] It vested Gazprom and its subsidiary 'Gazprom export' with exclusive right to export gas from Russia.

[23] Federal Law dated 18 July 2006 No.117-FZ 'On the Export of Gas' (Legal database 'Consultant').

This, in particular, effectively allows Gazprom to gain control over any new LNG projects constructed in Russia.

C. Comparison

To conclude this part it would be useful to make a short comparison between regulation of the Russian and European gas market. It appears that, as opposed to a widely accepted perception, in the midstream sector the instruments of reformation of Russian and European market conceptually coincide—they are focused primarily on liberalization of access to the natural monopoly segment of gas business (gas pipeline transmission) and apply more or less the same means:

- independent regulation (the functions of Russian national regulatory authority are performed by the Federal Tariff Service);
- third party access; and
- unbundling of vertically integrated entities (there are currently twenty specialized gas transmission operators in Russia which are wholly owned subsidiaries of Gazprom—this has much in common with the so-called 'legal unbundling' model required by the EU Second Gas Directive).

IV. Land Law Reforms in Russia and their Impact on Energy Networks

A. Reform process

In the Soviet times land was state-owned.[24] Only specific kinds of possession and lease rights were granted to individuals and legal persons.[25] After the break-up of the Soviet Union the stages of land reform were formalized by the Land Code 1991,[26] the Land Code 2001[27] (hereinafter the Land Code), and the federal law 'On the Enactment of the Land Code of the Russian Federation' 2001[28] (hereinafter the Law No.137). The latter two (having been amended many times since their entry into force) represent the result of the land reform in the contemporary Russia.

It should be noted here that there is a clear division in the Russian legal system between the land law and the civil law:

[24] Decree 'On Land', adopted on 27 October 1917 by the II All-Russian Congress of the Soviets of the Workers', Soldiers' and Peasants' Deputies; Art 3 of the Land Code of the Russian Soviet Federative Socialist Republic (RSFSR) dated 1 July 1970 (Legal database 'Consultant').

[25] Arts 3, 9, 11 of the Land Code of the RSFSR dated 1 July 1970 (Legal database 'Consultant').

[26] Land Code of the RSFSR dated 25 April 1991 No. 1103-1 (Legal database 'Consultant').

[27] Land Code of the Russian Federation dated 25 October 2001 No. 136-FZ (Legal database 'Consultant').

[28] Federal Law dated 25 October 2001 No. 137-FZ 'On the Enactment of the Land Code of the Russian Federation' (Legal database 'Consultant').

The land legislation governs the relationship regarding the use and protection of lands in the Russian Federation considering them as the basis for life and activities of the nations living on the respective territory (land relationship)... Property-related relationship regarding ownership, use and disposal of land plots, as well as regarding transactions therewith are governed by the civil legislation, unless otherwise provided by the land, forestry, water legislation, legislation on subsoil, on environmental protection, by specific federal laws.[29]

In line with the Russian Constitution 1993 the Land Code acknowledges variety of types of ownership on land—private and public. Since Russia is a federal state, there may be different forms of public ownership on land depending on the level of authorities—federal, regional, and local (municipal). As we will see below, such differentiation plays an important role for the network facilities.

B. Categories of land

The Land Code provides for seven different categories of land, depending on their purported use, such as agricultural lands, settlement lands, lands of protected territories and objects, lands of the forestry, etc. The electricity and gas networks may be located on the lands, which belong to the category 'lands of industry, energy, transport, telecommunications, informatics, lands for the purposes of the space activity, lands of defense, security and of other special purpose' (hereinafter the Industry and Other Special Purpose Lands). The Industry and Other Special Purpose Lands are in their turn divided into seven sub-categories, including the industry lands, the energy lands, the transport lands, etc.

It is interesting to note that the lands on which it is allowed to build the electricity networks fall into the sub-category of energy lands, whereas the lands on which it is allowed to build gas pipelines (as well as oil pipelines) fall into the sub-category of transport lands. Both sub-categories have their specific regulation.

The process of relocation of lands between categories and sub-categories is governed by the specific federal law (hereinafter the Law on Relocation of Lands) and is extremely formalized and full of uncertainties. The formalization, which is aimed at prevention of the abuse and the wasteful use of the lands, can represent a significant obstacle for the network development.

C. Federal energy systems

A very important limitation is imposed on the Industry and Other Special Purpose Lands, which include the so-called 'federal energy systems', federal transport, and other strategic objects, including nuclear energy objects and objects of defense industry. Such lands belong to the federal ownership only.

Whereas it is clear what kind of gas networks may be regarded as 'federal energy systems', the question is not so clear with regards to the electricity networks. In accordance with the federal law 'on Gas Supply' the entire federal gas supply system—consisting of (i) the Unified Gas Supply system; (ii) regional gas supply

[29] Art 3 of the Land Code.

systems; (iii) gas distribution systems; and (iv) independent organizations—is qualified as one of the federal energy systems.[30]

The Law on Electricity Sector does not clarify which components of the sector are included in the federal energy systems. One may reasonably assume that the Unified National Electricity Grid of Russia is indeed part of the federal energy networks. Nevertheless, there is no express provision about this in the law. Moreover, the Grid being 'the sum of production and other electricity assets connected by the united electricity production (including combined electricity and heating production) and transportation process within the auspices of the centralized electricity operational dispatch management'[31] does not include significant parts of the overall electricity network, such as technologically isolated territorial electricity systems, territorial networks, etc.

Lands occupied by gas networks can only be state-owned and then leased to the respective network owner. The owners of the electricity networks are not banned from acquiring ownership title to land. The absence of express qualification of the Grid as a federal energy system allows the owners of the elements of the Grid (be it the Grid Operator or other transmission network owners (the latter hereinafter called the TN owners) to avoid the application of those limitations.

D. Conversion of title, buy-out, and lease rates

In the Soviet times the land plots beneath electricity and gas networks were provided to network owners (all of which were state-owned) on the basis of the so-called 'permanent (with unlimited term) right of use'. In accordance with the Law No. 137 the said title should have been converted by the network owners into either ownership or lease title upon the entry into force of the Land Code. The right to decide, which title to choose as a result of the conversion, was provided to the network owners. The said right was subject to the abovementioned limitations concerning the federal ownership in lands underneath the 'federal energy systems'. The initial time limit for the conversion process was 1 January 2004.[32] It was subsequently changed to 1 January 2006,[33] then to 1 January 2008,[34] and so on.

Moreover, the Law No. 137 in its initial version provided for substantial exemptions granted to the network owners with respect to the lease payments. It was stated that the amount of the lease payments for the use of the public-owned land plots occupied by the 'objects of the transport systems of the natural monopolies' shall not exceed the rate of the land tax for the Industry and Other Special Purpose Lands. In practical terms this meant that the current payments would have been

[30] Art 5 of the Law on Gas Supply.

[31] Art 3 of the Law on Electricity Sector.

[32] Art 3 of the Law No. 137 (initial version).

[33] Federal Law dated 8 December 2003 No. 160-FZ 'On the Introduction of Change in the Article 3 of the Federal Law "On the Enactment of the Land Code of the Russian Federation"' (Legal database 'Consultant').

[34] Federal Law dated 27 December 2005 No. 192-FZ 'On the Introduction of Change in the Article 3 of the Federal Law "On the Enactment of the Land Code of the Russian Federation"' (Legal database 'Consultant').

the same irrespective of the title to land (ownership or lease) received by the network owner as a result of the conversion of title.

There could have been another question: what types of facilities should be qualified as 'transport systems of the natural monopolies'? There was no legal definition for that concept; theoretically, electricity networks could have been excluded from the list of 'transport systems of the natural monopolies', since that type of network is located on the electricity lands and not on the transport lands.

Nevertheless, in practice it was acknowledged that the owners of the electricity networks (eg the Grid Operator) could benefit from the rule about lower lease payments as well. Moreover, in 2010, three years after the said provision of the Law No. 137 was repealed, the Supreme Arbitration Court of the Russian Federation recognized that the words 'objects of the transport systems of the natural monopolies' have the same meaning as the 'linear facilities', which in turn include electricity networks.[35] For the purposes of the legal linguistics it may be noted that electricity networks are included by the Urban Planning Code in the category 'linear-cable constructions', whereas the 'linear facilities' include pipelines, roads, and railways.[36]

Substantial changes were introduced into the Law No. 137 in 2007 by the federal law 'On Introduction of Changes in the Legal Acts of the Russian Federation in the Part Clarifying Terms and Conditions of Acquisition of Rights to the Land Plots in the State and Municipal Ownership' (hereinafter—the Law No. 212).[37] The Law No. 212 provided for specific (and relatively favourable) conditions for buy-out of lands occupied by privately owned buildings and facilities, including energy networks.

The privilege granted by the said law is twofold. First, the buy-out price of the land plots can be easily defined, since it is tied to the land tax rate. The other multiplier is also very clear—it is the location of the land plot. This latter multiplier depends on whether the land plot is inside or outside a populated locality and on the number of inhabitants in that populated locality. For example, if a land plot occupied by a network (not being part of the 'federal energy systems') is located in a city with the number of inhabitants of more than 500,000, but less than 3,000,000, buy-out price may vary from five to 17 annual land tax rates, depending on the respective regional legislation in force. If no legislation is adopted on the regional level in this respect, then the buy-out price would be equal to the minimum level, ie five annual land tax rates.

Second and more commercially important, the buy-out (even if the maximum rate for the cities with more than 3,000,000 inhabitants is applied) would in any case turn out to be made at a discount to the cadastral price of a given land plot and

[35] Ruling of the Supreme Arbitration Court of the Russian Federation dated 28 April 2010 No. BAC-1557/10 (Legal database 'Consultant').

[36] Art 1 of the Urban Planning Code of the Russian Federation dated 29 December 2004 No. 190-FZ (Legal database 'Consultant').

[37] Federal Law dated 24 July 2007 No. 212-FZ 'On Introduction of Changes in the Legal Acts of the Russian Federation in the Part Clarifying Terms and Conditions of Acquisition of Rights on the Land Plots in the State and Municipal Ownership' (Legal database 'Consultant').

at an even bigger discount to the market price. Cadastral price is a price of a land plot calculated on the basis of the administratively approved methodology, primarily for the purposes of taxation; usually it is below the market price.

The favourable buy-out conditions with respect to land plots underneath the linear facilities only apply until 1 January 2015[38] (initially until 1 January 2013[39]), whereas the owners of other types of buildings and facilities may only enjoy such privileges until 1 January 2012 (initially until 1 January 2010). The above is, of course, true if no extensions would be further provided by the law.

The lease of land is the sole conversion option for the land plots, which fall within the exclusive federal ownership, and an alternative to the buy-out for the land plots, which do not. It is obvious that lease may entail less financial burden for the owner of the energy infrastructure in the short-term perspective. The owner of the linear facilities may use current payments to finance the lease, whereas to finance the buy-out a specific development programme or a third party financing would be required.

The current trend is to conclude long-term lease contracts for the land plots underneath the linear facilities, usually for 49 years. In such a long-term perspective the overall sum of the payments in case of buy-out (buy-out payment plus annual land tax) would in the majority of situations be definitely less than in cases of that kind of long-term lease.[40] In case of buy-out, the land plot may also be mortgaged by the owner for the purposes of financing development activities and will remain his property after the end of the 49-year period, whereas the tenant would have to renew the lease agreement.

Federal Law No. 212 has also changed the calculation rules for the lease payments. From October 2007 they are no longer equal to the land tax. The general rule is that the public-owned land should be leased to the private entities on the basis of the rates calculated in accordance with the administrative regulations of the respective federal, regional, or municipal government. This rule has, of course, its own exemptions. The exemptions consist in tying the lease rate to the (very favourable) percentage of the cadastral price of the land plots. The general rate is 2 per cent and it is even less in the agricultural lands—only 0.3 per cent.

The exemptions apply, first of all, to the land plots, the title to which had been converted from permanent (with unlimited term) right of use to lease. Certain parts of the land plots occupied by the electricity and gas networks fall within this category if the tenants are able to prove the conversion directly to the long-term lease without any intermediary contracts (eg short-term lease contracts).

As the wording of the Federal Law No. 212 was not entirely unequivocal, it was for some time unclear whether the said legally established rates could be applied to the leased land occupied by the network facilities.[41] In 2010 the Presidium of the Supreme Arbitration Court expressly stated that the relevant provisions of the

[38] Art 2 of the Federal Law No. 212 (current version).

[39] Art 2 of the Federal Law No. 212 (initial version).

[40] For more details see S.S. Seliverstov, *Land beneath Transportation, Communications and Utilities Lines: Rent or Buy-out*, (Moscow: Wolters Kluwer, 2008) No. 7 Corporate lawyer 14 (in Russian).

[41] Ibid.

Federal Law No. 212 relate to the land plots occupied by the specific facilities: electricity and telecommunications networks, pipelines, roads, railways, and other similar facilities. 'Taking into account specific character of those facilities, the said provision *de facto* provides for a specific privilege established with the aim to contain the price increase for the products and services of the natural monopolies'.[42]

In our opinion, the Supreme Arbitration Court has gone a bit further than the express wording of the law. At the same time, taking into account the importance of the network industries for the society and the share of the land payments in the overall amount of expenses, this may be considered absolutely reasonable.

The process of conversion is not over yet. It may be argued, however, that the existing legal and judicial framework of the land law provides for relatively comfortable conditions for the existing network owners. The lease rates for the public-owned lands are significantly lower than average. Moreover, it seems that in the cases of the buy-out of lands Russian network owners are experiencing a 'grace' period, whereby they are allowed to purchase land plots at a significant discount.

E. Public buy-out, reservation, and public servitudes

Russian land law also provides for a number of tools for the situations where the network facilities should be built or modernized using private lands. Those are public buy-out of land plots, reservation of land plots, and public servitudes.

Article 49 of the Land Code envisages that land plots may be subject to public buy-out for state and municipal purposes. The requisition may be effected 'in exceptional cases' connected *inter alia* with the placement of the facilities of the federal energy systems and energy systems of regional importance, linear facilities of federal and regional importance to be used in the activities of the natural monopolies, and objects of electricity and gas supply of municipal importance. The public buy-out may only be executed if there are no possible alternatives for the placement of such objects.

The public buy-out is executed on the basis of a decision of relevant state, regional, or municipal authority. The owner of the land plot subject to this procedure shall be notified one year in advance. The procedure of the public buy-out is governed by the Civil Code envisaging that the amount of compensation should be calculated on the basis of the market price. Both the Civil Code and the Land Code[43] stipulate that the compensation should include not only the direct loss, but the lost profit as well. In case the owner of the land plot is unhappy with the valuation made, it may be subject to the judicial review. The same compensation rules apply to the reservation and public servitudes.

It should also be noted that the long-term (registered) tenants should be compensated as well. The compensation rules only apply to the tenants who have lease agreements for more than a year. There is a mandatory requirement that such lease

[42] Decision of the Presidium of the Supreme Arbitration Court dated 2 February 2010 No. 12404/09 (Legal database 'Consultant').

[43] Art 281 of the Civil Code, Art 57 of the Land Code.

agreements are subject to public registration.[44] Thus, a clear interrelation may be seen between the one-year notification requirement in case of the public buy-out and the said registration rules.

Reservation of land plots may serve as a temporary alternative to the public buy-out. On some occasions reservation takes place before the public buy-out. It can be executed with respect to both private-owned and public-owned lands. In the former case reservation is made on the same grounds as the public buy-out. In case of the public-owned lands reservation may be made for the purposes of construction of engineering and transport infrastructure and of objects of federal, regional, and municipal importance.

The usual maximum reservation term is seven years. However, the reservation term may extend up to 20 years for the public-owned land plots used for the purposes of construction of linear facilities. In practice, reservation may take place if a given land plot will be used during the construction period, but should not be required, at least in its entirety, during the period of operation of the linear facility. After the construction is finished, the reservation of the part of the land plot may be terminated, and the part of the land plot necessary for the operation of the facility may either be bought-out by the state and then transferred to the owner of the facility or (if already owned by the state) leased to the latter.

Public servitudes may be imposed if the access to and the use of the land plot is necessary for the purposes of repair of electricity lines and networks and of transport infrastructure. It is imposed on the basis of a legal act of federal, regional, or municipal authorities taking into account the result of the public hearings. The owner of the land plot encumbered by the public servitude may require the compensatory payments or, if the servitude makes impossible the use of the land plot, the public buy-out of the land plot.

The variety of legal mechanisms in theory allows public entities and network owners relatively flexible policies with respect to the land owners. Moreover, there is a rule enabling the network owners to compensate the owner (or tenant) of the land plot subject to the public buy-out directly, without the necessity to make money transfer from the state (municipal) budget. The said rule allows the elimination of the public entities from the compensatory issues. Their involvement is nevertheless crucial in terms of political decision-making as to the application of the specific instrument.

V. Electricity Networks' Development Prospects

A. Commercial factors

The commercial factors, such as transportation tariffs and profit forecasts, surely represent the main incentives for the networks' development. At the same time, the land management system, the regulatory framework, and the ownership structure are the key factors that shape the development.

[44] Art 26 of the Land Code.

The electricity transportation tariffs, as well as the tariffs for the connection of the additional capacities to the network, are set down by the Federal Tariff Service (FTS), the central tariff regulator for natural monopolies in the energy, telecommunications and transport sectors, and regional authorities responsible for the tariffs' regulation. The tariffs are calculated on the basis of methodologies approved by the FTS, which are in turn adopted in accordance with the Government Regulation 'On the Price Formation with Respect to the Electricity and Heating Energy in Russian Federation'.[45] The tariffs are established for a five-year perspective, but may be recalculated on an annual basis.

B. Networks development executed by grid operators

Article 7 of the Law on Electricity Sector states that the Grid is a complex of electricity networks and other network-related facilities, which are owned or otherwise belong to the persons acting in the electricity sector. The Grid Operator (which is 75 per cent state-owned) is responsible for the management of all parts of the Grid, ie both the parts owned by him and the parts that belong to the TN owners. Thus, apart from the abovementioned legal unbundling provisions, the rights of the TN owners are also limited to the extent that they transfer the overall Grid management powers to the Grid Operator. The Grid Operator has the exclusive right to enter into electricity transmission contracts involving the Grid with any third parties and should approve the use and the phasing-out of the network facilities of other TN owners.

At the same time the Law on Electricity expressly envisages that any persons other than the Grid Operator may build and develop electricity transmission networks. Such persons are entitled to connect their networks on the basis of the non-discriminatory rules common for the producers, customers, and TN owners. The Grid Operator should enter with the TN owners into specific contracts defining the use of the relevant network facilities. The Law on Electricity Sector provides for non-discrimination rules regarding profit-sharing. The rate of return on the capital should be equal for the Grid Operator and the TN owners.[46]

One of the tasks of the Grid Operator (formalized by way of the 'soft law' requirement meaning that there is no remedy for the breach thereof) is the development of the Grid.[47] To that end the Grid Operator shall elaborate the long-term development programmes and schemes to be approved by the relevant authorities. The current investment programme of the FGC UES for the period of 2010–2014 for the overall amount of 952,4 billion Roubles (approximately €23.8 bn) was approved by the Ministry of Energy.[48] The investment programme shall be

[45] Regulation of the Government of Russian Federation dated 26 February 2004 No. 109 'On the Price Formation with Respect to the Electricity and Heating Energy in Russian Federation' (Legal database 'Consultant').

[46] Art 8 of the Law on Electricity Sector.

[47] Art 10 of the Law on the Electricity Sector.

[48] Order of the Ministry of Energy of Russian Federation dated 12 November 2010 No. 547 'On the Approval of the Investment Programme of OJSC "FGC UES" for 2010–2014', available at <http://www.fsk-ees.ru/media/File/stockholders/documents/docs/Prikaz_12.11.10.pdf> (accessed 2 October 2011).

financed by the federal budget, the funds of the Grid Operator, equity financing, loans, and connection payments.

The Grid Operator also has a significant number of agreements with the regional governments on the cooperation regarding the networks' development. The said agreements provide for a specific form of partnership between the public entities on the regional level and the state-owned electricity transmission monopoly. The regional governments undertake to coordinate the development plans, to facilitate the approval procedures with respect to the building and developing of the network capacities, and to provide different forms of co-financing. Thus, the governmental and the intra-corporate development programs traditionally have a lot of common elements.

C. Networks development executed by other network owners

The privately owned transmission networks do not seem a 'self-sufficient' business. The owner of the network being obliged to transfer the management rights to the Grid Operator will always be partially deprived of his assets. In exchange for that he would be able to count on the return rate, which would never be higher than the return rate of the Grid Operator. Therefore, the private investors would develop the transmission networks either in case the profit of the Grid Operator would be higher than the market average (which is hardly achievable bearing in mind the 'social' component in the tariffs), or in case the particular network would be necessary for the purposes of horizontal integration with other businesses. A direct line between third party electricity-generating capacities and a heavy machinery plant belonging to TN owner may be a good example of that.

As outlined above, the electricity sector reforms in Russia in the beginning of 2000s resulted, *inter alia*, in the creation of the state-owned electricity distribution champion—IDGC Holding—controlling the so-called inter-regional network companies. The existence of such holding does not preclude, however, the creation of the privately owned distribution networks' owners and operators. The restrictions imposed on the distribution networks' owners are less severe than those for the TN owners. Moreover, there are constant discussions about the privatization (at least partial) of the IDGC Holding.

The financing tools of the IDGC Holding are quite similar to the ones of the Grid Operator, including the agreements with the regional governments. Therefore, it may be said that the IDGC Holding would have substantial competitive advantage vis-à-vis its privately-owned competitors in this regard, since the business decisions adopted at the company level are politically supported by the federal government and the regional authorities. This relates, of course, to the land management and land relocation issues.

The creation of the privately owned distribution companies should not be ignored. It may serve as an alternative way of partial privatization for the distribution segment. There are a number of isolated markets in Russia, which theoretically may be supplied by the local supplier owning the local distribution network.

The prospects of formation of the independent network operators are very uncertain. Economically, the activities of the independent distribution companies may be profitable in the densely populated areas. However, the regulatory framework, including the provisions of the land and city planning laws, require substantial political and administrative support. Therefore, different forms of public–private partnerships look like the most viable form for that.

VI. Gas Networks' Development Prospects

As in the electricity sector, the development of networks in the gas sector (through creation of new infrastructure or expanding the existing one) primarily depends upon questions of ownership and financing.

As regards ownership, the principle of indivisibility of the UGS presupposes that all networks (gas transmission systems and gas distribution networks) which form part of the UGS (including their expansions) shall be owned by Gazprom irrespective of who has financed their creation or expansion. The UGS is being developed in accordance with the investment programme of Gazprom and gasification programs approved by competent state authorities. Although the independent producers and other third parties are not prohibited from financing the expansion of the UGS required for accommodation of their needs, it is suggested that they may not become owners of the expanded facilities due to the abovementioned principle.

Gas networks outside the UGS may currently be created, expanded, and owned by persons other than Gazprom, even by private entities without state participation, provided that such networks, if connected to the UGS, shall be centrally managed by Gazprom.

There are currently no legal restrictions for building and owing gas networks independent of the UGS. In reality, however, there are not many trunk gas pipelines owned by persons other than Gazprom (examples include trunk pipelines forming part of regional gas supply systems; a trunk gas pipeline built in the framework of Sakhalin 2 project and owned by Sakhalin Energy Investment Company; a trunk gas pipeline on Sakhalin owned by Rosneft).

The draft law on Trunk Pipeline Transport, which has been under discussion in the State Duma since 1999, proposes to establish a rule that any trunk pipeline in Russia must belong to the state-controlled (50 per cent +1 voting share) enterprises. This proposal is based on recognition of the strategic importance of this kind of infrastructure. The draft law also: (i) provides that the organizational form of the owners of trunk pipelines shall be open joint stock companies (which would ensure quire broad transparency requirements); (ii) develops TPA rules, including the definition of 'spare capacity' available for third parties; and (iii) regulates connection to the trunk pipelines and recognizes the possibility of both state and private financing of the development of the trunk pipelines.

As regards financing, the key problem is to define the proper source and mechanism of financing of the network development, particularly of the UGS, given that the owner of its elements may not change irrespective of who finances UGS's expansion.

In order for the third parties to get their gas transported through the networks they need to both get physically connected to the networks and conclude a relevant gas transportation agreement. The order of connection to the networks is defined by the Governmental Regulation No. 83 (based on the City Construction Code),[49] while the order of access to the GTS of Gazprom is established by the Governmental Regulation No. 858 (based on Law on Gas Supply) and the order of access to the gas distribution networks by Governmental Regulation No. 1370.[50]

The expenses for the connection to the networks shall be borne by the third party being connected, which is explicitly recognized in the law at least in relation to the gas transmission system.[51] It is suggested that these expenses shall be included in the connection fee (which is not subject to state regulation at the moment). However, if in order to make practical use of the connection the capacity of the networks needs to be expanded, how shall such expansion be financed?

The law on gas supply (article 17) provides that the government may apply special surplus payments (in excess of tariffs for transportation of gas through the distribution networks) in order to finance the gasification programmes. The government has adopted a Regulation No. 335 introducing these surplus payments but has dedicated these payments only for financing of gas distribution networks (and not of gas transmission systems).[52] It follows that neither the source nor mechanism of the financing of gas transmission systems development is currently clearly defined, it being understood that the owner of the gas transmission systems generally finances their development and third parties compensate these expenses through payment of transmission tariffs regulated by the FTS. The draft law on Trunk Pipeline Transport may become an appropriate instrument to efficiently address this issue.

VII. Conclusion

The transformation process in the Russian electricity sector during the past 20 years does not seem to be completely over. At their current stage the ownership in and the control over both transmission and distribution networks in the major part of the territory belongs to two state-owned companies.

[49] Regulation of the Government of the Russian Federation dated 13 February 2006 No. 83 approving Rules on defining and granting of technical conditions for connection of the object of capital construction to the engineer-technical networks and Rules on connection of the object of capital construction to the engineer-technical networks. Gas networks for the purposes of this Regulation form part of the engineer-technical networks (Legal database 'Consultant').

[50] Regulation of the Government of the Russian Federation dated 24 November 1998 No. 1370 'On Establishing of a Provision on Ensuring Access of the Organizations to the Local Gas Distribution Networks' (Legal database 'Consultant').

[51] Regulation of the Government of the Russian Federation dated 5 February 1998 No.162 'On Approval of Rules of Gas Supply in the Russian Federation' (Legal database 'Consultant').

[52] Regulation of the Government of the Russian Federation dated 3 May 2001 No.335 'On the Order of Imposing Special Surplus Payments to the Tariffs for Transportation of Gas by Gas Distribution Organizations for Financing of Programs of Gasification' (Legal database 'Consultant').

The land law provides for very significant benefits to the companies who are the 'historical' owners of the linear facilities. Those benefits consist in the low lease rates (which were equivalent to the amount of the land tax before 2007) and in the extremely favourable conditions for the purchase of land beneath the linear facilities. By 2015 (or probably by some later date) the said buy-out preferences will expire, leaving the state-owned champions with the 'historical' competitive advantages.

The development of the transmission networks, which are predominantly composed of the Unified National Electricity Grid, will be executed almost exclusively by the Grid Operator. It will be financed on the basis of the transmission tariffs, connection fees, and the public funds transferred by the federal and regional governments. The said governments will play the key roles in terms of facilitating the land allocation process.

The distribution networks' development may attract more interest from private investors. At this stage both the privatization of the national champion and the creation of the private-owned competitors may be expected. Nevertheless, the public entities will still play a very significant role in that process. Therefore the establishment of private–public partnerships may be expected in this segment.

Unlike the electricity sector, which was the subject of major structural reforms, the gas market is protected against radical unbundling measures by the principle of indivisibility of the UGS. This principle forms the basis for maintaining the unique position of Gazprom being the owner of the UGS on the Russian gas market, in particular in the sphere of network activities.

The basic legal issues relevant for further development of gas networks in Russia include the clear definition of ownership rights over the networks (whether non-state-controlled entities will be prohibited to own the trunk pipelines) and of rules for financing network development (mechanisms and sources of network investments by third parties).

PART IV

OVERVIEW AND CONCLUSION

22

The Role of Networks in Changing Energy Markets and the Need for Innovative Solutions

Martha M. Roggenkamp, Donald N. Zillman, Lila Barrera-Hernández, and Iñigo del Guayo

I. Introduction

Our 29 authors have provided a wide variety of insights into energy networks in 2012. Their observations remind us of both the similarities and the differences in energy network issues—technological, financial, legal, and political—from country to country and region to region.

National network policies are shaped by such factors as the presence or absence of raw energy resources, geography, political, and legal traditions, and relations between government and market economy. Network issues also connect with our previous AAG projects for Oxford University Press.[1] Energy security is often crucially dependent on reliable infrastructure. Definitions of property rights are essential in shaping networks. New forms of regulation have a considerable influence on network practices; and, 'moving beyond the carbon economy' implies new forms of network development and regulation. As previous AAG works have suggested, rarely does one size fit all when it comes to law and policy in networks. Equally important, academic theories that postulate perfection in economics or governance regularly yield to realities on the ground.

II. Network Promoters and Facilitators

The chapters identify the variety of promoters and facilitators needed to move a network project from inception to completion. Depending on the situation, one party may lead while others follow. In the ideal project, all parties are supporting the network project with equal zeal.

A first group of promoters and facilitators are the prospective customers of the completed network. The history of many electricity and natural gas networks has

[1] An overview of these publications can be found at the end of the Introduction to this book.

involved the expansion outward from the big-city, affluent customer base to isolated groups of citizens and small businesses in rural areas who are eager to share the benefits of electricity or natural gas service. Frequently, the network businesses providing the remunerative service to the populated area are less than enthusiastic about some expansions of the service. The extension of a network line 100m to capture a well-paying new customer is a profitable 'no brainer'. The decision to extend a network another 100km to pick up a small number of poor customers typically doesn't appeal to network companies' financial officers.

The second promoters and facilitators are the various businesses connected to the network project. These may include the resource developers (eg the oil and gas extractors, the hydroelectric promoters, or the solar or wind project developers). They will find common cause with the planning, financing, construction, and service businesses that will create the networks to move raw resources to the eventual customers. As suggested above, sometimes the projected economic returns on a network project may be compelling. In other situations, profits may be more speculative and partners or guarantees from governments may be required.

The third group of promoters and facilitators are governments at the local, state or provincial, or national level. Governments have a variety of ways to assist a network project. Most directly, government may be the planner, financier, builder, and operator. For good portions of the age of modern energy, some governments in both capitalist and communist nations played some or all of these roles. With the collapse of much of the communist world and the discontent with state capitalism in other quarters, governments' roles have become more complex. To be certain, government resources still fund energy and energy network projects around the world. Often, they are the difference maker in persuading the private sector to undertake the project. Government also plays a regulatory role in commanding, encouraging, or discouraging network ventures. This role for government is often prompted by the shortcomings of private enterprise systems in dealing with natural monopolies. The inherent nature of certain types of enterprise (among them network ventures) tends to result in single enterprise dominance. With that dominance comes the risk of consumer exploitation because of the lack of competition for the supply of essential services. Government is compelled to restore something that resembles the benefits of free competition.

Another important group of promoters and facilitators in network decisions are international institutions such as the World Bank and regional organizations. The several chapters on the European Union (EU), the Americas, and the one on Africa (Chad and Cameroon) suggest the potential for such groupings. The groups can either compel action by the individual nations or merely play an advisory role. Often, their positions regarding networks will be motivated by goals larger than simply the supply of an energy product to a body of consumers. Efforts to control climate change or to assure energy security or to broaden economic development to energy-disadvantaged citizens are examples.

Once the combination of promoters and facilitators for a network project has developed a vision, they face the three challenges discussed in the Introduction. Does science and technology allow such a project to succeed? The chapters in the innovative section of the book consider the challenges of offshore wind production and transmission, carbon capture and storage, and 'smart' energy technologies. The second challenge is financing. Are the dollars or euros or yuan available to see the project through to completion? Network project costs are notoriously front-loaded. Enormous investments must be made before the first shipment of natural gas or petroleum or electricity provides the first return to the investor on her investment. Increasingly, the model for such major network projects is a combination of private wealth and valuable government subsidies or guarantees. The third challenge is whether legal and public policy mandates or preferences will allow the project to go forward.

A century or more of land use and regulated industry law have set frameworks for how a network project can secure government support. But, network innovations often need to start by creating a new body of law for the new energy enterprise. Here, government may be adversary rather than supporter of the venture. This is especially true if the virtues of the new network venture are more visible at the national level than at the local level. The combination of new local legislation (often involving land use controls), opposing local citizens, and uncooperative media have brought down network projects that have state or provincial or national government on their side and that have garnered great wealth to make the project succeed.

III. Networks—New and Innovative Concepts Emerging

As the Introduction illustrated, energy and networks have gradually been evolving from man and horse power to modern energy sources like the fossil fuels, hydro and nuclear power, and modern renewables. Most of these modern resources are transported through pipelines and cables which can be constructed in the subsoil or above ground. These networks are part of an energy chain: from well head to burner tip or electric generator to the residential light switch. The energy chain begins at the level of production and high-pressure/high-voltage lines are connected to upstream pipelines and/or cables of the production facility. These pipelines or cables are then connected to the next level in the energy chain: distribution grids that are operated at a lower pressure or voltage level. Distribution grids are subsequently connected to the premises of final consumers—major industries, small businesses, public buildings, and households. This top-down approach is currently being augmented with a bottom-up approach. Small consumers are turning into producers ('prosumers') and inject some of the energy they produce from solar and/or wind into the distribution grid. In addition, facilities producing renewable energy are connected to the distribution grid instead of the transmission grid. Consequently new and innovative concepts of grid management emerge as the ones explained below.

In the past few decades, government and business decisions have separated production and generation from shipment of the energy product. That has dampened some of the earlier desire to invest more in infrastructure to increase profits to the company. A further modern trend has been to discourage expanding network facilities by attempting to limit demand. This concept of demand management does not result in new networks but it does create new and innovative network management. Paddock and Youngblood's chapter 9 illustrates the concept.

A second new concept, crucial to the success of modern demand management, is the 'smart grid'. What is a smart grid? Is this new type of grid different from the traditional 'dumb grid'? In order to answer this question, consider what the concept of 'grid' entails. Is a grid merely a set of steel pipes or cables or is it more? We know that transformer stations, converters, and compressor stations are considered as part of the grid. But, smart grids are also smart because of smart meters which are situated in premises. These new metering systems are crucial in operating the grids as they indicate when certain machines can start working, ie, washing machines that turn on at night when there is otherwise little electricity consumption and electric cars that can be recharged at night. However, are these smart meters part of the grid or just part of some sort of new network management? If the latter is the case, then there are no smart grids, but only smart grid mechanisms. Anita Rønne's chapter 8 explores these issues.

Another new grid concept involves offshore electricity grids. In Europe the development of offshore wind farms has been discussed for almost a decade. Several wind farms have been constructed in the North Sea, as well as in the Baltic Sea. In the United States, similar offshore developments are in the planning stage. These wind parks need to be connected to shore and this requires some sort of subsea cable. The cables can be either a direct park to shore cable or it may involve one major cable to which several offshore wind parks are connected. In other words, some sort of a transmission cable is developed. But can it be considered, treated, or classified as a transmission cable? Maybe one should qualify it similar to the upstream natural gas pipelines which are not qualified as transportation pipelines. So far a decision on the proper qualification of these offshore cables has not been made. Given the unpredictable nature of offshore wind energy production, these cables need to be smart or managed smart in order to guarantee regular energy supply. Chapter 10 by Olivia Woolley, Peter Schaumberg, and Graham St. Michel considers these matters.

In contrast to electricity cables, which all transport the same product, electricity (either green or grey), pipelines can transport a variety of products. In addition to 'traditional' products like crude oil, chemical products, and natural gas, pipelines can also transport other commodities like milk, beer, and more recently, CO_2. Does the product being transported have an impact on the regulatory framework? In other words, what is decisive: the infrastructure or the commodity being transported? This depends on whether there are alternatives. Oil, milk, and beer can be transported via other means of transportation such as trucks, trains, and ships. Natural gas, however, in its original form can only be transported via pipelines. These pipelines are therefore considered natural monopolies and consequently may need a special regulatory framework. The fact that gas pipelines are natural monopolies entails that producers, suppliers, and consumers depend on a regulatory framework enabling them to make use of that specific infrastructure. Access rules may be based on specific laws or the application of the essential facilities doctrine as, *inter alia*, described by Nigel Bankes and Rick Nilson in chapter 13.

The special status of pipelines in the gas industry is undisputed. However, the status of pipelines is not yet resolved in other *emerging* sectors, as illustrated in Bankes and Nilson's chapter (13) on CO_2 pipelines. Although CO_2 can be transported by ships and trucks, it is generally accepted that in the case of carbon capture and storage (CCS), pipelines will be the obvious type of transportation. Are CO_2 pipelines then to be considered natural monopolies and essential facilities? Bankes and Nilson's chapter shows that the status of these pipelines is not yet clear but that a comparison with the natural gas sector is apparent, and as a result CO_2 pipelines will require a specific regulatory framework. In other words, CO_2 pipelines may have a 'special status' like natural gas pipelines, and need to be treated differently from pipelines transporting other products.

IV. Jurisdiction and Networks

A. Impact of jurisdiction

The extent and manner of regulation is a matter of state sovereignty. As a manifestation of such sovereignty, states have exclusive jurisdiction within their territory and may thus prescribe and enforce all laws and regulations deemed necessary. However, many national states are subdivided into states or provinces and/or are part of a federation of geographical entities. Examples in the text are Australia, Brazil, Mexico, the United States, Canada, and Germany. Subunits are viewed as sovereign within the larger state and that sovereignty is not subject to change by act of the national legislature. Court decisions at the highest judicial level will adjudicate exactly where the lines between state and other sovereign entities will be drawn. Alastair Lucas's chapter (2) also highlights the fact that certain native or First Nations' communities partake of sovereignty over their land, people, and resources.

In addition to the sovereign entities (states, provinces, etc.), many lesser units of government also exercise powers of jurisdiction. These may be identified as cities, towns, counties, special purpose districts, etc. These entities are not sovereign in the sense of states or provinces. They are controlled by national, state, or provincial law and are not free to assert that their sovereign rights overrule control by the legislative bodies of their state or province. But, those bodies have often granted considerable powers to the local units by statute. A common example is the considerable control over land-use matters by local governments that are the closest to the land in question. As with national and state issues, jurisdictions may overlap. In case of such overlapping jurisdictions, there may be some shared or concurrent jurisdiction over certain matters. Where the overlap leads to conflict between local and state authority, state legislative or executive action may resolve the conflict. Alternatively, court decisions can settle the matter.

An increasingly common feature in network development and operation is the fact that more than one jurisdiction may be involved. Regardless of whether the borders crossed are international or local, as with different provinces or municipalities within the same national unit, different legal and regulatory regimes may apply to a network as a whole, with a variety of results. As illustrated by Alastair Lucas's chapter on Canadian law and practice, an exact delineation of jurisdictional attribution of power over network projects is not always possible. More often than not, the same undertaking will be subject to overlapping authorities, and lead to rivalries, tension, and uncertainty, placing constraints on network development. The ability to rely on some sort of mechanism or procedure to overcome conflict, contradiction, and unnecessary duplication may be central to future network development. In the Canadian example, a web of inter-jurisdictional agreements is in place to guide parties with a stake in network development. According to Lucas, '[i]nter-jurisdictional agreements providing for joint regulatory processes are a potent instrument for mitigating federalism constraints'. However, the author warns, cooperation and harmonization may come at a price, often reflected in the complexity of the resulting processes or their length. Overcoming those pitfalls may command much of lawmakers' and regulators' attention in the future, particularly in those regions that are actively promoting network interconnection as an integral part of energy integration, as is the case of the EU and the Americas.

The warning has a particular relevance for offshore networks, especially in the EU. In order to secure long-term energy supply and meet carbon reduction goals, coastal states are increasingly developing offshore electricity projects, ie offshore wind farms. In order to bring the electricity to shore, a network of cables is required. The European Commission pays special attention to the situation in the North Sea and encourages the establishment of a 'North Sea super grid', ie one North Sea grid to which all offshore wind parks can be connected. However, such a grid will involve the cooperation of all North Sea states, which all have exclusive rights to develop offshore wind energy and to construct and exploit the necessary installations and cables. So, what should the regulatory framework governing this super grid look like? When taking into account the experiences with onshore interconnections and regulatory processes, one common EU approach for developing such a grid could eventually solve problems of jurisdiction and harmonization which otherwise will exist. However, in the absence of any specific EU competences, such approach will need the cooperation of all coastal states involved.

B. Inter-jurisdictional coordination and intergovernmental agreements

Intergovernmental cooperation regarding the construction and use of cross-border networks is not a new phenomenon. The North Sea petroleum pipeline systems are all based on bilateral agreements between either the sending and receiving states (usually Norway being the sender state and Germany, the UK, Belgium, and

France the receiver states) or between the states linked by the interconnector pipeline (UK–Belgium and UK–the Netherlands).[2]

Inter-jurisdictional coordination also figures prominently in the chapters by Mohammed Bekhechi and Catherine Redgwell (chapters 5 and 6, respectively). International law in the form of intergovernmental agreements, host government agreements, loan agreements, or traditional international treaties, particularly on environmental issues and human rights, may be called upon to fill in gaps or smooth over disparities between jurisdictions. According to Redgwell, '[n]onetheless, there are a number of international treaties that may apply to condition where, and how, such projects may be carried out particularly with respect to environmental impact and human rights'. Moreover, also according to Redgwell, international agreements can serve to 'insulate the project from the application of ordinary domestic law and to internationalize the project by benchmarking conduct to international standards'. Both observations are confirmed in Bekhechi's Chad–Cameroon Pipeline case study. Bekhechi's chapter highlights the role played by international lenders such as the World Bank as standard-setters in environmental protection matters but warns readers that 'international ad hoc mechanisms are not a panacea and cannot replace government agencies responsible for protecting the environment and the people in any country'. Ideally, local capacity, including well-funded agencies and adequately trained personnel, to monitor project implementation should precede network development. Meanwhile, enhanced access to dispute resolution seems to gain in significance in terms of a country's and developer's ability to undertake a project and avoid disruptions to its progress.

On the other hand, as is made clear in the case studies included in the book, international law's role in insulating a project from domestic law can be problematic. Stabilization clauses inserted in concession agreements or host government agreements are as much welcome as they are vilified. So far, however, they seem to be a necessary precondition to investment in network projects in developing countries. Arguably, investment guarantees and adherence to the rule of law, and the consequent ability to set aside the stability clause, could derive from regional, ie multinational, cooperation agreements on energy integration. However, forging consensus at a regional level to the point of building a robust legal platform to facilitate network development seems to be easier said than done, as explained by Lila Barrera-Hernández in chapter 4 on energy integration in South America. Yet, the energy integration process in the EU offers an encouraging example. Since the Internal Energy Market was launched at the end of the 1980s, several attempts have been made to liberalize and integrate the national markets. Currently, the EU member states are faced with the Third Energy Package, which brings the integration process to another level as Iñigo del Guayo and Johann-Christian Pielow explain in chapter 19.

[2] These agreements have been discussed as part of an earlier AAG project 'Managing Jurisdictional Conflicts following Globalisation and Liberalisation' presented at the Cape Town biannual in 1998. See further Martha M. Roggenkamp, 'Petroleum Pipelines in the North Sea: Questions of Jurisdiction and Practical Solutions' (1998), 16 *J Energy & Nat Resources L* 92–109.

V. Developing Energy Networks in Changing Markets

A. Changing markets

Each national state may decide the regulatory framework governing the construction and use of energy networks. This book clearly illustrates that the role of networks is constant, as a network's main function is to transport a commodity from point A to point B. As a result, networks are a crucial element in securing energy supply. Although this role remains unchanged, the regulatory framework within which these networks operate has changed in recent decades for two important reasons: (i) market liberalization; and (ii) climate change, in combination with supply security resulting in an increased use of renewable energy sources. Both changes and the impact they have on energy networks can be noticed throughout the world and are illustrated by several chapters in this book.

This book shows that energy market liberalization is a global phenomenon. However, it also shows that the level of market liberalization differs. Whereas the process of energy market liberalization started around the 1980s and early 1990s in developed countries in Europe (EU member states), Canada, the US, New Zealand, and Australia, a similar process began some ten years later in Brazil, China, and Russia. Irrespective of the degree of market liberalization, all liberalized markets have in common that (i) networks and transportation activities are separated (or unbundled) from energy production and supply activities; and (ii) network access needs to be regulated. The degree of separation/unbundling may differ, as well as the way in which it is organized. A liberalized market thus changes the perspective on network development. It is the company owning and/or operating the network that is responsible for maintaining the networks it operates and extending/reinforcing these networks if necessary. Following this market change, the role of governments has changed. Governments no longer decide directly about the maintenance and operation of energy networks. However, they may do so indirectly through ownership in the grid and a wide range of approval mechanisms included in laws and regulations, ie approval of transportation/network tariffs, access rules, and a variety of permits in relation to the construction of new or additional grids. Within that regulatory framework the network companies need to operate and transport the commodity from the producer/supplier to the end consumer. However, due to the fact that networks are considered natural monopolies and any abuse of their dominant position should be prevented, the national legal frameworks provide for some regulatory boundaries within which network companies may operate. Those boundaries provide that the sole focus of the network company should be on the networks and that any investment decisions should be based on that notion.

How does the market liberalization process and the role of the network companies therein fit with another important policy goal: the need to combat climate change and secure long-term energy supply? The latter goal can be met by, amongst other means, increasing the level of renewable energy sources or organizing for a more efficient use of energy resources and thus limiting the use of networks. Are the latter measures in the interest of network operators? Why limit the use of networks if their sole business is operating the networks? A possible reason is that by limiting the use of networks a decision to invest in an expansion or reinforcement of the networks could be avoided. Network companies can be faced with a similar decision when governments promote the use of more renewable energy sources. Renewable energy sources can be located in a different part of the country where there are insufficient network connections or be injected at a different level of the existing grid. Again, network companies could oppose an investment decision, which would not result in a sufficient rate of return but from a government point of view is necessary. Who makes the final investment decision? The chapters of this book make clear that we need to distinguish between investments in electricity and gas networks and other networks. Pipelines transporting oil or CO_2 can be treated under a different investment regime. For example, the Chad–Cameroon pipeline has been the result of involvement of the World Bank and the development of CO_2 pipelines may benefit from light-handed regulation instead of heavy-handed rate-making rules, as usually applied in the natural gas sector. The regimes governing investments in electricity and gas networks will be discussed below.

B. Network investments

In a liberalized market network operators and users need to agree on the use of the networks. Third party access (TPA) is usually negotiated on the basis of predefined tariffs. A variety of tariff regimes apply. Within one country a distinction can even be made between regimes applying to the transmission and the distribution grid.

1. Regulated tariffs and network investments

As a general rule these regimes have been introduced and/or are meant to be sufficient to cover the costs of grid investment. The recent introduction of a liberalized gas market in Brazil discussed by Yanko Marcius de Alencar Xavier and Anderson Souza da Silva Lanzillo in chapter 14 is a clear example of the difficult balancing act between the new regulatory regime and the need for network investments. Brazil's 2009 Gas Law introduced a system of price cap regulation, so the tariffs are not necessarily based on the costs of transportation or on the profits as in a rate of return regime. The price cap is based on the market prices which are defined by the previous (and existing) monopolist. It is anticipated that the revenues from gas transportation will not be sufficient to make the necessary investments in grid expansion. The authors conclude that 'introducing rules of open access is not enough. Tariff regulation must take into account not only the aspects of the market or an annual income, but should also take into account the costs and the investments needed to build the networks'. This shows that Brazil is still in the early phases of market liberalization and probably will be confronted with several changes in the tariff and investment regime. Are there lessons to be learned from other regimes?

Moving to another region in the southern hemisphere we note examples in New Zealand and Australia, where energy market liberalization was introduced in the 1980s and 1990s. Since then the regulatory regime has gradually been intensified. New Zealand began with a tariff regime based on price cap regulation (CPI-X) but is now gradually moving away from this model and will apply a revenue cap instead. Barry Barton analyses how the transmission companies' needs have replaced the earlier efficiency goals. The physical characteristics of New Zealand have also led to another interesting phenomenon: the transmission pricing methodology (TPM), which allows for different users paying different tariffs. All users are categorized on the basis of the different use they make of different parts of the network and the tariff is set accordingly. It means that the regime allows for some degree of different treatment as long as users within the same category are treated similarly. The prices set under the TPM may not exceed the maximum allowable revenue set. These revenues can be used to invest in grid development. Such a decision is subject to a grid investment test, which is an instrument to find out whether the intended investment could be made at lower costs. In Australia the liberalization process started as early as the 1980s and aims to create an efficient energy market, including efficient network operations. Investments in networks are regulated and regulation is based on efficiency goals. All major network investments are subject to approval from the energy regulator (which then supplies some funds). As in New Zealand each project will also be subject to a regulatory test based on economically efficient investment criteria. This again is accompanied by a system of strategic investment planning which aims at linking investment decisions on electricity production with investment in the networks so that a situation is avoided where the networks are expanded/constructed without the generation facilities being built, and vice versa. Australia is currently faced with the need to invest in the replacement of networks constructed some 50–60 years ago. These old networks are no longer safe and may be the cause of bushfires. The current regulatory regime is not suited for encouraging these investments. Efficiency goals are contrary to investment needs. Lee Godden and Anne Kallies (chapter 16) conclude that currently there is no legal regime in place or a reform announced enabling network operators to meet these new goals.

The experiences in the EU mirror to some extent the above examples. EU member states have usually made use of the CPI-X tariff system. The UK is the cradle of the CPI-X regime. Aileen McHarg analyses the origin of this regime in chapter 17. The aim was that of introducing competition and thus creating more efficiency, which again is to the benefit of the consumer. Since its introduction, the regime has become more complicated and is not very transparent anymore. Nor is it suited for investing in large infrastructure projects. In 2010 the UK government proposed a new regime which will replace the CPI-X model. The Revenue using Incentives to deliver Innovation and Output (RIIO) model is a regime that particularly aims at creating revenues for long-term investments or innovative projects. The change in the regulatory framework in the UK from CPI-X to RIIO illustrates the need for a new mechanism to guarantee grid investments. The new RIIO formula clearly shows how crucial a proper remuneration scheme is to have the infrastructure constructed or expanded, as well as properly managed and to help achieve other policy aims through innovation, ie a carbon-free economy. Although it is too early to assess the impact of this new regime, it may be monitored closely by other jurisdictions struggling with long-term network investments.

Most EU member states make use of price cap mechanisms such as the CPI-X regime. The EU understands that this tariff regime is not sufficient to promote some specific network investments, in particular cross-border networks, which are crucial for the establishment of an internal energy market and long-term supply security. Therefore an exemption procedure is included in EU law, ie the possibility to be exempted from the principle of regulated access and the regulated tariff regime. This exemption regime applies mainly in relation to the construction of new infrastructure such as cross-border networks, eg electricity and gas interconnectors. Exemptions are granted by member states, but the EU Commission has the right of approval and may therefore require amendment or withdrawal of the exemption decision. Tjarda van der Vijver in chapter 18 discusses this exemption regime and assesses decisional practice on the basis of an analysis of a variety of exemption decisions on the development of gas and electricity interconnectors. The main element of assessment is the need for investment, ie if without an exemption the investment would have been made. The analysis shows that the EU Commission is getting stricter in its assessment and exemptions are increasingly awarded under specific conditions. The author recognizes that the exemptions show that a proper balance needs to be struck between the need to stimulate investments on the one hand and the need to strengthen competition on the other hand. He concludes with the positive assessment that the regime of third party access (TPA) is hopefully 'now firmly secured as the leading principle regarding energy infrastructure'. Time will show whether this is the case. It may depend on the judgement of investors, as the tariff itself has not necessarily been the stumbling block so far but rather the regulatory uncertainty following the introduction of new EU legislation every five years when most investment periods have a lead time of 20–25 years.

2. *Infrastructure planning and network investments*

Liberalization was introduced in the EU some 30 years ago and its sole aim was to introduce competition. It has led to an unbundling of network activities and production and supply activities, the latter companies not interfering in investment decisions of the former and vice versa. Unbundling combined with a system of regulated tariffs based on a price cap principle have proven to be a disincentive to invest in infrastructure. The EU Commission now acknowledges that there are some adverse effects: a decrease in proper maintenance and development of infrastructure following a lack of proper investment in infrastructure. That is why the liberalization process is giving way to a new policy of infrastructure planning, with binding investments components imposed upon companies. When introducing its last regulatory package in 2009 the EU recognized the need for more long-term planning and a stronger involvement of governments in planning, investing, and fostering investment in infrastructure. That move explains the background for chapter 19, the del Guayo and Pielow chapter on the EU planning provisions in the

2009 rules relating to the European internal energy market. Planning is now again central in the process to build a European energy market. It is an interesting development showing how planning must play a role within the regulatory and legal framework. This development should not be seen as a mere result of a pendulum-style motion, from free market in energy towards a new planned energy model. Rather, it opens a space for planning within liberalization and competition. It is a new way of planning, far from the old planning of the period 1945 to 1980.

In chapter 20 Gunther Kühne provides us with an example of the impact of the new EU policy. He analyses the situation in Germany, which is characterized by a direct need for large-scale development of energy networks, in particular electricity networks, following the decision to phase out nuclear energy and to switch to renewable energy sources instead. As the renewable energy sources are predominantly situated offshore (North Sea and Baltic Sea), the existing north–south extra-high-voltage grid needs to be expanded. The intended expansion not only requires investments but also some integrated planning. As to the latter element, the German law implementing the EU planning provisions is relevant. The German implementation law requires all transmission network operators to annually submit a joint national network development plan, which must contain an overview of all measures necessary to maintain and operate the grid effectively during the next decade, including measures relating to the extension and construction of grids. This joint plan is again the basis for the Federal Grid Requirement Plan, which aims at a further streamlining of procedures. The next issue is, of course, how to finance the extension of existing grids and/or new grid. Germany applies a tariff system based on incentive regulation, which again is based on efficiency goals. Investments in transmission grids need to be approved by the energy regulator and such approval is based on a 'necessity' test. How does one assess 'necessity'? Do the same criteria apply to planning and investment? These questions are still unresolved.

Similar questions are raised in China where, since the 1970s, a reform and opening-up policy is being implemented. Most recently, the electricity sector is subject to such reform. As a result, the direct influence of the government has declined as the sector has evolved. In addition, energy production has been separated from network activities. However, some sort of government planning still exists. In order to meet energy needs and to limit climate change, China also aims at introducing renewable energy sources on a large scale. A more 'modern' use of renewable sources will not only enhance the standard of living in rural areas but also secure energy supply in the already industrialized areas. In case of the latter, large-scale networks need to be constructed. The Renewable Energy Law of 2005 as amended in 2009 is a crucial tool in this process. However, the relevant plans are not always coordinated. The Medium & Long-term Development Plan for Renewable Energy of 2007 sets production targets but does not connect these targets with network connection plans. The network companies are under a separate obligation to develop network construction plans. Wang Mingyuan in chapter 12 illustrates the tension between both planning systems and the need for improvement by analysing the legal and policy issues relating to the development of wind energy in China.

3. *Regional integration and network investment*

Finally, according to Barrera-Hernández in chapter 4, network investment in South America could receive a boost from regional or sub-regional integration schemes which might be able to deliver the clarity, stability, and predictability that the individual countries' regulatory systems cannot provide. After an in-depth look at UNASUR, MERCOSUR, and CAN (ie the Union of South American Nations, the Southern Common Market, and the Community of Andean Nations), the author concludes that:

> [t]he demise of Washington Consensus policies and the enhanced presence of the state in the energy sector through newly empowered national energy companies may bring around a uniquely South American model of energy integration and network development. It is yet to be seen whether that integration will move in the direction of a free regional energy market where national oil companies (NOCs) are at a par with private companies or whether it will favour a strong presence of the state(s) in developing and operating the network(s).

C. Constructing the networks: approval procedures, land use, and the environment

Securing the necessary investments and providing for some sort of network planning is just part of the story. Other challenges arise when the networks are to be constructed. These challenges vary from public acceptance and transparency to land use and environmental issues.

1. *Public acceptance and transparency*

Public acceptance is closely connected to public participation, for which transparency is a pre-requisite. The basic principles of public participation have been discussed in a previous AAG project in 2002 (*Human Rights in Natural Resource Development—Public Participation in the Sustainable Development of Mining and Energy Sources*). Several chapters in this book highlight the role of public participation in creating public acceptance of major infrastructure projects, including energy networks.

It is interesting to note McHarg's observation in chapter 17 on British energy network regulation that Ofgem hopes that by introducing the new regulatory regime RIIO:

> enhanced engagement will increase the transparency and legitimacy of the price control process. It may also be hoped that greater participation at an early stage will help to reduce later opposition to new network infrastructure when it enters the land-use planning process, planning delays having been a major problem in securing the connection of new renewable generation.

In order to facilitate stakeholder engagement, Ofgem will provide more accessible documentation. As RIIO has not yet entered into force, it is too early to assess whether public acceptance will improve.

In Germany the intended expansion of the north–south extra-high-voltage grid may be delayed due to public protest. Throughout Germany local politicians and civil action committees are opposing many infrastructure projects. Apart from expensive technical solutions like burying cables, several legal measures have been introduced to speed up and streamline the administrative processes, ie the 2006 Infrastructure Planning Acceleration Act and the 2009 Electricity Grid Expansion Act. These Acts intend to streamline the licensing procedures and thus limit the extent of public participation. This has not led to the intended result. The 2011 legislative package implementing the Third Energy Package includes some further instruments which, amongst other things, aim at tighter planning mechanisms and interaction between undertakings and governments.

Ulf Hammer identifies similar problems in the Simna–Samnanger case involving the construction of a 410 kV transmission line from the Simna power station to the Samnanger transformer station, a distance of approximately 90km. The licence to construct and operate the cable was awarded in June 2006 by the competent authority and was followed by objections from 49 local and regional public and private organizations who are concerned about the biological diversity of the landscape and the population of wild reindeer. Although the decision was confirmed in March 2011, other offshore options are being assessed. Be that as it may, the case illustrates the problems relating to balancing two crucial interests: security of energy supply versus environmental protection. This balancing becomes even more complicated if one environmental goal (clean energy) needs to be balanced with another environmental goal (nature protection).

The need to balance competing interests—transparency versus security—is central to Martha Roggenkamp's chapter (chapter 7) on protecting energy networks in the EU. The author illustrates her discussion with examples from the Netherlands and Belgium, where a suitable compromise seems to be struck. The jury is still out, however, on whether local country solutions may be compatible with EU mandates on European critical infrastructure and the consequent need for enhanced confidentiality.

2. *Environmental impacts*

Managing environmental and social impacts due to network development to satisfactory levels for all persons and communities involved across the jurisdictions traversed by a network project is a recurring issue. The above section acknowledged the tension that may exist between networks and the impact they may have on the environment. The tension becomes even greater if the networks are designed to replace traditional 'grey' energy with clean 'green' energy, as it means balancing two environmental objectives as chapter 20 on Germany clearly illustrates. The Norwegian and German chapters (11 and 20) also show that environmental concerns can result in long-lasting procedures.

In the case of cross-border networks the duration of licensing procedures are usually substantially longer. The need for legal and regulatory harmonization to adequate standards becomes clear in a number of chapters, including José Juan Gonzalez's and Lila Barrera-Hernandez's writing on the Americas (chapters

3 and 4). Indeed, even in regions where there is substantial progress on harmonization of technical and economic issues, as in the case of the Andean Community of Nations, arriving at a consensus on minimum environmental and social impact management standards seems to be a distant goal, if a goal at all. Network development, however, does seem to have a beneficial impact by raising environmental standards and promoting capacity development, particularly in transboundary contexts. Such was the case in Mexico, where the negotiation of a free-trade agreement with the USA and Canada, prompted the enhancement of the country's environmental norms and institutional framework. The environment was specifically targeted in the agreements that made the development of the Chad–Cameroon pipeline possible. There, the World Bank's involvement ensured that some measure of protection and mitigation was applied. Most of all, the Bank's involvement raised the level of awareness of the international community regarding the environmental and social issues surrounding large network development projects in Africa, and called attention to the need to enhance developing countries' capacity to handle environmental and social impacts before large development projects take place.

3. Land use

Although land use featured prominently in the preceding AAG's book project *Property and the Law in Energy and Natural Resources*, it is addressed in some contributions to this book. It is obvious that developing energy networks requires land use, although a difference can be made between aerial cables, subsoil cables, and pipelines, as aerial cables have less impact on the land/subsoil. Any decision to bury aerial cables in the subsoil, as proposed in Germany and Australia, will require land-use agreements and thus additional costs. In addition, a distinction needs to be made between networks developed onshore and offshore, as the latter area is governed by a different regulatory regime where states have limited rights with regard to the construction and exploitation of networks.

The political reform in Russia led to a process of energy market liberalization but also to a reform in land ownership and use. In chapter 21 Sergey Seliverstov and Ivan Gudkov discuss this process of land reform based on the Land Code of 1991 and 2001. The Land Code acknowledges that there may be private and public land and that ownership rights can exist at federal, regional, and local levels. In addition, the Land Code identifies seven categories of land, depending on their use: agricultural land, settlement land, land with protected areas and objects and 'land of industries, energy, transport, and telecommunication [...] and other special purpose'. The latter category is again divided in seven sub-categories, amongst which are energy land and transport land. Whereas electricity networks can be built on 'energy land' only, gas pipelines fall in the category 'transport land'. Any relocation of land categories is governed by a special federal law. An additional complication is that gas networks are considered as 'federal energy systems' but the law does not clarify whether this also applies to the electricity networks. As a result gas networks can only be constructed on land which is state-owned and then leased to the network company; however, electricity networks can be constructed in privately owned land. This distinction between gas and electricity networks is also relevant with regard to payments to be made, ie operators of electricity grids may be in the position that they pay less compensation than operators of gas grids.

José Juan Gonzales pays some attention to land use issues in Mexico. As in Russia, it may be necessary to obtain an authorization for land use change if, for example, a transmission line crosses forest land. Such a land use change authorization is not easy to obtain. Therefore network developers can face lengthy and costly procedures. The same can apply to obtaining a right of way. Under Mexican law there are three mechanisms to secure the right of way: an easement, a purchase agreement, or a rental agreement. In all cases the land owner is entitled to receive economic compensation. Obtaining a right of way is often not easy, as property rights may conflict and thus lead to uncertainty regarding the identification of the person having legal capacity to grant the easement, sell, or rent the land. In Africa, the Chad–Cameroon pipeline was constructed on the basis of two concession agreements entered into by the government of Chad and the government of Cameroon, respectively. These agreements specify the rights and duties of each party. As to the issue of land use, it is notable that the agreements recognize that the construction of a hydrocarbons' pipeline is considered a public interest and that—if necessary—land can be expropriated under the countries' respective laws on expropriation of land in the public interest. However, as the compensation regimes did not fulfill the World Bank standards, additional conditions were included in the concession agreements. Both chapters highlight an important dimension at the intersection of network development and land use, ie social impacts and the need to weigh those impacts against the public interest in network development and security of supply.

D. Network protection

Developing reliable energy networks goes beyond the stage of construction. Energy networks—subsoil as well as above-ground networks—need to be protected. Such protection may be necessary as external circumstances can damage networks. Similarly, unsafe networks may also have an impact on health, safety, and the environment. Therefore usually some safety standards apply when cables and pipelines are constructed. Such standards will include instructions as to the distance between the networks and the nearest buildings and a minimum of ground cover (burial depth).

The consequences of unsafe networks have been illustrated in chapter 16 by Godden and Kallies on network developments in Australia. They analyse, amongst other things, the bushfires that took place in 2009, which were caused by an aerial electricity cable constructed in the 1950s. The cable had reached the end of its engineering life but had not been replaced. The subsequent inquiry concluded that 'major changes should be made to Victoria's electricity distribution infrastructure, and its operation and management, if there is to be a substantial reduction in the risks to human life posed by bushfires'. Basic instruments to mitigate such risks are included in the Electricity Safety Act 1998, which was amended in order to give the regulatory agency—Energy Safety Victoria—an extended mandate to intervene. However, the inquiry also concluded that the core problem was the need to replace the old electricity cable and to find the necessary funds to do so.[3]

Another example involves the protection of cables and pipelines in the EU and, in particular, in the Netherlands and Belgium. Both countries have similar subsoil and geographical conditions, with an extensive subsoil gas grid which can be easily damaged by construction activities. The accident near the town of Gellingen (*Ghislenghien*), Belgium, in 2004 damaging a subsoil gas transmission pipe, is a clear example of the devastative effects such construction activities can have: the damaged pipeline led to a major gas explosion and fire. As a result of the dangers inherent to subsoil networks, the Netherlands and Belgium have regimes which require all constructors to make use of an online registration system so that it is clear which network companies operate in a specific area and detailed information can be obtained to avoid damaging the subsoil network. Apart from their direct consequences, such damages can obstruct security of supply. This is also the main reason why the EU has introduced measures to protect 'critical energy infrastructure'. Such protection can be necessary in case of major environmental disasters, but especially in case of terrorist attacks. Such measures will only apply in case the damages have an impact on at least two member states. Given the fact that most member states are connected through several connections (or will be in the future), it remains to be seen what the impact of this protective measure is on energy networks.

VI. Conclusion: Changing Markets and Innovative Solutions

As we prepare the final lines of this Conclusion in August 2011, economic, social, and political turmoil are the order of the day. Political and economic traumas have placed rioters on the streets of the United Kingdom. The year 2011 has seen regime change, or threatened regime change, percolate throughout the Middle East. Final results are still uncertain for the geographic region containing most of the world's petroleum reserves. Japan and New Zealand are coping with major earthquakes, tsunamis, and nuclear plant accidents that also have an impact on energy supply issues. Evidence of a modest recovery from the recession of 2008–09 at the start of 2011 is now very much in question. National debt has moved from a rather boring subject for specialists to the hot topic on the news and in electoral politics. The United States Congress flirted with default. The EU weighs how to bail out debt-heavy members. Expensive, government-funded network projects force very hard decisions.

Making energy law and policy in this troubled time is not easy. It would be tempting to say: do nothing. Government dollars, euros, and yuan are in short supply. Political consensus on any matters is hard to reach. But, the energy picture of 2011 cautions against inactivity. Economic recovery relies heavily on secure and affordable access to energy. Concerns about climate change demand new patterns

[3] See for this above, section IV B.

of energy use or non-use. Development of the disadvantaged portions of the globe cannot wait. Inactivity will change the status quo in unfortunate ways.

Our authors have explored the many aspects of the legal system that impact energy networks. Law may sometimes stand as an impediment to optimal network development. Often, this reflects a body of law adopted before the world was aware of new energy technologies or network plans. Legislators, treaty writers, and agency heads tend to be reactive to developments in the energy world, rather than writing their laws for the 'next big thing' coming a decade in the future. New network developments and new development affecting the existing networks may need to bring the law along with it.

The chapters in this book indicate a possible reassessment of our current vision on networks. Three important developments can be noted.

First, fossil fuels and also renewable energy sources like hydropower or wind are increasingly located in remote areas within or outside national borders. The latter require networks to cross more than one border and involve several jurisdictions. In order to create legal certainty (investment protection and the use of common standards) states have entered into bilateral or multilateral agreements. The chapters have provided us with insight into such agreements from several parts of the world.

We note that multilateral treaties and bilateral agreements can smooth the way for working across national boundary lines or in international waters. Large network projects, like those case studies examined in the book, invite project specific legal agreements among the private entrepreneurs, the financiers, and the governments involved. These can certainly raise memories of unequal concession agreements in the early days of energy resource exploitation. All too often, the concession agreement left the division of profits heavily weighted to the businesses and paid little attention to such other host country issues as environmental protection, human rights, and national economic development. Today, the bargaining power and the law governing it (national and international) are more balanced. The best of such agreements will consider not just how to enable an economic venture to succeed, but to do it with proper attention to the environmental and human consequences of the development and operation of the network.

Second, in contrast to this development, we see that traditional renewable energy sources are being used in a modern fashion. Wind, solar, and biomass have been used for centuries and are still playing a role in local energy supply. However, these energy sources are today used in a different and more intelligent way. Consumers are increasingly making use of solar panels and wind turbines, but not only for their own use. Any surplus power generated is injected into the grid. The consumer is turning into a producer but still relying on the traditional energy supplier if private production is insufficient. This consumer has a dual role and has turned into a 'prosumer'. This is a process which demands a lot from the grid, as the flexibility of the prosumer has to be met by a flexibility of the grid and the operator of the grid. The grid has to become smart and this requires smart legal solutions. Chapter 8 by Anita Rønne suggests that such solutions are not yet available. Smart grids so far are a matter of smart technology but it is now up to the legislators to provide smart legal solutions. The challenge will be to find these solutions in a liberalized market where production/supply and network activities are unbundled activities, whereas these smart solutions seem to require a close cooperation between supplier and network company. Is there an innovative solution to be found in a liberalized market or is this a sign of a return to a more vertically integrated market?

These two approaches may also lead to a new concept of networks. On the one hand, extensive international systems linking several transmission pipelines bring oil and gas to consumption areas and, within those consumption areas, to consumers that are relying less on these traditional fuels and more on cooperation via smart technology. The transmission networks will therefore have a different role. They may act less as a network providing regular energy supply but more as a network of last resort. Who will pay for the operation, maintenance, and replacement of these networks if there are fewer parties making use of them and network income is based on the extent to which they are used? Many transmission grids are now either written off or unsafe and need to be modernized. In other words, governments and grid operators need to make decisions on long-term investments without knowing how these networks will operate in order to face new challenges. This requires innovative legislation.

The chapters have shown that in a liberalized and changing market, the reinforcement, extension, or expansion of a network can be difficult to finance. As a general rule these investments need to be made by the companies operating the networks and financed on the basis of the tariffs paid by system users. However, these tariff regimes are often based on efficiency goals and price cap mechanisms. Due to regulatory uncertainty investors may also be little tempted to make long-term investment decisions, which may be necessary as networks become old and unsafe, more customers (producers as well as consumers) require access or transportation capacity, or need to be extended as resources are located elsewhere. Our studies provide some example of possible solutions.

In the UK a new regulatory regime, RIIO, has been introduced, which also aims at investments in 'special' projects. A different example is the possibility under EU law to apply for an exemption from the principle of third party access and/or the regulated tariff. As a consequence, however, the use of these networks is often limited to the applicants of the exemption. So far the exemption regime only applies to a specific type of network: interconnectors. Would it be possible to use such an exemption regime for specific national networks, such as new cables necessary to transport renewable energy from far away locations not yet connected to the grid? Does such an extended application of the exemption regime require the introduction of a new type of grid as, for example, the 'closed distribution system' introduced in the EU following the Third Energy Package? Another alternative may be an investment test as used in Australia. On the basis of such a test a regulatory authority could assess whether an investment is needed and, if so, whether the income from the tariffs is sufficient or whether some additional government subsidy would be necessary. This would be a regime similar to subsidies given to renewable energy producers for the extra costs of producing 'green' instead of 'grey' energy. In the latter regime, it is still the network company taking the initiative and in practice network operators and government do not always have the same interest at heart. In that respect we note that even in liberalized markets there is a re-introduction of government planning. These plans usually focus on investment decisions and do not necessarily have a direct link with physical planning and on the duration of licensing procedures. Could an investment test and investment planning be integrated in the sense that an investment is assessed positively if the network is part of such a plan and would network companies therefore be more easily persuaded to make an investment?

Third and finally, we note that the energy chain is becoming integrated and more complex. The gas chain (production, transmission, distribution, and supply) is directly connected to the electricity chain as significant quantities of natural gas (and oil) can be used for generating electricity. So, the well head is in the end connected to the residential light switch. Moreover, this chain of networks has been extended with another network: CO_2 pipelines. The electricity generators using natural gas may in the future be required to capture and store the CO_2 emissions. The CO_2 will be transported via pipelines to an underground storage facility such as a depleted gas field. In this respect we see a cycle of networks bringing natural gas from a subsoil reservoir to consumers and then returning another gas (CO_2) to the same reservoir. All those in this extended energy chain need to be regulated and each set of regulations has an impact on another part of the chain. This requires innovative solutions and maybe a more integrated approach than we have had so far.

One overarching conclusion that can be derived from all chapters is that the law and regulation of network development and use is far from static. Shifts in economic policy, globalization, safety, climate change, and technological advancements, just to name a few potential influences, demand constant adaptation. Some of the features of network law and policy, such as the resurgence of planning, may be more familiar than others, which introduce entirely new ideas such as the need to regulate CO_2 networks. Innovation, however, permeates the entire field.

Index